CELLULAR COFFERDAMS

A Pile Buck® Production

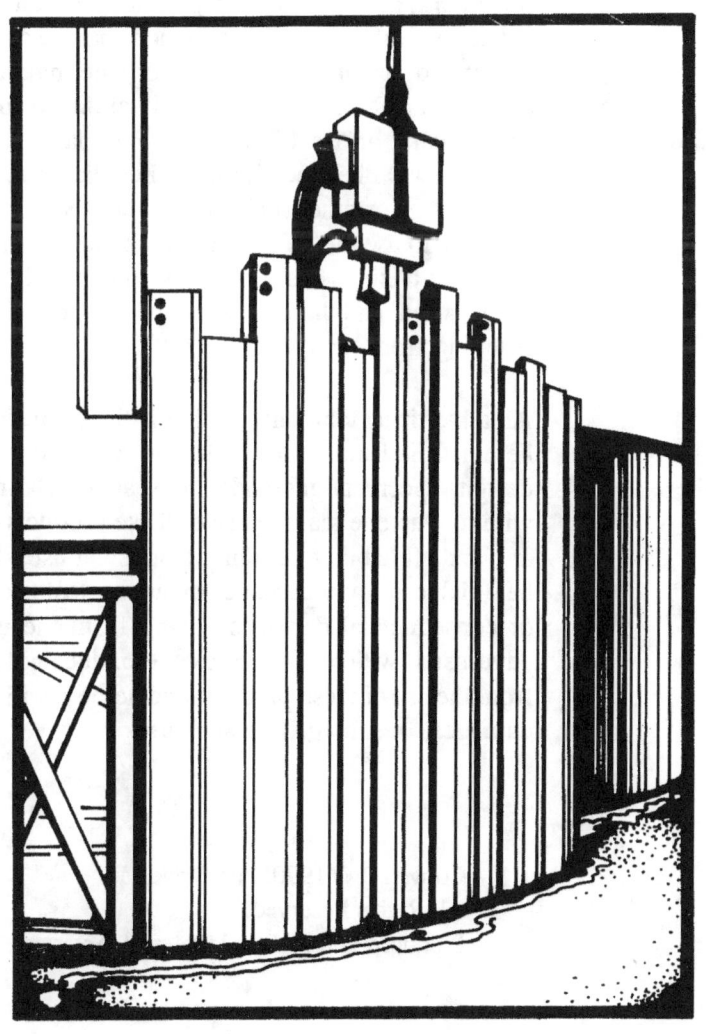

© COPYRIGHT 1990, PILE BUCK® INC.

CELLULAR COFFERDAMS

A Pile Buck® Production

NOTICE

"The information, including technical and engineering data, figures, tables, designs, drawings, details, procedures and specifications, presented in this publication has been prepared in accordance with recognized contracting and/or engineering principles, and are for general information only. While every effort has been made to insure its accuracy, this information should not be used or relied upon for any specific application without independent competent professional examination and verification of its accuracy, suitability and applicability, by a licensed professional.

This handbook is provided without warranty of any kind. Pile Buck®, Inc. hereby disclaims any and all express or implied warranties of merchantibility, fitness for any general or particular purpose or freedom from infringement of any patent, trademark, or copyright in regard to information or products contained or referred to herein. Nothing herein contained shall be construed as granting a license, express or implied, under any patents. Anyone making use of this material does so at his own risk and assumes any and all liability resulting from such use. The entire risk as to quality or useability of the material contained within is with the reader. In no event will Pile Buck®, Inc. be held liable for any damages including lost profits, lost savings or other incidental or consequential damages arising from the use or inability to use the information contained within. Pile Buck®, Inc. does not insure anyone utilizing this Handbook against liability arising from the use of this information and hereby will not be held liable for "consequential damages," resulting from such use.

All advertising contained within is the exclusive representation of those registered herein. Pile Buck®, Inc. makes no representation as to the accuracy, performance, design, specifications and/or any such "claims" made by advertisers, contained within. Anyone making use of these products does so at his own risk and assumes any and all liability resulting from such use. In no event will Pile Buck®, Inc. be held liable for any damages including lost profits, lost savings or other incidental or consequential damages arising from the use or inability to use the products advertised within. Pile Buck®, Inc. does not insure anyone from liability arising from the use of these products and hereby, will not be held liable for "consequential damages," resulting from such use.

Copyright © 1990, Pile Buck, Inc.
All Rights Reserved.
Produced in the United States of America.

Published By:
Pile Buck® Inc.
P.O. Box 1056
Jupiter, FL 33468-1056
PH: 407-744-8780
FAX: 407-575-9748

ACKNOWLEDGEMENTS

The publishers of Cellular Cofferdams, A Pile Buck® Production gratefully acknowledge the work of the Waterways Experiment Station, Vicksburg, MS., and the authors of Theoretical Manual for Design of Cellular Cofferdams, Mark Rossow, Edward Demsky and Reed Mosher, as well as the Tennessee Valley Authority, Knoxville, TN, the authors of Steel Sheet Pile Cellular Cofferdams on Rock.

Also, thanks to Bill Lovelady of W.E.S. as well as Alan Carmichael and Dean Robinson of TVA.

Additional thanks to Harry A. Lindahl, P.E., who was instrumental in the compilation of the enclosed works.

And finally to Don C. Warrington of Vulcan Iron Works Inc. who once again has devoted his time to the study and further understanding of the operation of pile driving equipment.

ADDITIONAL BOOKS, PUBLICATIONS AND SPECIFICATION CHARTS AVAILABLE FROM PILE BUCK®

In addition to this book, our Pile Buck® "Series" of (6) Marine Construction Handbooks consist of:

COASTAL CONSTRUCTION

■

HARBORS, PIERS & WHARVES

■

BULKHEADS, MARINAS & SMALL BOAT FACILITIES

■

PROTECTION, INSPECTION & MAINTENANCE OF MARINE STRUCTURES

■

MATERIALS & EQUIPMENT FOR MARINE CONSTRUCTION

Also available:

PILE BUCK® NEWSPAPER *(Published Twice Monthly)*
PILE BUCK® STEEL SHEET PILING DESIGN MANUAL
PILE BUCK® PILE HAMMER SPECIFICATIONS CHART
PILE BUCK® STEEL SHEET PILING SPECIFICATIONS CHART
PILE BUCK® PRODUCT DIRECTORY

For an up-to-date descriptive brochure regarding the available Pile Buck® Publications please call, FAX or write:

Pile Buck®, Inc.
Attn: Publications
P.O. Box 1056
Jupiter, Florida 33568-1056

PH: (407) 744-8780   FAX: (407) 575-9748

MANUFACTURERS, SUPPLIERS AND CONTRACTING SERVICES

Customer Number	Company Name	Ad Location (Page No.)
1	American Underwaters Contrs. Inc.	153
2	Bethlehem Steel Corporation	5
3	East Coast Pile	129
4	L.B. Foster Company	67
5	H & M Vibro, Inc.	79
6	Hercules Machinery Corp.	45
7	International Construction Equip. Inc.	21
8	Nestor - Merrick Materials	91
9	Penn State Fabricators	139
10	Pennco Pipe & Tube Corp.	29
11	Pipe & Piling Supplies (U.S.A.) Ltd.	13
12	Seaboard Steel Corporation	115
13	Shugart Manufacturing Inc.	55
14	Skyline Steel Corporation	37
15	Vulcan Iron Works Inc.	103

CUSTOMER #1

AMERICAN UNDERWATER CONTRS. INC.
219 Foerster Road   314/739-5235
St. Louis, MO 63042   EVES. 314/739-3939

CUSTOMER #2

BETHLEHEM STEEL PILING PRODUCTS
701 East Third Street   800/521-0432
Bethlehem, PA 18016   FAX: 215/694-2640
                         TELEX: 84-7417

CUSTOMER #3

EAST COAST PILE
87 Barnyard Lane   516/579-PILE
Levittown, N.Y. 11756   FAX: 516/735-8753

CUSTOMER #4

L.B. FOSTER COMPANY
415 Holiday Drive   412/928-3400
Pittsburgh, PA 15220   FAX: 412/928-7891

ADDITIONAL OFFICES:
Two Perimeter Park South   205/870-5413
Suite 210W
Birmingham, AL 35243

250 South 59th Avenue   602/269-5804
Phoenix, AZ 85043

Long Beach Corporate Square 1   213/422-6616
4300 Long Beach Blvd.
Suite 250
Long Beach, CA 90807

1837 Whipple Rd.   415/489-7000
Hayward, CA 94544

1051 Winderley Place   407/660-1222
Suite 105
Maitland, FL 32751

6455 Old Peachtree Road   404/662-7700
Norcross, GA 30071

1111 East Touhy Avenue   708/299-4450
Suite 285
Des Plaines, IL 60018

700 Bob-O-Link Drive   606/278-6926
Suite 220
Lexington, KY 40504

1921 Corporate Square Bldg. 504/649-7721
Suite 4
Slidell, LA 70458

Point West Place   508/879-8691
111 Speen St., Suite 423
Framingham, MA 01701

One Executive Drive   201/560-9790
Somerset, NJ 08873

15628 S.E. 102nd Avenue   503/657-7711
Clackamas, OR 97015

7015 Fairbanks-North Houston Rd.   713/466-2700
Houston, TX 77040   TELEX: 166122

7400 Beaufont Springs Drive   804/320-3345
Suite 225
Richmond, VA 23225

CUSTOMER #5

H & M VIBRO, INC.
P.O. Box 224   616/538-4150
Grandville, MI 49418   FAX: 616/532-8505

CUSTOMER #6

HERCULES MACHINERY CORPORATION
3101 New Haven Avenue   219/424-0405
P.O. Box 5198   U.S.   800/348-1890
Fort Wayne, IN 46895   IN   800/552-4848
                             FAX: 219/422-2040

ADDITIONAL OFFICES:
P.O. Box 6045   314/441-1871
St. Charles, MO 63301

03896-444 Picciola Road   904/787-7648
Fruitland Park, FL 32731

CUSTOMER #7

INTERNATIONAL CONSTRUCTION EQUIPMENT INC.
301 Warehouse Drive   704/821-8200
Matthews, NC 28105   800/438-9281
                             FAX: 704/821-6448
                             TELEX: 572385 ICE INTL

ADDITIONAL OFFICES:
83 Carmel Hill   203/266-0774
Woodbury, CT 06798   800/438-9281
                             Car (NY,NJ): 201/259-1510
                             Car (CT):   203/592-0952
                             Car (MA):   617/571-2720
                             FAX: 203/266-4440

Route 15   201/361-7494
North Wharton, NJ 07885   800/438-9281
                             Car: 201/308-6963
                             Truck: 201/303-2028
                             Truck: 201/394-3132
                             FAX: 201/361-8270

337 Oakdale Place   800/438-9281
Springfield, PA 19064   Car: 215/350-0403
                             FAX: 215/544-2757

5350 Beliveau   416/295-3007
Brossard, PQ J4Z-2C6   704/821-8200
                             FAX: 514/445-6919

8311 Mundare Cresent   416/295-3007
Niagara Falls, ON L2G-7M6   704/821-8200
                             Car: 416/687-4783
                             FAX: 416/295-3003

310 Winston Creek Parkway   813/680-1508
Lakeland, FL 33809   800/541-8244
                             Car: 813/297-8447
                             Car: 813/623-9018
                             FAX: 813/680-1540

5920 Cunningham   713/466-9781
Houston, TX 77041   800/438-9281
                             FAX: 713/466-9829

2222 Engineers Road   504/392-4131
Belle Chasse, LA 70037   800/438-9281
                             FAX: 504/392-0107

100 Stewart Street   707/643-8898
Vallejo, CA 94590   Car: 707/483-0743
                             FAX: 707/643-7960

8101 Occidental Ave. South   206/764-4787
Seattle, WA 98108   800/247-4335
                             FAX: 206/764-4985

19 Tuas Cresent   65/861-3773
Singapore 2263   FAX: 65/861-3043
                             TELEX:786/29252

CUSTOMER #8

NESTOR - MERRICK MATERIALS
P.O. Box 95   516/542-9100
Merrick, N.Y. 11566   516/546-7900
                             FAX: 516/542-9087

CUSTOMER #9

PENN STATE FABRICATORS
124 Newton Street   718/388-1890
Brooklyn, N.Y. 11222-4898   FAX: 718/388-4741

CUSTOMER #10

PENNCO PIPE & TUBE CORP.
1100 East Hector St.   215/834-6310
Conshohocken, PA 19428   FAX: 215/834-6318
                             TELEX:244894

ADDITIONAL OFFICES:
Daingerfield, TX   214/645-3944
Salt Lake City, UT   801/264-8011
Montreal, CN   514/621-7702

CUSTOMER #11

PIPE AND PILING SUPPLIES (USA) LTD.
244 Kincheloe Road   906/495-2245
Kincheloe, MI 49788   800/874-3720
                             FAX: 906/495-5754

ADDITIONAL OFFICES:
Shipyard Road   800/874-3720
Seneca, IL 61360

14110 Giles Road   800/874-3720
Omaha, NE 68138

5025 Ramsay Street
St. Hubert, Quebec
Canada J3Y 2S3
514/445-0050
FAX: 514/445-4828

Brampton, Ontario, Canada
416/796-6800

1835 Kingsway Avenue
Port Coquitlam, B.C.,
Canada V3C 1S9
604/942-6311
FAX: 604/941-9364

5515 40th Street, S.E.
Calgary, Alberta
Canada T2C 2A8
403/236-1332
403/279-4588

220 5008 86th Street
Edmonton, Alberta,
Canada T6E-5S2
403/465-0501
FAX: 403/469-4971

99 Rocky Lake Drive, Unit 17
Bedford, Nova Scotia,
Canada B4A 2T3
902/835-6158
FAX: 902/835-6079

CUSTOMER #12

SEABOARD STEEL CORPORATION
P.O. Drawer 3408
Sarasota, FL 34230
813/355-9773
FAX: 813/351-7064

CUSTOMER #13

SHUGART MANUFACTURING, INC.
P.O. Box 748
Chester, S.C. 29706
803/581-5191
FAX: 803/581-1080

CUSTOMER #14

SKYLINE STEEL CORPORATION
405 Murray Hill Parkway
East Rutherford, NJ 07073
201/933-7070
FAX: 201/933-0787

ADDITIONAL OFFICES:

Suite 345
11550 Fuqua
Houston TX 77034
713/484-4000
FAX: 713/484-2782

Suite 202
60 E. Sir Francis Drake Blvd.
Larkspur, CA 94939
415/461-4900
FAX: 415/461-1624

17W703G Butterfield Road
Oakbrook Ter., IL 60181
708/620-6300
FAX: 708/620-6394

Suite 506
3169 Holcomb Bridge Rd.
Norcross, GA 30071
404/447-1600
FAX: 404/447-1151

303 West Main Street
Freehold, NJ 07728
201/409-2800
FAX: 201/409-2828

Suite 230
5641 Burke Center Pkwy.
Burke, VA 22015
703/978-2500
FAX: 703/978-2908

Bell Realty Building
Highway 231 South
Pell City, AL 35125
205/338-7774
FAX: 205/338-7748

Suite 900
3 Monroe Parkway
Lake Oswego, OR 97035
503/636-9633
FAX: 503/636-9273

CUSTOMER #15

VULCAN IRON WORKS
2909 Riverside Drive
P.O. Box 5402
Chattanooga, TN 37406
615/698-1581
800/742-6637
FAX: 615/698-1587
TELEX: 558412

SERVICES AND SERVICES # IN ADDITION TO THE CORRESPONDING NUMBERS OF THE COMPANIES THAT SUPPLY THE SERVICE

SERVICE # AND SERVICE	ENGR./CONTRACTING SERVICE #
Barge, Workboat Rentals	
Blasting	
Bulkhead Design	
Concrete Repair	
Construction Claims Assistance	
Dock/Pier Design	8
Drystack Storage Design	
Environmental Studies	
Groundwater Control	
Harbor Design	
Hydrographic Survey	
Marina Design	8
Marine Insurance (General)	
Marine Insurance (Hulls & Equip.)	
Marine Structure Design	
Permitting	
Pile Driving	8
Pile Driving Studies	8
Pile Restoration	
Pressure Grouting	
Site Investigations	
Underpinning System Designs	8
Underwater Construction	1,8
Underwater Inspections	1
Underwater Welding and Burning	1
Vibration/Noise Monitoring	
Workers Compensation Insurance	

PRODUCT AND PRODUCT # IN ADDITION TO THE CORRESPONDING MFR./SUPPLIER NUMBERS OF THE COMPANIES THAT SUPPLY THE PRODUCT

PRODUCT # AND PRODUCT	MFR./SUPPLIER #	PRODUCT # AND PRODUCT	MFR./SUPPLIER #
Air Compressors	3	Auger Cutterheads & Tools	3,6,7,12
Air Hoists/Winches		Auger Flighting & Accessories	3,6,7,12
Air Hose & Access.	3,12	Auger Teeth	3,6,7,12
Air/Steam Pile Hammers	3,6,9,12	Augers (Augered Piles)	3,6,7,12
Air Tuggers		Augers (Test/Soil Boring)	6
Aluminum Channels, Bar & Plate		Augers (Truckmounted)	6
Aluminum Decking		Barge Fittings	
Aluminum Dock Systems		Barge Mats	
Aluminum Gangways		Barge Movers (Winches)	
Aluminum Ladders		Barges, Crane	13
Aluminum Sheet Piling		Barges, Deck	
Amphibious Work Vehicles		Barges, Derrick	
Anchor Chain		Barges, Hopper	
Anchor/Mooring Systems		Barges, Pontoon	13
Anchors (Barge, Boat, etc.)		Barges, Sectional	13
Anchors (Bulkhead)	14	Batch Plants	
Anchor Winches/Hoists		Beams, Aluminum	

PRODUCT # AND PRODUCT	MFR./SUPPLIER #
Beams, Concrete	
Beams, Fiberglass	
Beams, Steel	3,6,8,10,11,12,14
Beams, Timber	
Blocks, Sheaves & Tackle	
Boat Hoists	
Boat Lifts/Davits	
Boat Ramps	
Bollards	
Bolts, Washers, etc.	
Boring & Sounding Equip.	
Bottom Braces/Spotters	3,6
Breakwaters	3
Bridges (Floating)	13
Bridges (Portable)	
Bridges (Pre-Fab. Steel)	6,8
Bridges (Pre-Fab. Timber)	
Buckets (Clamshell, Dragline, Grapples, etc.)	
Bumpers/Fenders, Barge	
Bumpers/Fenders, Boat	
Bumpers/Fenders, Bow & Stern	
Bumpers/Fenders, Bulkhead	3,14
Bumpers/Fenders, Camels	3
Bumpers/Fenders, Dock	3
Bumpers/Fenders, Dolphin	3
Bumpers/Fenders, Foam	
Bumpers/Fenders, Marina	3
Bumpers/Fenders, Pier	3,14
Bumpers/Fenders, Pneumatic	
Bumpers/Fenders, Pushboat	
Bumpers/Fenders, Rubber	
Bumpers/Fenders, Timber	
Bumpers/Fenders, Tugboat	
Buoys	
Caisson Pipe	3,4,6,8,10,11,14
Caissons	3,6,8,11,14
Camels	3
Capstans	
Car Floats	
Casing	3,4,8,10,11,14
Cathodic Protection Devices	
C.C.A. Piling	6
C.C.A. Timbers	
Centrifical Pumps	7
Chain (Anchor/Mooring)	
Chain (Crawler)	
Channel Caps	3,8,14
Channel, Aluminum	
Channel, Steel	3,8,14
Cherry Pickers	
Clamshell Buckets	
Cleats	
Clevis Assemblies	
Coal Tar Epoxy	3,4,8
Coatings/Coaters (Corrosion Control)	3
Column/Pier Forms	
Computer Software (Dredging)	
Computer Software (Misc.)	
Computer Software (Piling)	
Concrete Batch Plans	
Concrete Cribbing	
Concrete Decking	
Concrete Forms	13
Concrete Hose & Accessories	
Concrete Pile Cutters	3,6
Concrete Pile Splicers	6
Concrete Piling	
Concrete Pumps	
Concrete Sheet Piling	
Containment Systems (Floats)	
Corrosion Protection	3
Couplings (Dredge Pipe)	
Couplings (Hose)	
Couplings (Pile)	3,6

PRODUCT # AND PRODUCT	MFR./SUPPLIER #
Crane Barges	13
Crane Booms, Jibs, etc.	
Crane Mats	
Crane Parts & Access.	
Cranes, Crawler	
Cranes, Derrick	
Cranes, Hydraulic	
Cranes, Truck	
Creosoted Piling	6
Creosoted Timbers	
Crewboats/Supply Boats	
Cushion Material (Pile Hammer)	3,6,7,9,15
Cushion Material (Piling)	3,6,9
Cutterheads (Augers)	3
Cutterheads (Dredge)	
Davits/Boat Lifts	
Deck Fittings	
Deck Winches	
Decking, Aluminum	
Decking, Concrete	
Decking, Fiberglass	
Decking, Treated Timber	
Decking, Untreated Timber	
Derricks	
Dewatering Pumps & Equip.	
Diesel Pile Hammers	3,4,6,7,12
Diving Equip./Supplies	1
Dock Boxes	
Dock Fingers	
Dock Hardware	
Dock Lockers	
Dock Power Units	
Docks (Fixed/Stationary)	
Docks (Floating)	13
Dock Washers	
Docking/Boarding Systems	
Dolphins	3
Draglines	
Dragline Buckets	
Drainage Fabrics	7
Drainage Pipes	10,11
Dredge Frames	
Dredge Pipe	3,4,8,10,11,14
Dredge Pipe Couplings/Connections	4,8
Dredge Pipe Flotation	
Dredge Pumps	
Dredges (Bucket)	
Dredges (Cutterhead)	
Dredges (Hydraulic)	
Dredges (Portable)	
Dredges (Suction)	
Drill Bits & Access.	6
Drill Rigs (Crawler Mounted)	6
Drill Rigs (Soil Testing)	
Drill Rigs (Truck Mounted)	6
Drop Hammers	3,6,7
Dry Docks	
Dry Stack Storage Systems	
Earth, Rock, Soil Anchors	6,8
Electrical Dock Components	
Electric Winches/Hoists	
Epoxy Products	
Erosion Control Blocks	
Erosion Control Fabrics	
Excavators	8
Excavators, Amphibious	8
Extractors (Pile)	3,4,5,6,7,8,12,15
Fairleads	6
Fasteners (Hardware)	
Fenders/Bumpers, Barge	8
Fenders/Bumpers, Boat	
Fenders/Bumpers, Bow & Stern	
Fenders/Bumpers, Bulkhead	8,14

PRODUCT # AND PRODUCT	MFR./SUPPLIER #
Fenders/Bumpers, Camels	3
Fenders/Bumpers, Dock	3,8
Fenders/Bumpers, Dolphin	3
Fenders/Bumpers, Foam	
Fenders/Bumpers, Marina	3,8
Fenders/Bumpers, Pier	3,14
Fenders/Bumpers, Pneumatic	
Fenders/Bumpers, Pushboat	
Fenders/Bumpers, Rubber	
Fenders/Bumpers, Timber	
Fenders/Bumpers, Tugboat	
Fiberglass Decking	
Filter Cloth	
Finger Docks/Piers	
Fixed (Stationary) Dock Systems	
Flap Gates	
Float Balls	
Float Drums	
Floating Dry Docks	
Floats	
Floats, Containment	
Floats, Pipe Line	
Floating Breakwaters	
Floating Bridges	
Floating Dock Systems	
Foam Flotation	
Foundation Testing Equip.	6
Gabion Wire	
Gabions	
Gangways	
Generators	
Geotextile Fabrics	7
Grapples	
Greenheart Piling	6
Grout Hose & Accessories	
Grout Pumps	
H-Piling	2,3,4,6,8,10,11,12,14
HZ-Sheet Piling	4,8,14
Hairpins	3,6,7,12,15
Handrail Hardware	
Handrails	
Handrails (Aluminum)	
Hardware (Galvanized)	
Hardware (Marine)	
Headsheaves	6
Hoists/Winches (Air)	
Hoists/Winches (Electric)	
Hoists/Winches (Hydraulic)	6
Hoists/Winches (Mechanical)	6
Hoists/Winches (Mooring)	6
Hoists/Winches (Spud)	6
Hydraulic Jacks	6
Hydraulic Power Packs	6,7
Jacks	6
Jacking Equipment	6
Jet Pumps, Hose & Access.	6,7
Ladders	
Ladders (Aluminum)	
Leads & Accessories	3,6,7,12,15
Line Pipe	3,4,6,8,10,11,14
Load Testing Equipment	6
Lubricators	3
Lumber, Treated	
Lumber, Tropical	
Lumber, Untreated	
Marine Electronic Equip.	
Marine Engines	
Marine Generators	
Marine Hardware (Bolts, Tie-rods, etc.)	
Marine Propellers & Units	
Marine P.T.O. Assemblies, etc.	
Marsh Buggies	

PRODUCT # AND PRODUCT	MFR./SUPPLIER #
Mats, Crane	
Metal Docks	
Mooring/Anchor Systems	
Mooring Arms	
Mooring Buoys	
Mooring Docks/Piers	
Mooring Dolphins	
Mooring Hooks	
Mooring Lines	
OCTG Pipe	4,11
OGee Washers	
Penta Piling	6
Pile Beavers (Concrete Pile Cutters)	3,6
Pile Caps	6,7,8
Pile Clamp Assemblies	3,6,7
Pile Cushion Material	3,6,9,15
Pile Driving Analyzers	8
Pile Forming Equipment	
Pile Guides (Floating Dock)	6
Pile Guides (Lead Systems)	3,6,7,15
Pile Hammer Analyzers	6
Pile Hammer Cushion Material	3,4,6,7,9,12,15
Pile Hammer Extractors	3,4,6,7,12,15
Pile Hammer Leads & Access.	3,4,6,7,12,15
Pile Hammers, Air/Steam	3,6,9,12,15
Pile Hammers, Diesel	3,4,6,7,9,12
Pile Hammers, Drop	3,6,7
Pile Hammers, Hairpin	3,4,6,7,12,15
Pile Hammers, Hydraulic	3,4,5,6,7,12
Pile Hammers, Vibratory	3,4,5,6,7,12,15
Pile Jackets	6
Pile Points	3,4,6,7,14
Pile Restoration Materials	
Pile Shackles & Threaders	3,4,6,12
Pile Splicers	3,6,7,10,14
Pile Testing Equipment	6
Pile Wraps	6
Piling, Aluminum Sheet	6
Piling, CCA	6
Piling, Concrete	6
Piling, Creosoted	6
Piling, Greenheart	6
Piling, HZ	4,8,14
Piling, H-Pile	2,3,4,6,8,10,11,12,14
Piling, Penta	6
Piling, Plastic	6
Piling, Pipe	3,4,6,8,10,11,14
Piling, Spiral Weld	3,4,6,8,10,11,14
Piling, Steel Sheet	2,3,4,6,8,11,12,14
Piling, Treated Timber	6
Piling, Tropical	6
Piling, Untreated Timber	6
Pipe, Aluminum	
Pipe, Steel	3,4,6,10,11,14
Pipe Couplings (Dredge)	
Pipe Couplings (Piling)	3,4,6,7,14
Pipe Line Floats	
Pipe Struts	3,8,14
Plastic Piling	
Plastic Sheet Piling	6
Plate, Aluminum	
Plate, Stainless Steel	
Plate, Steel	3,8
Points, Pile	3,4,6,7,14
Polyurethane Coated Piling	3,4
Pontoons	
Pontoon Barges	13
Post Tensioning Equip.	
Post Tensioning Jacks	
Power Lifts (Boat)	
Power Units, Dock	
Prestressed Concrete Decking	
Prestressed Concrete Beams, Girders, etc.	
Prestressed Concrete Piling	

PRODUCT # AND PRODUCT	MFR./SUPPLIER #
Pumpout Systems	
Pumps, Concrete	
Pumps, Dewatering	6,7
Pumps, Grout	
Pumps, Jet	6,7
Pumps, Sludge	
Pushboats/Workboats	
Rigging (Shackles, etc.)	
Rock & Soil Anchors	6
Rock Drilling Equipment	6
Rough Terrain Cranes	
Sectional Barges	13
Shackles	3,4
Sheaves, Blocks & Rigging	
Sheet Pile Shackles	3,4,6,12
Sheet Pile Threaders	3,4,6,12
Sheet Piling, Aluminum	6
Sheet Piling, Concrete	
Sheet Piling, HZ	4,8,14
Sheet Piling, Lightweight Steel	2,3,4,6,8,12,14
Sheet Piling, Plastic	6
Sheet Piling, Steel	2,3,4,6,8,11,12,14
Sheet Piling, Timber	6
Shoring	3,6,8,14
Sluice Gates	
Soil/Rock Anchors	6
Soil Stabilization Fabrics	7
Splicers, Pile	3,4,6,7,14
Spotters (Bottom Braces)	3,6,7
Spike Grids	
Spuds (Barge)	3
Steam Hose	3,15
Steel Fabricators	6,8
Steel H-Piling	2,3,4,6,8,10,11,12,14
Steel Pipe	3,4,6,8,10,11,14
Steel Sheet Piling	2,3,4,6,8,11,12,14
Structural Steel, Plate, Wide Flange, etc.	3,4,8,14
Submersible Pumps	7
Swamp Buggies	
Tampers (Vibratory Compactors)	
Test/Soil Boring Equipment	6

PRODUCT # AND PRODUCT	MFR./SUPPLIER #
Throttle Valves	6,15
Tie Rods	3,4,6,14
Tieback Machines	6,7
Timber Billets	
Timbers, CCA	6
Timbers, Creosoted	6
Timbers, Greenheart	6
Timbers, Penta	6
Timbers, Treated	6
Timbers, Tropical	6
Timbers, Untreated	6
Tongs (Log)	
Tongs (Rock)	
Trench Boxes	
Tropical Lumber	
Tropical Piling	6
Turnbuckles	4
Tugboats	
Tunneling Equipment	
U-Bolts	
Underpinning Equipment	6,8
Underreamers	
Underwater Const. Equip.	6,7
Utility Boxes (Dock)	
Vibration Measuring Devices	6
Vibratory Pile Drivers/Extractors	3,4,5,6,7,12,15
Wale Systems	3,4,14
Washers, Bolts etc.	
Wave Attenuators	
Well Point Equipment	
Wide Flange Beams	3,6,14
Winches/Hoists, Air	
Winches/Hoists, Electric	
Winches/Hoists, Hydraulic	6
Winches/Hoists, Mechanical	6
Winches/Hoists, Mooring	6
Winches/Hoists, Spud	6
Windlesses	
Wire Rope & Accessories	
Wire Rope Pullers	
Wire Rope Tension Monitoring Equip.	
Workboats/Pushboats	13

TABLE OF CONTENTS

STEEL SHEET PILE CELLULAR COFFERDAMS ON ROCK
Technical Monograph No. 75

Chapter	Page
1. Introduction	3
Cofferdams for TVA Projects	3
The Steel Sheet Pile Cellular Type	3
Considerations Influencing Selection	4
Adaptability	4
Scope of Report	4
2. Cells	4
Circular-Type Cell	4
Advantages of the Circular-Type Cell	4
Disadvantages of the Circular-Type Cell	10
Diaphragm-Type Cell	11
Advantages of the Diaphragm-Type Cell	11
Disadvantages of the Diaphragm-Type Cell	16
Diaphragm-Type Versus Circular-Type Cell	16
Cloverleaf-Type Cell	16
Other Types of Cell	18
Other Uses of Circular Cells	18
Factors Affecting Selection of Cell Sizes	18
3. Steel Sheet Piling	19
Types	19
Web Thickness	19
Fabricated Piling	19
Interlock Strength	22
Piling for Kentucky Cofferdam	23
Field Driving Test	23
Specifications for New Steel Sheet Piling for Kentucky Cofferdam	23
Interlock Failure--Kentucky Cofferdam	24
Length of Piling	24
Splicing Piling	24
Splicing Procedure--Chickamauga Cofferdam	24
Splicing Procedure--Kentucky Cofferdam	25
4. Fill Materials	25
General Requirements for Cell Fill	25
Cell Fill	25
Silt	25
Mud	25
Clay	25
Sand	25
Sand and Gravel	25
Other Materials	26
Placing Cell Fill	26
Berms	
Placing Berms	26
Saturation and Drainage of Cell Fills and Berms	26
Saturation Line	26
Tests for Saturation Line--Guntersville and Pickwick	27

Tests for Saturation Line--Chickamauga Lock Cofferdam	28
Tests for Saturation Line--Kentucky Cofferdam	28
Location of Saturation Line for Design Purposes	28
Cell Drainage	28
Cell Drainage at Kentucky Cofferdam	28

5. Design ... 29
The Development of Cellular Cofferdams ... 29
Fundamentals of Design ... 32
Cells ... 32
- Sliding Resistance ... 32
- Overturning ... 32
- Shear ... 32
- Revised Method for Computing Shear Resistance in Cell Fills ... 33
- Frictional Resistance of Piling on Fill ... 34
- Average Width of Cofferdam ... 35
- Fill Pressure ... 35
- Interlock Tension ... 36
- Resistance of Piling to Sliding in Interlocks ... 38
- Overburden ... 38
- Berms ... 38
  - Outside Berms ... 38
  - Inside Berms ... 38
  - Berm Design ... 38
  - Reactive Pressure for Berms with Negative Unbroken Backslopes ... 38

Construction Features Affecting Design ... 40
Notations and Formulae ... 40
Design Procedure ... 41
- Cofferdams Without Berms ... 41
- Cofferdams With Berms ... 42

Design Example ... 42
- Case I ... 44
- Case II ... 44
- Case III ... 46
  - Conclusions from Investigations, Case III ... 46
- Case IV ... 47
  - Comparison Check with Case I ... 47
- Case V ... 48

Examples of Actual Cofferdam Design Used on TVA Projects ... 48
- Early Cofferdams ... 48
- Kentucky Cofferdam ... 48
- Hales Bar Cofferdam ... 49

6. Final Layout ... 66
Size of Cofferdam Enclosure ... 66
Auxiliary Cofferdam Structures ... 66
Cell Alignment ... 66
Selection of Cell Size ... 66
Cell and Connecting Cell Layout and Details ... 66
- Deflection Angles in Cell Alignment ... 71
- Tie-In Detail ... 71

7. Construction ... 71
Template Design and Use ... 71
- Welded 1-Piece Pipe Template ... 71
  - Alignment and Setting of 1-Piece Pipe Template ... 71
  - Connecting Cell Template ... 78

Pipe Template Data 78
Wood Ring Template 78
Sectional All-Welded Pipe Template--Kentucky Cofferdam 78
   Spreaders 78
   Template for Connecting Cells 78
   Template Removal 78
Sectional All-Welded Outside-Type Pipe Template--Hales Bar Cofferdam 78
   Outer Template Section 79
   Inner Template Section 81
   Special Features 81
   Template Setting 81
Cloverleaf Template--Kentucky Cofferdam 81
Timber Templates 81
   Timber Templates--Kentucky Cofferdam 81

Handling and Setting Sheet Steel Piling
Pile Setting 86
Order of Setting Piling 86
Pile Setting Procedure on TVA Projects 86
Pile Setting Crews 87

Timber Trestle
Kentucky Cofferdam Trestle 87
   Timber Pile Template 87
   Timber Pile Driving 87
   Trestle Deck Section 87

Steel Sheet Pile Driving and Driving Equipment
Marking Piling 87
Guys for Steel Sheet Piling 87
   Order of Driving 91
Experience on TVA Projects 91
   Guntersville Cofferdam 91
   Chickamauga Cofferdam 91
   Kentucky Cofferdam 91
Driving Specifications for Steel Sheet Piling 95
Special Job-Built Driving Rig for Kentucky Cofferdam 95
Equipment Summaries 95

Load Bearing Piling
Test Piles 95
Pile Driving Formulae 99
   Engineering News Formula 99
   Comprehensive Pile Driving Formula for Practical Use 99
   Static Formula 100
   Formulae for Analysis of Energy Losses and a Check of Computations 100

Pile Hammers and Their Selection
Limitations Due to Type of Pile 101
Types and Sizes of Hammers 101
   Drop Hammers 101
   Single-Acting Hammers 101
   Double-Acting Hammers 101
   Differential-Acting Hammers 102
Data for Selection of Proper Size Hammer 102
Jetting 102
Hammer and Driving Data for TVA Projects 102
   Fishtail Guides 102

Tie-Ins
Tie-Ins to Riverbank 102
Tie-Ins to Completed Dam Structure 103

   Tie-Ins at Kentucky Cofferdam 104
 Closure 105
  Closure Procedures 109
   Chickamauga Cofferdam, Stage 3 109
   Guntersville Cofferdam, Stage 3 109
   Kentucky Cofferdam, Stage 2, Upstream Arm 109
   Kentucky Cofferdam, Stage 2, Downstream Arm 109
 Cell Filling 109
  Hydraulic Dredging 109
   Dredging Specification 114
   Special Hydraulic Placing Method--Chickamauga Cofferdam, Stage 3 117
   Special Hydraulic Placing Method--Hales Bar Dam 117
  Truck and Clamshell Filling 117
 Floodgates and Sluiceways 121
  Kentucky Cofferdam Floodgates and Sluices 121
  Floodgates for Other TVA Projects 121
 Foundation Treatment 121
  Grouting Equipment 121
  Foundation Treatment at Kentucky Cofferdam 121
   Before Cell Construction 121
   After Cell Construction, Stage 1-B 125
   During Stage 2 125
   River Arm Steel Sheet Pile Seal, Stages 1-B and 2 125
   Left Bank Rock and Overburden Treatment, Stage 2 125
  Foundation Treatment on Other TVA Projects 129
   Guntersville Cofferdam, Stage 2 129
   Chickamauga Cofferdamk, Stage 1 131
   Chickamauga Cofferdam, Stages 2 and 3 131
   Fort Loudoun Cofferdam, Stage 1 131
   Fort Loudoun Cofferdam, Stage 2 132
   Effect of Grouting on Pile Removal 132
 Unwatering and Maintenance 132

8. Removal 132
 Pile Extraction 132
  Extraction Procedure 133
  Extractor Size 133
 Removal Problems on TVA Projects 133
  Guntersville Cofferdam 133
   Special Pile Puller for Guntersville Dam 134
  Pickwick Cofferdam 134
  Kentucky Cofferdam 134
   Removal Equipment 134
   Sluicing Barge 134
   Extractor Units 134
  Special Methods of Extracting "Difficult" Piles 138

9. Tests 138
 Fill Material Characteristics for Kentucky Cofferdam 139
 Coefficient of Internal Friction 139
  Tests by TVA for Coefficient of Internal Friction 141
 Fill Material for the Hales Bar Upstream Cofferdam 145
 Test Cell at Kentucky Dam 145
 Model Studies 145
 Location of Saturation Line in Cofferdam Cells 145
 Cell Deflection 145

Tests for Strength of Piling Interlocks 145
Photoelastic Model Studies 145

Acknowledgments 145

Bibliography 146

Footnotes 146

THEORETICAL MANUAL FOR DESIGN OF CELLULAR SHEET PILE STRUCTURES (COFFERDAMS AND RETAINING STRUCTURES)

Page

PREFACE 148

Conversion Factors, NON-SI to SI (Metric) Units of Measurement 148

PART I: Introduction 149
    Purpose 149
    Scope 149
    Design Methods Presented in Rational Form 149
    Criteria and Design Procedures 149
    Limitations of the Manual 149
    Basic Combinations 149
    Soil-Structure Interaction in a Cellular Cofferdam 149
    Two Possible Analogies 149
    State of the Art 150
    Hypothesized Failure Modes 150

PART II: Analysis of Failure Modes 150
    Conventional Simplifications and Equivalent Layout 150
    Critique of Simplifications 150
    Failure Modes and Example Problems 153

PART III: Bursting 153
    Effects of Internal Lateral Stresses 153
    Critical Loading Cases 153
    Considerations in Interlock-Tension Calculations 155
    Alternate Method of Locating Plane of Fixity 157
    Interlock-Tension Calculations in Crosswall 157
    Rational Design Procedure to Avoid Bursting 158

PART IV: Slip on Vertical Center Plane in Fill 160
    Effects of External Lateral Forces 160
    Cell Foundation 160
    Considerations in Analysis of Failure by Vertical Shear on Center Plane 162
    Overturning Moment 162

PART V: Slip on Horizontal Planes in Fill (Cummings' (1957) Method) 167
    Horizontal Plane Sliding Due to Lateral Forces 167
    Considerations in Horizontal Shear Calculations 167
    Representative Integrals 171

PART VI: Slip Between Sheeting and Fill 173
    Vertical Sheeting Slip from Overturning Moment 173
    Considerations in Calculations for Slip Between Fill and Wall 174
    Contribution of Cell Bulging 175

PART VII: Pullout of Outboard Sheeting 177
    Rotation about the Toe 177
    Considerations in Calculations for Pullout 181

PART VIII: Penetration of the Inboard Sheeting (Plunging) 181
    Effects of Friction Downdrag 181
    Considerations in Calculations for Penetration 182
    Comments on the Design Procedure for Preventing Penetration 182

PART IX: Bearing Failure of Foundation 183
    Effects of Lateral Forces on Bearing Capacity 183
    Considerations in Calculations for Avoiding Bearing Failure of Foundation 183

PART X: Sliding Instability 183
    Effects of Lateral Force on Sliding 183
    Considerations in Calculations for Sliding Instability 183

PART XI: Slip on Circular Failure Surface (Hansen's Method) 185
    Alternative Mode of Failure 185
    Considerations in Calculations for Hansen's Method 186
    Failure Modes for Cofferdams on Sand 186
    Comments on Hansen's Method 187

PART: XII: Overturning 188
    Cause of Overturning 188
    Considerations in Overturning Calculations 188

APPENDIX A: Example Problems 190

APPENDIX B: Notation 210

REFERENCES 212

VIBRATORY PILE DRIVERS/EXTRACTORS

INTRODUCTION 215

HISTORICAL DEVELOPMENT 215

OVERVIEW 216

EQUIPMENT OPERATION 219

THEORY OF OPERATION 220

DRIVABILITY AND CAPACITY PREDICTION 222

CONCLUSION 227

ACKNOWLEDGEMENTS 227

REFERENCES AND FURTHER READING 228

PILE HAMMER SPECIFICATIONS CHART

Diesel Hammers 230
Hydraulic Vibratory Drivers/Extractors 232
Air/Steam Hammers 233

STEEL SHEET PILING SPECIFICATIONS

General Notes 236
"Z" Shaped Sheet Piling 239
Larssen Type and other "U" Shaped Sheet Piling 242
Arch and Light Gauge Sheet Piling 245
Flat Shaped Sheet Piling 247
Obsolete Sections 248

CREDITS/REFERENCES

PUBLICATION ONE:

Title	Steel Sheet Piling Cellular Cofferdams on Rock (Technical Monograph No. 75)
Publication Date	December, 1957
Document/Report Number	Technical Monograph No. 75
Author(s)	T.V.A. Tennessee Valley Authority
Courtesy of	Tennessee Valley Authority, Knoxville, TN

PUBLICATION TWO:

Title	Theoretical Manual For Design Of Cellular Cofferdams On Rock (Cofferdams And Retaining Structures)
Publication Date	May, 1987
Document/Report Number	Technical Report ITL- 87-5
Author(s)	Mark Rossow, Southern Illinois University; Edward Demsky, U.S. Army Engineer, District St. Louis; Reed Mosher, U.S. Army Engineer, Waterways Experiment Station
Courtesy of	Waterways Experiment Station, Vicksburg, MS.

PUBLICATION THREE:

Title	Vibratory Pile Driving Equipment
Publication Date	December, 1989
Document/Report Number	N/A
Author(s)	Don C. Warrington, Vulcan Iron Works Inc.
Courtesy of	Don C. Warrington and Vulcan Iron Works Inc., ©Copyright 1989, Don C. Warrington.

PUBLICATION FOUR:

Title	Pile Hammer Specifications Chart
Publication Date	August, 1989
Document/Report Number	N/A
Author(s)	Pile Buck®, Inc.
Courtesy of	Pile Buck®, Inc. , ©Copyright 1989, Pile Buck®, Inc.

PUBLICATION FIVE:

Title	Steel Sheet Piling Specifications Chart
Publication Date	September, 1989
Document/Report Number	N/A
Author(s)	Pile Buck®, Inc.
Courtesy of	Pile Buck®, Inc., ©Copyright 1989, Pile Buck®, Inc.

EDITORS NOTE

There has rarely been a better, more complete coverage of a specific, construction-related subject, than the manual assembled by TVA engineers and originally published in December 1957. While cellular sheet piling structures are only one of many types of sheet pile useages, it has been and will continue to be, an important one. Filled cells are used to build large temporary cofferdams and also for permanent structures such as deep water piers, breakwaters, mooring cells, dams, guide and guard walls and bridge protection devices.

Sheet pile cells were first used to cofferdam the site of the lock on the Black River Canal in Buffalo N.Y. in the early 1900's. When TVA began to plan for it's series of more than twenty hydro-electric dams to be built in the Southeastern United States in the 1930's, available design techniques were still based on rough assumptions and experience. Installation techniques (if any had been perfected) were concealed in the minds of contractor's superintendents.

With many years of cofferdam construction ahead, TVA decided to completely research the subject. Developing new theories regarding failure modes, which were independently confirmed by Karl Terzaghi in his famous paper "Stability and Stiffness of Cellular Cofferdams" (A.S.C.E. Transactions Vol. 110-1945), TVA designed and constructed seven large multistage sheet pile cofferdams. This experience was then assembled into Technical Monograph 75 and became available to anyone interested. The first printing (1957), was followed by a second printing in late 1966.

Subsequent to Professor Terzaghi's paper, (1945), other theories of design were proposed by Cummings and by Hansen. Cumming's approach to failure models has been used by various engineers including TVA in some of their later cofferdams. This is discussed briefly in the Forward to the 1966 reprint (See page 3). Our readers however, are encouraged to research these newer methods for themselves. The second publication in this book - "Theoretical Manual for Design of Cellular Sheet Pile Structures" contains a review and analysis of both the Cummings and Hansen Methods. A single, universally recognized, approach to design which will always produce safe and economical coffer-cells still seems out in the future, awaiting further research on full size structures.

This reprint of Technical Monograph 75 is offered in it's original form with the exception of some photographs which could not be reproduced. Beyond it's discussion of design, the manual would be of extreme interest for it's coverage of pile driving, template design, inspection, filling, maintenance, extraction and other subjects. Pile Buck has foot-noted some of the areas where the reviewer should be aware of changes since the manual was published. In particular, the reader should be aware that sheet pile sections offered when the TVA cofferdams were being built, may no longer be available. More efficient sections have taken their place. Also, pile driving equipment and techniques, which receive generous coverage in TVA's text, have also changed for the better. To bring our readers up to date on these subjects, Pile Buck has included its two most recent specifications charts as well as a chapter on Vibratory Drivers/Extractors.

The Pile Buck® Steel Sheet Piling Specifications Chart lists over 200 sheet piling sections offered by 14 different producers and includes Z-type, U-shape, arch and flat web shapes, their properties for design and specification and other descriptive information.

The Pile Buck® Pile Hammer Specifications Chart lists over 250 diesel, steam/air and vibratory type pile hammers along with their specifications and descriptional information.

In addition to our 2 specifications charts we have included a chapter on Vibratory Pile Drivers/Extractors, the primary tool utilized today in the driving and extraction of steel sheet piling for cellular cofferdam construction. Written by Don C. Warrington of Vulcan Iron Works Inc., this excellent chapter is the most up-to-date information on Vibratory Drivers/Extractors available today.

As a companion piece relating to the general subject, Pile Buck has also reprinted a recent (1987) U.S. Government publication, "Technical Report ITL-87-5 Theoretical Manual for Design of Cellular Sheet Pile Structures" for reviewers interest and comparison with the design methods contained in TVA Manual 75. This work, sponsored by the Waterways Experiment Station U.S. Army Corps of Engineers Vicksburg, Mississippi, was authored by Mark Rossow, Edward Demsky and Reed Mosher. This report reviews each of the possible failure modes affecting cellular cofferdams, analyzes historical and current methods of designing against these possibilities, and finally expresses some opinions and conclusions regarding what might constitute a proper approach to design. Some of the design steps inTVA 75 are found to be either unnecessary or possibly mis-applied. The Cummings and Hansen theories regarding failure of the fill are seriously questioned. This paper is an important summary of the current state of the art and covers many questions which may have bothered designers of cellular structures in the past. The need for further full scale tests of cells is made apparant in preparation for probable future design solutions by finite element analysis.

Pile Buck agrees with the general consensus that cellular design is still very much an art which should be based on experience as well as sound engineering judgement. In re-printing these papers, Pile Buck does not necessarily endorse or recommend any of the methods proposed for design or construction. A study of past problems is of great value in designing for the present. Most cofferdam failures have been caused by construction or material shortcomings rather than sizing of the cells during the design phase. It seems rather foolish now to be concerned about small differences in input values for soil fill when a poor installation practice or out-of-specification material is the more likely cause for a failure.

All three chapters contain bibliographies listing most of the available discussion of this subject up until 1987.

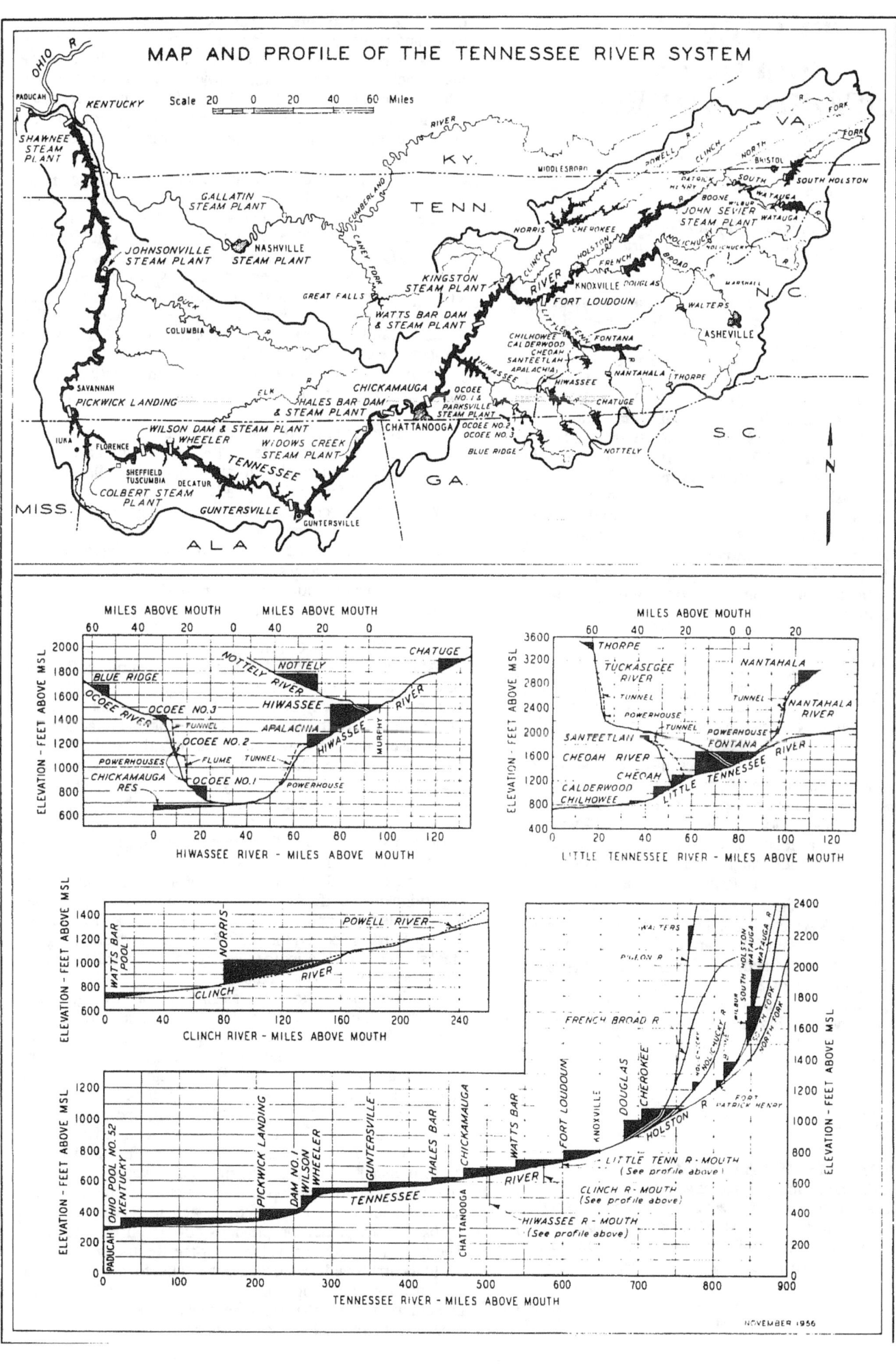

STEEL SHEET PILE CELLULAR COFFERDAMS ON ROCK

(December 1957)

(Technical Monograph No. 75)

*Tennessee Valley Authority
Divisions of Engineering and Construction*

The information comprising this monograph ("Information") was most recently reviewed for publication by the Tennessee Valley Authority in December 1957 and has not been revised since then.

Information is provided "AS IS" WITHOUT WARRANTY OF ANY KIND, EITHER EXPRESSED OR IMPLIED, including, without limitation, the IMPLIED WARRANTIES OF MERCHANTABILITY AND FITNESS FOR A PARTICULAR PURPOSE. The United States, TVA, and their agents and employees assume no risk as to the accuracy, quality, and performance of Information. In no event shall the United States, TVA, and their agents and employees assume any liability whatsoever to any party for any damages, direct, indirect, incidental, or consequential, including, without limitation, any lost profits or lost savings arising out of the use or inability to use Information.

FOREWARD

This second reprint, June 1990 of Tennessee Valley Authority Technical Monograph No. 75, Volume 1, Steel Sheet Piling Cellular Cofferdams on Rock, is a reproduction of the original issue of December 1957.

In the decade after 1957, TVA designed and constructed several major cellular steel cofferdams and has, in each case, taken advantage of the latest applicable developments in this field of heavy construction. However, TVA's 24 years of experience to 1957, as related in this monograph, continues to be a valuable reference for designers and constructors who may find therein some applications to their cofferdam problems.

A new theory of design has gained favorable attention and has been used by TVA since 1959. The new theory is presented in Paper No. 1366, entitled "Cellular Cofferdams and Docks," by E.M. Cummings, published in the Journal of the Waterways and Harbors Division of the ASCE, Vol. 83, No. WW3, September 1957. Subsequent discussion by Messrs. Samuel Heyman and Cevdet Z. Erzen is published in the Journal of the Waterways and Harbors Division of the ASCE, Vol. 84, No. WW2, March 1958.

CHAPTER 1

COFFERDAMS FOR TVA PROJECTS

Stream diversion and the care of water play a greater part, perhaps, in shaping a hydro project construction program than any other single factor. This is reflected in the size, design, and cost of the cofferdam structures, the construction equipment and construction operations involved, and the arrangement of the construction plant and other construction facilities to best serve the construction project. See Figures 1, 2, 3, 4 and 5. This is not always strictly true for small tributary dams where the streams are narrow and shallow, cofferdams small and relatively simple in design, and needed preparation against possible floods relatively minor. Stream diversion is a major factor, however, on larger projects such as Fontana Dam where earth dike cofferdams and diversion tunnels were used; or at Kentucky Dam where earth dikes and large cellular steel sheet pile cells were used, involving an estimated expenditure of $3,000,000.

Cofferdams at Kentucky, Pickwick, Guntersville, Hales Bar, Chickamauga, Watts Bar, and Fort Loudoun project--all on the main river--were principally of cellular steel sheet pile construction. Earth dikes were used extensively at Kentucky and Fort Loudoun Dams, and to a lesser extent at some of the other projects for inshore enclosures where local conditions and depth of water permitted. The cofferdams at Wheeler (also on the main river), the river arm of stage 1 cofferdam at Fort Loudoun, and a number of tie-in structures at some of the other projects were of timber construction.

In the design of a cofferdam, careful consideration must be given to preliminary investigations which cover:

1. Selection of cofferdam type.(The information presented in this volume is based on the assumption that the steel sheet pile cellular type cofferdam has already been selected.)
2. The number and arrangement of stages to suit construction sequence.
3. Determination of the cofferdam height based on flood frequency studies.
4. Site investigation, using core borings.
5. Available materials for cell fill, berms, etc.

The Steel Sheet Pile Cellular Type

At the time TVA started its first steel sheet pile cofferdam, little written information was available on design and construction of such structures. The development of design procedure had been left almost entirely to manufacturers of steel sheet piling and was available mostly in the form of empirical formulae. Construction details were sketchy and not readily available, and a relatively few publications dealt with this subject. Existing records covered the design and construction of a number of types of steel sheet pile cofferdams, but the information was conflicting and a minimum of conclusive data were available. With data that could be obtained, and aided by information gained from observing other cofferdams under construction, TVA designed and built its first steel sheet pile cofferdam at Pickwick in 1935. This cofferdam formed the basis for all subsequent design of steel sheet pile cells prior to the design and construction of the Kentucky cofferdam, started in 1939.

The Kentucky cofferdam was almost twice as high as any cofferdam yet built by TVA, and it was obvious, therefore, that careful consideration would be required in the design and construction of such a structure. Since there had still been no significant advances in the cofferdam design and construction field--at least insofar as the availability of publications was concerned--during the 3 years following the start of the Pickwick cofferdam, TVA engineers were required to develop the design and construction methods to be followed for the much higher Kentucky cofferdam. The investigations made and the procedure developed were much more thorough and rational than any yet undertaken and were materially enhanced by experience gained on previous cofferdam construction.[2] The design procedure developed for the Kentucky cofferdam led to the improved design methods discussed in chapter 5 and later used in the design of the cofferdam at Hales Bar for construction of the powerhouse extension for additional units No. 15 and 16. Much of the discussion in this volume is based on experiences in the design and construction of the Kentucky cofferdam, but information and procedures developed at other TVA construction projects are also included.

Considerations Influencing Selection--Considerations which influenced the selection of the steel sheet pile cellular cofferdam were:

1. Depth of water, rate of flow, or both.
2. Excessive hydrostatic heads.
3. The greater security and watertightness offered in the case of heavy overburden.
4. Simplified design.
5. Economy in construction.
6. High salvage value of piling for reuse.
7. Reduced channel restriction possible with this type of structure.

Adaptability--The ideal condition under which the steel sheet pile cellular cofferdam may be built is apparently a solid rock foundation covered with sand and gravel overburden free from boulders, and of sufficient depth to support the cell until the cell fill has been placed. This type of structure was used successfully on TVA projects, however, overburden covering the rock foundation ranged from zero on some projects to a maximum of 95 feet in depth at Kentucky cofferdam. In the case of extremely deep overburden, experience indicates that excessive interlock stresses with some interlock failure may be expected. (See "Interlock Failure--Kentucky Dam," page 24, for procedure followed in handling this condition.) In setting piling where no overburden exists, the piling must be set and the cell fill material placed before complete removal of the cell template. (See chapter 7, "Construction.")

Normally, the steel sheet pile cellular cofferdam is not readily adaptable or economical where the area to be enclosed or the required height of the cofferdam is small. Local conditions may alter this, however, where piling and driving equipment are readily available and can be used more economically than other methods or materials.

> ***Editors Note***
> Gravity cells are often the only way to cofferdam even small areas where rock may prohibit penetration of single wall sheet piling. Cells provide open area, free of bracing.

The adoption of the steel sheet pile cell for cofferdam construction does not necessarily exclude the use of other types of cofferdams as a supplemental type of construction. Earth dikes and other types of cofferdams are often used to fill in short gaps between groups of cells. They may be used for tie-ins and inshore cofferdams to enclose small areas and thus facilitate construction while the large steel sheet pile cellular cofferdam is being built. This is common practice and is of significant importance economically.

SCOPE OF REPORT

The subjects treated in the following chapter of this report fall under the following headings"

Cells
Chapter 2, "Cells," discusses the advantages and disadvantages of the various types of steel sheet pile cells used in cellular cofferdam construction.

Steel Sheet Piling
Chapter 3, "Steel Sheet Piling," describes the characteristics of steel sheet piling and the methods followed in its use for cofferdam cells.

Fill Materials
Chapter 4, "Fill Materials," is devoted to discussions of the various materials available and their use for filling cells, and the construction of berms, together with factors affecting this use.

Design
Chapter 5, "Design," covers fundamentals of design, design procedure, design examples, and examples of actual TVA design of sheet steel pile cellular cofferdams driven to rock.

Final Layout
Chapter 6, "Final Layout," discusses the considerations involved in the layout of the cofferdam system.

Construction
Chapter 7, "Construction," is a comprehensive coverage of the many and various steps involved in TVA's sheet steel pile cellular cofferdam construction program, including equipment and foundation treatment.

Removal
Chapter 8, "Removal," discusses mainly the problems involved in the removal of these cofferdams together with procedures and equipment required for extracting the piling.

Tests
Chapter 9, "Tests," includes descriptions of the tests and model studies made to determine characteristics of the materials and the type of cofferdams discussed in the preceding chapters.

CHAPTER 2

CELLS

Circular-Type Cell

The circular-type cell with connecting arcs was adopted for the first steel sheet pile cellular cofferdam built by TVA at Pickwick Dam in 1935. This same type was also used for all subsequent cofferdam construction on the main Tennessee River (see figure 6 and table 1). It should be noted here that the cells on all projects, except Kentucky, employed 90-degree tee sections in the connection between the main cell and the connecting arcs. At Kentucky a 30-degree wye was used to replace this tee. This was due to the unusual depth of penetration of the piling (from 50 to 95 feet) and to the possible high interlock stress which might develop during cell filling. The 30-degree wye gave a longer connecting arc to provide more flexibility during driving, and it was thought that this would result in less interlock rupture due to boulders and other causes. It was further considered that the 30-degree wye would transmit less stress to the main cell than the tee pile. The tee piles, however, gave satisfactory service in all cases where they were used. (For a complete discussion of interlock stresses, see chapter 3, "Steel Sheet Piling," and chapter 5, "Design." Also see figure 7 for typical circular cell used in the Kentucky cofferdam.)

Advantages of the Circular-Type Cell--The advantages of using such a cell are:

1. It is a self-supporting unit which may be used singly (as for temporary bridge piers), in small groups, or as the end cell for other types of cofferdams.

2. In the event of cell failure, due to split interlocks or other causes, the adjoining cells will not be affected.

3. Circular cells may be filled immediately after being completed regardless of the relative height of fill in adjoining

For cellular structures, specify:
BETHLEHEM STEEL SHEET PILING

Strength • Economy • Availability

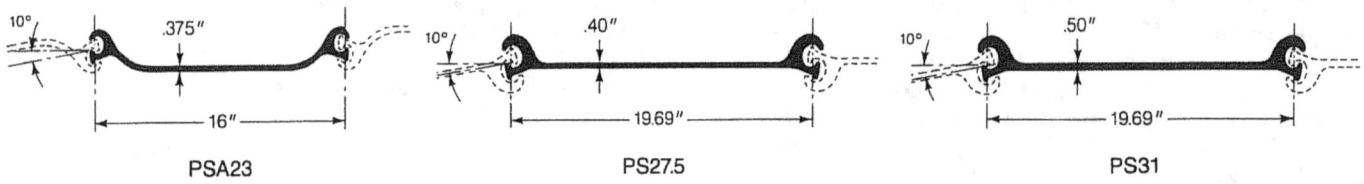

PSA23 PS27.5 PS31

Properties and Weights

Section Designation	Area sq in.	Nominal Width, in.	Weight in Pounds		Moment of Inertia, in.^{4}	Section Modulus in.^{3}		Surface Area sq ft per lin. ft of bar	
			Per lin. ft of bar	Per sq ft of wall		Single Section	Per lin. ft of wall	Total Area	Nominal Coating Area*
PSA23	8.99	16	30.7	23.0	5.5	3.2	2.4	3.76	3.08
PS27.5	13.27	19.69	45.1	27.5	5.3	3.3	2.0	4.48	3.65
PS31	14.96	19.69	50.9	31.0	5.3	3.3	2.0	4.48	3.65

* Excludes socket interior and ball of interlock.

Interlock Strength

Section PSA23

PSA23, when correctly interlocked, develops a minimum ultimate interlock strength of 12 kips per inch. Excessive interlock tension results in web extension for section PSA23. Therefore, the interlock tension for this section should be limited to a maximum working load of 3 kips per inch.

Sections PS27.5 and PS31

PS27.5 and PS31, when correctly interlocked, provide the minimum ultimate interlock strengths in kips per inch as shown in the following table.

Section	Steel Grade		
	A328	A572-Grade 50 or A690	A572-Grade 60
PS27.5	16	20	N/A
PS31	16	20	24

Bethlehem Steel Sheet Piling is readily available from frequently scheduled mill rollings in your specified lengths. These sections are generally available for immediate shipment from inventories at dealer locations.

If you need additional product data, or information more specific to your particular project, please call our Piling Product Sales and Marketing Office in Bethlehem direct: (800) 521-0432.

Piling Products
Bethlehem Steel Corporation
Bethlehem, PA 18016

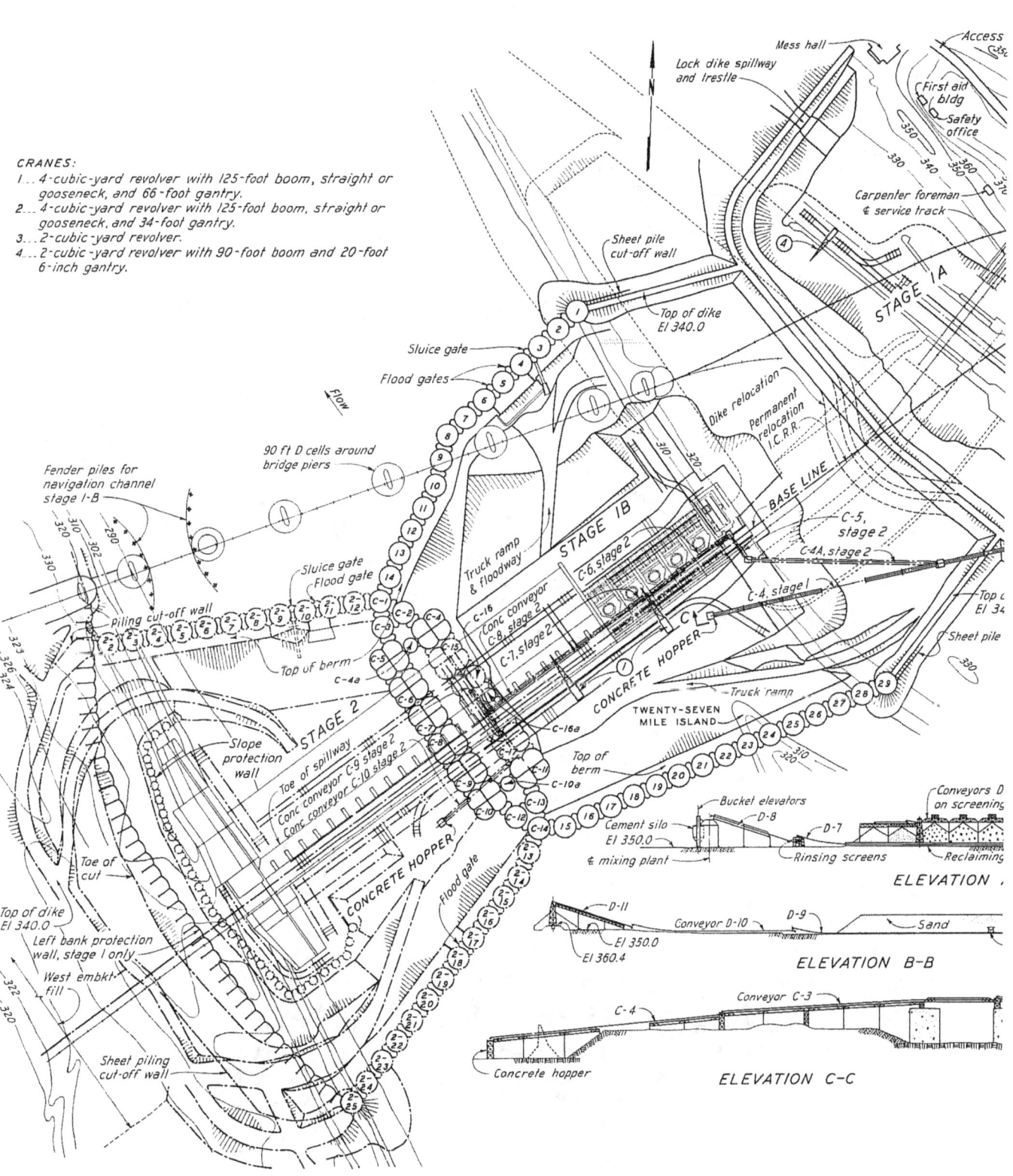

**Figure 1
General Plan of Construction Plant and Cofferdam
Arrangement - Kentucky Dam**

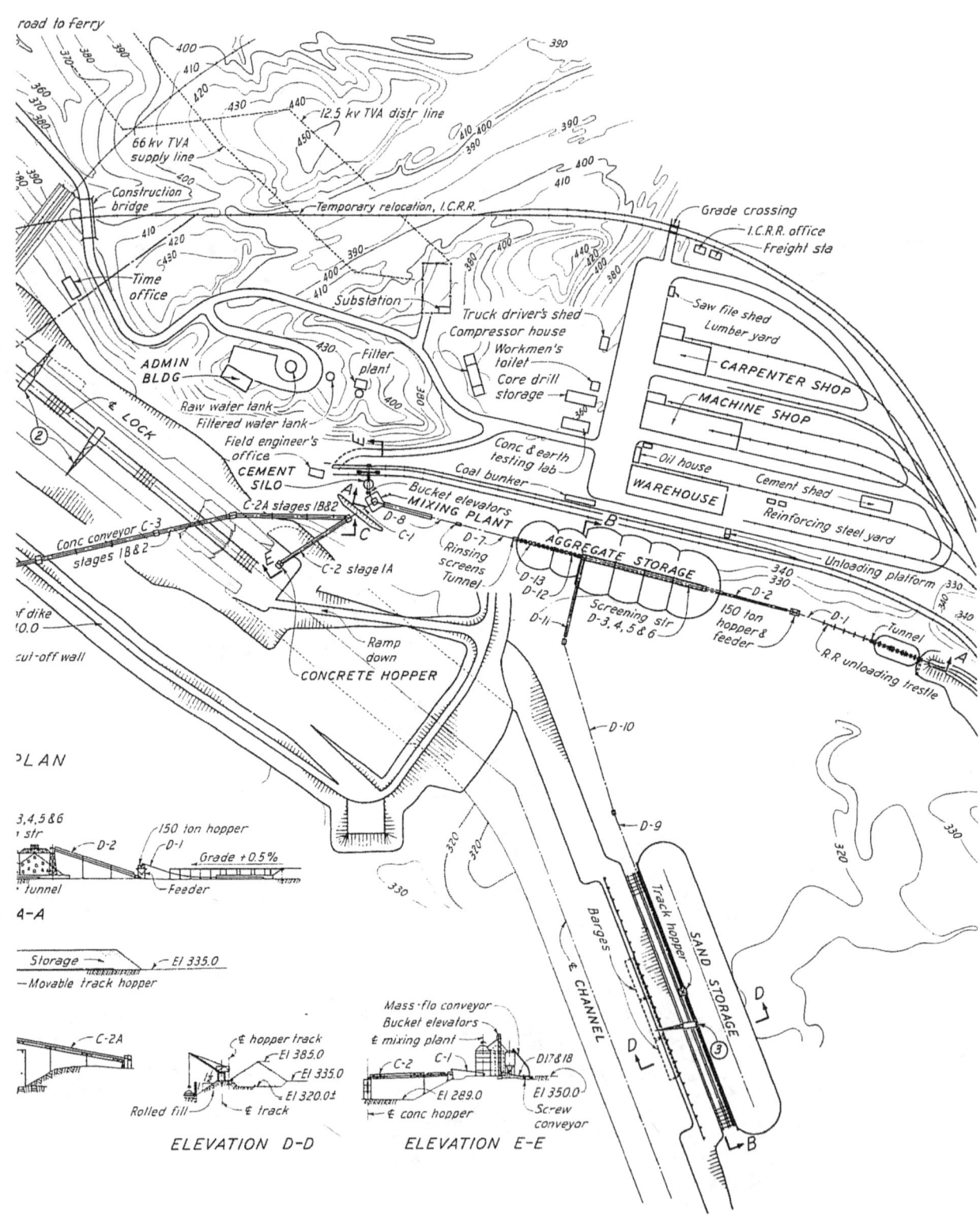

**Figure 1 (Continued)
General Plan of Construction Plant and Cofferdam
Arrangement - Kentucky Dam**

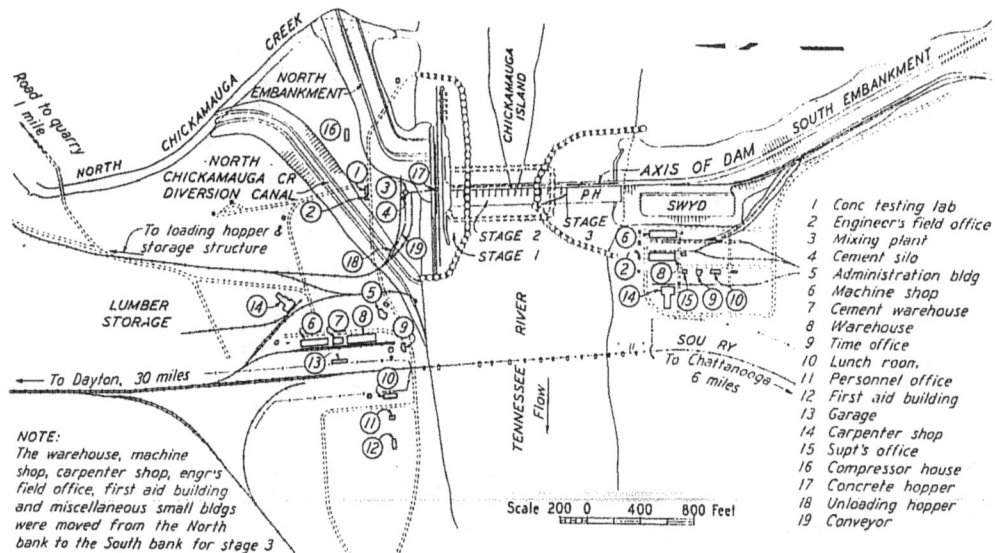

**Figure 2
General Plan of Construction Plant and Cofferdam
Arrangement - Chickamauga Dam**

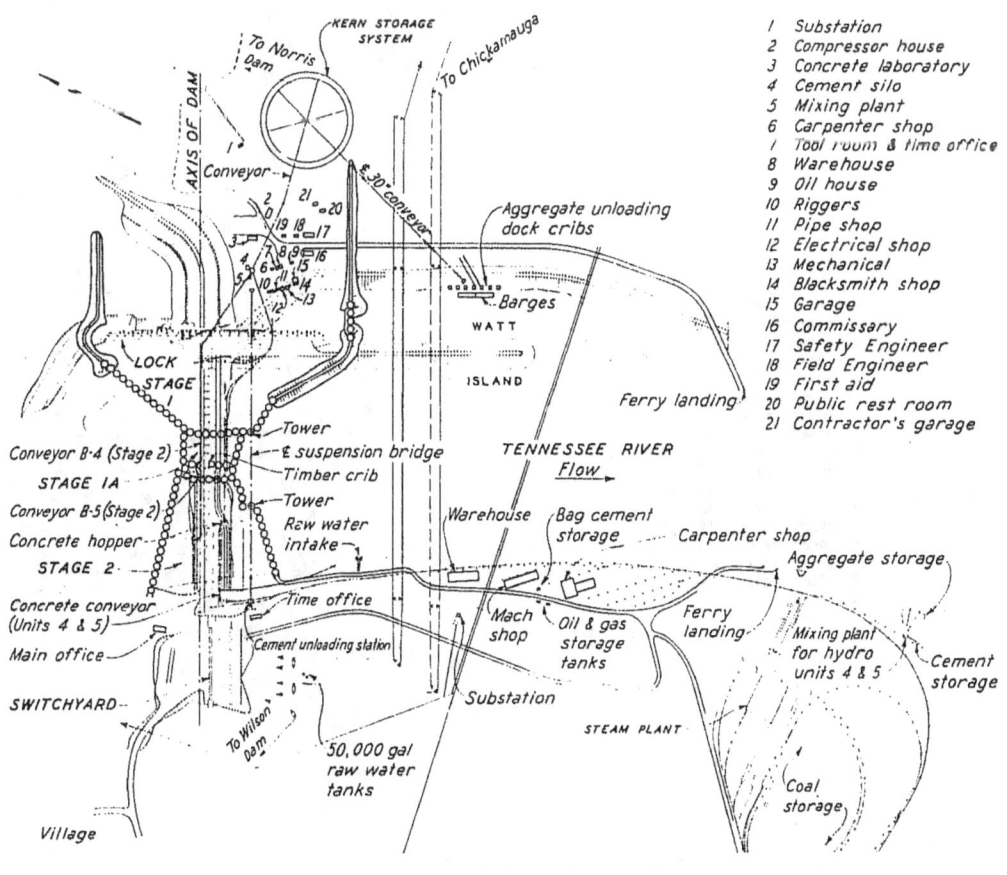

**Figure 3
General Plan of Construction Plant and Cofferdam
Arrangement - Watts Bar Dam**

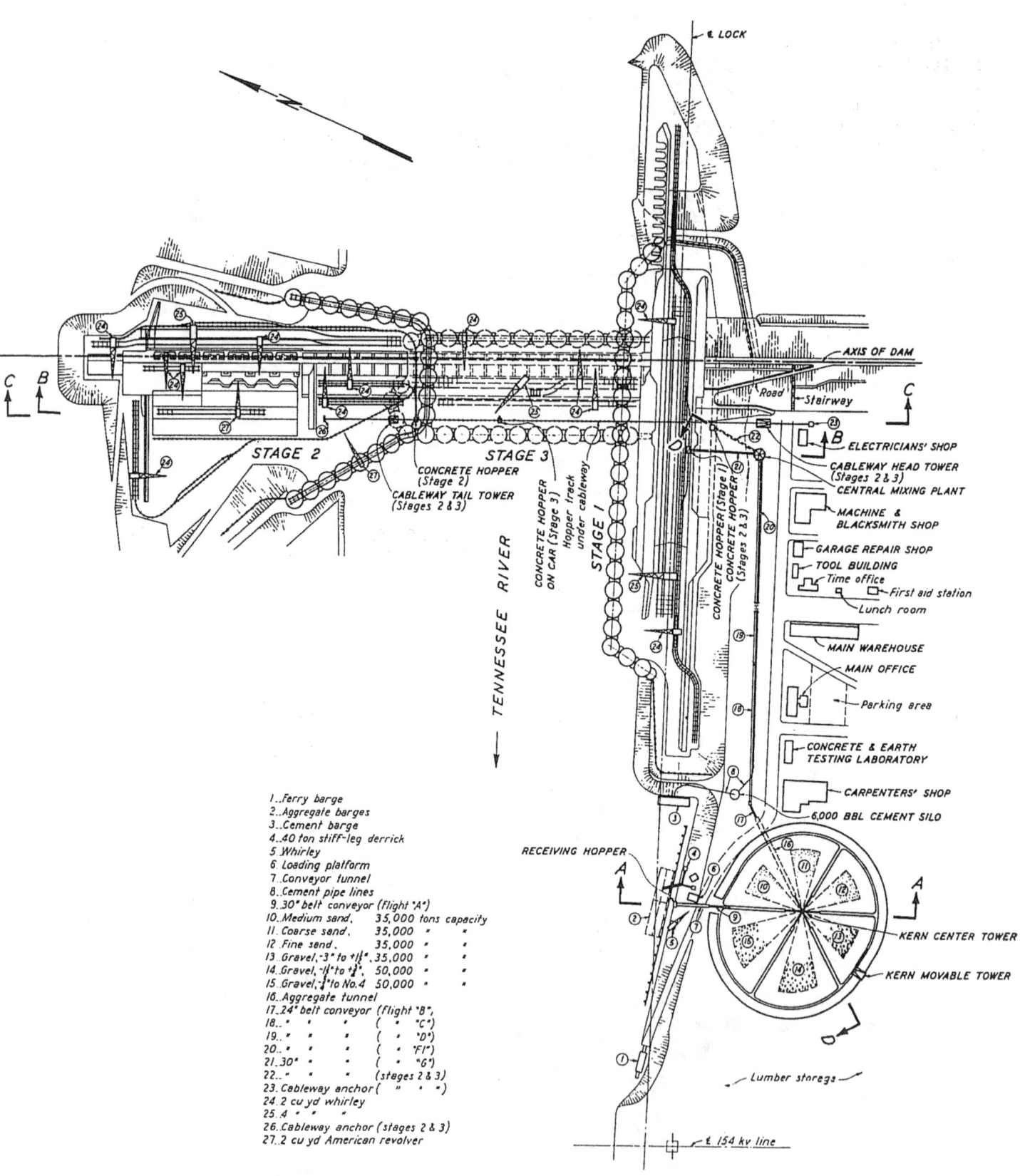

Figure 4
General Plan of Construction Plant and Cofferdam Arrangement - Pickwick Landing Dam

4. Construction of the cofferdam can be facilitated by working from the fill at the top of the completed cells.

5. Circular cells are especially adapted to hydraulic filling since they may be filled in one continuous operation.

6. The circular cell gives greater security against failure from flash floods, particularly in the case where cells are being built across the current of the stream. The security feature is due to the fact that the cells can be kept filled as fast as constructed and none are left unfilled to collapse under the pressure from the swift stream or floodwaters.

7. The circular cell is easily adapted to a tie-in at any point along the cofferdam. This is accomplished by inserting a tee pile at the point of tie-in at the time the cell is constructed.

8. Fewer piles per lineal foot of cofferdam are required than for the diaphragm-type cells. (See "Diaphragm-Type Versus Circular-Type Cells," page 16.)

Disadvantages **of the Circular-Type Cell**--This type cell has two principal disadvantages, which are:

1. For the same radii, the interlock stresses in a circular cell will be slightly higher than in the diaphragm-type cell (due largely to the additional pull from the connecting arc in the circular-type cell). See "Interlock Tension," page 36.

2. The diameter of the cell, and thus the effective net width of the cofferdam for resisting high heads, is limited to avoid exceeding the allowable interlock stress.

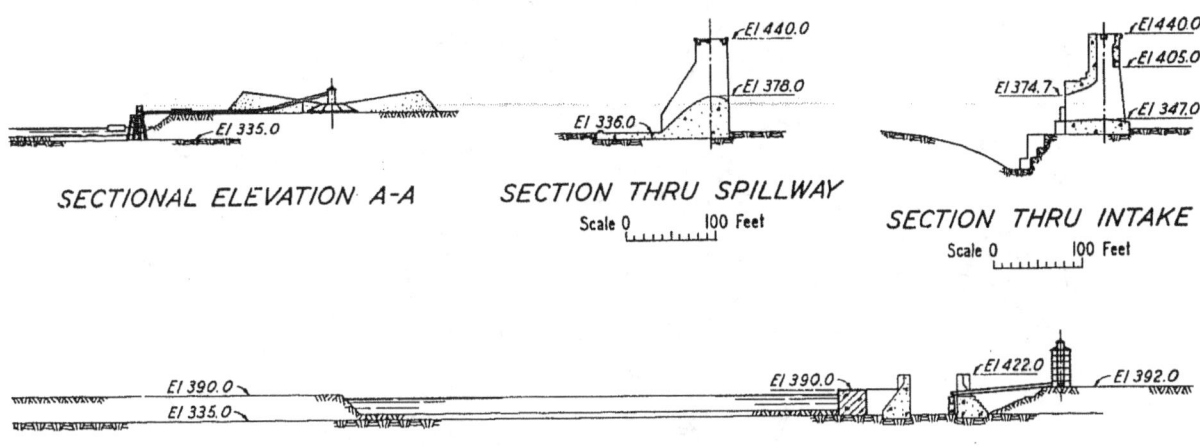

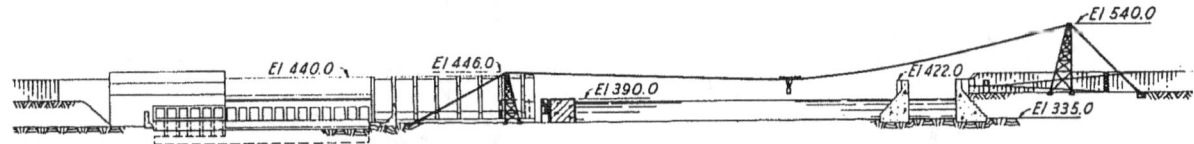

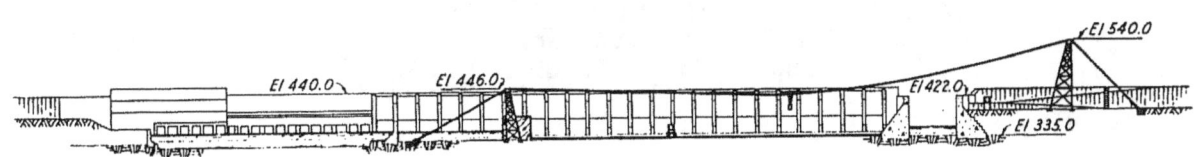

Figure 4 (Continued)
General Plan of Construction Plant and Cofferdam Arrangement - Pickwick Landing Dam

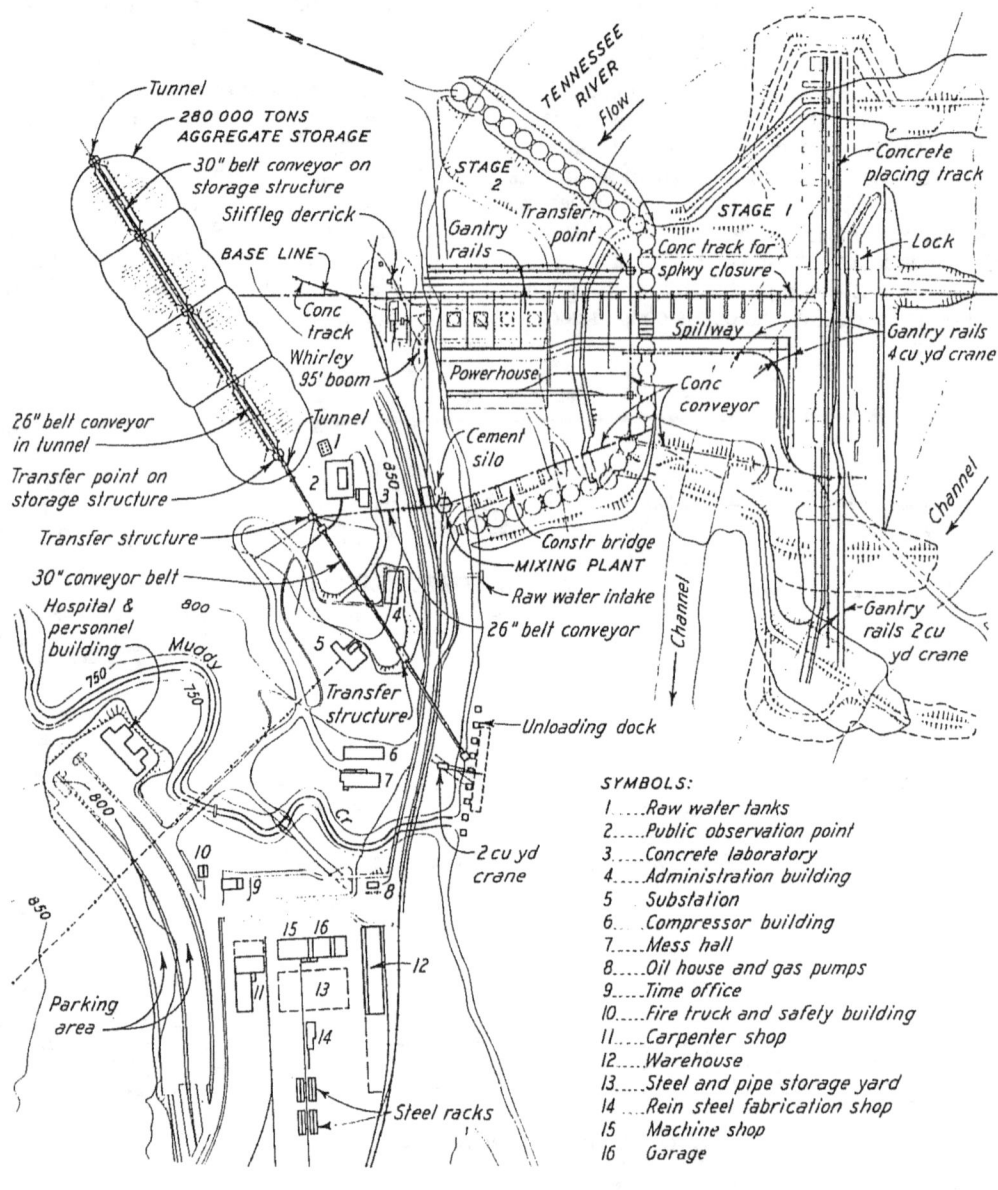

**Figure 5
General Plan of Construction Plant and Cofferdam
Arrangement - Fort Loudoun Dam**

Diaphragm-Type Cell

The diaphragm-type cell has been widely used and recommended by engineers and construction organizations. Its principal features are indicated in figure 8. Here the width (W) is usually taken equal to the radius (R). This requires the use of a fabricated wye with 120 degrees between legs for connecting arcs to the diaphragm. With these proportions in design it is easy to show that the interlock stresses at a given elevation will be the same magnitude at any point along either arc or diaphragm.

At least one cofferdam has been built where the width (W) was made greater than the radius (R). This required the fabrication of a special wye, and the interlock stresses were higher in the diaphragms than in the arcs. The principal advantage of this type of cell, perhaps, is that the diaphragm may be made any length (it should be in multiples of the width of the pile being used, plus allowances for wyes), as it is independent of the width and radius of the arcs. A careful analysis, however, will reveal that the disadvantages far outweigh the advantages.

Advantages of the Diaphragm-Type Cell--The two principal advantages of this cell are:

1. The interlock stresses are uniform and less than the maximum interlock stresses in the circular type cell for a condition where the radii and height of the two types are the same and where the cell fill materials and filling conditions are similar.

2. The effective width of the cofferdam may be increased to resist high heads by increasing the length of the diaphragm (The radius of the arcs being limited by the maximum

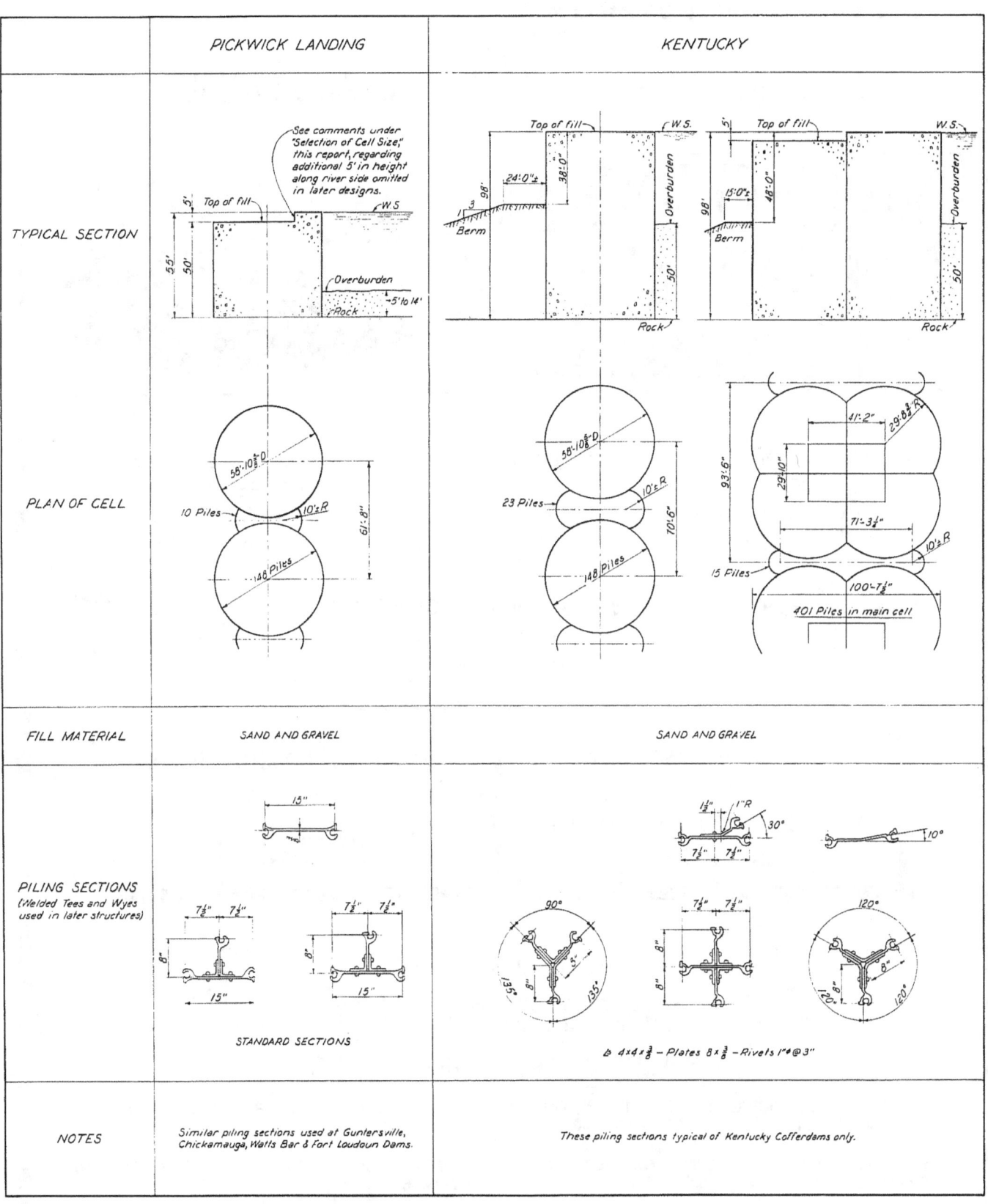

**Figure 6
General Arrangement of Circular and Cloverleaf
Cells Built Along the Tennessee River by TVA**

PIPE and PILING SUPPLIES

STOCKING LOCATIONS IN THE U.S. AND CANADA

PILING PIPE
- MILL SECONDS
- NEW SEAMLESS A252-3
- NEW SPIRAL WELD

NEW H-BEAMS
- 8"
- 10"
- 12"

SHEET PILING
- LARSSEN 2N
- LARSSEN 3N
  (NEW & USED)

SEAMLESS MILL SECONDS
2⅜" OD to 12¾" OD
- LINE PIPE SIZES
- OCTG SIZES
- MILL LACQUERED
- BEVELED & SQ. CUT ENDS

AVAILABLE IN ILLINOIS AND NEBRASKA
#1 RECONDITIONED STEEL PIPE
UNCLEANED OR CLEANED AND BEVELLED
24" x .375

AVAILABLE IN HOUSTON, TEXAS
SURPLUS SPIRAL
42" X .530

IN U.S. CALL TOLL FREE
1-800-874-3720

SALES OFFICES AND YARDS

KINCHELOE, MICHIGAN
906-495-2245

BRAMPTON, ONTARIO
416-796-6800

PORT COQUITLAM, BRITISH COLUMBIA
604-942-6311

ST. HUBERT, QUEBEC
514-445-0050

EDMONTON, ALBERTA
403-465-0501

BEDFORD, NOVA SCOTIA
902-835-6158

CALGARY, ALBERTA
403-236-1332

OTHER STOCKING LOCATIONS: SENECA, ILLINOIS • HOUSTON, TEXAS
OMAHA, NEBRASKA

FOR THESE LOCATIONS CALL TOLL FREE 1-800-874-3720

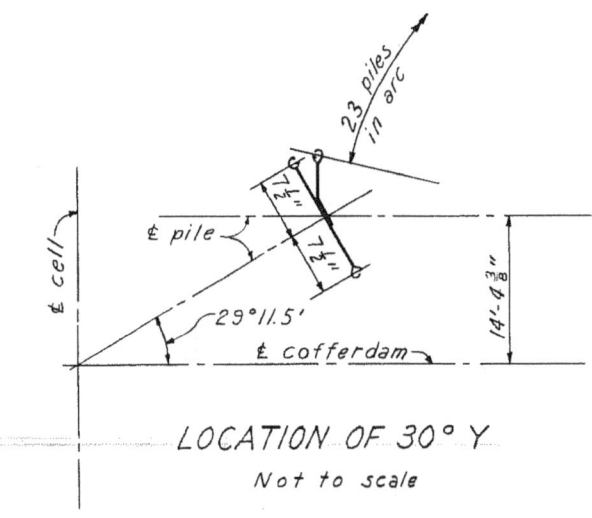

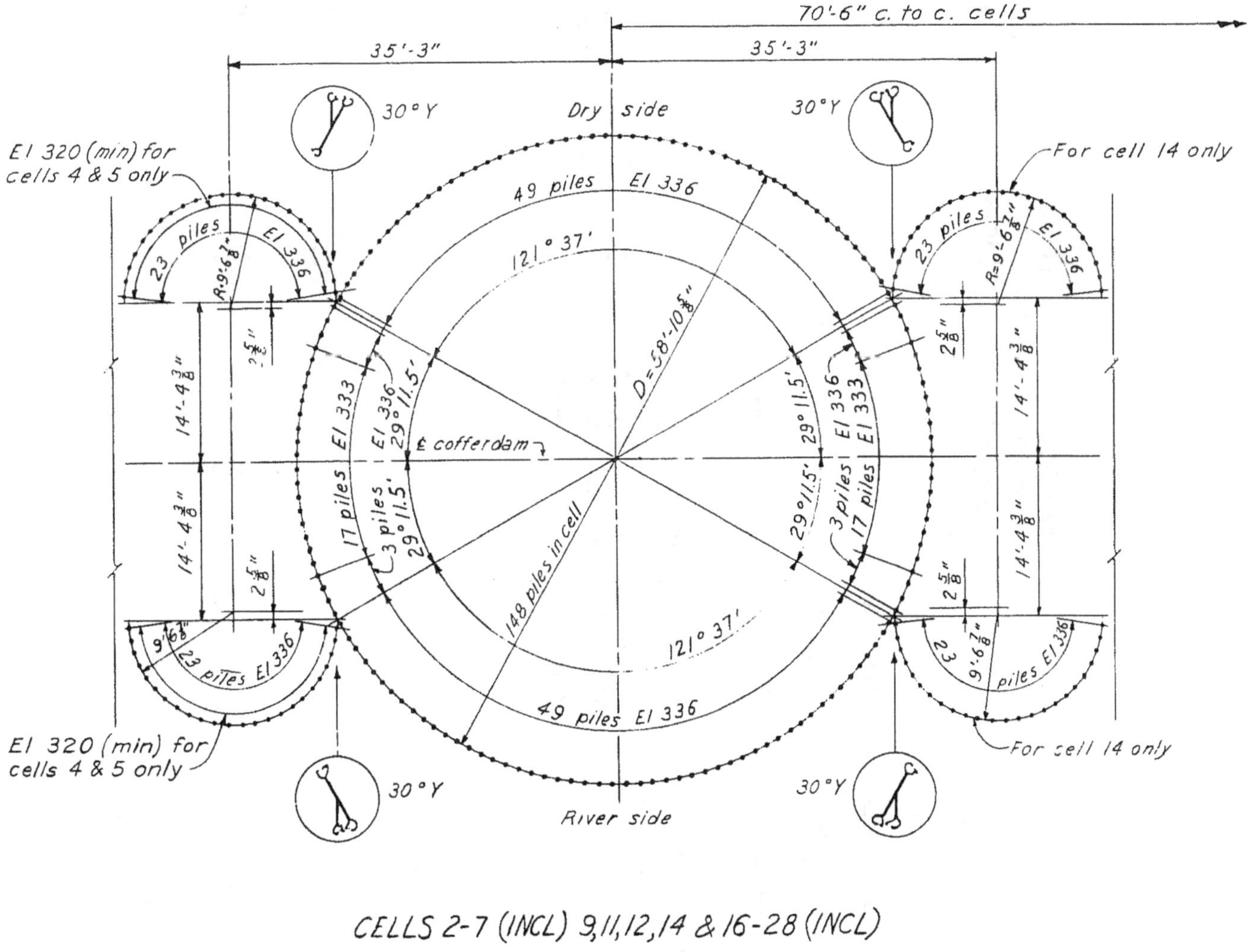

**Figure 7
Typical Details of Circular Cell - Kentucky Dam**

Table 1
Summary - Steel Sheet Pile Cellular Cofferdam Construction Along the Tennessee River

	Pickwick Landing Dam	Guntersville Dam	Chickamauga Dam	Watts Bar Dam	Fort Loudoun Dam	Kentucky Dam	Hales Bar Dam (Upstream Cofferdam Only)
Cell diameter Cell height	58' 10-5/8" 55'	54' 1-1/4" 51'	47' 9" 56'	42' 11-3/4" 42'	47' 9" 53'	See figure 8	70' 58'
Fill material (all hydraulic)	Sand and gravel	Sand and gravel with small amount of silt and clay	Sand and gravel with some clay	Sand and gravel	Sand and gravel with some clay	Sand and gravel	Sand and gravel with content of No. 4 material less than 15%
Provisions for drainage	1-1/2"-dia. weep holes at 3-ft centers, every third pile on dry side.	Main cell: 1-1/2"-dia. weep holes on dry side in each fifth pile (lower one-half only). Connecting cell: Six 1-1/2" weep holes 3'0" on centers in each of two piles.	Main cell: 1-1/2"-dia. weep holes at 3-ft centers on dry side in each fifth pile (lower half only). Connecting cell: Six 1-1/2" weep holes at 3-ft centers in each of two piles.	Main cell: Six 1-1/2"-diameter weep holes at 3-ft centers in each fifth pile (lower half only). Connecting cell: Weep holes in two piles.	Main cell: Six 1-1/2"-diameter weep holes at 3-ft centers on dry side in each fifth pile (lower half only). Connecting cell: Weep holes in two piles.	Circular cells: Two horiz. rows of nine 4"-dia. weep holes cut in each main and connecting cell on dry side (one row at river bed and other at top of berm). Cloverleaf cells: One horiz. row of twelve 4"-dia. holes was cut in main cell at river bed, and one horiz. row of three 4"-dia. holes was cut in each connecting cell at river bed level on dry side.	Main cell: 3"-dia. weep holes in each third pile on dry side. Starting 2' from bottom, space 3' for second, 4' for third, 5' for the fourth, and 6' for the fifth, with total of 5 holes. Connecting cell: 3"-dia. weep holes spaced as above in two piles on dry side of cell.
Size of cofferdam enclosure (a)	Stage 1 Cells: 26 Acres: 14.38	Stage 1 Cells: 37 Acres: 8.7	Stage 1 Cells: 38.5 Acres: 13.5	Stage 1 Cells: 50 Acres: 34.3	Stage 1 Earth dike Acres: 42.75	Stage 1 Cells: 35 cir. (std.) 2 small 8 cloverleaf Acres: 71.1	Stage 1 Cells: 4 (+2 small) Acres: (See figure 50
(b)	Stage 2 Cells: 22 Acres: 5.65	Stage 2 Cells: 38.5 Acres: 6.95	Stage 2 Cells: 36 Acres: 6.36	Stage 2 Cells: 37 Acres: 9.58	Stage 2 Cells: 30 Acres: 12.84	Stage 2 Cells: 31 cir. (std.) 4 small 6.5 cloverleaf Acres: 32.85	
(c)	Stage 3 Cells: 29 Acres: 5.78	Stage 3 Cells: 28 Acres: 12.25	Stage 3 Cells: 35 Acres: .14.8				

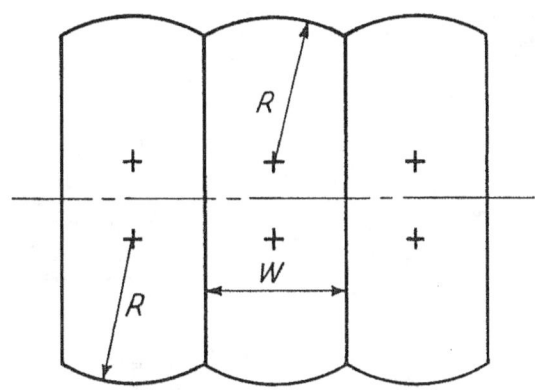

**Figure 8
Diaphragm-Type Cells**

allowable stress for the piling interlocks). This is not true in the case of the circular cell, as the effective width of the circular cell is directly proportional to the radius.

Disadvantages of the Diaphragm-Type Cell--The four main disadvantages are:

1. The collapse or failure of one cell will cause the failure of a number of adjoining cells. The danger from this is especially present during flash floods where the cells are being constructed at right angles to the flow of the stream. (See discussion on danger of failure from ice in Cofferdams.[1])

2. During filling operations the fill level in any one cell must not vary more than a few feet above or below the fill level in an adjoining cell. Too great a variation will cause distortion of the diaphragm. To fill these cells hydraulically would require the use of an excessive amount of pipe work.

3. Because of the stair-step method of filling, the template and supporting piling must be left in place in the cells until filling is completed to a point of stability. As many as 10 to 12 templates may be required at one time, and it may be impossible to salvage much of the piling or template material.

4. More piles per lineal foot of cofferdam are required for the diaphragm cell than for the circular cell. (See following paragraphs).

Diaphragm-Type Versus Circular-Type Cell

A study of the relative advantages and disadvantages of the two types will readily show the reasons why TVA adopted the circular-type cell at the very start of its cofferdam construction program on the main river and why it has continued to use it in preference to the diaphragm-cell type.

It has been stated that the cost of the diaphragm-type cell per lineal foot of cofferdam is less than the circular-type in low cofferdams, but the difference in cost decreases as the height of the cofferdam increases until the cost of the circular-type cell becomes less for the very high cofferdam. Computations show, however, that on a strictly comparable basis the circular-type cell normally requires less piling per lineal foot of cofferdam for all cases up to a cell diameter of something over 50 feet (see figure 9), and that the difference decreases as the size of the cell increases. The circular cells were all computed with the tees set at 30 degrees from the centerline of the cofferdam and with eight piles in each connecting arc. All diaphragm cells were computed with an arc radius equal to that of the corresponding circular cell (so that the interlock stresses would remain about the same), with the distance between diaphragms equal to this same radius. The average width of the diaphragm cells, at right angles to the axis of the cofferdam, was also made equal to the average width of the circular cell and connecting cells so that the two would have comparable stability. The Kentucky cofferdam cell (figure 7) required more piling than usual because of the longer connecting arcs, amounting to 2.78 piles per lineal foot of cofferdam, but was still less than the 2.85 piles per lineal foot for an equivalent diaphragm cell. Computed material requirements on a comparable basis for the Pickwick cofferdam were:

Diaphragm Cells	Circular Cells
5078 tons	4857 tons

Since the rest of this volume is based on TVA experience, the discussions are confined almost entirely to the circular cell-type cofferdam. The cloverleaf-type cells used in the Kentucky cofferdam are included, however.

Cloverleaf-Type Cells

The cloverleaf-type cell (figure 10) is normally used where height and mass, required for stability, prohibit the use of the single-radius-type circular cell. A number of variations are possible with the cloverleaf-type cell, such as increasing the number of compartments by increasing the number of diaphragms. Any size cell may be built up to meet stability requirements, but the radii of the arcs will be controlled by the maximum allowable interlock stress in the piling. The cloverleaf cell will require more piling than the standard-type

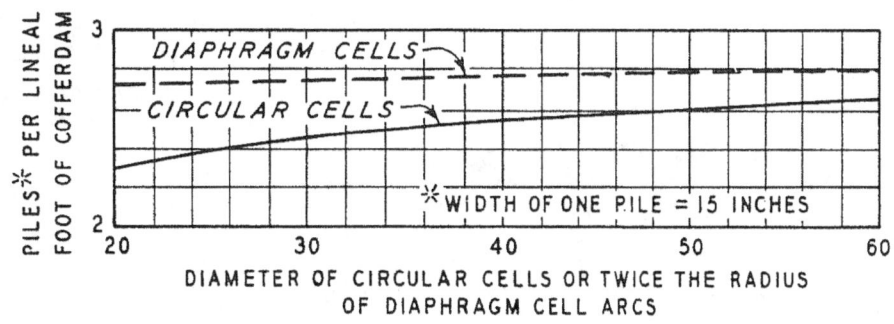

**Figure 9
Piling Required per Linear Foot of Cofferdam**

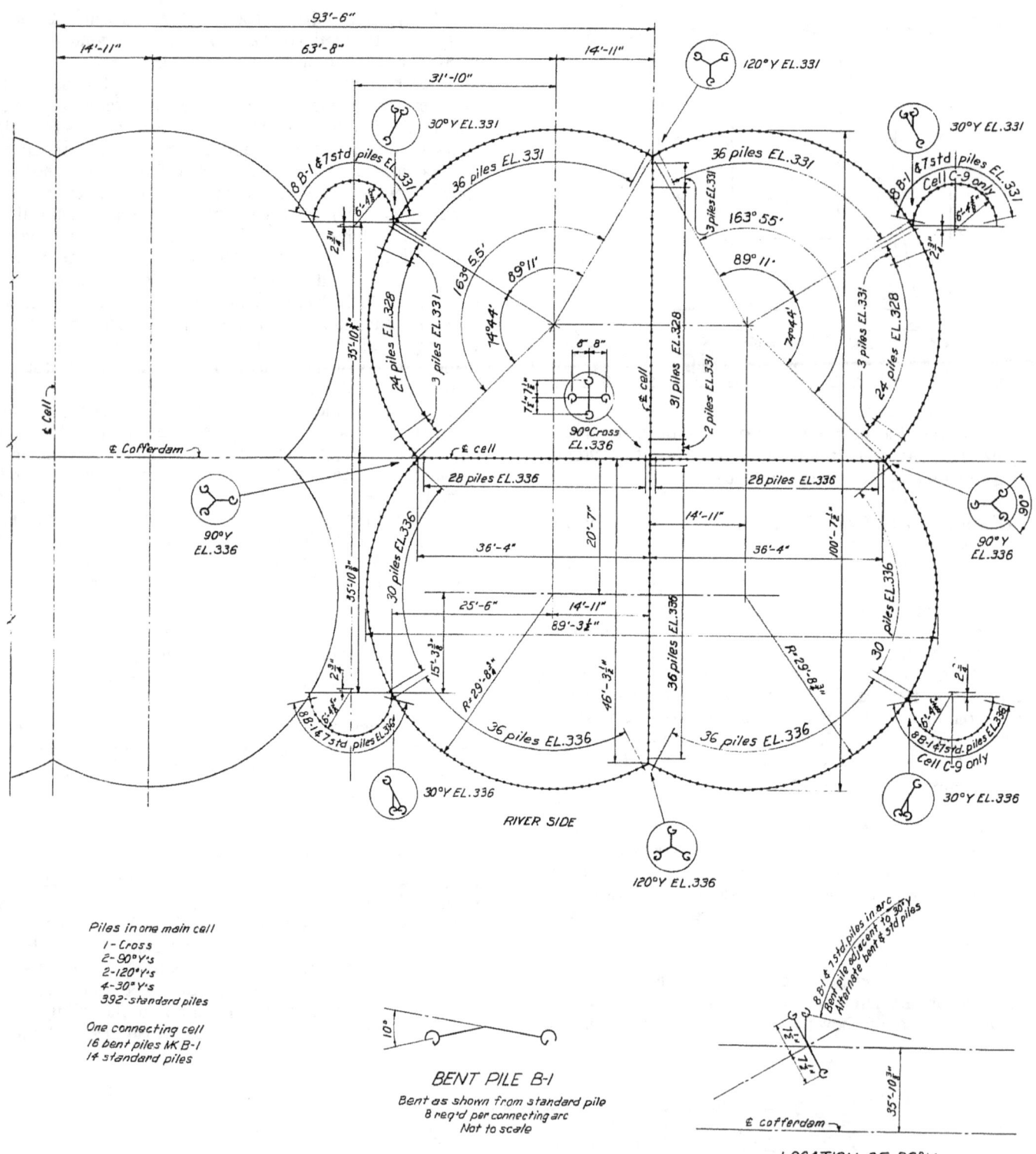

**Figure 10
Typical Details of Cloverleaf Cell - Kentucky Dam**

cell but, like the circular cell, it has the advantage of being a self-sustained unit. In filling a cloverleaf cell, no compartment should be filled to more than 5 feet above the fill in any adjoining compartment to avoid distorting the diaphragm wall.

In the construction of the diaphragm cellular-type cofferdam, the cloverleaf cell is, because of its stability, often used for end cells, corner cells, tie-ins, and anchors. Where the supporting berm along the dry side of the cofferdam is objectionable during construction, the cloverleaf cell may be used to give the desired height and stability without the use of a berm. This is particularly advantageous where the cells must be built close to the inside structure to reduce channel restriction or afford maximum operating efficiency to floating equipment which may be working on the inside structure.

Figure 1 shows cloverleaf cells used along the river arm of stages 1-B and 2, Kentucky Dam. Here is was necessary to place the concrete in the dam as close to the cofferdam as possible to avoid a large overlap in construction stages. These cells were designed for stability without the aid of the supporting berm to meet this need.

Other Types of Cell

Circular cells connected by a single arc make excellent retaining walls. Such a structure, marked "Slope Protection Wall," may be seen in figure 1, stage 2, Kentucky cofferdam.

> *Editors Note*
> Some retaining walls and marine bulkheads have been built with the rear connecting arc omitted. This practice has been considered but not recommended for cofferdams.

Another type, while not strictly cellular, was built during stage 1-B, Kentucky cofferdam to protect the left bank against scour (see figure 1). It is similar to the diaphragm-type cellular cofferdam except that it has arcs only along one side of the structure. The straight walls (normally used as the diaphragm walls for the diaphragm-type cell) are extended far enough back into the riverbank to develop, through friction, the necessary resistance to pull. A similar structure was used by TVA at Watts Bar Dam in the construction of a permanent coal unloading dock.

Other Uses of Circular Cells

Circular cells have been used by TVA during permanent construction operations for a number of uses other than the conventional cofferdam discussed so far.

At Kentucky Dam a circular cell was driven around each pier of the Illinois Central Railroad bridge (figure 1) for protection against scour caused by the river current during construction operations. Here the cells were filled with gravel and capped with concrete just above low water elevation.

In constructing the piers for the relocation of the N.C. & St. L. Railroad bridge in the Kentucky Reservoir area, a steel sheet pile circular cell was driven to rock for each pier around a series of laminated timber template rings. The overburden inside each cell was removed with clamshell and replaced with tremie concrete. The upper part of the cell was concreted in the dry after the cell was unwatered. The template rings were left in place and covered with concrete.

At Watts Bar Dam, single circular cells filled with gravel were used in the river to support the towers for the temporary suspension bridge. At Johnsonville Steam Plant the raw water intacke pumps for construction were supported on a single gravel-filled cell, the intacke being inside the cell with the water entering the pumps through the gravel fill material.

Circular cells have also been used by TVA in permanent construction as a substructure for lock training walls. At a number of TVA steam plants a series of isolated circular cells were used in connection with the construction of the permanent coal unloading docks.

Factors Affecting Selection of Cell Size

A careful study of flood frequencies, relation of stages to construction schedules, the relative economy of various arrangement schemes, etc., is necessary before reaching a decision as to the size of the area to be enclosed and the elevation of the tope of the cofferdam.

Upon selection of the elevation for the top of the cofferdam, the height of the cells can then be determined. The diameter of the cofferdam cell is, of course, a function of the cell height. Where the elevation of the rock foundation varies greatly, it may be found advisable to use more than one size of cell. Although the amount of piling required for a given length of cofferdam varies only slightly regardless of the diameter of the cell (see figure 9), a material saving in the cost of cell fill material can be effected by using smaller cells. Smaller cells should be used then, where possible, provided it is economically feasible. To use more than one size of cell will require more than one size of template, thereby resulting in a reduction in the saving that would otherwise be effected through use of smaller cells.

For estimating purposes, the required diameter for the circular cell may be taken as approximately 1.25 times the cell height. This corresponds to the 0.85H for the average width of the cofferdam suggested by Karl Terzaghi.[2] (See "Average Width of Cofferdams," page 35, for final recommended diameter).

The width, or diameter, of the high cells is often limited by the maximum allowable interlock stress (this was 12,000 pounds per lineal inch of interlock for piling used by TVA in cofferdam construction). Where there is room on the dry side of the cofferdam for a supporting berm, stability requirements may be met without increasing the size of the

> *Editors Note*
> Guaranteed interlock strength for new standard flat sheet piling is generally at least 16,000 pounds per inch of interlock. Higher strengths are available from most producers.

cell. Costs incurred in making the cofferdam enclosure larger to accommodate construction operations without interference from the inside berm should be balanced against the cost of a smaller area enclosure using larger cells which do not require a supporting berm. Cells driven through heavy overburden may be supplied with a natural berm by leaving part of the overburden in place at the time that excavation is being done inside the cofferdam.

The cloverleaf cell offers the most practical alternate in special cases where stability requirements cannot be met by the single-type cell.

> *Editors Note*
> Cloverleaf cells were the only solution to high-head cofferdams when this manual was written. New higher strength interlocks now permit the design of larger diameter cells.

All cellular steel sheet pile cofferdams built by TVA prior to the construction of the Kentucky cofferdam were designed with the piling along the river side of the main cells and connecting cells extending 5 feet above the rest of the piling in the cell. The piling was strong enough to act as a cantilever and support a head of water extending 5 feet above the cell fill, and it was felt that this arrangement would result in a worthwhile saving in piling cost. The careful design analysis of the Kentucky cofferdam, however, indicated that this procedure was of doubtful value; the less amount of fill for this type of cell made it necessary to increase the cell diameter to provide sufficient mass for stability. This complicated both design and construction procedure. It was therefore decided to consider the top of the piling and the top of the fill to be at one elevation, except in the case of the cloverleaf cells where the inner compartments were left 5 feet low. Based on TVA experience, then, it is now felt that the circular cell in most cases should be designed with a uniform top elevation. The piling of the circular cell inside the connecting arcs--except for the first three piles immediately inside the connecting tees--should also be left approximately 3 feet below the top of the other piling outside the connecting arcs, however, so as not to interfere with the construction of a roadway along the centerline of the cofferdam. (See figures 31 and 32).

Hydraulic computations may indicate a sufficient drop in head between the upstream and downstream limits of the cofferdam to warrant making the downstream cells lower than the upstream cells. The difference in elevation, however, should be somewhat less than the computed drop in head.

CHAPTER 3

STEEL SHEET PILING
Types

The development of sheet piling from timber to modern rolled steel sections is covered very thoroughly in textbooks and handbooks on the subject. The rolled steel pile of the straight web type has been used by TVA in all its steel sheet pile cofferdams, and the discussions in this chapter pertain primarily to this type and TVA's experience with it.

Three types of steel sheet piling are almost universally used throughout the United States. These are (1) the straight web type (figure 11), (2) the arched web type, and (3) the Z type. Differences in interlock details are minor, making it possible to interlock the piling manufactured by one company with that manufactured by another. This is not desirable, however, where the developed stresses are high since this arrangement may develop eccentricity and perhaps early failure.

> *Editors Note*
> In the past, sheet piling of different manufacture was routinely interlocked, contrary to the producers recommendation. Interlock guarantees are generally based on new material, same manufacturer, same section.

The interlocks are designed for 3-point contact between the "fingers" of the interlocks where the piling is driven in a straight line, for 2-point contact where the angle between adjoining piling does not exceed 10 degrees, and 1-point contact where the angle between adjoining piling is greater than 10 degrees. It is advisable, therefore, to limit this deflection angle to 10 degrees or less, particularly where interlock stresses are high. Bent piles may be used where it is necessary to have a large angle of deflection, as in the connecting arc in the Kentucky cofferdam (see bent pile B-1, figure 10).

> *Editors Note*
> Large deflection angles between sheets are usually encountered only in connecting arcs between cells or in very small diameter cells. The producer should be consulted regarding his "swing" guarantees.

Certain European piling sections are designed for 1-point contact for deflection angles up to 15 degrees. Claims are made that such piling has decided advantages over multiple-point contact piling. (For a technical discussion of this see papers by Bae's[2] and Descans.[3])

> *Editors Note*
> Most worldwide producers have settled on three-point contact interlocks. There is only one producer of a single point contact interlock at this time.

Arch web piles are adapted for use in straight walls supported by wales and tie rods. The high section modulus provides resistance to longitudinal bending between wales (see figure 12). Z-piles may be used where an even higher section modulus is required. Arch web piling has been used in cellular construction employing circular arcs but is not well adapted to such use since high ring tension in the arc will tend to flatten the arch out into a straight line. This will allow the cell to expand. Furthermore, the arch pile interlocks are designed for lower interlock stresses more or less in line with the ability of the section to resist flattening out.

The straight web piling interlocks are designed for much higher interlock stresses than the interlocks of other type piling and are, therefore, much better suited for the circular and disaphragm wall stresses where tension is the principal stress. In the case of the circular arcs, the resistance to longitudinal bending is of little importance (except in the case of handling extremely long sections), as each section is supported along the interlock by the adjoining pile. Diaphragm walls present no problems in longitudinal bending except during cell filling operations, where care should be exercised to keep pressure from fill material in adjacent cells equalized to prevent distortion in the wall alignment. (The difference in fill level in adjoining diaphragm-type cells should not exceed 5 feet.) TVA has used the straight-web type of pile in all of its cofferdams and has found it to be eminently satisfactory.

Piling used by TVA was furnished by three manufacturers[4] and, with the exception of slight differences in interlock details, is identical. These manufacturers and piling designations are: U.S. Steel Corporation, sections MP 101 and 102; the Inland Steel Company, section I-28-S; and the Bethlehem Steel Corporation, sections SP-6a and SP-7a. Additional data, such as the recommended driving radii for circular cells, details of fabricated sections, and design of retaining walls, etc., may be found in the manufacturers' catalogs and information bulletins.

> *Editors Note*
> USX Corp. (formerly U.S. Steel) and Inland Steel Corp. no longer produce steel sheet piling. Bethlehem Steel Corp. is the only domestic producer of hot-rolled sheet piling. Several foreign producers also offer flat sheet piling shapes. The properties and dimensions will differ from those described in the TVA Manual 1957).

Web Thickness--The web thickness of most standard steel sheet piling is either 3/8 or 1/2 inch thick. The 3/8-inch-thick web has been used exclusively throughout TVA cofferdam construction with good results. The 1/2 inch thick web might give better performance, however, in certain cases of heavy driving.

Fabricated Piling--Special piling used at the junction of cell arcs, straight walls, etc., is fabricated from whole piling and combinations of piling sections either welded or riveted together and reinforced with plates, angles, etc. (See figure 11. See also figure 6 for the Kentucky tees, crosses, and 120-degree wyes that were fabricated from piling sections, 3/8 inch thick angles and plates, and 7/8 inch diameter rivets.) These are standard sections. Because of the heavy overburden and driving conditions at Kentucky cofferdam, however, the plate and angle thickness (figure 6) was increased to 5/8 inch in some cases. Special sections, succh as 90- and 30-degree wyes, are fabricated by the manufacturer on request. Bent pilings are also furnished on request to meet specifications set

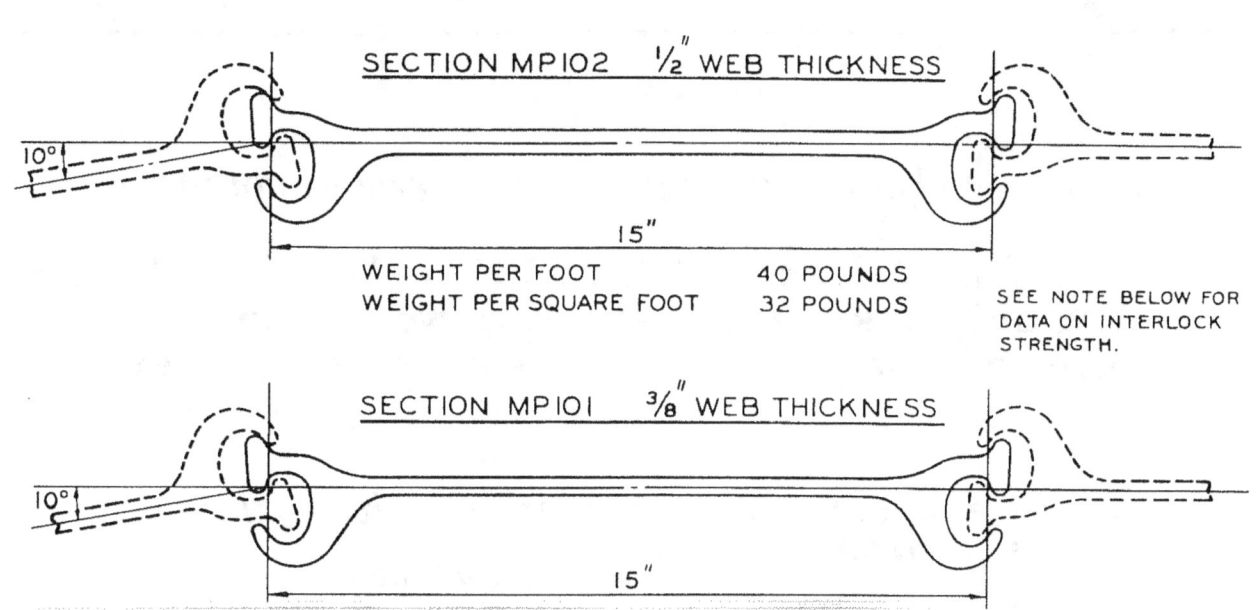

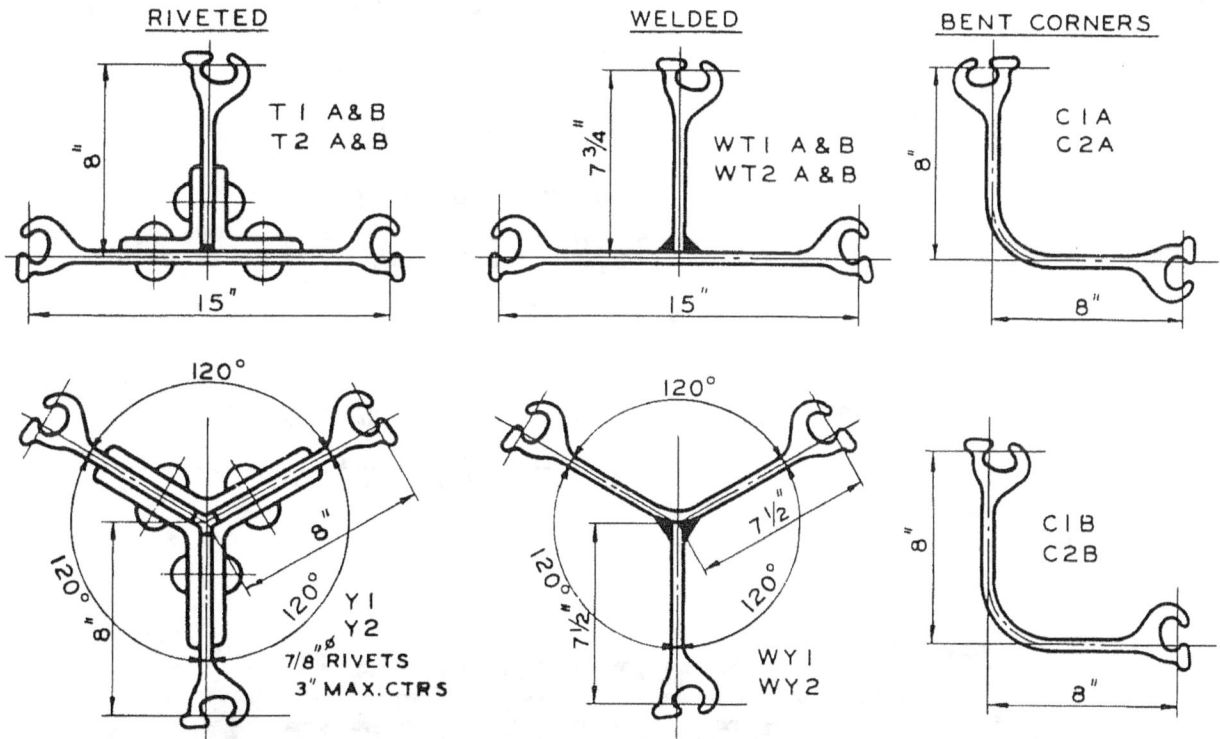

Editors Note
Rivets have generally been replaced by high strength bolts.

CONNECTIONS USING CODE MARK "1" ARE MADE FROM MP101, CODE MARK "2" MADE FROM MP102

WEIGHTS IN POUNDS PER LINEAR FOOT INCLUDING ANGLES & 7/8"⌀ RIVETS

CODE MARK	T1A	T2A	T1B	T2B	WT1A	WT2A	WT1B	WT2B	*Y1	*Y2	WY1	WY2	C1A	C2A	C1B	C2B
WEIGHT	75	82	75	82	53	61	53	61	95	115	53	61	35	40	35	40

* Y1-5"x5"x3/8" ANGLES MAXIMUM RECOMMENDED WORKING STRESS 6,000 POUNDS PER LINEAR INCH. — Y2-5"x5"x1/2" ANGLES MAXIMUM RECOMMENDED WORKING STRESS 8,000 POUNDS PER LINEAR INCH.

NOTE:—RECOMMENDED MAXIMUM WORKING STRESS FOR INTERLOCK TENSION 8,000 POUNDS PER LINEAR INCH, BASED ON AN AVERAGE ULTIMATE STRENGTH OF 16,000 POUNDS.

Editors Note
USX Corp. (U.S. Steel Corp.) no longer produces sheet piling.

Figure 11
U.S. Steel Corporation Piling Sections MP101 and MP102

When time and money are important, experience counts...

DEEP FOUNDATION EQUIPMENT AND PRODUCTS
are our only business. That's why we do more of it and have to do it better than anyone else in America.

We manufacture, sell and rent	*Vibratory pile driver/extractors*
	Diesel pile hammers
	Predrill & auger-cast augers
	Tie-back installation equipment
	Pile leads & spotters
	Pile points & splicers
	Jet & water supply pumps
	Submersible pumps
	Hydraulic power units
We sell and rent	*Menck offshore hammers*
	Mitsubishi diesel hammers

Corporate offices: 301 Warehouse Dr., Matthews, NC 28105 USA
800 438-9281 & 704 821-8200 FAX 704 821-6448 Telex 572385

Matthews NC - Springfield PA - Woodbury CT - Niagara Falls ONT
Brossard PQ - Lakeland FL - Belle Chasse LA - Houston TX
Vallejo CA - Seattle WA - Ft. Wayne IN - Sterling VA - Jessup MD

up by the user. In bending piling, however, care should be exercised to ensure that piling is set in correct relation so that the Interlock fingers will mesh properly.

Welded connections (see figure 11) are now more widely used. This type of connection may not be sufficiently strong, however, to resist high tension, and a careful investigation should be made before deciding on which type to use. The riveted type of connections will be stiffer for hard driving.

> ***Editors Note***
> Some cell failures have occured where welded connectors were used. The manufacturer should be consulted regarding his current recommendation.

Interlock Strength

Leading manufacturers will guarantee a maximum working interlock stress of 8000 pounds per lineal inch for standard 3/8 and 1/2 inch thick straight web piling. For the Kentucky cofferdam this was raised to 12,000 pounds per lineal inch of piling. This corresponds to a (tensile stress) tension of 32,000 pounds per square inch for the 3/8 inch web and 24,000 pounds per square inch for the 1/2 inch web. Web failure during interlock tests or in actual use, except under difficult driving conditions, has been a rare occurrence. It is therefore considered safe to base the design of the cell on the piling interlock strength. An allowable maximum interlock stress of 8000 pounds per lineal inch of piling was adopted for the 3/8 inch thick straight web pile used in all TVA cofferdams except Kentucky cofferdam, thereby giving a factor of safety of 1.5. The design of the Kentucky cofferdam was a special problem and is discussed in chapter 5, "Design."

> ***Editors Note***
> Manufacturers generally guaranteed a minimum ultimate interlock strength for a representative sample of 16,000 pounds per inch of interlock. The figure 8000 pounds was a recommended maximum design tension to provide a safety factor of two.

PROFILE		Section Index	Driving Distance Per Pile	Weight		Web Thick-ness	Section Modulus	
				Per Foot	Per Square Foot of Wall		Per Pile	Per Foot of Wall
			Inches	Pounds	Pounds	Inches	Inches³	Inches³
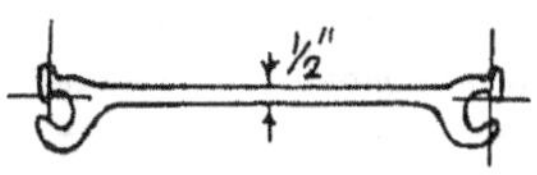	INTERLOCK WITH EACH OTHER	MP-102	15	40.0	32.0	½	2.4	1.9
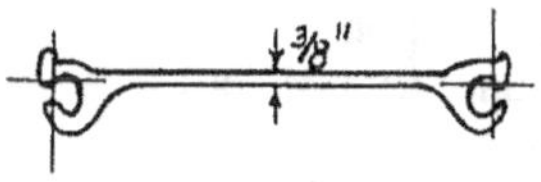		MP-101	15	35.0	28.0	⅜	2.4	1.9
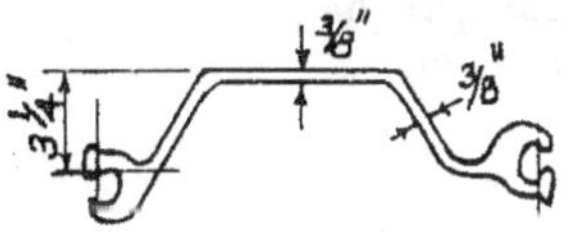		MP-117	15	38.8	31.0	⅜	8.9	7.1
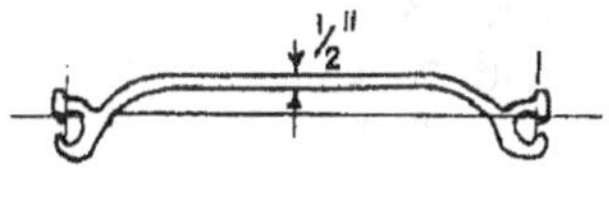	INTERLOCK WITH EACH OTHER	MP-113	16	37.3	28.0	½	3.3	2.5
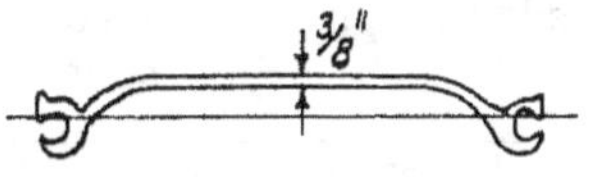		MP-112	16	30.7	23.0	⅜	3.2	2.4
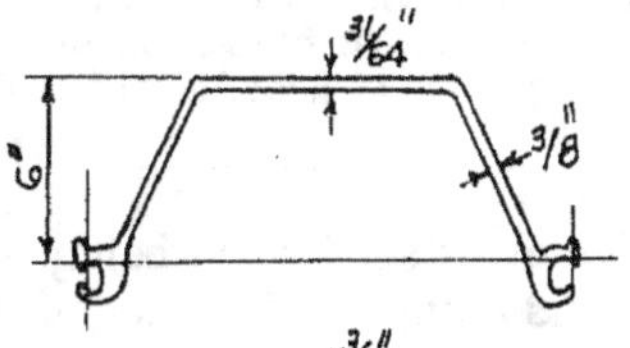		MP-110	16	42.7	32.0	21/64	20.4	15.3
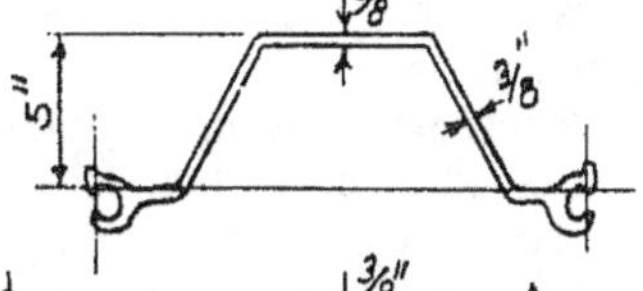		MP-116	16	36.0	27.0	⅜	14.3	10.7
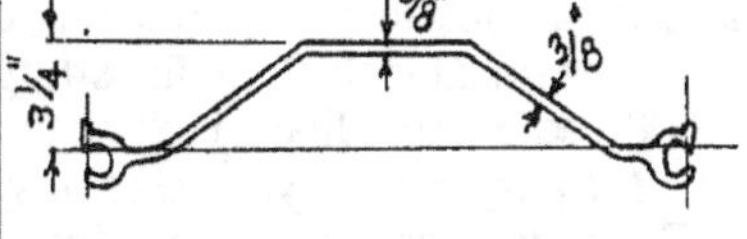		MP-115	19⅝	36.0	22.0	⅜	8.8	5.4

> ***Editors Note***
> This chart is now obsolete (1990) but has been included since it is referenced in the text. For a complete listing of all major steel sheet piling shapes currently produced - see pages 236-248.

Figure 12
Straight and Arch Web Piles

As intimated above, the inherent weakness of the steel sheet pile cofferdam is the vulnerability of the piling interlocks. Every consideration involving both design and construction which will affect the stresses in the piling interlocks should be carefully investigated. In addition to the type and characteristics of the cell fill, these considerations include:

1. Maximum allowable interlock stress.
2. Maximum computed interlock stress for the cells.
3. Depth and condition of the overburden.
4. Condition of the rock foundation.
5. Results from practical tests to determine driving conditions and the resulting effect on the piling interlocks.

Where conventional principals of soil mechanics and carefully determined data are used in conjunction with the best available information gained through practical experience, there seems to be little reason for failure due to faulty design. Practically every instance on record of failure from bursting can be traced directly to piling being driven out of interlock. (See "Interlock Failure--Kentucky Cofferdam," page 24, and footnote 3).

> *Editors Note*
> Other possible bursting failure causes: failure of connector piles, improper interlocking, accidental change of steel properties due to torch cutting or welding.

Piling for Kentucky Cofferdam

The 3/8 inch thick web for the 15 inch straight pile adopted by TVA was satisfactory for all its cofferdam needs. But because of the heavy sand and gravel overburden at Kentucky Dam, ranging from 50 to approximately 85 feet in depth, it was questionable whether or not the standard 3/8 inch web would stand up under the heavy driving conditions to which it would be subjected. At the time the piling was purchased for Kentucky, therefore, manufacturers were requested to supply a piling with higher guaranteed interlock strength than was guaranteed for the standard pile. This was accomplished through special supervision at the mills during the manufacturing process without making any changes in the standard pile section. An allowable interlock working stress of 12,000 pounds per lineal inch was used in designing the cells, with a guarantee from the manufacturers of 16,000 pounds per lineal inch maximum ultimate.

In analyzing the strength of the 3/8 and 1/2 inch webs, the guaranteed ultimate of 16,000 pounds per lineal inch was permitted for both sections of piling. This, then, was equivalent to a unit strength of 42,700 pounds per square inch for the 3/8 inch web and 32,000 pounds per square inch for the 1/2 inch web. For stresses of 12,000 pounds per lineal inch for the piling interlock, the unit stress in the piling webs was 32,000 and 24,000 pounds per square inch respectively. Since a 70,000 pound per square inch ultimate strength steel was being used, it was apparent that the interlocks would fail long before the piling web. (The only web failures noted have apparently been due to foundation faults, boulders, etc., and not to a weakness of the web or interlock.) Tests of the 3/8 inch piling interlocks indicated failure at 20,000 to 25,000 pounds per lineal inch or up to 66,600 pound per square inch web stress, proving again the strength of the web over the interlock. Based on the indicated strength of the 3/8 inch piling, a deicsion to use the 15 by 3/8 inch pile instead of the 15 by 1/2 inch pile was made in order to effect a material saving in steel tonnage. (See specifications starting on page 23 for piling bought for the Kentucky project. See also figure 6.)

Field Driving Test--Because of the depth of overburden, expected driving difficulties, and magnitude of the cofferdam to be built, a practical pile driving test was made at the site of the Kentucky cofferdam. Here a circular cell made up of 3/8 and 1/2 inch piling was driven through 95 feet of overburden (job average was approximately 50 feet) and was then unwatered, excavated, and the piling inspected. The results obtained with the 3/8 inch piling were considered satisfactory and it was therefore adopted in preference to the 1/2 inch piling to effect the saving in steel tonnage mentioned above. (For description of test see "Test Cell for Kentucky Cofferdam," page 145.) A set of Kentucky piling specifications follows:

Specifications for New Steel Sheet Piling for Kentucky Cofferdam-- The following specifications give the requirements for sheet piling for the Kentucky cofferdam.

Section 1--Service requirement. Steel sheet piling to be furnished under these specifications will be used in the construction of circular and arc-diaphragm cell cofferdams and cutoff walls.

Section 2--General. All piling shall be new and free from defects that would impair its strength, durability, and suitability for the use intended throughout its entire length.

All work shall be done and completed in a thoroughly workmanlike manner and shall follow the best modern practice.

Section 3--Type and size. Steel sheet pile sections shall be straight web and shall have the following characteristics:

> *Editors Note*
> Manufacturers current dimensions may not agree.

Normalwidth	15 inches
Weight per foot of pile	38.8 pounds
Weight per square foot of wall	31.0 pounds
Webthickness	3/8 inch
Section modulus (per pile)	3.7 inches
Guaranteed full interlock	16,000 pounds per lineal inch (Min.)

The steel shall be made by the open-hearth process.

Chemical composition
Phosphorous shall not exceed	.06 percent
Sulfur shall not exceed	.06 percent

Strength
Minimum tensile strength	70,000 pounds per square inch

> *Editors Note*
> Minimum yield point is now a more important specification than "minimum tensile strength" (ultimate strength.) Sheet piling is normally ordered to meet ASTM A-328 (latest issue) specification.

Bend test specimens shall stand being bent cold through 180 degrees around a pin, the diameter of which is equal to twice the thickness of the specimen without cracking on the outside of the bent portion.

Section 4--30 degree wye sections. Each wye section to be furnished shall be constructed of two 15-inch, straight web sheet piles built in accordance with drawing No. 502-B-9 and arranged to develop a minimum guaranteed interlock strength of 16,000 pounds per lineal inch along the straight section and not less than 12,000 pounds per lineal inch on the angle section.

Section 5--120--degree wye sections, tees, and crosspieces. Each wye section to be furnished shall be constructed of three 15-inch, straight web sheet piles joined by three bend plates and riveted, the entire section to develop the full interlock strength of the section (16,000 pounds per lineal inch).

Each crosspiece to be furnished shall be constructed of

15-inch straight web sheet piles joined by angles and riveted, the entire section to develop the full interlock strength of the 15-inch pile section.

Each tee piece to be furnished shall be constructed of 15-inch straight web piles joined by angles and riveted, the entire section to develop the full interlock strength of the 15 inch pile section.

Section 6--Handling holes. Two handling holes shall be provided at the top of each of the straight web piles. Holes shall be 2-1/2 inches in diameter. All wye sections shall have handling holes at each end.

Section 7--Marking. All piling shall be marked for length at the ends and sides of the piling and wyes.

Section 8--Shipping. Piling shall be stacked (and blocked) in lifts of five (5) piles each to allow slings to be attached for facilitating unloading, each stack to contain piling of the same length, and with oak strips separating each lift.

The provisions of section 6, 7 and 8, above, should be noted. Handling holes are valuable for handling during erection and can be used later for attaching pile extractors. Marking the piling length on each piece greatly facilitates handling and placing on the job. Proper scheduling of shipments, so that the piling can be handled directly from the cars to the work, will often effect important economies.

Interlock Failure--Kentucky Cofferdam--Experience at Kentucky cofferdam indicated that piling driven through as much as 40 to 50 feet of sand and gravel overburden will be subject to excessive stresses, resulting in considerable interlock failures. Some of these failures were exposed when overburden was removed along the dry side of the cloverleaf cells. Repairs to prevent further failure wer made as excavation progressed. More serious ruptures were repaired in 3-foot vertical stages by using longer metal straps covered with a concrete face wall to prevent further leakage and possible loss of cell fill.

Where overburden could not be removed to expose broken interlocks, the overburden was consolidated through grouting to increase the security against leakage and further failure. (See "Foundation Treatment," page 121). Although the overburden at Kentucky Dam was, in general, free from boulders, a program of foundation investigation was carried on simultaneously with pile driving operations. Where piling reached refusal at an elevation higher or lower than was indicated for bedrock by adjoining piling, a core drill hole was drilled as close to the pile affected as possible to determine whether a break in the interlock had occurred at a point high enough above bedrock to threaten the stability of the cell. At times a hole on either side of the piling was necessary before a definite conclusion could be reached.

Editors Note
Interlocks of both 3/8" and 1/2" web sections were identical and would have had no effect on driving. The slightly stiffer half-inch section might have driven better with less web damage.

In spite of all precautions taken, there was a high percentage of lost piling during first stage driving due to twisted piling and ruptured interlocks. This damage to the piling was apparently due, for the most part, to excessive interlock stresses caused by driving through the heavy overburden and in some cases to overdriving after bedrock had been reached. Over 25 percent of the piling pulled from stage 1-B was unfit for reuse, resulting in the purchase of approximately 1500 tons of extra piling for use in the construction of stage 2. If all of the piling in stage 1-B had been pulled, this percentage of unusable piling might have been even higher.

Because of this large loss of piling for reuse, it is apparent that careful consideration should again be given to the choice between the 3/8 and 1/2 inch web thickness. The thicker web will doubtless add stiffness to the piling, which should reduce to some extent the damage occurring during driving. The saving in steel tonnage at Kentucky cofferdam effected through the use of 3/8 inch web instead of 1/2 inch web was only 5 percent and in the final analysis should be balanced against the cost incurred in repairing interlocks, foundation grouting, and pulling crooked piling before making a decision on the thickness of web to be adopted.

Length of Piling

Accurate information on the elevation of bedrock will be needed in determining the required lengths of the steel sheet piling to be purchased. This is obtained from core borings and preliminary foundation investigations. It is essential that this exploration data be complete and reasonably accurate in order that proper lengths may be determined. It is not advisable to prepare the bill of material too carefully, however, since it is practically impossible to determine all of the high and low spots in the bedrock.

The longest "non-premium" pile length is 80 feet, and where required over-all lengths exceed this amount the pile should be purchased in two lengths. The bottom section should be long enough to reach to above-normal low water, and alternate piles should differ in length by not less than 5 feet so that the splices will be staggered.

Splicing Piling

Splices may be made before, during, or after the piling has been driven, depending on local conditions. An example of each is given below.

At Fort Loudoun and Watts Bar Dams the piling for the cofferdams was supplied largely from used piling from other projects. It was therefore necessary to build up piling to the desired lengths by splicing. This was done, for the most part, before the piling was driven.

At Chickamauga Dam the piling for the cofferdam was spliced during driving. Here the foundation was filled with faults, crevices, and solution channels, and it was impossible to tell in advance just how far the pile would penetrate. If the first length did not reach sound rock, a second length was added by splicing and driving continued. As many as three sections were used in some instances.

At Kentucky Dam the cofferdam cells were so high that more than one length of piling was required to complete the full height of the cells. Splicing in this case was completed after the first section of the piling was driven. Splicing of the long lengths before driving introduces a difficult handling problem and therefore is not advised.

Splicing Procedure--Chickamauga Cofferdam--Three methods of splicing were tried at this project:

1. Butt-weld with no splice plates.
2. But-weld with an 8 inch by 3/8 inch by 1 foot 8 inch long splice plate on each side of the pile and welded all around.
3. Butt-weld with one 8 inch by 3/8 inch by 1 foot 8 inch long splice plate added to the outside of pile and welded all around.

The butt-weld in method No. 1 failed repeatedly in tension during pulling and broke during handling of piling. Methods No. 2 and 3 were used throughout most of the driving operation and were satisfactory.

Editors Note
While TVA does not mention any specific problems, todays contractor should verify proper procedures with his supplier prior to welding or burning.

Splicing Procedure--Kentucky Cofferdam--Three methods of splicing were used also at this project:

1. Butt-weld with no splice plate. This splice was used on the river side of the cell only for watertightness in piling which was not to be pulled.
2. One splice plate cut from an 18 inch section of piling or equal, placed on outside face of pile and welded all around except for a 4 inch space on each side of the joint. This 4 inches of welding was omitted to permit buckling in the splice plate during driving without breaking weld. This type of splice was used along the dry side of cell in which the piling was to be pulled.
3. Same splice as in No. 2 above except that the plate was welded all around. That part of the joint not covered by the splice plate was also welded from one side for watertightness. This type of splice was used on piling along the river side of cells in which the piling was to be pulled.

In general, splices of types No. 2 and 3 were used only in cases where it became necessary to drive the top of the first, or bottom, length of piling to an elevation below normal low water. Welding of the splices on the dry side of the cells was usually deferred until after the cofferdam had been unwatered. Welding along the river side of the cells was accomplished from a scaffold attached to a work barge. Each method of splicing was satisfactory for the purpose for which it was used and, in general, required less welding than other methods previously used by TVA.

CHAPTER 4

FILL MATERIALS

The general requirements for the material to be used in filling the cells and in the construction of berms, a description of the various materials, their availability, the factors affecting their use, tests made to determine their characteristics, and related matters concerning the filling of cells and the construction of berms are discussed in this chapter.

General Requirements for Cell Fill

The general requirements for a fill material for cellular steel sheet pile cofferdams resting on rock foundation are:

1. High coefficient of internal friction.
2. Free draining.
3. Resistance to scour.
4. Weight in mass to resist sliding.
   Overturning is no longer considered a primary factor in the design of this type of cofferdam. (For further comments see "Other Materials," page 26).

Cell Fill

It is highly desirable that the cell fill material be taken from the immediate vicinity of the cofferdam. This is often done, and as a result the fill material will be found to vary widely from project to project. If no suitable fill material can be found within a radius from which it can be transported economically, then it may be advisable to use another type of cofferdam.

The cell fill material should have certain inherent qualities, which must not be ignored. For projects where the cells are small these qualities may be determined by using sound judgment and past experience. For large cells, however, it becomes increasingly important that the fill material be subjected to laboratory or practical field tests in order to determine its characteristics and suitability.

Silt--Silt offers resistance to leakage but no resistance to scour and little resistance to internal shear. such a fill, unless covered with a layer of scour-resisting material, will suffer heavily in the event that cells are overtopped by floodwaters. High interlock stresses are also apt to develop during hydraulic filling because of the liquid nature of the material. Fill materials with a fairly large silt content will tend to develop similar characteristics to that of pure silt and are, in general, to be avoided.

Mud--Mud has the same general characteristics as silt and should likewise be avoided.

Clay--Clay is a material which is extensively used in cofferdam construction because of its impervious qualities and resistance to scour. However, it is not suitable as a cell fill material because of its low frictional value, when wet, in resisting sliding and internal shear. Where the supply of a more suitable material is limited, however, clay may be used satisfactorily for that part of the cell fill above the line of saturation, or as an aid in sealing leaks.

"Sand piles" were used successfully to drain and stabilize diaphragm cells filled with marl in the construction of a ship construction basin (see Engineering-News Record, October 21, 1943, pages 38-40). It is possible that this method could be used just as efficiently to increase shear resistance in cells filled with clay. The method might be too costly for temporary cofferdam construction, however. The procedure followed in this method consisted essentially of driving 12 inch diameter closed end pipes to the full depth of the compelted cell fill, filling them with sand and gravel, and then pulling the pipe, leaving the sand and gravel in place. To facilitate removal, the pipes were capped and compressed air applied to force the sand and gravel out as the pipe was removed.

Clay, thoroughly saturated, will become a soupy liquid weighing up to 115 pounds per cubic foot and will exert full liquid pressure for the full height of the cell. Great care should be exercised, therefore, in the investigation of interlock stresses. Where the cells must be driven through a stratum of mud or soft clay which is unsuitable for fill material because of its low shear resistance, it is advisable to remove the objectionable material before driving the cells. If this is not feasible, the objectionable material inside the cell must be replaced by sand and gravel, by a carefully controlled process of displacement, to avoid collapsing the cell due to external pressures.

Sand--Clean coarse sand has been suggested by Terzaghi[1] as a suitable fill material. Since it is seldom (if ever) found readily available in sufficient quantity to meet the need of such proportions, it is hereby classed as impractical. It may be used successfully, however, in isolated cases.

Sand and Gravel--A natural mixture of sand and gravel offers one of the most satisfactory materials for cell fill and is often found along the bed of large streams where the cellular type of cofferdam is frequently used. It was available from the bed of the Tennessee River and was used, for the most part, as a fill for all of the cellular steel sheet pile cofferdams built by TVA. (Typical specifications for sand and gravel fill for circular steel sheet pile cells built at Guntersville Dam are noted below.) This type of material is ideally suited to placement by hydraulic methods and was therefore the method used by TVA. (See page 115 for specifications on furnishing fill materials for filling cofferdam cells at Guntersville Dam which was built in 1936-37.)

In constructing the upstream cofferdam at Hales Bar in connection with the installation of generating units No. 15 and 16, cells 70 feet in diameter were required. Because of the great height of these cells (58 feet) and the impracticability of using a supporting berm, it was very necessary that the cell fill material be free drainage. Tests were made of sand and gravel along the riverbed in the vicinity of the cofferdam but all of the material had a high content of fines. It was then decided that the fines would have to be screened out. A sand bar located downstream from the dam was selected and the material was pumped hydraulically to the upstream cells. All fines less than size No. 4 were removed with a revolving screen and dumped back into the river. The final product contained less than 15 percent of undesirable material and gave an excellent fill material that permitted saturation line to be held at a very low level. Test observations revealed that the upper two-thirds of the cells were well-drained, even before the cofferdam was completely unwatered. After unwatering was completed and the water table had been stabilized, the saturation line dropped even lower.

Other Materials--The cell fill materials just described, as well as combinations of these and other materials not mentioned, have no doubt been used successfully. In any case, however, the material to be used should be subjected to laboratory tests to determine whether or not it can be expected to meet minimum requirements listed on page 25. Before proceeding further, and to avoid a misunderstanding with regard to the determination and use of laboratory data, let it be said here that such data should be used with the greatest care. The weight of the materials, dry, loose, wet, saturated, etc., can of course be determined with a fair degree of accuracy. Frictional coefficients, however, will vary widely for the same material with successive tests and test methods. Test results of doubtful value should be repeated. Reference to similar tests conducted on the cell fill material for Kentucky cofferdam ("Fill Material Characteristics for Kentucky Cofferdam," page 139) will indicate the variations that might be expected. In this case where results were doubtful the tests were repeated. It is urgently recommended, therefore, that laboratory test results be used with very conservative values for the soil constants.

Placing Cell Fill--The method used and the time at which the cell fill is placed are of great importance. Hydraulic handling of the sand and gravel fill, aside from providing the greatest economy, has another distinct advantage. During filling the cell fill will be saturated to the top, thereby inducing higher interlock stresses than if dry fill had been used. The pressure of the water and the saturated fill will tend to change the shape of the cell into the familiar barrel form. As the water drains out of the cell fill the cell will probably maintain most of its changed shape. The reactive pressure will be substituted for the water pressure as the cell tends to contract, thereby maintaining a high pressure in the cell fill. This combination of high interlock stresses and high pressure in the cell fill will provide high shearing resistance for the cell as a whole. If the cells are filled while the water surface is near the top of the cells, however, this advantage gained through hydraulic filling will be minimized. It is apparent, then, that the method and time of filling should be studied and carefully coordinated in working out the design for the cells. A more complete discussion is found in chapter 5, "Design."

Berms

Two types of toe fills, or berms, are generally used in cofferdam construction. These are:
1. A fill placed along the river side of the cell to resist leakage.
2. A supporting fill, or berm, placed along the dry side of the cells, when necessary, to increase the stability of the structure.

In the first case the fill should be an impervious material, preferably one while will resist scour. Clay is the most widely used material for this purpose although other materials are frequently used because of their availability. Where scour from streamflow is high, rock is usually placed as a protective coating for the fill. The additional pressure exerted against the cofferdam by this type of fill, however, should be taken into account in designing the cells. Toe fills along the river side of the cofferdam are generally used with the crib-type cofferdam but were used by TVA in the construction of stage 2 cofferdam at Fort Loudoun Dam and in the protection of small groups of cells on other projects. In the second case, however, where the berm is built along the dry side of the cell, it becomes an integral part of the design of the cell and must be selected with regard to both material and size. A material giving weight in mass and resistance to sliding and internal shear is most desirable.

Placing Berms--Berms built along the river side of the cells to resist leakage may be placed during low water by hauling the fill material in wagons or trucks and spreading with bulldozers.

Where little or no overburden exists, the placement of a supporting berm along the dry side of the cells should receive careful consideration, as it should be able to develop the full reacctive pressure depended upon to resist deflection in the cells. If this berm is placed during high water its effectiveness will be greatly reduced. The best time for placement is during low water and before the cofferdam is unwatered. (For complete discussion see chapter 5, "Design.")

Where sufficient overburden exists, the berm may be partly or even completely formed by leaving part of the overburden in place at the time the working area inside the cofferdam is excavated. Where little or no overburden exists the berm may be formed by first unwatering the cofferdam and then building up a berm from rock and other material excavated in connection with the permanent work. This, of course, should be done during a low-water stage.

Saturation and Drainage of Cell Fill and Berms

Saturation Line--The saturation line, or elevation of saturation, exists to some degree in every steel sheet pile cell which is resisting a head of water. The height to which the saturation line will rise depends largely upon the characteristics of the fill material. In designing the cell, cognizance is taken of the full hydrostatic pressure exerted by the water contained in the saturated fill and of the pressure exerted by the submerged fill. It is therefore important that a rational conclusion be reached regarding the probably location of the line of saturation in the cell to be designed. However, little information is available that will help to approximate its location in every instance. For the most part its location is still approximated, based on previous experience.

Starting with the first cofferdam designed by TVA, it was realized that hydraulic filling of cells would require interlock stresses to be computed on the basis of full hydrostatic pressure, plus pressure from the submerged fill. Early analysis was based on the tacit assumption that weep holes would almost completely drain the sand and gravel fill (see figure 13-a). Some doubt existed, however, as to the correctness of this assumption and as a result a series of tests was made, first at Pickwick cofferdam and later at Guntersville, Chickamauga, and finally Kentucky cofferdam. These tests are described starting on page 27.

Professor Karl Terzaghi presents a theoretical explanation of the factors affecting the location of the saturation line,[2] and a brief discussion is inserted at this point:

Figure 13 represents vertical sections through cells resting on an impermeable base. The inner row of sheet piling is provided with weep holes. If the cells are filled with a clean, coarse sand or gravel (figure 13-a) the quantity of water that will leak through the joints at the outer wall is small. The sand offers little resistance to flow and there is, therefore, a big drop in head at the outer wall, leaving only the bottom of the cell to be covered with water. If the cell is filled with silt (figure 13-b), the quantity of water flowing through the outer wall is still small, since the fill offers another obstacle to its passage. The low permeability of this fill requires a steep hydraulic gradient to discharge even a small quantity of water. The saturation line will therefore assume a high elevation. The cells filled with clay (figure 13-c) offer more resistance to flow than silt and nearly as much as the outer row of piling, thereby resulting in a negligible drop in head at this point. Water will percolate through the cofferdam fill, then, as if the outer row of piling were nonexistent.

In practice, the condition illustrated in figure 13-a has rarely been encountered. The mixture of sand and gravel found at most cofferdams contains enough silt and fines, and sometimes clay, to make the saturation line approach that shown in figure 13-b or even figure 13-c. The saturation line is apt to be raised and flattened out if the weep holes fail to function properly. Extreme care should therefore be taken to maintain the maximum possible drainage for the cell fill. (See previous discussion dealing with screened fill for Hales Bar cofferdam under "Sand and Gravel," page 25.) There the saturation line closely approached the condition illustrated figure 13-a.

Tests for Saturation Line--Guntersville and Pickwick--Two inch diameter pipes perforated with 1/4 inch holes were sunk in a line across three cells in Pickwick lock cofferdam and across seven cells in the Guntersville lock cofferdam. Only one cell in the Pickwick lock cofferdam revealed reliable information, however, Results obtained from these two sets of tests varied widely although similar cell fill material was used in each case. A sieve analysis of samples taken from the fill used in the cells at Pickwick indicated the following properties:

Gravel larger than 1/2 inch	36 percent
Coarse sand	28 percent
Fine sand	36 percent
Total	100 percent
Material passing a 3/8 inch standard sieve	50 percent

At Pickwick the saturation line varied with the rise and fall of the river level. It remained nearly level over approximately two-thirds of the width of the cells, with the highest observed elevation at the center of the cells (when the river was near the top of the cells) being 0.71 times the height of the river above the base of the cells. This ratio varied, however, but the exact cause is unknown. Water level in the test wells lagged behind the rise and fall of the river level. This was thought to be caused, at least in part, by the 1/4 inch holes in the test pipes being plugged up. It is possible that this condition had some effect upon the variation in the ratio mentioned above (See figure 14 for the general trend the saturation line followed at Pickwick).

At the Guntersville lock cofferdam the saturation line was nearly level for the full width of the cells and stood at a height of only a few feet below the level of the river, averaging approximately 2 feet 6 inches for cells having no weep holes. After 1-1/2 inch weep holes were burned at 3 feet on centers in every fifth pile (along the dry side of the cell) the saturation line dropped to an average of approximately 8 feet below the level of the river.

The saturation line at Pickwick lock may be classed as following approximately the theoretical (see figure 13-b), due to the high content of fines in the fill material. The deviation from the theoretical may be explained as being due to the method by which the fill was placed. The swirling of the material entering the cell from the dredge pipe had a tendency

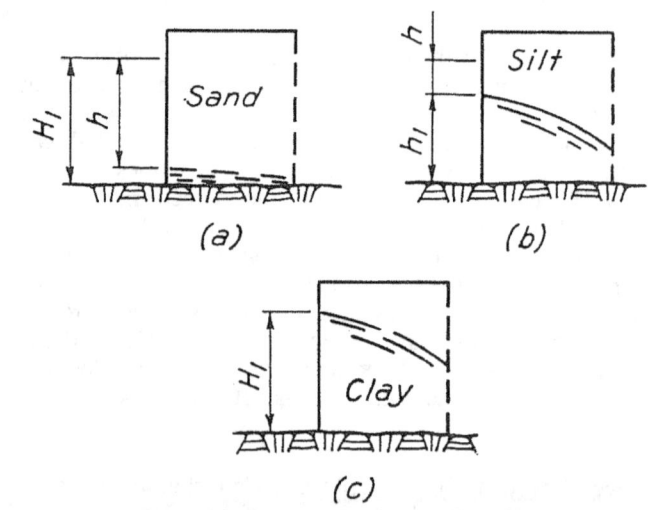

Figure 13
Saturation Line in Steel Sheet Pile Cells Resting on Rock Foundation

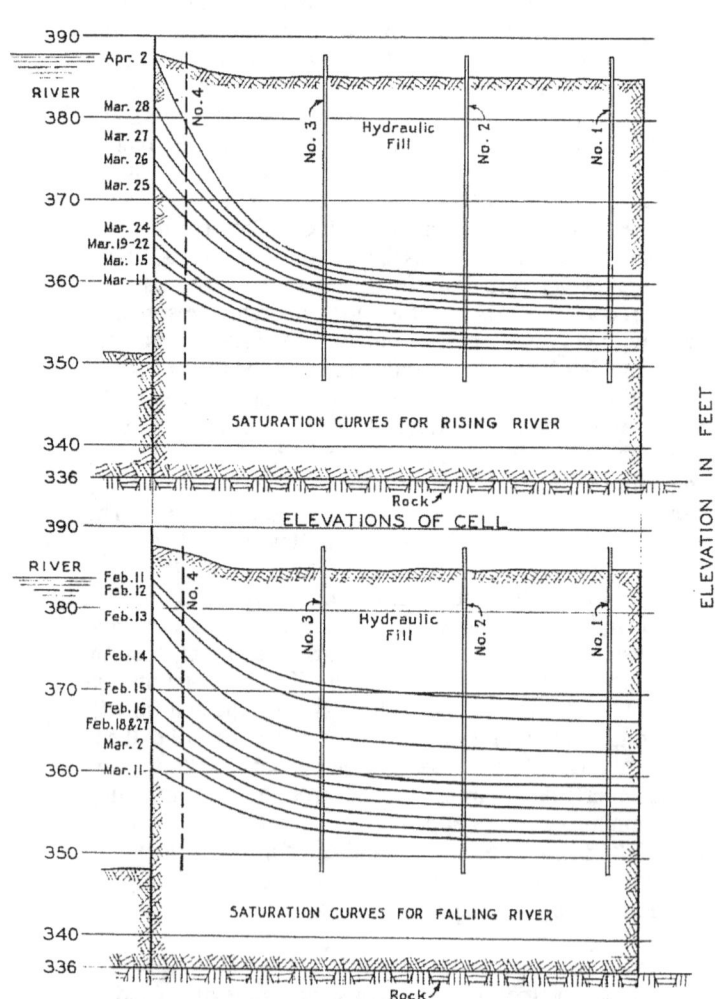

Figure 14
Test Pipe for Water Level and Saturation Line in Steel Sheet Pile Cell - Pickwick Landing Dam

to float the fines to the outer rim of the cell, thereby partially sealing the piling interlocks with fines and at the same time forming a rim of semi-impervious material along the outer

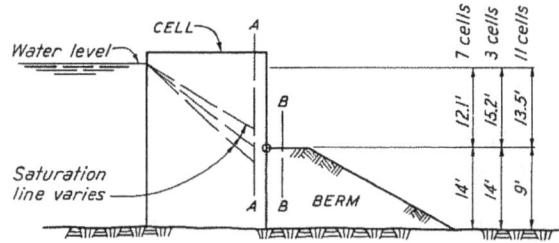

**Figure 15
Results of Tests for Saturation Line in Steel Sheet Pile Cell - Chickamauga Dam**

edges of the cell and leaving a core of more or less pervious material at the center. Also, weep holes in the Pickwick lock cofferdam were plugged during filling operations to stop the heavy loss of fines held in suspension in the waste water. This further increased the fine content of the cell fill and no doubt partially blocked future drainage to the weep holes after the holes were unplugged. Had the weep holes been rodded out to get better drainage, the saturation line might possibly have been lowered.

The saturation line at Guntersville lock cofferdam, when using weep holes, approximated the condition shown in figure 13-a. Before the weep holes were burned through the piling, however, a high head was required to force the water through the inner row of piling. After the weep holes were burned in the piling along the dry side, the seepage line dropped to near the bottom of the cells. The fill material no doubt contained a relatively small percentage of fines and silt. This, together with careful rodding of the weep holes, was perhaps responsible for the better drainage condition. No actual sieve analysis of the fill material was made.

Tests for Saturation Line--Chickamauga Lock Cofferdam--At Chickamauga cofferdam during third-stage construction, auger holes instead of perforated piping were used for locating the saturation line (See figure 15). Holes drilled in 21 cells, along plane A-A, showed the elevation of the saturation line to be very nearly the same elevation as the top of the berm at point P. A number of holes were then drilled just outside the cells along line B-B to a depth of 10 feet but no water was found except in one instance. No weep holes were drilled below the elevation of the top of the berm, however, indicating a rather high resistance to flow through the piling along the dry side of the cells.

Test for Saturation Line--Kentucky Cofferdam--An elaborate drainage system was devised for this cofferdam. Weep holes were provided at the riverbed (elevation 290±) and at the top of the berm (elevation 300 and 310), and connected to drainage heads (see "Cell Drainage at Kentucky Cofferdam," opposite column. For a check on the efficiency of the drainage system, observation wells were placed in all cells. The saturation line at the inner side of the cells was found to remain very constant (about 4 to 5 feet below the top of the berm) in spite of the 32-foot variation in the river level.

Location of Saturation Line for Design Purposes--An analysis of observations made on TVA projects seems to indicate that it is possible to keep the saturation line quite low, but that it should not be so assumed for design purposes unless it is certain that care will be taken to see that the drainage system functions properly. Where a berm is used the saturation line in the cell fill should not be assumed to be below the top of the berm, however.

In the absence of better information, TVA for design purposes considered the saturation line to be straight and to extend downward on a 2:1 slope from the top of the cell on the river side to the dry side, where no berm was used. Where a berm was used, as at Kentucky cofferdam, it was taken as a stright line from the top of the cell on the river side to the top of the berm on the dry side. These assumptions appear to be on the safe side. It seems, however, that other assumptions might be made where a careful analysis of the fill material is available. For example, the saturation line for a free draining, granular material might be safely assumed to exist along a horizontal line at a depth of one-half the diameter of the cell below the top. The horizontal position of the line for design computation purposes should produce results just as accurate and at the same time simplify the design procedure.

Little is known with regard to the location of the saturation line in berms. At Kentucky it was assumed to lie along a straight line from where the berm touched the cell to its toe. The stability of the cell was also checked for a condition where the berm was completely drained. A thorough knowledge of the type of material being used, and good judgment, should be employed in deciding what location to use.

Cell Drainage--In general, the supervision to maintain drainage through the weep holes in all TVA cofferdams except Kentucky and Hales Bar was poor. (See table 1.)

Cell Drainage at Kentucky Cofferdam--At Kentucky cofferdam the cells were high and drainage was a critical feature. Here a more elaborate drainage system was installed than had been used on any previous cofferdam (see figure 16). One horizontal row of nine 4 inch diameter holes, equally spaced, was burned in each circular cell and its connecting cell. These were placed along the top of the berm. A similar line of holes was placed along the riverbed. At riverbed elevation 4 inch drain pipes were welded into the weep holes at one end and connected to a 15 inch corrugated pipe header at the other end. The corrugated pipe ran parallel to the cells (See typical layout and B-B figure 16.) This header, in turn, led to 24 inch diameter collecting headers spaced at approximately 300 foot intervals along the inside of the cofferdam. The collecting headers led out under the berm at riverbed elevation and discharged into open flumes laid along the toe of the berm. At the top of the berm the 4 inch pipes, welded into the weep holes, discharged into an open flume laid along the top of the berm. This open flume was connected to the 24 inch header of the lower drainage system at intervals through a 15 inch corrugated riser. A small amount of fill material was removed from the cell around the weep hole opening and was replaced with 3 inch gravel to ensure free drainage. As a result the saturation line along the inside of the cells was kept at approximately 4 to 5 feet below the top of the berms.

In the cloverleaf cells only one row of twelve 4 inch diameter holes was actually burned in each main cell and three 4 inch diameter holes in each connecting cell along the riverbed. This varies to some extent from the instruction shown on figure 16.

CHAPTER 5

DESIGN

All of the steel sheet pile cofferdams constructed by TVA were either built on rock where little or no overburden existed or were driven to rock through sand and gravel overburdens ranging up to 50 feet in thickness. This discussion on design, therefore, is confined to the design of steel sheet pile cofferdams driven to rock. For a discussion of cellular cofferdams resting on deep sand or clay foundations see the paper by Professor Terzaghi[1] and Cofferdams by White and

H-PILES
CAISSON
LINE PIPE
PILING PIPE
STEEL PLATE
DREDGE PIPE
STRUCTURAL PIPE
STRUCTURAL STEEL
WIDE FLANGE BEAMS

*With stocking locations throughout the United States and Canada, we are your "one stop" source of supply for **IN STOCK** new and used steel for Pile Driving, Foundation and Marine Contractors throughout North America.*

Pennco

Pennco Pipe & Tube Corporation • 1100 East Hector Street, Suite 395 • Conshohocken, PA 19428
PH: (215) 834-6310   FAX: (215) 834-6318   TELEX 244894

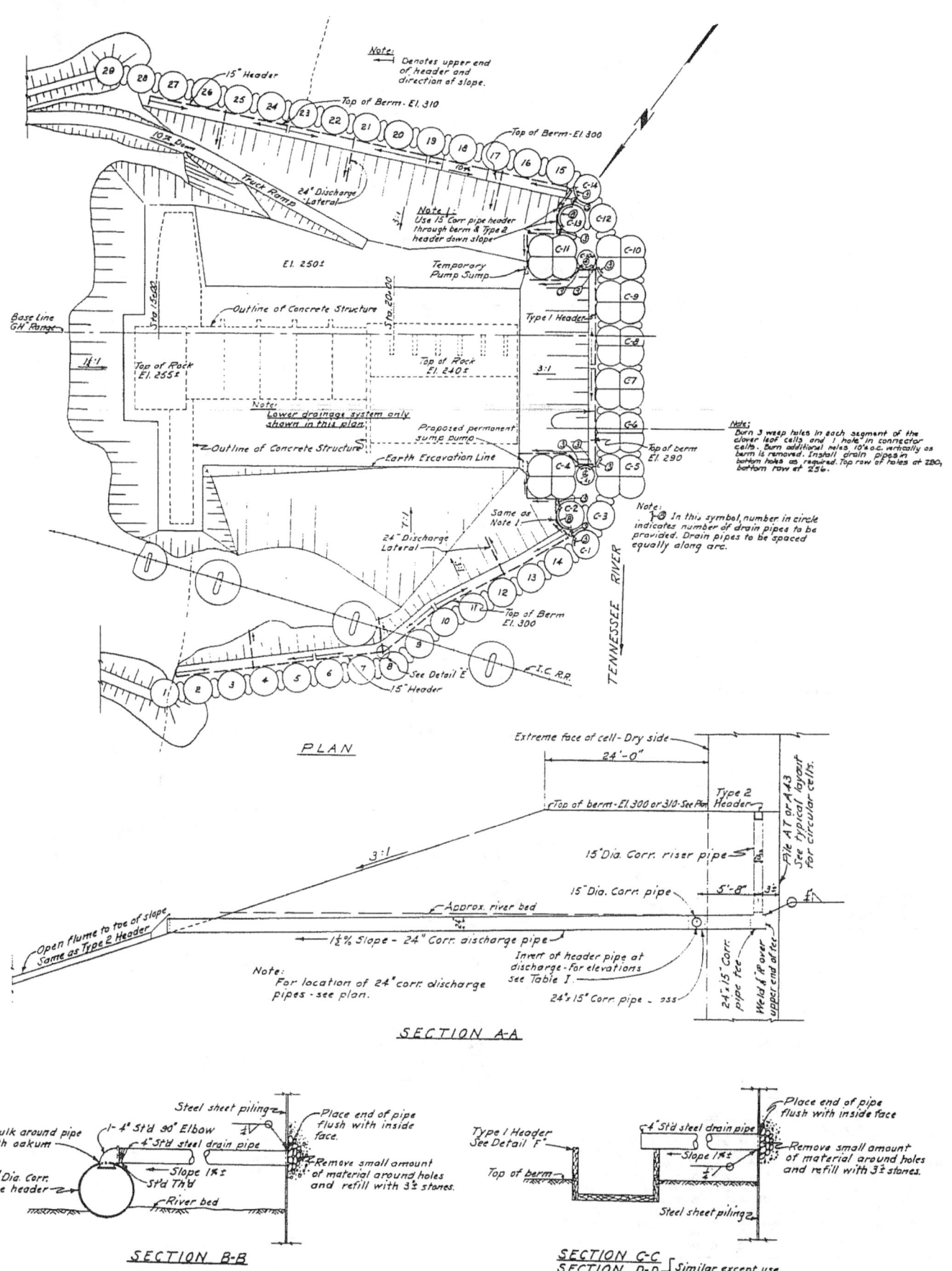

**Figure 16
Drainage System for Steel Sheet Pile Cofferdam - Kentucky Dam**

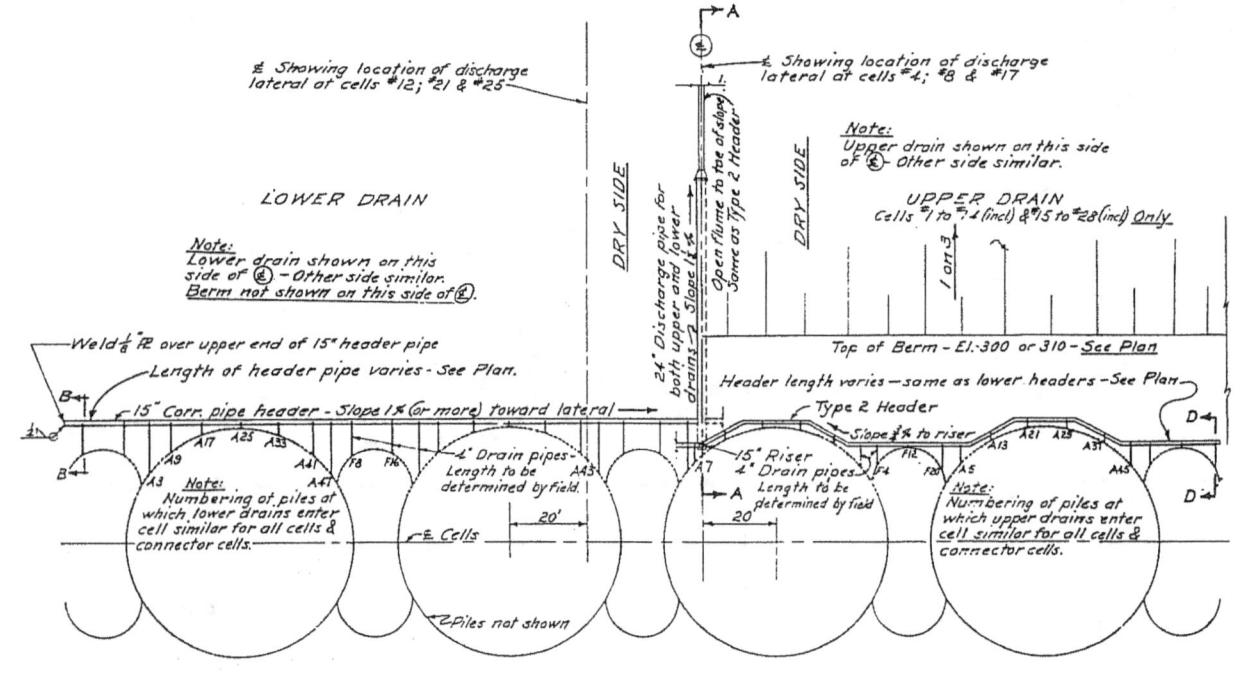
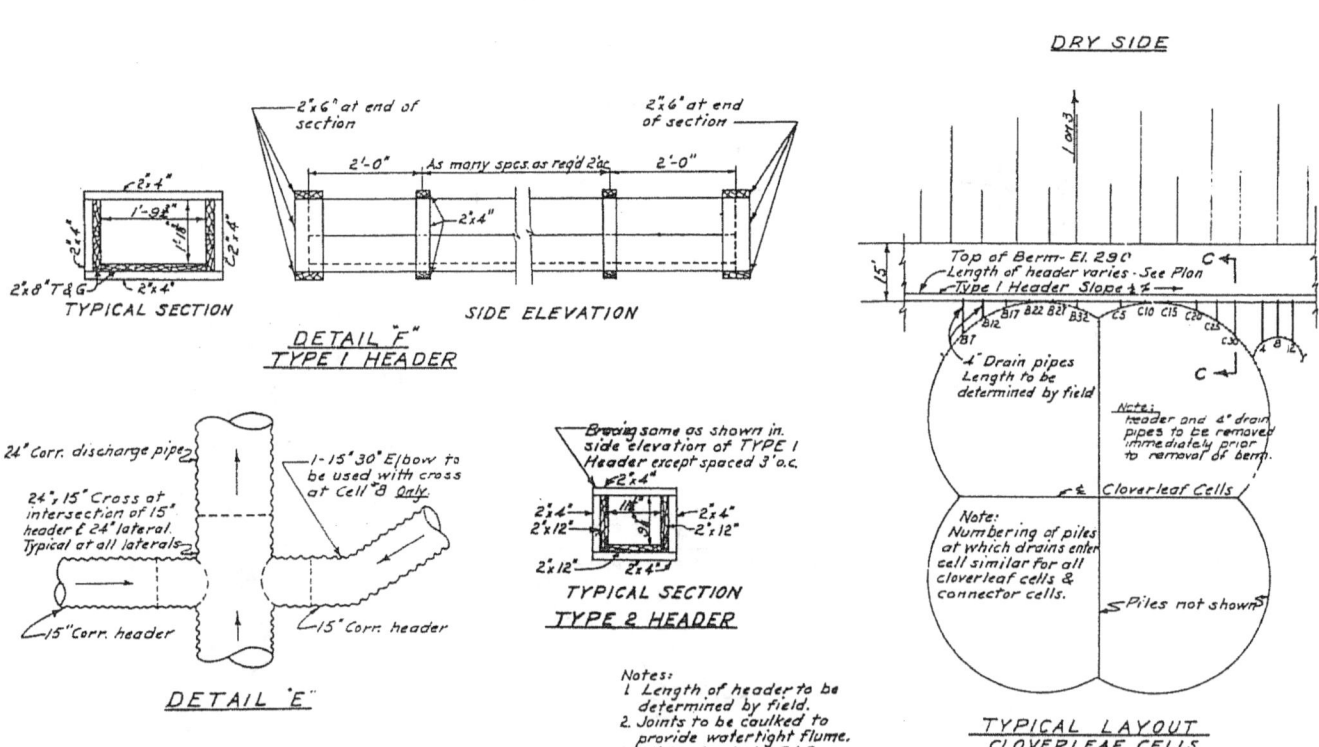

**Figure 16 (Continued)
Drainage System for Steel Sheet Pile Cofferdam –
Kentucky Dam**

Prentis.[2]

The Development of Cellular Cofferdams

The development of the design of cellular cofferdams, from the building of the first one at black Rock Harbor, Buffalo, New York, in 1908 or 1909 to the present day, is ably covered in the paper presented before the American Society of Civil Engineers in 1944 by Professor Karl Terzaghi.[3]

Although early designs were based on many faulty assumptions, the cofferdams stood up under actual service, due no doubt to their having more actual resistance to failure than was originally anticipated. This is discussed later. After a brief resume of early cofferdam construction, Professor Terzaghi continues his paper by presenting a rational method for cell design. This method had already been anticipated by TVA engineers in the design of the Kentucky cofferdam, however, and is also discussed later.

FUNDAMENTALS OF DESIGN

Cells

The paper by Professor Terzaghi[4] includes a comprehensive discussion of the fundamental principles of cellular cofferdam design and should be studied carefully by the designer. A brief summary is as follows:

The construction of a steel sheet pile cell involves the use of two distinctly different materials--soil and steel. Its design involves the use of the basic laws of statics and the fundamental laws of soil mechanics. Also, since the cell consists of two very different materials, its elastic properties may be compared to those of a composite material such as steel and concrete. These properties are governed chiefly by those of the soil in the cell, but the steel wall definitely acts to stiffen the structure. The cell should be classed, therefore, as a plastic structure and should not be treated as a gravity wall.

Several essential design considerations are discussed below, in what is felt to be their logical sequence.

Sliding Resistance--Early design methods included this as fundamentally important. Resistance to sliding was computed by multiplying the total weight of the cell by the coefficient of friction of the fill material on rock. One error was generally made, however. This was to consider that fill material was completely drained, computations being based on the dry weight of the material. Actual observations indicate that this condition is practically impossible to attain (see discussion on page 26).

Water standing in the cell fill has an uplifting effect, and the resulting weight is actually the sum of the submerged weight of material below the saturation line plus the dry weight of material above the saturation line. During early cofferdam design TVA neglected the effect of submergence but in later designs came to the conclusion that for design purposes a saturation line should be used. The line adopted in TVA design extended downward on a 2:1 slope from the water surface on the river side of the cell toward the dry side of the cell. A saturation line should not be assumed at a lower elevation unless extreme care is exercised in maintaining proper drainage. (See previous discussion under "Cell Drainage," page 28, and subsequent discussion on "Hales Bar Cofferdam," page 49; and "Fill Material for Hales Bar Upstream Cofferdam," page 145).

In spite of failure to consider saturation in earlier designs, there seems to have been no record of cell failure due to sliding. Such failure was doubtless prevented by a much rougher surface of rock foundation than was anticipated or than was indicated by the customary value of 0.5 for the coefficient of friction used for sand sliding on smooth rock. Where the rock surface is only moderately rough, the coefficient of sliding of fill material on rock will approach a value equal to that of the coefficient of internal shear of the fill material. In almost every case the resistance of the cell to sliding will be further increased by piling being driven into the rock surface, pockets, or cracks and crevices in the rock. The piling along the inner, or dry, side of the cell will also be pressed downward due to the action of the friction between the fill material and steel piling. This is discussed throughout the remainder of this chapter. It is wise, then, to consider the cells' resistance to sliding in preparing the designing. It will be found in most cases, however, that the coefficient of sliding can be taken as at least equal to the coefficient of internal friction of the fill material.

Overturning--The second step in early design methods was to consider the cofferdam as a gravity wall and to check its resistance against overturning. If the resultant of the forces fell within the middle third of the base, then it was considered stable against overturning. Here again the erroneous consideration that the cell fill was completely drained led to a resulting factor of safety which was much larger than actually existed. But the cofferdam cannot be considered to act as a gravity wall, and computation indicating its resistance to overturning do not give a true measure of its stability.

Shear--The magnitude of the Kentucky cofferdam led engineers to a very careful analysis of all accepted design methods, and the conclusions reached were that stability of the cell is a function of the shear resistance of the cell fill and of the frictional resistance of the piling to sliding in the interlocks. Professor Terzaghi had also arrived at this same conclusion,[5] not being aware at the time that TVA engineers had already done so.

TVA engineers computed the amount of vertical shear in a cell by a different process from that used by Professor Terzaghi, however, although results were essentially the same in each case. The method used by TVA is illustrated in figure 17 and the procedure is as follows: First the pressure on the base of the cell is computed, using a straight line variation. The difference in weight of the cell fill and the pressure on the base, to the left of any section AA, is then computed and gives the value of the shear on the section. This value is then used to plot one point ("a") on the curve, figure 17. A repetition of the process produces other points to complete the shear curve. It may be shown that, mathematically, the resulting shear curve is a parabola; which, incidentally, is analogous to the horizontal shear in a beam. Thus, the total shear S_v on the centerline is $3/2\ M/b$, in which M equals $1/3\ P_wH$. (See figure 17). This formula will apply in all cases similar to that shown in figure 17, whether the saturation line is level or inclined. The principal advantage of the method outlined in the first part of this paragraph is that it is applicable for any type of pressure diagram or other condition such as a cofferdam with a berm.

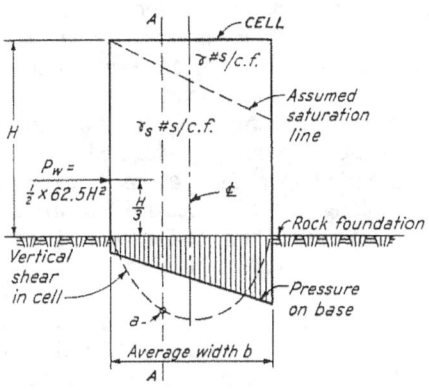

**Figure 17
Vertical Shear in Cell Fill**

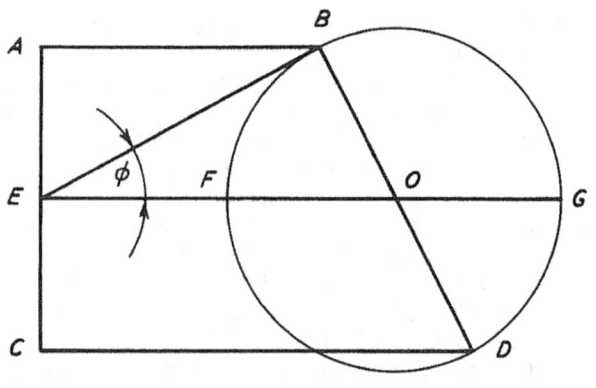

**Figure 18
Mohr's Circle**

The problem of evaluating shear resistance of the cell proved more difficult, however. Professor Terzaghi shows[6] that by multiplying the active pressure in the cell (as computed by the Rankine formula) by the coefficient of internal friction of the fill material, the resulting resistance to shear (even when combined with the resistance of the piling to sliding in the interlocks) gives a value for the factor of safety of less than 1. Previously, however, the width of the cofferdam cells constructed on a rock foundation had averaged about 0.85 of the cell height; still none had failed. From this Professor Terzaghi concludes that the horizontal pressure inside a cell must be considerably greater than the active Rankine pressure. During the design studies conducted by TVA, this same conclusion was reached with two possible explanations. These were: (1) when a cell is filled hydraulically, at a time when the water outside the cell is at a low stage, a high interlock tension will be developed in the piling ring due to the addition of water pressure to the submerged fill; (2) after the fill has drained, the tendency of the cell is to resume its original diameter but is restrained from doing so by the extremely compact fill material. This, then, will develop a high reactive pressure in the fill and may be equally as great as that caused by the water and fill material during filling operations. The method used in filling the cells, therefore, becomes of considerable importance. In designing the Kentucky cofferdam, where a high inside berm was used, TVA engineers felt that the cell would be squeezed between the pressure along the river side and the resistance of the berm along the dry side, with the resulting pressure on the inside of the cell considered to vary in a straight line between these two forces (see figures 35 to 38. However, Professor D. P. Krynine in his discussion of Professor Terzaghi's paper, points out an error into which everyone had fallen in the computation of shear resistance. The correction of this error seems to be of great importance in working out a rational method for cell design, so it is discussed in detail in the following paragraphs.

Revised Method for Computing the Shear Resistance in Cell Fills--Professor Krynine shows, in his discussion of Professor Terzaghi's paper[7] that past methods of computing shear resistance along vertical planes had been in error because of an incorrect use of the Rankine formula. He states that in assuming shear failure will occur along vertical planes (as proposed by Professor Terzaghi and used by TVA in the design of the Kentucky cofferdam), the use of the Rankine coefficient ($C = \tan^2(45° - \phi/2)$) for determining the ratio between the horizontal and vertical pressures at the point in question is incorrectly used. The value of C, he points out, is the ratio between the minor and major principal stresses at a point in a cohesionless mass just immediately preceding failure of the mass in shear; and that since failure is expected to occur along a vertical plane this direction is not a principal one because, by definition, there are no shears along principal axes. Therefore, the above value of C cannot be used to compute the horizontal pressure on the vertical failure planes. Professor Krynine shows instead that the well-known Mohr's circle may be applied in finding the ratio "K" between the horizontal and vertical pressures at a point on a plane (as AA in figure 17) at a depth Z from the top of the fill just immediately prior to failure in shear. To demonstrate, simply draw a horizontal line and a circle of arbitrary radius, with its center at 0, on the horizontal line (see figure 18). Next draw a tangent line to the circle that will describe an angle ϕ intersecting the horizontal at a point E (ϕ being the angle of internal friction for th cell fill). Further, draw the diameter of the circle through the point of tangency B, and then the vertical line AEC. Now, the ratio of the distances EF and EG is the ratio of the principal stresses at the point in question. AB is proportional to the normal pressure on the plane of failure (section AA, figure 17), and the distance CD is proportional to the pressure on the plan perpendicular to that plane of failure--that is, in the given case, the horizontal plane. Hence the ratio of the distances AB and CD multiplied by γZ (which, incidentally, is the vertical pressure at the point, γ being the weight of the material per cubic foot, and Z being the depth of the point from the top of the fill) is the horizontal pressure on the plane AA at the given point. Simple algebra will show that the value of $K = \cos^2\phi/(2 - \cos^2\phi)$. (This constant is designated as "K" to distinguish it from the constant "C" as determined by the normal Rankine formula.)

Now, using an angle of internal friction of 28° 50' as determined for Kentucky cofferdam, "K" determined by Mohr's circle method would be 0.623, and "C" determined by the Rankine formula would be 0.29, for active pressure used in computing shear resistance. The resistance shown by Mohr's circle method is over twice as great as that actually computed in designing the Kentucky cofferdam cells while using the Rankine formula method, indicating that the berms perhaps could at least have been reduced in size.

Professor Terzaghi agrees[8] that Professor Krynine has offered convincing proof that shear failure on vertical planes is inevitably preceded by, and associated with, a considerable increase in normal pressure on these planes. However, Professor Terzaghi takes exception[9] to Professor Krynine's suggestion of possible failure along shear planes other than vertical ones.[10] Professor Krynine suggest that perhaps failure would occur along a plane (or curved surface) extending from the heel of the cell to the top of the cell at the dry side, accompanied by a rise in the surface of the fill along the dry side of the cell. In an attempt to verify these contentions the simple test described below was performed.

The top and one side were removed from a thin cigar box to construct the appartus shown in figure 19. The top was replaced with glass to facilitate observation of the fill particles, and the box was set on its edge with the open side up. Two thin wood strips as shown in figure 19 were then inserted through the open side to simulate a section through a double-wall cofferdam. The tops of the two wood strips were tied together with a tie-rod to hold the fill pressure, and the space between the strips was filled with fine sand to simulate cell fill. Although this apparatus did not duplicate the circular cell, it was felt that fill action in either case should be somewhat similar. As a force was applied at P, moving the strips from a to a_1 and from d to d_1, the fill (as shown in figure 19) piled up at the river side. This was just opposite to the results envisaged by Professor Krynine. The distance a_1b became greater than H, and the distance cd_1 became less than H. As the deflection increased the indicated plane of rupture a_1c became more and

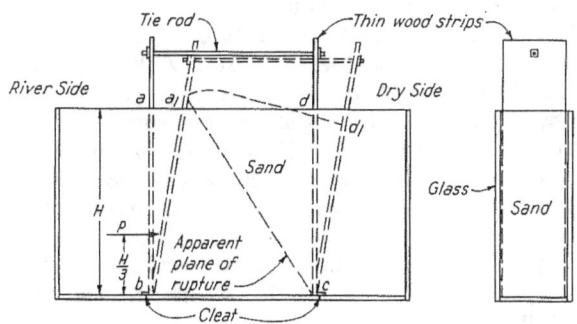

**Figure 19
Test Apparatus for Checking Plane of Rupture in Cell Fill**

more apparent. Also the fill in the triangle a_1bc became highly compressed with a marked tendency for the strip ba_1 to ride upward, while the sand in the triangle a_1cd_1 remained noticeable loose. This apparently indicated that the pressure on cd_1 was only equal to the active pressure of the sand while that on a_1b was equal to the applied lateral load.

Professor Terzaghi points out[11] that failure occurs simultaneously on vertical and inclined planes, the latter sloping upward toward the river side similar to the direction of the plane ca_1 and therefore considers it justifiable to evaluate the shear resistance on the basis of vertical planes. In the above model test the movement of the sand particles along the indicated plane of rupture ca_1 only became apparent after considerable deflection had taken place. There was also some indication of vertical movement of the sand particles during the initial stages of deflection. It is quite possible, then, that shear resistance along vertical planes may give a measure of the stability of the cell for small deflections.

It should be noted that the results of the test agreed very closely with the action of a double-wall cofferdam as described by Mr. Savile Packshaw, with the British Steel Pile Company, Ltd., London, England, M. Am. Soc. C. E., in his discussion of Professor Terzaghi's paper.[12] The question now arises as to why a cellular cofferdam should not be investigated in the same manner as outlined by Mr. Packshaw for the double-wall cofferdam. It would seem that a cellular cofferdam of the same average width as the double-wall cofferdam should be at least as safe without considering the added resistance of the piling in the cross-walls. However, computations based on Mr. Packshaw's method for a cellular cofferdam resting on rock, without berm and having an average width of 0.85 times the height (and head of water), give a factor of safety for the fill resistance to the water pressure of 1.77 (without considering the added resistance to shear offered by the resistance of the piling to sliding in the interlocks). The resistance to shear along a vertical plane through the centerline of the cell, even when using K = 0.623 (see "Shear," page 32), gives only a factor of safety of about 1. During the course of the test illustrated in figure 19, however, another feature was noted which might be a controlling factor in cofferdam design. This is the frictional resistance of piling on fill, which is discussed in detail below.

Frictional Resistance of Piling on Fill--As the pressure applied it became necessary to hold down the side ab (figure 19) firmly to prevent its riding up on the fill. When not held down it moved up and allowed the cell fill to run out at the heel. There was still some question, however, as to whether the cell made up of interlocking piling would act in the same manner. It was decided, therefore, to make a small model cell (figure 20-A), using interlocking piling, and compare the resulting

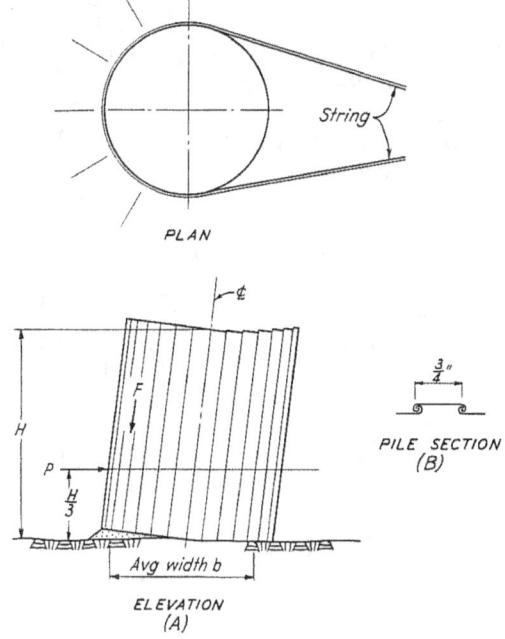

**Figure 20
Sliding Test With Model Sheet Pile Cell**

action with that of the double-wall model. A number of small interlocking sheet piles were made from 1 1/2 inch by 8 inch strips of 25 gage aluminum by bending the edges to form the interlocks (see figure 20-B). Twenty-six piles were made and placed to form a cell approximately 6 1/2 inches in diameter. The cell was then filled with fine sand to a depth of about 6 inches. Water pressure was simulated by pulling on a string looped around the cell. This gave an even distribution of the pressure over the river half of the cell. Test procedure was as follows.

The cell was first placed on a paper surface and the pressure applied at the 1/3 point of the cell height. Failure in this case occurred through sliding. A thin layer of sand adhered to the paper with the rest of the fill sliding over this, bearing out a previous suggestion that the coefficient of sliding, even on smooth surfaces, will be at least as great as the coefficient of internal shear.

Next, resistance to sliding was provided by restraining the cell at the toe and pressure again applied. A tendency for the piling along the pressure side to creep up on the fill was at once noted. As the pressure continued the fill material began to run out at the heel of the cell, and the cell assumed the shape shown in figure 20-A. Note here that the piling along the dry side of the cell had slipped in the interlocks while along the river half there was no noticeable slip. In fact, the piling on the river side remained in one undisturbed group. Slipping in the interlocks started along the dry side at the extreme inside piling and continued progressively to about to centerline of the cell. At this point a large part of the cell fill had escaped at the heel and, with increased pressure against the cell, it collapsed completely by overturning. The piling along the river side, upon which pressure had been applied, was still in one undisturbed group with the ends of the piling remaining flush one with the other. However, slipping in the interlocks of the other piling had increased quite markedly. A possible explanation of why the piling along the river side of the cell did not slip in the interlocks is given below.

First, assume:

Cell diameter = D
Cell height = H
Coefficient of friction for piling on fill = 0.4
Coefficient of friction in piling
  interlocks = 0.3
Coefficient of lateral fill pressure = 0.29
Weight of fill material (dry) = 110 lb per cu ft

The following equations may now be developed:

Water pressure $P_w = 1/2 \times 62.5 \times H^2 = 31.25 H^2$
Active cell fill pressure $= 1/2 \times 0.29 \times 110 \times H^2$
$= 15.9 H^2$

Tension in piling $= 15.9 H^2 \times 1/2 D$
Interlock resistance $= 2(0.3 \times 1/2 D \times 15.9 H^2)$
$= 4.77 DH^2$ (2 sides)

Friction of fill on piling $= 0.4 \times 31.25 H^2$
$= 12.5 H^2$

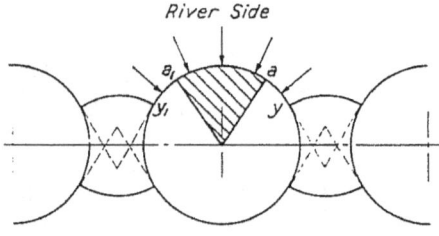

**Figure 21
Pressure Diagram for Cell in Sliding**

The interlock resistance, then, will balance the friction of the fill on the arc aa_1 (figure 21), with length equal to $4.77 DH^2 \div 12.5 H^2$ or $0.38 D$. The rest of the piling in the river side of the cell arc, a_1y_1 and ay (figure 21), will then be in unstable equilibrium and might move with the rest of the piling or slip in interlock, depending upon slight variations in the frictional resistance.

From the tendency of the river side piling to slip up on the fill, it would appear that the frictional resistance of the piling to sliding may be a more critical point in design than the shear resistance of the fill, especially where no overburden exists and where the fill material is mainly sand and gravel. Professor Krynine[13] suggest that this frictional resistance will add to the safety of the cell but the results of the above test seem to indicate clearly that it is not additive and may be a weak point. Of course, where the piling has been driven into a deep strata, the resistance to withdrawal will add to the stability but this is not the case for cofferdams founded on rock with little or no overburden.

A method of evaluating the frictional resistance of the piling against rising on the fill is indicated in figure 20. The piling on the river side of the cell will be pressed against the cell fill by the water pressure P_w. It is possible, however, that the pressure inside the cell may be even greater than this for the case where fill has been placed hydraulically, at a time when the outside water level was low (as explained previously under "Placing Cell Fill," page 26). In this case the tension in the piling ring is induced by the full head of the water in the cell fill plus the pressure exerted by the submerged fill. Then as the water is drained from the cell fill, relieving the cell of this inside pressure, the cell will tend to resume its original shape. The reactive pressure set up in the cell fill, or in other words the resistance of the cell fill to this contraction, may be as great as the inside pressure which existed while the cell was being filled. In the case where the fill is very compact (as when placed hydraulically), this reactive pressure will last throughout the life of the cell. No matter what method of filling the cell is used, however, the pressure of the river side piling against the cell fill will be at least as great as the outside water pressure. Therefore, unless it is certain that all cells will be filled hydraulically, it would be safest to consider that the maximum pressure will be that due to the outside water pressure only.

Accordingly, if:

$\tan \alpha$ = the coefficient of friction of piling on fill, and
b = the average width of cofferdam (figure 20),

then the frictional force resisting the upward movement of the piling is F, where:

$F = P_w \tan \alpha$
Then for a factor of safety of 1.25:
$P_w b \tan \alpha = 1.25 \times H/3 \times P_w$
$b = 0.417 H / \tan \alpha$

Where $\tan \alpha = 0.4$, then $b = 1.04 H$ and for $b = 0.85 H$,[14] the factor of safety will be only 1.02.

Average Width of Cofferdam--The design of cellular cofferdams may be simplified considerably by using the average width of the cells and then considering a section of the cofferdam 1 foot long, rather than working with the entire cell and connecting cell.

Another method is to compute the theoretical width of a rectangular cross section having a section modulus equal to that of the circular cell and connecting cell. The average width based on areas will vary up to as much as 6 percent greater than that based on section moduli. However, TVA has used the average width based on areas, and Professor Terzaghi agrees[15] that it is sufficiently accurate for all practical purposes. For the standard type of cofferdam, with 60 degrees between tees and 7 to 10 piles in each connecting arc (see figure 45), the average width is very nearly equal to the area of one circular cell divided by its diameter, or $b = 1/4 \pi D = 0.785 D$. When the cell diameter is within a range of 30 to 60 feet, the maximum variation between the above value and the actual average, allowing for the connecting arcs, is only 2 percent. For cofferdams with 90 degrees between tees (see figure 39), the average width is very nearly $0.875 D$ for cells with diameters ranging from 30 to 70 feet. However, if special connecting arcs or cloverleaf cells are used, as in the Kentucky cofferdam, the actual average width should be determined on the basis of a section of cofferdam of length equal to the distance from center to center of cells.

Fill Pressure--In computing the horizontal pressure of the cell fill against the piling, it is necessary to determine the ratio C of the horizontal to vertical pressure in the fill. Professor Terzaghi[16] points out that the value of C is very uncertain and may vary from point to point in the fill. From the results of his computations he suggests an empirical value of 0.4 to 0.5. However, in view of Professor Krynine's discussion,[17] in which he shows that a value of $K = \cos^2 \phi / 2 - \cos^2 \phi$ should be used in computing shear resistance, it would seem that the customary value of C for ordinary active pressure may safely be used for computing interlock tension. The value of C is usually determined from Rankine's or Coulomb's well-known formulae which give essentially the same value. Coulomb's formula was used by TVA in the Kentucky cofferdam design (see figure 33).

Interlock Tension—The method customarily employed by TVA for computing interlock stress was to compute the active earth pressure inside the cell (using the submerged weight of the fill), to which was added full water pressure. Full water pressure was added only in the case where the cells were filled hydraulically. A maximum pressure was considered to occur at the base of the cell or at the top of the overburden, whichever was highest. Piling stresses were determined by the ring tension formula in figure 34. Such a method implies that there will be no expansion in the cell at the top and a maximum at the bottom where no overburden exists. (See line a_1b, figure 22). Actually, however, the condition indicated by the line a_1b can occur only where the rock surface is perfectly smooth, or for a condition such as a tie-in cell resting on the smooth concrete of the dam spillway apron. The cell piling will nearly always "bite" into the rock foundation sufficiently to prevent any movement of the piling at the base. The cell in this case, then, will assume the familiar barrel shape indicated by the curved line ab_1, figure 22. Note here that the piling ring is free to expand at the top also, and will move from b to b_1. It is also obvious that the maximum piling tension will occur at the point of maximum expansion. Field observations show this to occur at a point ranging from 1/4 to 1/3 the height of the cell above the base. Where the piling is driven through deep overburden to reach rock foundation, however, the lower part of the piling will be held in a vertical position and the point of maximum expansion will occur at a higher elevation than in the case where no overburden exists.[18] In short, it seems safe to compute the maximum interlock stress at a point approximately one-fourth the height of the cell above the base or one-fourth of the height above the top of the overburden. Where the cells have been filled hydraulically the procedure should be to compute the unit pressure at the one-fourth point based on the sum of the inside water pressure plus the pressure due to the submerged cell fill. A deduction should then be made of the water pressure on the outside of the cell for a height equivalent to the minimum expected river state at which the cell fill is to be placed.

In computing the maximum interlock stress, the effect of the pull from the connecting cell arcs has customarily been

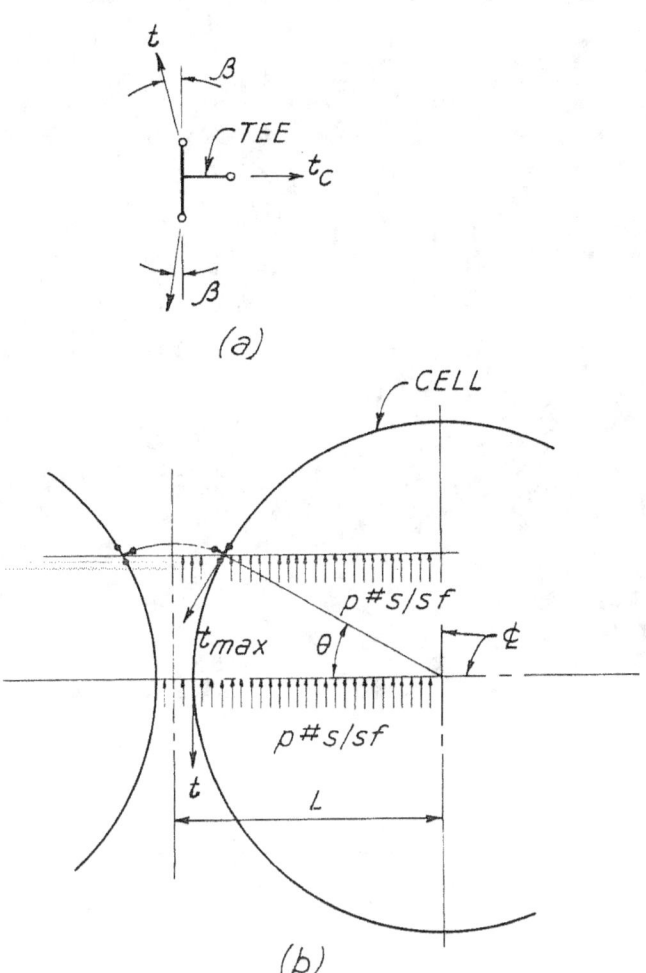

Figure 23
Interlock Stress Diagram

neglected. However, TVA began to make allowances for this with the design of the Kentucky and Watts Bar cofferdams (see figure 23-a). Here t_c represents the tension from the connecting arc, t the tension in the main cell, and β the deflection angle per pile. Then $2t \sin\beta = t_c$ and $t = t_c/2 \sin\beta$. Due to the smallness of angle β, t becomes quite large and to it must be added the ring tension in the main cell. Actually the tension (t_c) in the connecting arcs is probably much smaller than computed due to the bin action in the relatively small connecting cells. Furthermore, a tendency of the tee (figure 23-a) to move in the direction of pull (t_c) will be resisted by the reactive pressure in the fill of the small, or connecting, cell. In an effort to reduce the effect of the connecting arc pull on the main cell (and also to provide more flexibility through the use of longer arcs), special wye piles were used for Kentucky cofferdam in place of the conventional tees. (See figure 34 for analysis of the tensions at the wye.) A similar analysis for a tee connection should show comparable results. However, the maximum tension in the cell may be computed as indicated in figure 23-b in which L is one-half the distance center to center of cells and p is the unit horizontal pressure at the elevation under consideration. Then, on a plane through the centerline of the cell it is obvious that $t = pL$, and on a plane through the tees, $t_{max} = pL\sec\theta$. The difference in t and t_{max} must be taken, by friction, into the cell fill. This value of t_{max} is much smaller than that computed by combining the ring tensions in the large and small cells but appears to be adequate in view of the fact that there has been no record of interlock failures

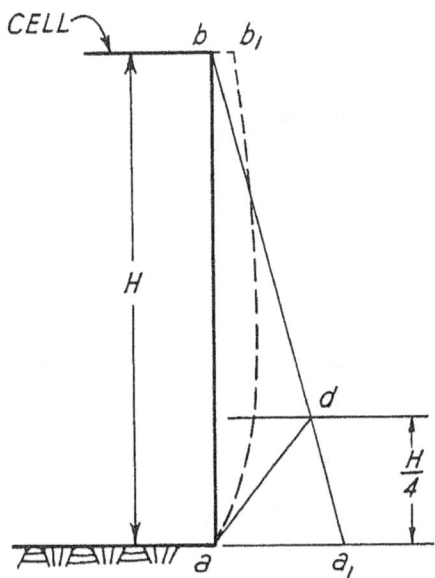

Figure 22
Piling Deflection

ARBED/UNIMETAL INTRODUCES 3 NEW STEEL SHEET PILING SECTIONS THAT ARE
- **STRONGER**
- **LIGHTER**
- **MORE ECONOMICAL**

AZ - Section	DIMENSIONS				WEIGHT		SECTION AREA	SECTION MODULUS	MOMENT of INERTIA
	W Width	H Height	A Flange	E Web	Per Lin Ft. of Pile	Per Sq. Ft. of Wall	Per Lin Ft. of Wall	Per Lin Ft. of Wall	Per Lin Ft. of Wall
	In.	In.	In.	In.	Lbs.	Lbs.	In.²	In.³	In.⁴
AZ 13	26.38	11.93	0.375	0.375	48.38	21.92	6.47	24.2	144.3
RU +0.02	26.38	11.95	0.394	0.394	50.25	22.86	6.75	25.0	149.4
RD −0.02	26.38	11.91	0.355	0.355	46.62	21.21	6.26	23.4	139.2
AZ 18	24.805	14.96	0.375	0.375	49.99	24.17	7.09	33.5	250.4
RU +0.02	24.805	14.98	0.394	0.394	52.00	25.16	7.38	34.6	258.8
RD −0.02	24.805	14.94	0.355	0.355	48.03	23.36	6.85	32.5	242.4
AZ 26	24.805	16.81	0.512	0.480	65.72	31.75	9.35	48.4	406.5
RU +0.02	24.805	16.83	0.532	0.500	67.69	32.75	9.64	49.6	417.1
RD −0.02	24.805	16.79	0.492	0.460	63.85	30.89	9.10	47.2	396.3

Once again, ARBED leads the industry with their new AZ sections; a perfect combination of the outstanding characteristics of a new rolling program, and the proven performance of the world renowned "LARSSEN" interlock.

The new AZ sections are the direct result of findings derived from a worldwide survey* of engineers to determine their needs in Z piling sections. These results were part of an intensive R & D program to redesign the interlock geometry so that when combined with the new rolling technology the sections had good swing with tighter tolerances.

Extensive testing in the laboratory as well as the field have proven ARBED's new AZ sections to be superior to other Z sections.

*Details of this private survey are available upon written request.

Skyline Now Has 23 Stocking Facilities Strategically Located Across the Country

There's One Near You!

Corporate Office:
Skyline Steel Corporation
405 Murray Hill Parkway
East Rutherford, NJ 07073
(201) 933-7070
FAX: 201-933-0787

Skyline Steel Corporation
Suite 202
60 E. Sir Francis Drake Blvd.
Larkspur, CA 94939
(415) 461-4900
FAX: 415-461-1624

Skyline Steel Corporation
Suite 506
3169 Holcomb Bridge Rd.
Norcross, GA 30071
(404) 447-1600
FAX: 404-447-1151

Skyline Steel Corporation
Suite 345
11550 Fuqua
Houston, TX 77034
(713) 484-4000
FAX: 713-484-2782

Skyline Steel Corporation
Suite 230
5641 Burke Center Pkwy.
Burke, VA 22015
(703) 978-2500
FAX: 703-978-2908

Skyline Steel Corporation
Suite 900
3 Monroe Parkway
Lake Oswego, OR 97035
(503) 636-9633
FAX: 503-636-9273

Skyline Steel Corporation
303 West Main St.
Freehold, NJ 07728
(201) 409-2800
FAX: 201-409-2828

Skyline Steel Corporation
17W703G Butterfield Road
Oakbrook Ter., IL 60181
(708) 620-6300
FAX: 708-620-6394

Skyline Steel Corporation
Bell Realty Building
Highway 231 South
Pell City, AL 35125
(205) 338-7774
FAX: 205-338-7748

except those resulting from driving damages. (For a highly technical study of piling stresses see footnote references 2 and 3. Baes's paper gives the results of photoelastic tests on one-point contact piles [Belval], and Descan's paper is a detailed study of the stressing in circular walls.

Resistance of Piling to Sliding in Interlocks--Professor Terzaghi[19] shows how the piling resistance adds to the shear resistance of the cell fill. In designing the Kentucky cofferdam this resistance of the piling to sliding in interlocks (see figure 36) was included by TVA.

It is obvious that such resistance is a function of the interlock tension and the frictional resistance in the interlocks. As pointed out above, the method of filling the cells may induce a high interlock stress which will remain throughout the life of the cell. Consequently, the use of such a value in computing interlock resistance to slippage is justified. However, as was pointed out previously, the interlock tension is proportional to the cell enlargement if in figure 22 triangle aa_1b represents the total active pressure (submerged fill plus water) in the cell, only part of this causes tension in the piling ring. It appears reasonable to use as the total pressure the area of the triangle abd, which will be called P_1. Then the shear resistance of the piling on the centerline of the cell will be $2LP_1/2L \times f = P_1f$ per lineal foot of cofferdam, where $2L$ = the distance center to center of cells and f = the coefficient of friction in the interlocks.

Overburden--Shallow overburden of only a few feet in depth may, as a rule, be neglected without appreciably affecting the design of the cells. In the case of deeper overburden, the active pressure of the submerged material along the river side of the cell will be added to the total water pressure (considered as acting for the full depth to rock foundation). In addition, the added value resulting from frictional resistance of the piling against withdrawal should be considered. In some instances the overburden adjacent to the dry side of the cells is left in place to act as a berm.

Berms

Inasmuch as the required amount of piling per lineal foot of cofferdam increases only slightly as the diameter of the cells is increased (see figure 9), it is, in general, advisable to increase the diameter of the cells until the design shows it to be sufficiently stable without the use of a berm. The absence of an inside berm can result in a reduction in the size of the cofferdam, maintenance pumping, and a net saving in construction costs. Other advantages lie in easier inspection for driving damage to the piling along the dry side, and easier cell drainage maintenance.

In high cofferdams where the diameter of the cells is limited by the maximum allowable interlock stress for the piling (as for Kentucky cofferdam), an inside berm may be necessary to meet stability requirements. Also, in certain instances cells approaching the required size may be built with existing templates to avoid designing and fabricating new templates. In this case a berm may be required. Where the piling is driven through a deep overburden the berm may be partially or wholly developed by leaving the overburden in place, adjacent to the inside of the cells, as excavation for the permanent structure progresses.

Outside Berms--Outside berms are not widely used in connection with the design and construction of cellular cofferdams, but a berm or clay blanket may be required in certain instances to prevent or reduce leakage. If not provided for in the original design, a check should be made to see that the cofferdam can safely withstand the added pressure due to the outside berm. Otherwise, an inside berm may have to be added to preserve the desired factor of safety.

Inside Berms--Some discussion of berms has been

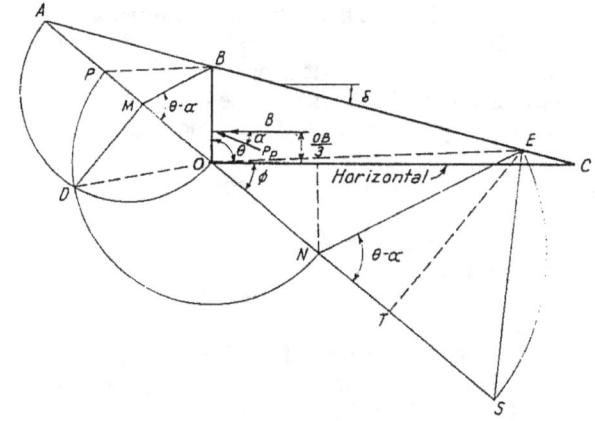

**Figure 24
Graphical Determination of Reactive Pressure in Berms**

included in chapter 4, "Fill Materials," but further discussion is also included here as it relates to design. The proper placement of a berm is important. Professor Terzaghi[20] discusses the deflection of a cofferdam in which a large rock berm was provided for in the original design. In spite of this the deflection in the cofferdam at the crest became quite large as the unwatering progressed. Since the cofferdam was being built to facilitate bridge pier construction, it was constructed while the river level was near the top of the cells. During construction then, both the berm and the cell fill were in an almost completely submerged state, and as unwatering of the cofferdam progressed it took some time for the berm to drain and develop its full reactive pressure. In view of this, then, it would seem highly desirable that inside berms (and also cell fill) be placed at low water stage, if at all possible. Overburden left in place to form a berm will, due to natural compactness, form a more resistant berm than one built up of loose materials.

Berm Design--The first consideration in determining the size of a berm is to determine the amount of reactive pressure that can be developed by the berm. For the berm with an unbroken negative back slope (figure 24), the value of the coefficient of reactive (or passive) pressure C_p may be determined by Coulomb's equation:[21]

(1) $C_p = \dfrac{\cos^2\phi \cos\alpha}{\cos\alpha \left[1 - \sqrt{\dfrac{\sin(\phi + \alpha)\sin(\phi - \delta)}{\cos\alpha \cos\delta}}\right]^2}$, and for $\alpha = 0$, this becomes,

(2) $C_p = \dfrac{\cos^2\phi}{\left[1 - \sqrt{\dfrac{\sin\phi \sin(\phi - \delta)}{\cos\delta}}\right]^2}$. Also, for $\alpha = 0$ and $\delta = 0$

(3) (level backslope) $C_p = \dfrac{1 + \sin\phi}{1 - \sin\phi}$.

The total pressure $P_p = 1/2\, C_p\gamma H^2$, acting at an angle α below the normal to the contact face OB and at $H/3$ above the base in every case (see figure 24).

Reactive Pressure for Berms with Negative Unbroken Backslopes--Following is the graphical solution shown in figure 24:

Given: The berm OBC with unbroken backslope BC in which

φ = The angle of internal friction
∝ = The angle of friction between berm material and wall OB
δ = The angle of the berm backslope
θ = The angle that OB makes with the horizontal
γ = The weight per cubic foot of berm material

Required: The magnitude of the reactive pressure Pp
Construction:
1. Draw AS through O, making the angle φ with the horizontal
2. Extend the backslope CB to intersection with OA at A
3. Draw BM to form the angle BMO = (θ - ∝)
4. With OA as a diameter, construct the semicircle ADO
5. Erect the ⊥ to OA at M and extend it to intersect semicircle at D
6. Lay off ON = to OD along AO extended
7. Draw EN parallel to BM, intersecting BC at E. (OE, then, is the plane of rupture of the berm.)
8. Next, lay off NS = NE along AS.
9. Now, the magnitude of Pp = γ x the area of the triangle NES. Also, for the condition θ = 90°, the normal pressure B = $1/2 \gamma \overline{ET}^2$ (where $\overline{ET}$ is the ⊥ to NS through E).
10. With OD as a radius and O as a center, extend the semicircle NDP to intersection with OA at P. (PB, then, is parallel to OE)

It can also be proved that the area of the triangle OEB = OEN. This being true, then, the direction of the plane of rupture OE for berms with broken backslope may be determined by making the area of the two triangles equal, the location of the line OE being determined by trial. The magnitude of the reactive pressure Pp may then be determined by proceeding with the construction and computations outlined in 7 through 9 above.

For certain values of δ, the plane of rupture OE will coincide with the horizontal base of the berm OC. Based on the constants determined for the berm used for the Kentucky cofferdam, this will occur when the berm is on an approximate 3 to 1 slope. The resistance of the berm to sliding in this case is equal to the weight of the entire berm multiplied by the coefficient of friction existing between the berm and the rock foundation along OC. Where the coefficient of friction is considered to be of lower value than the coefficient of internal shear for the fill, however, it will be necessary to compute the normal component of the reactive pressure, B (figure 24), and the resistance to sliding along the base, and use the lower value. Although it is obvious that δ must be less than φ, in some cases δ may be of such magnitude as to cause the plane of rupture OE to fall below the horizontal or base of the berm OC. It cannot fall below the base of the berm, however, since the plane of rupture will not penetrate the rock foundation. It will, in this case, lie along the base of the berm; and the resistance to sliding will be the sliding resistance of the fill on rock.

Professor Terzaghi, in chapter VII of Theoretical Soil Mechanics (see footnote 21), discusses passive (reactive) pressure in detail and shows that Coulomb's theory is only an approximation of the exact theory and that values of the passive pressure may be determined graphically by the logarithmic spiral method with an error of less than 3 percent. Professor Terzaghi also states on page 107 of Theoretical Soil Mechanics that for φ = ∝ = 30°, δ = 0°, and θ = 90° (level backslope), the value of Pp determined by the graphical method indicated in figure 24 or by Coulomb's formula (see "Berm Design," page 38), will give values up to 30 percent greater than those

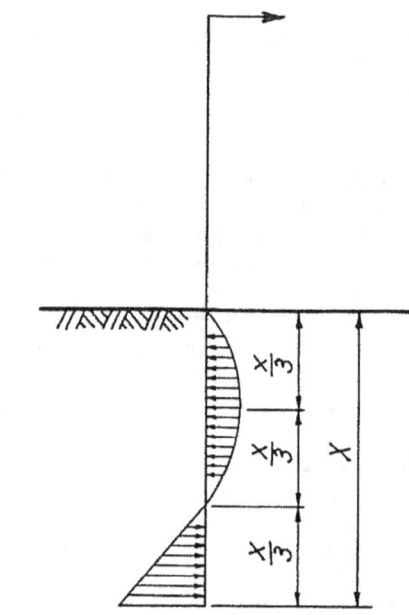

**Figure 25
Moment Resistance in Piling Wall**

obtained through the use of the exact method. Where ∝ is less than φ/3, however, the results obtained through use of the graphical method will be accurate enough for all practical purposes. Professor Terzaghi did not, however, make clear just what effect the negative backslope would have on the accuracy of results obtained by the graphical method. A check was therefore made using a berm with negative backslope of δ = 18°-26' (3 to 1 slope), φ = 28°-50', ∝ = 21°-50', θ = 90°, the value of Pp determined by the graphical method (Coulomb's theory), figure 24, and by the logarithmic spiral method. The latter method gave values approximately 6 percent less than those determined by the Coulomb method. From this, then, it would seem safe to use Coulomb's theory for slopes of 3 to 1 or steeper; but the error would increase for flatter slopes up to approximately 30 percent for the level backfill.

Another problem arising in the design of berms is that of determining the position of the resultant Pp. Professor Terzaghi, in Theoretical Soil Mechanics, shows this to be at h/3 for noncohesive soils. In other words, if a sufficiently large force were applied to a wall at this height above the base, and to the side of the wall opposite the berm, then the wall would move without rotation and the berm would fail along the surface of rupture as OE in figure 24. However, as in the case of the cofferdam cell, the berm must resist a moment as well as a lateral force. For instance, the single sheet pile wall driven into the ground and carrying a lateral load is sometimes considered to rotate about a point at one-third the depth of penetration above the bottom of the wall, with the distribution of resistance shown in figure 25. For the cell resting on rock foundation, however, the point of rotation is probably at the base with berm resistance considered to have a parabolic distribution and the resultant acting at the mid-point of the berm. The resistance offered under this condition probably could not be as great as the full passive resistance to horizontal movement offered by the berm. Consequently, although the resultant used in the design of the Kentucky cofferdam was considered to act as the mid-point of the berm, it would seem advisable to consider it as acting at the one-third point above the base of the berm for subsequent designs. It should be noted here, however, that the value of the berm resistance used in the design of the Kentucky cofferdam was considerably less than that considered as being full maximum, thereby making the design well on the safe side.

The exact amount of berm resistance mobilized under every condition is difficult to determine. Where the berm is made large enough to provide the required factor of safety against movement, then its full possible resistance, if applied against the cell, may cause the resultant of all forces acting on the cell to fall on the river side of the centerline of the cell rather than on the other side of the centerline, as is normal. Obviously, however, this could not occur as the resistance of the berm is passive and not active. But some assumption must be made, and the most reasonable one seems to be to determine the berm size which will give zero pressure under the cell at the river side (using the straight line distribution). Even more conservative assumptions could be made, but in any case the limiting condition would be one in which there was uniform pressure under the cell.

A typical section of the Kentucky cofferdam (figure 26) shows an analysis of the effect of a berm on the shear and shear resistance of a cell. Three cases have been shown, all being based on a full head of water. In the case of a low river stages, however, the minimum value of B (due to active pressure of the berm fill plus pressure from the water in the berm) may be sufficient to provide stability. As the river level rises the pressure from the river side will increase, thereby mobilizing more and more of the potential berm resistance until a point is reached where the shear curve becomes tangent to the shear resistance line. At this point deflection in the cell will cease. It is obvious that the value of B at which the cell will stop deflecting lies somewhere between the minimum value of B and a value of B required to give zero heel pressure under the cell. The use of the value of B which will give zero pressure at the heel of the cell, then, will provide some factor of safety. It should be noted here that if the berm is drained, the maximum value of B will be greater than for a saturated berm, and the shear computed on this basis would be less and the resistance greater.

CONSTRUCTION FEATURES AFFECTING DESIGN

Many of the construction features affecting design have already been discussed preceding chapters. However, their importance should again be stressed. Design should be based, in general, on the worst "in-service" and construction conditions. Where more favorable conditions have been selected as a basis for the cofferdam design, great care should be exercised to see that these conditions are maintained during actual construction and service. In the even that planned construction procedures must be changed, then the effects of the changes should be checked by the designer and steps suggested which will offset these effects.

As already pointed out, the most desirable time for filling cells and placing berms is at a period when the river level is at low stage. Cofferdam construction should therefore be scheduled for periods when, normally, low flow is to be expected. Low water is also desirable from a construction standpoint. Cells filled during high water will have a low initial tension in the piling rings, since most of the pressure inducing this tension is due only to the submerged fill. As a result, a large deflection in the cells may take place as the cofferdam is unwatered unless the cells are designed to meet this condition. It will be impossible to avoid this condition where cells are being designed for service in water which has only a small fluctuation in surface elevation. In this case, due

designed.

NOTATIONS AND FORMULAE

The following letter sumbols used in this discussion conform essentially with "Soil Mechanics Nomenclature" adopted by ASCE's Board of Direction in 1941, except as noted by an asterisk (those indicated by asterisk have been adopted for the convenience of this text):

A = area

*B = total effective horizontal resistance of a berm section 1 foot wide, assumed to act at the 1/3 point of the vertical contact face between berm and cell wall, and perpendicular to it. (B_{max} = Pp Cos $\propto$.)

b = average width of cellular cofferdam = equivalent width of a cofferdam with straight walls = 0.785D (approx) for 60° between tees, or 0.875D (approx) for 90° between tees. (See previous discussion on "Average Widths of Cofferdams," page 35.)

C_A = coefficient of active horizontal earth pressure, and may be computed by either Rankine's or Coulomb's formula.

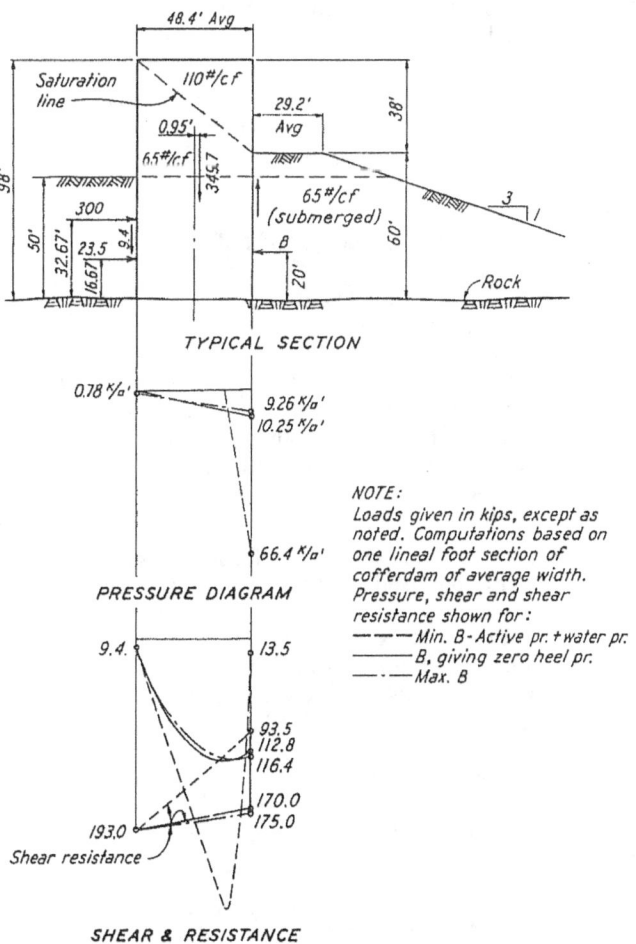

**Figure 26
Analysis of Berm Resistance**

Rankine's $C_A = \tan^2(45° - \phi/2)$ ... (C = 0.35 for Kentucky cofferdam - not used).

$$\text{Coulomb's } C_A = \frac{\cos^2 \phi}{\cos\alpha \left[1 + \sqrt{\frac{\sin(\phi + \alpha)\sin\phi}{\cos\alpha}}\right]^2}$$

(C = 0.29 used at Kentucky Dam).

Coulomb's formula includes the effect of friction on the piling wall and has been used in TVA design. The two formulae give the same result when such friction is neglected.

C_p = coefficient of reactive horizontal earth pressure (see "Berm Design," page 38, for formula).

D = diameter of circular cells.

*d = depth of water above base of cofferdam (during cell filling).

F = total frictional resistance on a surface of contact per unit width.

f = coefficient of interlock friction ..... (f = 0.3 used at Kentucky Dam).

G_s = factor of safety ..... (a minimum value of 1.25 used by TVA).

H = total height of cofferdam from rock to top of cells. The water surface is assumed to be level with the top of the cells for design purposes.

*h = depth of overburden, height of berm, etc.

*J = section modulus of base ..... (1-foot section of average width b).

*K = coefficient of active horizontal pressure in fill, used in computing shear resistance (See "Revised Method of Computing Shear Resistance in Cell Fills," page 33.)

$K = \dfrac{\cos^2 \phi}{2 - \cos^2 \phi}$ ..... (K = 0.623 for Kentucky Dam - not used in original design.)

L = half the spacing, center to center, of circular cells. See figure 30-b. It is also used as the distance between cross walls in the diaphragm-type cells.

M = moment.

P = resultant of all horizontal forces acting on a 1-foot long section of cofferdam.

P_A = total active pressure = $1/2\, C_A\, \gamma H^2$

P_w = total water pressure = $1/2 \times 62.5\, H^2$
    = $31.25\, H^2$

P_p = total reactive pressure = $1/2\, C_p \gamma H^2$

*P_o = active pressure from overburden

p = active horizontal earth pressure per unit area, at depth Z below the top of the fill (= $C \gamma Z$).

q = pressure per unit area of base of a cofferdam (= γH).

*R = resultant of all vertical forces.

S = total shearing resistance on a 1-foot-wide vertical plane.

S' = that part of S due to shearing resistance of the fill

Editors Note
This symbol (S') may appear throughout text as either S' or S'.

S'' = that part of S due to resistance to slip in piling interlocks.

*S_v = total vertical shear on any plane 1 foot wide.

T = total tension in one piling interlock.

t = tension per unit length of interlock at depth Z.

W = total weight of cell or berm (weight of piling neglected).

Z = depth to point under consideration as measured from top of fill.

*α = angle of friction between fill and steel piling (α = 21° 50' used at Kentucky Dam).

$\tan \alpha$ = coefficient of friction ..... ($\tan \alpha$ = 0.4 used at Kentucky Dam).

γ = unit weight of drained fill in cell or berm (γ = 110 lbs per cu ft used at Kentucky Dam).

*γ_s = unit weight of fill when submerged ..... (γ_s = 65 lb per cu ft used at Kentucky Dam).

*δ = angle that sloping surcharge makes with horizontal, as the angle of the negative backslope of a berm.

ϕ = angle of internal friction of the fill material (ϕ = 28° 50' used at Kentucky Dam).

$\tan \phi$ = coefficient of internal friction ..... ($\tan \phi$ = 0.55 used at Kentucky Dam).

θ = angle between centerline of cofferdam and tees for connecting arcs. (See figures 23 and 45)

DESIGN PROCEDURE

Cofferdams Without Berms
In view of past experience in the design of TVA cofferdams and of information offered by Professor Terzaghi,[22] (already

discussed in preceding paragraphs), the following is suggested as the procedure to be followed in design of a steel sheet pile cofferdam where no berm is to be used:

1. **Average width of cofferdam, b.** Base all stability computations on a section of cofferdam 1 foot long, and

for 60° between tees, b = 0.785D

for 90° between tees, b = 0.875D

(See discussion under "Average Width of Cofferdam," page 35.)

2. **Weight of fill.** Compute the weight of the fill, allowing for reduction in weight due to submergence below the assumed saturation line. (See discussion under "Sliding Resistance," page 32.)

3. **Sliding resistance.** Consider full water pressure acting against full height of the cofferdam, plus the horizontal pressure due to submerged overburden. Use a coefficient of sliding equal to the coefficient of internal shear in the fill material except for cells resting on a smooth surface such as very smooth rock or concrete of the dam spillway apron, etc. In the latter case a value for the coefficient of at least 0.5 may be used. (See discussion under "Sliding Resistance," page 32).

4. **Frictional resistance between piling and cell fill.** Compute the value of b as a function of the resistance of the river side piling to sliding up on the cell fill. The pressure between the piling and the cell fill will be at least equal to the water pressure along the river side of the cell, but may be somewhat greater where the cells are filled hydraulically (see discussion on "Frictional Resistance of Piling on Cell," page 34). Where the piling has been driven through a layer of overburden there will be a resistance to upward movement of the piling along the river side of the cell. This is due to the active pressure from the overburden on the outside of the cell. This frictional resistance should be added to the frictional resistance to upward movement resulting from pressure of cell fill on the inside of the piling.

5. **Shear resistance of the cell fill.** Select the maximum value of b as determined in steps 3 and 4 above and use this value in computing the shear resistance of the fill. Compute the shear in the cell by the method set forth in previous discussion on "Shear". In computing the shear resistance, the coefficient "K," determined by Mohr's circle, should be used. (See previous discussion of Mohr's circle, page 33).

6. **Interlock slip resistance.** Compute the frictional resistance to slip in the piling interlocks as outlined in previous discussions, "Resistance of Piling to Sliding in Interlocks," page 38.

7. **Total shear resistance.** Combine the resistances obtained in steps 5 and 6 above. To determine the factor of safety, divide the combined resistance from steps 5 and 6 by the shear on the plane in question (as determined by methods set up in step 5 above). If this gives a value less than the required factor of safety, then the value of b must be increased. It should be noted here that the shear will be a maximum on a vertical plane passed through the centerline of the cell, while the shear resistance will vary but slightly across the cell. An investigation of the shear at the centerline will be sufficient, therefore, as a rule.

8. **Interlock tension.** Interlock tension should be computed as already outlined. In the case of large diameter cells, where the maximum allowable interlock tension may be the controlling factor, it will be advisable to determine the interlock tension immediately after step 4 above.

9. **Cell diameter.** For cofferdams with 60° between tees, D = 1.27 times the value of b as found in steps 3 and 4 above. For cofferdams with 90° between tees, D = 1.14b.

Cofferdams with Berms

The following design procedure is suggested for steel sheet pile cofferdams with which a berm is to be used:

1. **Average width of cofferdam, b.** Where the average width b is not already known, it should be assumed.

2. **Weight of fill.** Proceed as suggested in step 2, opposite column.

3. **Sliding resistance.** Determine the weight (and size) of berm required so that the combined resistance of the cell and berm will give the desired factor of safety. The slope of the berm should be determined in advance. (See "Reactive Pressures for Berms with Negative Unbroken Backslope," page 38).

4. **Frictional resistance between piling and fill.** This procedure is the same as that outlined under step 4, opposite column, for cofferdams without berms, except that the moments should be taken about a point at a height of h/3 above the base of the berm along the contact face between the berm and the cell wall. It must be noted here that location of the cell wall is that based on the average width of the cell, b, and not the full diameter of the cell.

5. **Shear resistance of fill.** Determine a height h for the berm which will give zero heel pressure along the base of the cell (see discussion under "Berms," page 38). Select the largest value of h, determined from computations in steps 3, 4, and 5. Determine the berm resistance B (as outlined under "Berm Design," page 38, and figure 31). Now determine the vertical shear in the cell in a manner similar to that outlined in step 5 of "Cofferdams Without Berms," opposite column. Note here that the value of B used for this computation should not be greater than the value required to give zero heel pressure along the base of the cell. In other words, use the value of B based on the height of berm h determined by the computations made in the first part of step 5 above, regardless of the values of h determined in steps 3 and 4. This is necessary since in some instances berms giving greater values of h determined in steps 3 and 4. This is necessary since in some instances berms giving greater values of h and B are actually used as a result of leaving existing overburden in place to form natural berms. And, should the full value of B be mobilized for these higher berms, the total forces acting on the cell could result in an indicated cell deflection in the direction of the river. This, of course, would be impossible. In computing the shear resistance in the cell fill, the value of K, determined by means of Mohr's circle, should be used. As an alternate method the fill may be considered as being squeezed between the water pressure on the river side and the berm resistance on the other side, with a straight line variation through the cell. The greatest shear resistance resulting from the above method may be used.

The remaining procedure to be followed is essentially the same as that outlined under items 6, 7, 8, and 9, opposite column, for a cofferdam without a berm.

DESIGN EXAMPLES

A few typical design computations will now be presented to demonstrate the application of the principles already discussed.

Case I

A circular-type cofferdam of height h rests on a rock foundation with no overburden (see figure 27).

Assume:

1. The value of the required coefficients and constants to be the same as used for the design of the Kentucky cofferdam and noted under "Notations and Formulae," page 40, and figure

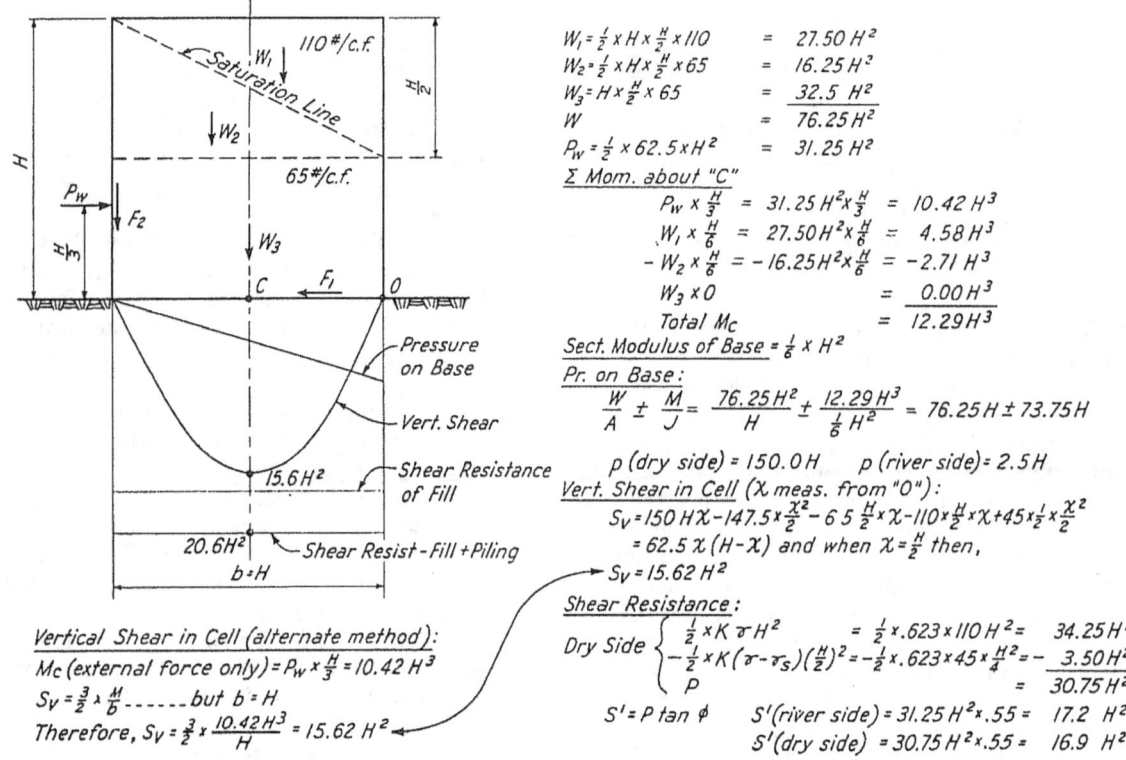

**Figure 27
Vertical Shear and Shear Resistance**

33.
2. 60° between tees (see figure 45).
3. Cells filled dry.
4. No water in the river at the time the cells are filled.
5. Saturation line as shown in figure 27.

Editors Note
The "60 degrees between tees" layout used by TVA on some of their cofferdams results in a smaller area between cells. Later practice used 30 degree "wyes" rather than tees (with 60 degrees between) a well accepted practice which reduces the pull on the main wall sheets from the connecting arc.

Solution:
1. $b = 0.785D$

2. $W = 65bH + (110 - 65) \, 1/2 \, b \, H/2$
   $W = 65bH + 11.25 \, bH$
   $W = 76.25 \, bH$

3. $P_w = 31.25H^2$
   $F_1 = W \tan \emptyset = 76.25bH \times 0.55 = 42.0 \, bH$

Selecting a factor of safety $G_s = 1.25$, then

   $F_1 = 1.25 P_w$ or $42.0 \, bH = 1.25 \times 31.25H^2$, and
   $b = 0.93H$.

4. $M_o = P_w \times H/3 = 31.25H^2 \times H/3 = 10.4H^3$
   $F_2 = P_w \tan \alpha = 31.25H^2 \times 0.4 = 12.5 \, H^2$

P_w is used in computing F_2, as the pressure in the cell during filling will be less due to the material being dry, but the force pressing the piling against the fill will be at least as great as P_w.

For $G_2 = 1.25$, $F_2 b = 1.25 \, M_o$, or
$12.5 \, H^2 b = 1.25 \times 10.4 \, H^3$, and
$b = 1.04 \, H$.

5. Next, compute the shear and resistance of the cell fill using $b = 1.00 \, H$. (If the weight of the piling were included, b would be reduced approximately 0.1 H.) See computations in figure 27.

In computing the shear resistance, the pressure against the river side of the cell will be at least as great as P_w and on the dry side will be at least as great as the pressure of the cell fill, computed using K instead of C (see "Revised Method for Computing the Shear Resistance in Cell Fills," page 33). These pressures were used in computing the shear resistance shown in figure 27. Actually the pressures may be greater, depending on the method used in placing the cell fill.

6. $P_1 = C\gamma \, 3H/4 \times H \times 1/2$ ..... (corresponding to area abd, figure 22).
   $P_1 = 0.29 \times 110 \times 3H^2/8$
   $P_1 = 11.96 \, H^2$
   $S'' = fP_1 = 0.3 \times 11.96H^2 = 3.6H^2$.
   (See "Resistance of Piling to Sliding in Interlock," page 38).

7. Total shear resistance $S = S'$ (figure 27) $+ S''$.
   S (river side) $= 17.2H^2 + 3.6H^2 = 20.8 \, H^2$
   S (dry side) $= 16.9H^2 + 3.6H^2 = 20.5 \, H^2$
   S (centerline of cell) $= 20.65H^2$
   $G_s = 20.65 \, H^2/15.62H^2 = 1.32$

8. $t_{max} = pL \sec \theta$
   For a cell filled hydraulically

$P = (62.5 + 0.29 \times 65)\ 3H/4 = 61.0\ H$

For a cell with a 58.9-foot diameter (148 piles) and 10-pile connecting arcs as at Pickwick:
$L = 30.83$ feet, $\theta = 34°\text{-}03'$, and
$t_{max} = 61.0\ H \times 30.83 \times 1.207 = 2270\ H$
(pounds per foot)
But, for $b = H$, then $H = 0.785D = 46.2$ ft.
Then t_{max} (per lineal inch of interlock) $= 1/12 \times 2270 \times 46.2 = 8750$ lbs.
This value of t is slightly over the usual allowable interlock stress of 8,000 pounds per lineal inch.

9. $D = 1.27\ b$ and $b = H$
Therefore, $D = 1.27\ H$
Now, select the diameter of cell to fit this condition from the column "Theoretical Cell Diameter D," table 2.

Case II

A circular-type cofferdam of height h rests on a rock foundation with no overburden, with the river assumed to be at some depth, d, during the times that the fill is being placed in the cells. The design procedure will be the same as for Case I except the computations for steps 6, 7, and 8, in which the pressure acting inside the cells will be reduced by the amount of the outside water pressure. It is obvious that if d is less than 1/4 H, the total shear resistance and interlock stress will be the same as in Case I. And, if d is greater than 1/4 H, then they will both be smaller.

Case III

A circular-type cofferdam of height h rests on a rock foundation with a sand gravel overburden overlying the bedrock. All overburden inside the cofferdam enclosure to be removed.
Assume: Depth of overburden - 0.4 H.
Solution:
1. $b = 0.785D$

2. $W = 76.25\ bH$ ..... (same as in Case I).

3. $P_w = 31.25H^2$
   $P_o = 1/2 \times 0.29 \times 65(0.4H)^2 = 1.51H^2$
   P (total) $= 32.76H^2$
   $F_1 = 42.0\ bH$ (Same as in Case I)
   For $G_s = 1.25$ then, $42.0\ bH = 1.25 \times 32.76\ H^2$ and,
   $b = 0.975\ H$.

4. $P_w \times H/3 = 10.417H^3$
   $P_o \times 0.4H/3 = 1.51H^2 \times 0.4H/3 = 0.201H^3$
   $M = ... = 10.618H^3$
   $F_2 = P \tan \alpha = 32.76H^2 \times 0.4 = 13.10H^2$
   $F_o = P_o \tan \alpha = 1.51H^2 \times 0.4 = 0.60H^2$ *
   $F_{2+o} = 13.70H^2$
   *friction of overburden on piling.
   For $G_s = 1.25$, then $13.70H^2 b = 1.25 \times 10.62H^3$, and
   $b = 0.969H$.

Note that the added friction of the overburden on the piling reduces the value of b from that determined for Case I, and that the values of b in steps 3 and 4 above are approximately equal for a depth of overburden of 0.4H.

5. The additional pressure from the overburden will increase the shear in the cell, so we will therefore try $b = 1.00H$ for shear computations.
   $W = 76.25H^2$ (Same as in Case I)
   $Mc .... = 12.29H^3$ (From Case I, figure 27).
   $Mo = 1.51H^2 \times 0.4H/3 = 0.20H^3$ (Additional moment due to overburden
   $Mr = -0.60H^2 \times H/2 = -0.30H^3$ (Additional moment due to friction on overburden).
   $Mc = ... 12.19H^3$

Pressure on base:
$R = W + Fo = 76.85H^2$
$R/A \pm M/J = 76.85H^2/H \pm 12.19H^3/1/6\ H^2 = 76.85H \pm 73.14H$
Then p (dry side) $= 149.99H$ and p (river side) $= 3.71H$.

Vertical shear at centerline:
$+149.99H \times H/2$ $= +75.00H^2$
$-(149.99 - 3.71)\ H/2 \times H/2 \times 1/2$ $= -18.28H^2$
$-65 \times (H/2)^2$ $= -16.25H^2$
$-110 \times (H/2)^2$ $= -27.50H^2$
$+45 \times H/2 \times H/4 \times 1/2$ $= +2.81H^2$
S_v $= +15.78H^2$

Shear resistance:
$S' = P \tan \phi$
S' (river side) $= 32.76H^2 \times 0.55 = 18.0H^2$
S' (dry side) $= 30.75H^2 \times 0.55 = 16.9H^2$
(Case I, fig. 27)
S' (at centerline) .. $= 17.45H^2$

6. Now, assume cell to be filled dry and overburden saturated (the overburden not to be removed before the cells have been filled).
   $p = 0.29 \times 110 \times 0.6H = 19.13H$ .. (at top of overburden)
   The pressures due to the overburden, inside and outside of the cell, will balance, and so:
   $p = 19.13H$ .. (at base of cell also)
   Then, in figure 28, area abcd represents the pressure contributing to the tension in the piling ring, and:

$19.13H \times 0.6H \times 1/2$ $= 5.74H^2$
$19.13H \times 0.15H$ $= 2.87H^2$
$19.13H \times 0.25H \times 1/2$ $= 2.39H^2$
$P = $ area abcd $= 11.00H^2$, and
$S'' = fP = 0.3 \times 11.00H^2$ $= 3.30H^2$

7. Total shear resistance at centerline:
   $S = S' + S'' = 17.45H^2 + 3.30H^2 = 20.75H^2$
   $G_s = 20.75H^2/15.78H^2 = 1.31$
   The resulting factor of safety of 1.31 is slightly greater than the required 1.25, indicating that a value for b of less than 1.0

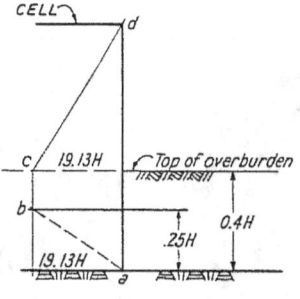

**Figure 28
Pressure in Cell During Filling**

HERCULES MACHINERY
CORPORATION

Pile Driving Equipment Specialists Since 1963

COMPLETE LINES OF PILE DRIVING EQUIPMENT
(Sale or Rent)

ICE "LINKBELT"
- Vibratory Hammers
- Double and Single Acting Diesel Hammers
- Auger Cast Equipment
- Fixed/Swinging Leads

VULCAN
- Air/Steam Hammers

H&M VIBRO
- Vibratory Hammers

BIG FOOT PAVEMENT BREAKER

ALSO
- Steel Sheet Piling for Sale or Rent
- H-Bearing Pile
- Pipe Piling

"If you break it, We can fix it!"

A COMPLETE SERVICE AND PARTS DEPARTMENT TO SERVE YOU:

Call Toll Free:
1-800-348-1890

IN INDIANA: 1-800-552-4848
(219) 424-0405
FAX: (219) 422-2040
IN ST. LOUIS: (314) 441-1871
MAIN OFFICE
3101 New Haven Ave.
P.O. Box 5198
Fort Wayne, Indiana 46895

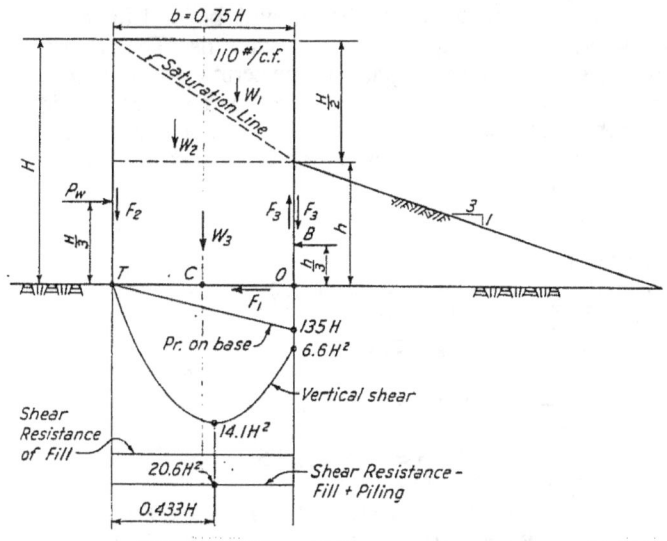

**Figure 29
Circular Cell Cofferdam Resting on Rock - With Berm and No Overburden - Case IV**

but not less than 0.975 as indicated in step 3 might have been used.

8. The maximum interlock tension during filling operations would occur at some point above the top of the overburden and would thus be less than in Case I. The maximum interlock tension after the overburden inside the cofferdam has been removed will occur at a lower point but would still be less than in Case I.

Conclusions from Investigations, Case III--Overburden, when not left in place as a berm, has little effect on the stability of cofferdam for depths up to approximately 0.4H.

Case IV

A circular-cell-type cofferdam has an average width $b = 0.75H$. Problem: To determine size of berm required for stability (see figure 29). Values for constants are the same as for Case I.

Solution:
1. $b = 0.75H$ (given)
2. W (for cell) = $65 \times 0.75H^2 + (110 - 65) \times 1/2 \times 0.75H \times 0.5H = 57.20H^2$
3. Let W_B = Weight of the berm (complete saturation). Then,
   $P_w = 31.25H^2$ and
   $F_1 = (57.2H^2 + W_B) \times 0.55$
   Now, for $G_s = 1.25$:
   $(57.2H^2 + W_B) 0.55 = 1.25 \times 31.25H^2$ and
   $W_B = 13.8H^2$.
   Assuming h = height of berm and a 3 to 1 backslope, then
   $W_B = h \times 3h \times 1/2 \times 65 = 97.5h^2$
   $97.5h^2 = 13.8H^2$
   $h = 0.376H$ .. (for sliding)

4. Assume berm resistance to act at h/3 or 0.125H above the base, and taking the sum of the moments about a point h/3 above "O":

$P_wH(0.333 - 0.125) = 31.25H^2 \times 0.208H = 6.50H^3$
$F_2 = 12.5H^2$ (same as in Case I)

and $G_s = 12.5H^2 \times 0.75H/6.50H^3 = 1.44$
Therefore, in this case sliding controls.

5. The value of the shear in the cell is dependent on how much of the berm resistance B is mobilized (see figure 24). So, determine the berm height h, for a heel pressure = 0 (see discussion in figure 26).

$P_w \times H/3 + W_1 \times 0.75H/6 - W_2 \times 0.75H/6 = B \times h/3 + F_3 \times 0.75H/2 + R \times 0.75H/6$
$W_1 = 1/2 \times 0.75H \times 0.5H \times 110 \quad = 20.6H^2$
$W_2 = 1/2 \times 0.75H \times 0.5H \times 65 \quad = 12.2H^2$
$W_3 = 0.75 \times 0.5H \times 65 \quad = \underline{24.4H^2}$
W_c (cell) .. $\quad 57.2H^2$
$F_3 = 0.4B$ and $R = (57.2H^2 - 0.4B)$
Taking the sum of the moments about c:
$P_w \times H/3 = 31.25H^2 \times 0.333H \quad = 10.42H^3$
$W_1 \times 0.75H/6 = 20.6H^2 \times 0.125H \quad = 2.58H^3$
$-W_2 \times 0.75H/6 = -12.2H^2 \times 0.125H \quad = \underline{-1.52H^3}$
M_1 .. $\quad = 11.48H^3$

Then $0.333Bh + 0.15BH + 7.15H^3 - 0.05BH = 11.48H^3$
$0.333Bh + 0.10BH = 4.33H^3$
but $B = (W_B + F_3) \tan \phi = (97.5h^2 + 0.4B) 0.55$
$B = 68.8 h^2$
Then $22.9h^3 + 6.88 Hh^2 - 4.33H^3 = 0$.

Now solve the cubical equation by the following trial method: Dividing by 22.9, the equation becomes $h^3 + 0.3Hh^2 - 0.189H^3 = 0$.

Solving:

	h^3	h^2	h	Absolute Term
	1	+0.3H	0.0	$-0.189H^3$
Try 0.490H......	+0.49H	+0.387H^2	+0.18968H^3
	+0.79H	+0.387H^2	+0.00068H^3

Note that the sum in each column is multiplied by the assumed root; that is, $0.79H \times 0.49H = 0.387H^2$ and $0.387H^2 \times 0.49H = 0.18968H^3$. As the difference in absolute terms approaches zero, the assumed root approaches its true value. Repeat the steps above by assuming a new root until the desired accuracy is reached.

h = 0.490H, the assumed root for the above equation, is sufficiently close to the true value and will be used.
$h = 0.490H \quad h = 0.376H$ (sliding) and so controls.
$B = 68.8h^2 = 16.50H^2$, and $F_3 = 6.60H^2$

Sum of moments about c:
$-B \times h/3 = -16.5H^2 \times 0.163H \quad = -2.69H^3$
$-F_3 \times 0.75H/2 = -6.60H^2 \times 0.37H \quad = -2.48H^3$
M_1 .... $\quad = \underline{\pm 11.48H^3}$
M_2 .... $\quad 6.31H^3$

Pressure on the base:
Sectional modulus of base, $J = 1/6 (0.75H)^2 = 0.0937H^2$
$W_c - F_3/b \pm M/J = (57.2H^2 - 6.60H^2)/0.75H \pm 6.31H^3/0.0937H^2 = 67.5H \pm 67.5H$
P(heel) = 0, P(toe) = 13.50 H

Vertical shear in cell: (location at x distance from T, figure 36).

$S_v = 65Hx + [(0.5H/0.75H)(x^2/2)(45)]$
$\quad\quad - [(135H/0.75H)(x^2/2)]$
$\quad\quad = 65Hx + 15x^2 - 90.0x^2$
$S_v = 65Hx - 75.0x^2$ (plotted on figure 29)
$d/dx = 65H - 150x = 0$.
$x = 0.433H$ = point of maximum shear measured from T (figure 29)
And $S_{v\,max} = 14.10H^2$

Shear resistance:
S' is the same as for Case I (see figure 27)
S' at the point of maximum shear = $17.0 H^2$

6. $S'' = 3.6H^2$ (same as for Case I).

7. $S = S' + S'' = 17.0H^2 + 3.6H^2 = 20.6H^2$
    $G_s = 20.6H^2/14.1H^2 = 1.46$

8. Next, the piling interlock stresses should be checked for a condition where cells are filled at minimum low water and prior to placing the berm. Computation procedure is similar to that followed in Case I.

Comparison Check with Case I—Upon inspection of figure 9, it will be noted that the cofferdam in Case IV will require approximately 2.5 percent to 5 percent less piling than Case I, depending of course on the value of D. However, the fill required for the cells and berm in Case IV is 11 percent more than required for the cells alone in Case I.

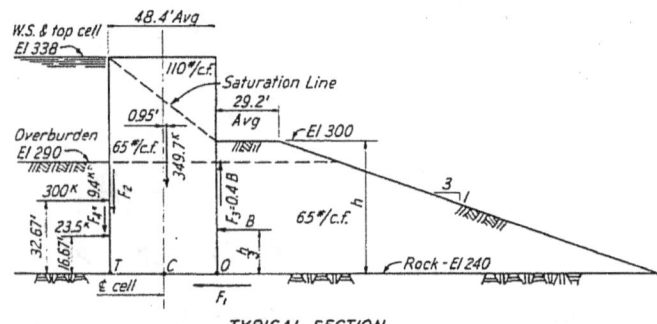

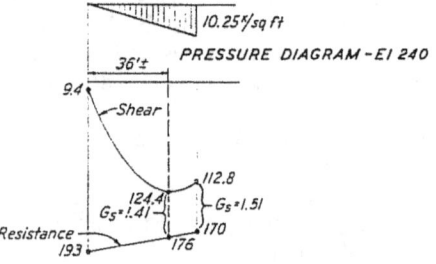

Figure 30
Berm To Give Stability to Typical Circular Cell Driven Through Heavy Overburden to Rock
Case V

Case V
A circular-cell type of cofferdam is driven through deep overburden and rests on a rock foundation.
Required: To determine the size of the berm which will give stability. (See figure 30).
In this case the same size cells have been used as were used in the design of the Kentucky cofferdam. The same shape of berm will also be assumed, but it will be considered as completely saturated to simplify computations. The problem now is to determine the height, h, of the berm by following the procedure previously outlined, and to compare the results with those for the original design of the Kentucky cofferdam shown in figures 33, to 38. Values of the constants to be used will be taken from figure 33.

Solution:
1. b = 48.4 (see figure 33)
2. W_c (cell) = 349.7 (see figure 35)
3. W_B = weight of berm
   P = $P_w + P_o = 323.5^k$ (see figure 35)
   $F_1 = (349.7 + 9.4 + W_B) 0.5$
And for $G_s = 1.25$:
   $(359.1 + W_B) 0.5 = 1.25 \times 323.5$
   $W_B = 449.66^k$
But $W_B = 0.065 [29.2h + (1/2 \times 3h \times h)] = 0.0975 h^2 + 1.898h$, and $0.0975 h^2 + 1.898h - 449.66 = 0$
   h = 58.9 feet

4. Assume berm resistance B to act at h/3, or 20 feet above a base (see figure 30), and taking the sume of moments about a point 20 feet above 0:

$300 \times (32.67 - 20)$ = 3,800 ft kips
$23.5 \times (16.67 - 20)$ = -78 ft kips
M_1 = 3,722 ft kips
$F_2 + F_4 = 323.5 \times 0.4 + 9.4 = 138.8$
$M_2 = 138 \times 48.4 = 6,720$
$G_s = 6720/3722 = 1.80$
Therefore, for sliding controls, see step 3.

5. Determine berm height h for heel pressure = 0.
Taking the sum of moments about C:
300×32.67 = 9,800 ft kips
23.5×16.67 = 392 ft kips
349.7×0.95 = 332 ft kips
$-9.4 \times 1/2 \times 48.4$ = -228 ft kips
M_1 = 10,296 ft kips

Net overturning moment
$= 10,296 - B h/3 - (0.4B \times 1/2 \times 48.4)$
$= 10,296 - 0.333Bh - 9.68B$

Now, for zero heel pressure the vertical resultant R must act at a point 1/6 b to the right of C.
$R = 349.7 + 9.4 - 0.4B = 359.1 - 0.4B$, and
$10,296 - 0.333 Bh - 9.68B - 48.4(359.1 - 0.4B)/6 = 0$
$10,296 - 0.333Bh - 9.68B - 2900 + 3.23B = 0$.
$7,396 - 0.333Bh - 6.45B = 0$.................(1)

Assuming that the maximum value of B is equal to the sliding resistance of the berm, then
$B = (W_B + 0.4 B) 0.5$
$\quad = (0.0975h^2 + 1.898h + 0.4B) 0.5$ .....see step 3.
$\quad = 0.0487h^2 + 0.949h + 0.2B$
$B = 0.0609h^2 + 1.185h$
Substituting in (1) above:
$7396 - 0.0203h^3 - 0.395h^2 - 0.393h^2 - 7.65 h = 0$
$7396 - 0.0203h^3 - 0.788h^2 - 7.65h = 0$.
$h^3 + 38.8h^2 + 377h - 364,000 = 0$.

Now, solving the cubical equation by the following trial method:

	h3	h2	h	Absolute Term
	1	+38.8	+377	-364,000
Trial root =		+59.1	+5780	+364,000
		+97.9	+6157	00000000 (check)

Now $h = 59.1$ is greater than $h - 58.9$ (sliding) and so controls.

$B = 0.0609h^2 + 1.185h = 282^k$
$0.4B = 282 \times 0.4 = 112.8^k$
$R = 349.7 + 9.4 - 112.8 = 246.3^k$

Taking the sum of moments about C:
M_1 = 10,296 ft kips
$-B \times h/3 = -282 \times 59.1/3$ = 5,550 ft kips
$-0.4B \times b/2 = -112.8 \times 24.2$ = -2,730 ft kips
M = 2,016 ft kips

Pressure on base:
$R/b \pm M/J = 246.3/48.4 \pm 2016/391 = 5.10 \pm 5.15$
$P(heel) = -0.05^k$, and $P(toe) = 10.25^k$.

This checks closely enough with the assumption of zero heel pressure.

Make berm height h = 60 feet and check the maximum possible value of B:
$W_B = 0.0975h^2 + 1.898h = 465^k$ see step 3.
$B = (465 + 0.4B) \, 0.5$
$B = 291^k$

The maximum value of B can be determined graphically (see discussion in figure 24). From the reactive pressure diagram in figure 30, then:
$B = 1/2 \, \gamma \, s \, \overline{CD}^2 = 1/2 \times 0.065 \times 101^2 = 332^k$

Therefore the value of $B = 282^k$ is safe as it is less than the values 291^k and 332^k above and will be used for checking the shear in the cell.

Vertical shear in cell (x distance from T):
$S_v = 9.4 + (98 \times 0.065x) + (1/2 \times 38/48.4 \times 0.045x^2) - (1/2 \times 10.25/48.4) \, x^2$
$= 9.4 + 6.36x + 0.0177x^2 - 0.1059x^2$
$= 9.4 + 6.36x - 0.0882x^2$
$dS_v/dx = 6.36 - 0.1764x = 0$
$x = 36.0 \pm$ (point of maximum shear measured from T), and $(S_v)_{Max} = 124.4^k$

Shear resistance:
Consider the cell as squeezed between the pressure from the water and overburden on the river side and the berm resistance on the dry side. Then,
S' (river side) $= 323.5 \times 0.55 = 178.0^k$
S' (dry side) $= 282.0 \times 0.55 = 155.0^k$

If the pressure on the inside of the cell is determined by means of the Mohr's circle relationship (see "Revised Method of Computing Shear Resistance in Cell Fills," page 33),
S' (river side) $= 1/2 K \, \gamma_s \, H^2 = 1/2 \times 0.623 \times 0.065 \times 98^2 = 194.3^k$
S' (dry side) $= 1/2 K \, [\gamma H^2 + (\gamma - \gamma_s) Z^2]$.
$= 1/2 \times 0.623 (0.065 \times 98^2 + 0.045 \times 38^2) = 214^k$

These values of S' may be even greater if the cells are filled hydraulically during a low river stage. However, to be on the safe side, the first values of S' (178.0^k and 155.0^k) will be used. From these:
S' (at the point of maximum shear) $= 161.0^k$.

6. If we consider the area abd in figure 22, (with (a) at the top of the overburden) to be the pressure contributing to the resistance of the piling to slip in interlock, then we will be on the safe side. It is possible that there will be some cell expansion below point a. The cells are also assumed to be filled dry, with the river level at elevation $302.0\pm$.

The pressure area abd(river side) $= 3/8 \, C \, \gamma \, ab^2$
$= 3/8 \times 0.29 \times 0.110 \times 48^2$
$= 27.6^k$

7. The pressure on the dry side will be as great, since the top 10 feet of the berm will not be placed until after the cofferdam is unwatered.
$S'' = 27.6 \times 0.55 = 15^k$
$S = S' + S'' = 161 + 15 = 176^k$
$G_s = 176/124.4 = 1.41$ (note that, due to the resistance being on slope, the minimum G_s will be slightly less).

8. The interlock stresses during hydraulic filling should now be computed for a point $48/4 = 12$ feet above the top of the overburden, with the river elevation at the lowest stage to be expected during filling operations. These computations will be similar to those already presented, and the stress will be lower than shown in figure 36.

EXAMPLES OF ACTUAL COFFERDAM DESIGN USED ON TVA PROJECTS

Early Cofferdams

Several TVA design sheets (figures 31-42, have been included here as a basis for comparison with the recommended procedure already outlined under "Design Procedure," page 41. These plates are discussed below.

Figure 31 covers the design at Pickwick Dam of the first circular-type cofferdam constructed by TVA. It will be noted that in the stability analysis of the cofferdam no allowance was made for submergence of the cell fill. It was, however, considered to be saturated up to an assumed line for the purpose of checking the interlock stress. If the design of this cofferdam had been checked by means of the foregoing recommended design procedure, results would have indicated that the cofferdam was unstable. No failure occurred during the actual life of the cofferdam, however, even under extreme conditions. This was due, no doubt, to the great frictional resistance between the piling and rock foundation (caused by a slight penetration by the piling) and to the great shear resistance of the cell fill and piling induced by the hydraulic filling methods.

Figure 32 shows the procedure used in the design of a later TVA cofferdam. In designing these cofferdams, design allowance for submergence of fill was used for the first time by TVA in checking for stability. A berm was then added to provide safety against sliding. This cofferdam would have figured safe, no doubt, had the design procedure recommended in this volume been used as a check. Note in this design that the pull from the connecting arc was included in determining the maximum interlock stress. the recommended procedure to be followed in subsequent designs, however, has been outlined previously under "Interlock Tension," page 36.

Kentucky Cofferdam

Due to the magnitude of the Kentucky cofferdam, extensive studies in cell design were made, giving careful attention to each condition which might possibly exist during construction and the life of the cofferdam. For some of the conditions considered and details of the design procedure followed, see figures 33-38. It will be noted here that for the first time in the design of TVA cofferdams, the theory of vertical shear in the cell fill was used. Figure 33 shows the design assumptions and data used. figure 34 shows the computation procedure followed in determining the interlock stress as previously discussed, and figure 35 and 36 show the design procedure for a typical circular cell with saturated berm. The design outlined in figures 35 and 36 should be compared with Case V, page 47, and figure 30 for revisions in computations and procedure based on the latest recommendations outlined in this chapter. In making the computations included in Case V, however, the berm was assumed to be completely saturated to simplify computations; and the berm resistance was considered to act at h/3 above the base rather than split up into a force B acting at h/2 and water pressure acting at h/3 above the base, as shown in figure 35. It is now felt that water pressure should not be considered as a separate force in conjunction with reactive pressures.

In considering pressure distribution over the base of the cell (see "Shear," page 32), some assumption must be made prior to determining the size of the berm. The choice of a zero heel pressure for the cells then (as in Case V), seems more logical than assuming a berm height (as indicated in figure 35) thereby limiting the value of B to one giving uniform pressure on the base as shown in figure 37.

Figures 37 and 38 show the design details for a condition where the berm is considered to be completely drained. Note here that the factor of safety against vertical shear in the cell is 1.10, while for the case of the saturated berm (figure 35) the factor of safety is 1.78. This is obviously illogical, for the cofferdam must be at least as safe with a berm completely drained as when the berm is saturated. And, had the two cases been analyzed on a comparable basis (that is, with the same pressure distribution on the base), the factors of safety would have been identical. As a matter of fact, the drained berm should have an added factor of safety in that only a small part of its potential resistance had actually been mobilized. Note also that in Case V no attempt was made to evaluate the vertical shear and shear resistance in the berm, as this appears to give no indication of the safe value of B.

In figure 37 the factor of safety in the berm, on this basis, would be 1.37. This is obtained by dividing the angle of internal friction of the fill material by the angle of friction between the fill and steel sheet piling, or tan ϕ/Tan $\propto$ = 0.55/0.40. The same factor of safety could have been obtained for the saturated berm if the water pressure had not been considered separately (the friction on the cell thus being smaller). In fact, an analysis of the problem shows that the factor of safety would be equal to tan ϕ/Tan $\propto$ for any berm of any size. So if tan $\propto$ decreases, then the factor of safety will increase, becoming infinite for perfectly smooth piling. This, of course, is absurd. In short, it would seem safe to use a value of B, which is equal to, or less than, the maximum berm resistance as determined by one of the methods described under "Berm Design."

Similar design analyses were made for the cloverleaf cells, the only essential difference being the presence of the straight pile diaphragm along the centerline of the cofferdam. The shear resistance at this point will naturally be lower than in the fill and is obtained by multiplying the lateral pressure at the centerline of the cell by tan$\propto$ instead of Tan ϕ. It can be shown that the resistance of the piling to slip in interlock will be P_1f per lineal foot of cofferdam, the same as for a circular cell, even though there is the extra cross wall in the cloverleaf cell. (See "Resistance of Piling to Sliding in Interlocks").

These cloverleaf cells were used in the Kentucky cofferdam to form the arm paralleling the river flow (figure 1) since concrete had to be placed to within 20 feet of the inside piles. This type of cell was used in preference to the circular cell in order to get stability without the use of a berm. A berm was used during the early stages of construction, however, but had to be removed during final stages to permit placing concrete close to the cofferdam. Accordingly, a design analysis was made to determine the height to which the river could rise without endangering the stability of these cells.

During excavation inside the cofferdam enclosure, it was discovered that cloverleaf cells had been driven through a thick layer of blue clay. A study was made, therefore, to determine its effect on the stability of the cells. Results showed that the stability of the cells was not seriously affected.

Hales Bar Cofferdam

The upstream cofferdam used at Hales Bar Dam during installation of generating units 15 and 16 presented an unusual problem. This work required that a section of the old dam be cut out to make room for new powerhouse construction to house the two additional units. Since it was necessary to maintain normal pool level and uninterrupted power generation, the cofferdam had to be built under special conditions. In addition to protecting the work, it was necessary:

1. That the cofferdam be unquestionably safe against failure.
2. That it be designed to withstand maxium head of 58 feet.
3. That piling be placed in deep water.
4. That generating capacity of existing units not be interfered with.
5. That not more than one spillway bay be obstructed during construction operations. (This meant that the cells had to be restricted to such a small area that a supporting berm could not be used.)

The above conditions led to the adoption and design of a cell 70 feet in diameter, the largest cofferdam cell ever built by TVA. So far as is known, larger cofferdam cells had been built in only one other instance by builders outside TVA. This, too, was a limited operation involving only a few cells. At Kentucky Dam single cells 95.5 feet in diameter were placed around each pier for the Illinois Central Railroad bridge that crossed the river in the vicinity of the dam. These, however, were built as a protection against scour around the pier footings resulting from changes in river flow conditions during construction and did not act as cofferdam cells.

Several studies were made before adopting the general arrangement and design shown in figures 39 through 43. This was the first cofferdam to be designed and built by TVA using the "revised method," (page 33) and design examples commencing on page 42, and it should be stated here that results in this instance were highly satisfactory.

The tees for connecting arcs were placed 90 degrees apart (at the quarter-points of the cell). This resulted in an increase in interlock stresses at the tee over that which would have been the case had the tees been placed 60 degrees apart as shown in figure 45. By figuring the maximum stress at the quarter-point above the base, it was found to be within the allowable limits (figure 40). The advantage in using 90 degrees between tees instead of 60 degrees was the increase in average width of cell thus obtained. In other words, had the tees been place at 60 degrees instead of 90 degrees, a cell 78 feet in diameter would have been required to produce an average width of cell equal to that produced by the 70-foot cell used, and the maximum interlock stresses would have been almost as great at the tees.

As pointed out in Chapter 4, "Fill materials," the riverrun gravel was screened to obtain a free-draining fill for the Hales Bar cofferdam. It might also be pointed out here that the position of the saturation line in a cell filled with a free-draining material should be approximately horizontal.[23]

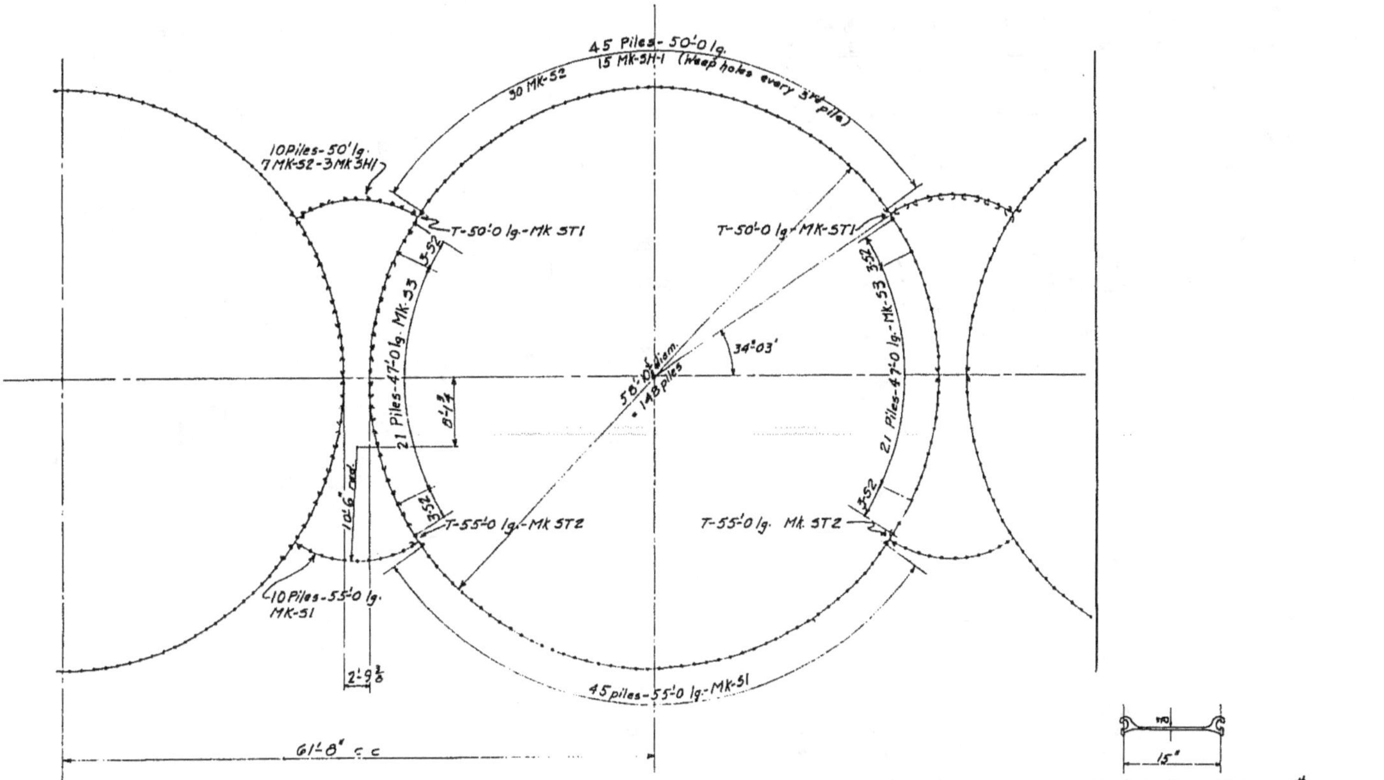

PLAN OF CELLS

PILING SECTION - 15" @ 38.8#
Min. tension in interlock - 12000#/lin.in. ult.
S.M. = 3.7 in³ = 3.0 in³/ft. of wall.
Weight = 31.0 #/sq.ft. of wall.

CASE I - STRESS DURING FILLING
(CONSIDER FILL IN CIRCULAR CELLS ONLY)

Assume: Fill by hydraulic method giving pressure of both water and submerged earth to bottom.
Back pressure up to normal low river level - El. 355

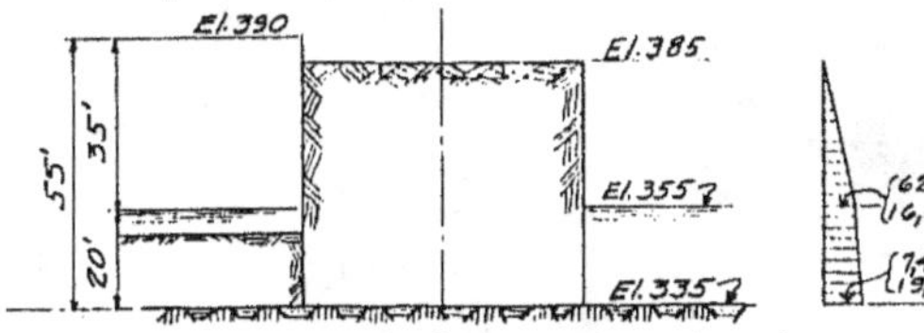

Pressure at bottom = $(62.5 + 60 \times .382) \times 50'$ = 4270 #/☐'
Back pressure = 62.5×20 = 1250 #/☐'
Resultant pressure = 3,020 #/☐'

Max. tension in interlock = $3020 \times \frac{58.2}{2} \times \frac{1}{12}$ = 7400 #/lin. in. (Allow 8000#/lin.in.)
" " in piling web = $7400 \div .375$ = 19750 #/☐" (Allow 24000#/☐")

Stress at El. 355 -
Tension in interlock = $85.4 \times 30 \times \frac{58.2}{2} \times \frac{1}{12}$ = 6,280 #/lin.in.
" in piling web = $6280 \div .375$ = 16,600 #/☐"

CASE II - STRESS AFTER UNWATERING

Assume: Saturation line as shown to 10 feet from bottom, with added superimposed load of 80 tons from movable whirley crane.

Load = $\frac{16,0000}{4 \times 4 \times 18}$ = 3620 #/Lin.ft.

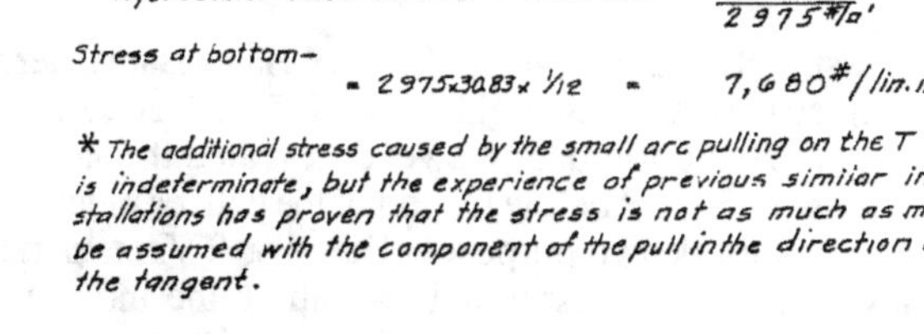

Pressure at El. 335:
Due to superimposed load - $\frac{3620}{2} \times .382$ = 60 #/☐'
Fill to El. 345 = $135 \times .382 \times 40$ = 2060 #/☐'
Fill below El. 345 = $60 \times .382 \times 10$ = 230 #/☐'
Hydrostatic below El. 345 = 62.5×10 = 625 #/☐'
Total = 2975 #/☐'

Stress at bottom -
= $2975 \times 30.83 \times \frac{1}{12}$ = 7,680 #/lin.in.*

* The additional stress caused by the small arc pulling on the T is indeterminate, but the experience of previous similar installations has proven that the stress is not as much as might be assumed with the component of the pull in the direction of the tangent.

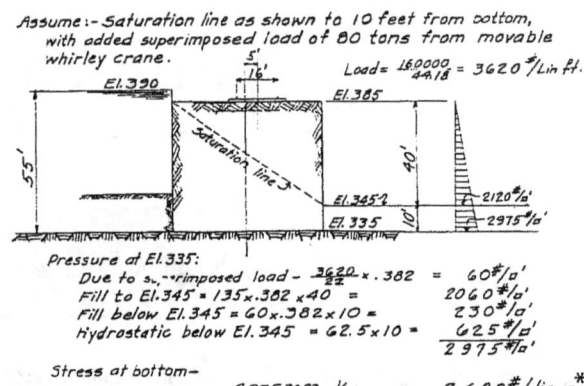

**Figure 31
Design Analysis for Circular Cell Type Cofferdam -
Pickwick Landing Dam**

STABILITY ANALYSIS

Overturning Moment:
 Water - 62.5 × 55 × 55/2 = 94,530# × 18.33' = 1,733,000'#
 Earth - 60 × .382 × 15 × 15/2 = 2,580# × 5.0' = 13,000'#
                               97,110# × 17.98' = 1,746,000'#

Per cell - 61.67' cc -
                        5,989,000# × 17.98' = 107,672,000'#

Flexural stress = $\frac{107,672,000}{.0982 \times 58.3^3}$ = ± 5,365 #/☐"

Direct pressure = 110# × 50' = + 5,500 #/☐"

Resultant pressure = + 135 #/☐" at river toe
                   = + 10,860 #/☐" at inside toe

Factor of safety against sliding =
    $\frac{.7854 \times 58.3^2 \times 110 \times 50}{5,989,000}$ × .5 = 1.25

Average shear at base = $\frac{5,989,000}{.7854 \times 58.3^2}$ = 2,200 #/☐"

Shearing resistance of fill material = w tan φ
    50 × 110 × 0.5 = 2,750 #/☐"

LOAD DIAGRAM

TYPICAL SECTION

PRESSURE DIAGRAM

ASSUMPTIONS –

1. Steel Sheet pile, Cellular type cofferdam, River wall top El. 390, Inside wall & fill top El. 385.
2. Rankine's Theory of earth pressure - $p = \frac{1}{2} wh^2 \left(\frac{1-\sin\phi}{1+\sin\phi}\right)$ For φ = 26°34', p = .382 $\frac{wh^2}{2}$.
3. Cofferdam driven thru pervious material - giving water pressure to rock.
4. Cofferdam to be flooded before overtopping.
5. All material removed from dry side.
6. Weight of submerged material - 60 #/cu.ft. - φ = 26°-34' (Slope 1V:2H)
    - See Merriman's American Civil Eng'r's Handbook - 5th Ed., Page 893.
7. Weight of filling material - Min. = 110 #/cu.ft. - φ = 26°-34' (Use for stability)
    - See Merriman & Ketchum's Struct. Eng'rs. Handbook - 3rd Ed. Page 309.
   Max. = 135 #/cu.ft. - φ = 26°-34' (Use for stress in piling.)
    - See Merriman & Ketchum.
8. Coefficient of friction - wet gravel on smooth rock = 0.5 - see C.E. Magazine, June, 1934, p. 302
9. Due to pervious fill (sand & gravel) & 1½"φ holes every three feet of every third inside pile, hydrostatic pressure is assumed at 10 ft., except during filling.
10. Allowable stresses :
    Normal condition of loading :–
    Tension in piling interlock - 8000 #/lin. in. (Specify 12,000 #/lin. in. ultimate.
    Tension in piling web - 24,000 #/☐" (High carbon steel - 70,000 #/☐" ultimate)
    Bending on extreme fiber - 25,000 #/☐" (20,000 #/☐" + 25%) (High carbon steel - 70,000 #/☐" ultimate)
    Bearing on power driven rivets - Double shear - 37,500 #/☐"; Single shear - 30,000 #/☐".
    Shearing on   "   "   - 16,900 #/☐"
    → A.I.S.C. - Building Code - allowable stresses +25% for structures of this temporary nature.

**Figure 31 (Continued)
Design Analysis for Circular Cell Type Cofferdam -
Pickwick Landing Dam**

On this assumption the saturation line for Hales Bar cofferdam was assumed to be at mid-height of the 58-foot-high cell. Tests on the completed cofferdam, however revealed the saturation line to be about 50 feet below the top of the cell, or much lower than assumed for design purposes. Based on TVA experience it is now felt that the horizontal saturation line more nearly approaches actual conditions than does the sloping saturation line that was used in the design of earlier TVA cofferdams. The use of a horizontal saturation line also simplifies design procedures. It is also felt that the assumption of a saturation line as being at one-half the height of the cell seems reasonable and quite adequate for design purposes, and the sloping line as shown in figure 27 seems unduly conservative.

The rest of the design of Hales Bar cofferdam, as shown in figure 39 through 43, in general, followed the procedure discussed under "Design Procedure". It should be noted, however, that the shear resistance developed during filling of the Hales Bar cells was considerably less than was the case where cells were filled hydraulically with the river level at a much lower elevation. (See "Placing Cell Fill,").

As an added precaution, a small concrete berm was placed along the inside of the cofferdam (see general note in figure 42). this was done just as soon as the cofferdam unwatering was completed. Pieces of railroad rail were used to dowel the concrete berm to the rock foundation and a gutter along the top of the berm served as a means of carrying the cell drainage to the sump pump.

ASSUMPTIONS:-

1. Steel sheet pile, Circular Cell Type Cofferdam.
2. Cofferdam driven through pervious material, giving water pressure to rock.
3. Cofferdam to be flooded before overtopping.
4. Rankine's & Coulomb's theories of earth pressures:
   (Rankine's) Active pressure (fill level at top) - $P_a = \frac{wh^2}{2}(\frac{1-\sin\phi}{1+\sin\phi})$; When $\phi = 26°-34'$ (slope 1:2), $P_a = .382 \frac{wh^2}{2}$
   When $\phi = 36°-52'$ (slope 1:1.33), $P_a = .25 \frac{wh^2}{2}$
   (Coulomb's) Reactive pressure (fill on slope δ) - $P_r = \frac{wh^2}{2}[\frac{1}{1-\tan\delta\tan\phi}]$; When $\phi = 40°-0'$ (slope 1:1.19) & $\delta = 33°41'$ (1:1.5), $P_r = 1.2 \frac{wh^2}{2}$.
5. Berm to be of free draining granular material.
   Stabilizing force determined by resistance to sliding or reactive pressure. (Use minimum figure.)
   Berm material - weight = 90#/cu.ft., $\phi = 40°-0'$, $\delta = 33°-41'$.
* 6. Cell filling material to be sand and gravel from river bed.
   Dry fill (placed by clamshell) - weight = 100#/cu.ft., $\phi = 36°-52'$.
   Hydraulic fill --------------- weight = 110#/cu.ft., $\phi = 36°-52'$.
   Submerged fill -------------- weight = 65#/cu.ft., $\phi = 26°-34'$.
** 7. Coefficient of friction for cell filling material and berm material sliding on smooth rock = 0.5
8. Due to pervious fill (sand & gravel) and 1½" weep holes @ 3'-0" c-c in lower half of every fifth inside pile, saturation line is assumed at slope of 1:2.5, except during filling. (See CASE II for added assumption for Guntersville)
9. Allowable stresses-
   Tension in piling interlock - 8000#/lin.in. (12,000#/lin.in. ultimate.) ⎤ Allow 50% increase in stresses for main cell piling
   Tension in piling web - 24000#/sq.in. (High carbon steel - 70000#/sq.in. ultimate.) ⎦ when considering effect of connecting arc pulling on Tee.
   Bending in extreme fibre - 25000#/sq.in. (20000#/sq.in. + 25%) (High carbon steel - 70000#/sq.in. ultimate.)
   Bearing on power driven rivets - Double shear - 37500#/sq.in.; Single shear - 30,000#/sq.in.
   Shearing on power driven rivets - 16,900#/sq.in.
   → A.I.S.C Building Code - Allowable stresses + 25% for structures of this temporary nature.

REFERENCES:-
* Merriman's American Civil Engineer's Handbook - 5th Edition, Page 893.
* Ketchum's Structural Engineers Handbook - 3rd Edition, Page 309.
** Civil Engineering Magazine - June, 1934, Page 302.

TANGENT LOADING AT TEE   STRESS DIAGRAM AT POINT "A"

T = Interlock stress in connecting arc.
d = Angle of deflection per pile in main cell.
S_c = Stress developed in main cell piling due to effect of connecting arc pulling on the Tee = $\frac{T}{t} \div \sin d'$.
S_M = Stress developed in main cell piling, computed from main cell pressures and radius.
Then S = Combined theoretical stress adjacent to Tee = $S_c + S_M$

ANALYSIS OF INTERLOCK STRESSES AT "TEE"

DESIGN DATA
Weight of one complete cell (during high water) = 4,202,800#
Section modulus of one complete cell = 8,280 ft³
Weight of berm per lineal ft. = 32,917#

STABILITY ANALYSIS

Water Pressure per cell = 62.5 × $\frac{40^2}{2}$ × 50,000 × 47.93' --------- 2,396,500 lbs.
Sliding Resistance of cell = 4,202,800 × 0.5 --------- 2,101,400 lbs.
Sliding Resistance of berm = 32,917 × 47.93 × 0.5 --------- 788,900 lbs.
Total Sliding Resistance = 2,890,300 lbs.
Factor of safety against sliding = 2,890,300 ÷ 2,396,500 --------- 1.21
Average shear at base of cell = (2,396,500 - 788,900) ÷ 1595# --------- 1008 lbs./sq.ft.
Average shearing resistance of fill material = w × tanϕ = 2635 × 0.5 --------- 1318 lbs./sq.ft.

Overturning Moment per cell = 2,396,500 × $\frac{40}{3}$ = 31,953,300 ft.lbs.
Resisting Moment of berm = 788,900 × $\frac{19}{3}$ = -4,996,000 ft.lbs.
Net Overturning Moment per cell = 26,957,300 ft.lbs.
Flexural Stress = 26,957,300 ÷ 8280 --------- ±3256 lbs./sq.ft.
Pressure at heel (river side) = -3256 + (35 × 65) --------- -981 lbs./sq.ft.
Pressure at toe (dry side) = +3256 + (16 × 110) + (19 × 65) --------- +6251 lbs./sq.ft.

Resisting Moment of cell about toe = 4,202,800 × 20.75 = 87,208,500 ft.lbs.
Factor of safety against overturning about toe =
$\frac{87,208,500 + 4,996,000}{31,953,300}$ --------- 2.89

CASE I - STRESS DURING FILLING

Assume - Fill to be placed by hydraulic method, giving pressure of both water and submerged earth to bottom of cell.
River level assumed at El.555² during time of filling.

A.- Hydraulic Fill to El.565².
Pressure at bottom of cell = 29(62.5 + 65 × .382) = 2532 lbs./sq.ft.
Back pressure = 19 × 62.5 = -1188 lbs./sq.ft.
Resultant pressure = 1344 lbs./sq.ft.
Interlock stress in main cell piling at Elev. 536.0 from:-
Pressure in main cell = 1344 × 21.49 × ½ = 2408 lbs./lin.in.
Pressure in connecting cell = 1344 × 2.55 × ½ ÷ .0581 = 9205 lbs./lin.in.
Total = 11,613 lbs./lin.in.
Tension in piling web = 11,613 ÷ .375 = 30,968 lbs./sq.in.

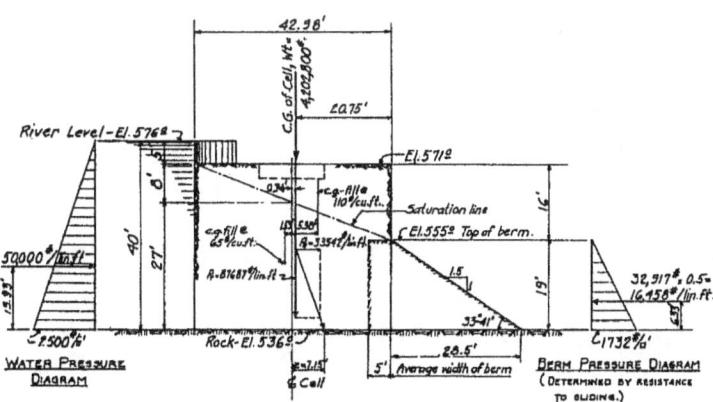

TYPICAL SECTION — WATER PRESSURE DIAGRAM / BERM PRESSURE DIAGRAM (Determined by resistance to sliding.)

TYPICAL SECTION — STRESS DIAGRAM (Tension in Interlock)

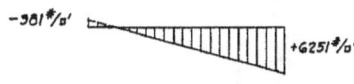

PRESSURE DIAGRAM

CONTROLLING FACTORS IN STABILITY ANALYSIS:-

Elevation of berm used at Guntersville was determined by limiting factor of safety against sliding to minimum of 1.21.

Elevation of berm used at Chickamauga was determined by maintaining zero uplift at heel of cell. Factor of safety against sliding exceeded 1.25 in all cases.

NOTE:-

Fill above elevation of initial fill was assumed to be placed slowly and freely drained to prevent developing excessive hydrostatic pressures in cell.

When interlock stress exceeded 12000#/lin.in with cell filled to top and drained to low water level, initial berm was used as req'd. (See STRESS DIAGRAM for river side - CASE II)

**Figure 32
Typical Design Analysis for Circular Cell Type
Cofferdam - Guntersville and Chickamauga Dams**

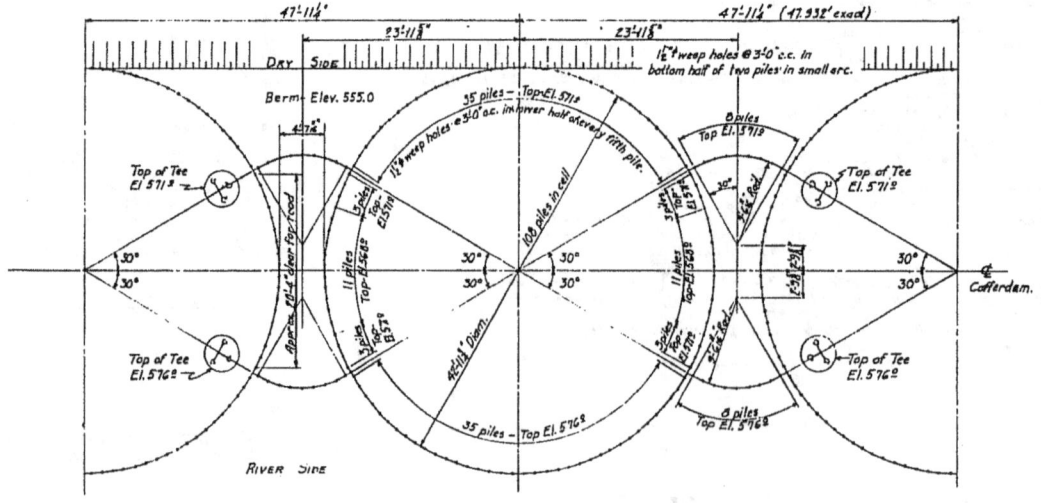

PLAN OF CELLS (Guntersville)
TYPICAL FOR DOWNSTREAM ARM

CASE II.- STRESS AFTER UNWATERING - BERM IN PLACE.

Assume - Saturation line as shown to top of berm. Field to provide sufficient weep holes
in piling on inside face of cofferdam to insure that all main and connecting cells
will be kept dry above El. 555º.   (This assumption
applied to Guntersville design only.)

Pressure on inside piling wall at El. 536.0 - (River level 571.0+)

Pressure at El. 555º = 110 × .25 × 16 = 440 lbs./sq.ft.		
Pressure at bottom of cell = 440 + (62.5 + 65×.362)×19	=	2099 lbs./sq.ft.
Back pressure (from Berm) = 91.2 × 19	=	-1733 lbs./sq.ft.
(Determined by resistance to sliding) Resultant pressure	=	366 lbs./sq.ft.

Pressure on outside piling wall at El. 536.0 (River level El. 555.0)

Pressure at bottom of cell	=	2099 lbs./sq.ft.
Back pressure = 62.5 × 19	=	-1188 lbs./sq.ft.
Resultant pressure	=	911 lbs./sq.ft.

Interlock stress in main cell piling from:-

Pressure in main cell = 911 × 21.49 × ½	=	1638 lbs./lin.in.
Pressure in connecting cell = 911 × 2.55 × ½ × ½ ÷ .0581	=	6239 lbs./lin.in.
	=	7877 lbs./lin.in.

Tension in piling web = 7877 ÷ .375   = 21,005 lbs./sq.in.

GENERAL NOTES-

Quantities and properties for one complete cell include
one main cell and one connecting cell.

Cofferdam section used for analysis is typical for
downstream arm of Guntersville cofferdam.

Design analysis is typical for both Chickamauga and
Guntersville 3rd Stage cofferdams except as noted.

Cells to be constructed of 15" @ 38.8#/ft steel sheet
piling and standard Tee connections.

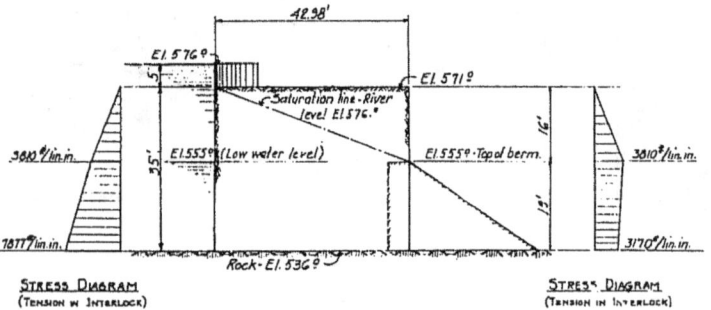

STRESS DIAGRAM   STRESS DIAGRAM
(TENSION IN INTERLOCK)   (TENSION IN INTERLOCK)

TYPICAL SECTION

NOTE:-
Conditions were also investigated with coffer unwatered to determine the
maximum permissible river level before berm is placed, without exceeding
the controlling factors in stability analysis or allowable interlock stresses.

**Figure 32 (Continued)
Typical Design Analysis for Circular Cell Type
Cofferdam - Guntersville and Chickamauga Dams**

ASSUMPTIONS:

Water pressure to rock.
Weight of dry material = 110 #/c.f.
Weight of submerged material = 65 #/c.f. } from Tests at Guntersville
Coeff. of friction – gravel on rock (sliding) = 0.5
         "   "   – gravel on steel ($2\alpha = 21°30'$) = 0.4
         "   "   – gravel on steel (piling interlock) = 0.55
Angle of internal friction, $\phi = 28°-50'$; $\tan \phi = 0.55$
         (for dry or saturated material)
Coeff. of friction – steel on steel (piling interlock) = 0.3
Slope of berm = 3:1 ; $\delta = 18°-26'$
Horizontal resultant of soil pressure computed according to Coulomb's theory:

Active pressure, $P_A = \tfrac{1}{2} w h^2 x \dfrac{\cos^2 \phi}{\cos \alpha \left(1 + \sqrt{\dfrac{\sin(\phi + \alpha)\sin\phi}{\cos\alpha}}\right)^2}$ $C = 0.29$
(level surface)

Reactive pressure, $P_P = \tfrac{1}{2} w h^2 x \dfrac{\cos^2 \phi}{\cos \delta \left(1 - \sqrt{\dfrac{\sin(\phi - \delta)\sin\phi}{\cos \delta}}\right)^2}$ $C = 1.58$
(negative slope, unbroken)   (for 3:1 slope, $\delta = 18°26'$)

Reactive pressure for berms with flat tops computed graphically.
(see sheets #7 and #8)

Allowable stresses in piling:
  Tension in web   32000 #/☐"   (for 70,000 #/☐" ultimate)
  Interlock tension   12,000 #/lin.in. (for 16,000 #/lin.in. min. ultimate)

Factor of safety 1.25 minimum – for sliding, shear, & interlock.
Pressure distribution:
  Active pressure: triangular; resultant at $\tfrac{1}{3}h$ above base.
  Reactive   "   :   "   " $\tfrac{1}{2}h$ "   "
       "   "   : parabolic
Water surface El. 338 (for Stage #2 Coffer.) has been used as a maximum.
Rock elevation – average El. 240.
Saturation line as indicated on sections. Field to provide sufficient drain holes on the inside of cofferdam cells to assure such drainage.
Field to grout foundation rock to insure tight base.

DETAILS OF CELLS:

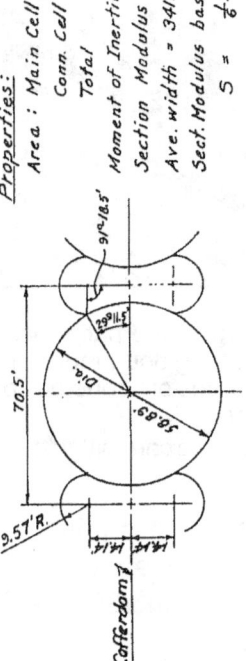

TYPICAL CIRCULAR CELLS
Upstream & Downstream Arms

Properties:
  Area: Main Cell   = 2,725 s.f.
        Conn. Cell   =   687 s.f.
        Total   = 3,412 s.f.
  Moment of Inertia   = 721,735 ft.⁴
  Section Modulus   = 24,500 ft.³
  Ave. width = 3412 ÷ 70.5 = 48.4 ft.
  Sect. Modulus based on ave. width:
    $S = \tfrac{1}{6} \times 48.4^2 = 391$ ft.³

TYPICAL CLOVERLEAF CELLS
Crosswall

Properties:
  Area: Main Cell   = 7,936 s.f.
        Conn. Cell   =   728 s.f.
        Total   = 8,664 s.f.
  Moment of Inertia   = 6,322,238 ft.⁴
  Section Modulus   = 125,666 ft.³
  Ave. width = 8664 ÷ 93.5 = 92.8 ft.
  Sect. Modulus based on ave. width:
    $S = \tfrac{1}{6} \times 92.8^2 = 1,435$ ft.³

**Figure 33
Design Analysis - Sheet 1 - Assumptions and Data
 - Kentucky Dam**

Reaching Out

For 25 years, Shugart employees have reached out to the construction industry with a commitment.

We're committed to understanding your needs, so we can manufacture quality products to serve you.

We're committed to identifying your problems, so we can design product features that save you money.

As a result, we offer a line of quality products produced by dedicated, experienced people.

We have a full line of barges and barge accessories, hydraulic barge pushers, floating doughnut cranes, utility tug boats, concrete screeds, custom concrete forms, and special hydraulic equipment.

When you're looking for people committed to quality and innovation, reach out to Shugart.

Shugart Manufacturing, Inc.
Post Office Box 748 • Chester, SC 29706
803/581-5191 • FAX: 803/581-1080

Where The Commitment To Quality and Innovation Continues

INTERLOCK STRESSES During filling: (Hydraulic fill)
River at El. 305

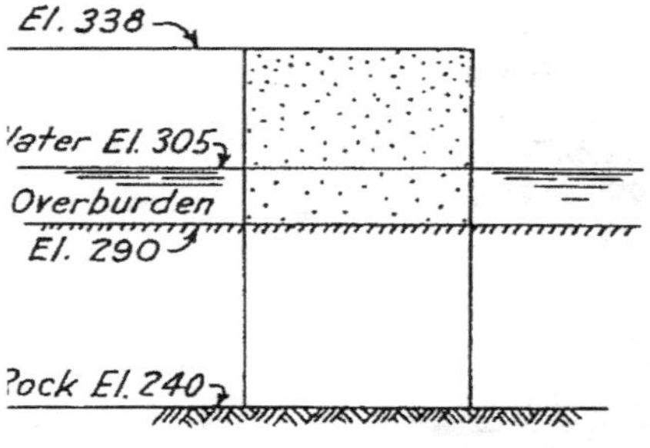

Section through Cell

Pressure at El. 290

Water = 62.5 x 48 = 3000

Subm. fill = .29 x 65 x 48 = 904

3904

−62.5 x 15 = −937

2967 #/□'

Tension (main cell) = $2967 \times \frac{29.44}{12} = 7280$ #/lin. in. = T_m

Assuming that the main cells have drained before filling adjacent conn. cells:-

Pressure at El. 290

Dry fill = .29 x 110 x 33 = 1051

Subm. fill = .29 x 65 x 15 = 283

1334 #/□'

Tension (main cell) = $1334 \times \frac{29.44}{12} = 3275$ #/lin. in.

Conn. cell saturated:-

Tension (conn. cell) = $2967 \times \frac{9.57}{12} = 2365$ #/lin. in. = T_c

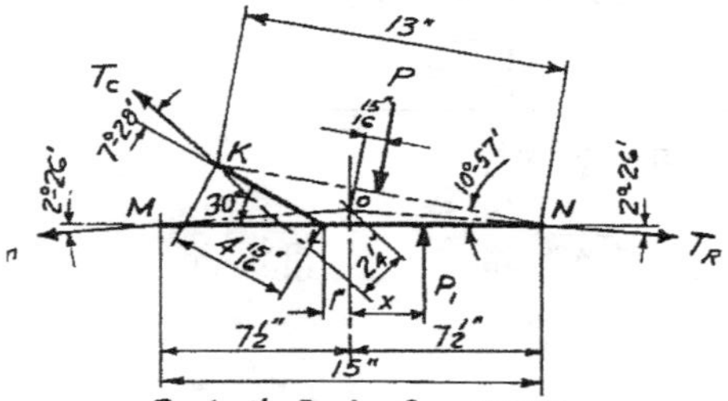

Detail Data for 30° Y

P = 1.083 x 2967 = 3220# or

P = 1.36 x T_c

P_1 = 1.25 x 2967 = 3710#

In order to calculate maximum interlock stresses which may obtain, the use of the 7280 #/lin. in. interlock tension in the main cell during filling is recommended, since it is conceivable that a permanent set is obtained which remains even after the cell has drained. To this figure must be added the effect of the pull of the Y-pile joining the connecting cell to the main cell. The computation of the interlock tension in the connecting cell is based on a condition during filling. It is also assumed that for this condition $P \geqq 3220$# and $P_1 \geqq 3710$#

Projection of forces on line MN must equal zero.

$T_m \times \cos 2°26' + T_c \times \cos 37°28' + P \times \sin 10°57' = T_R \cos 2°26'$

$7280 \times .999 + 2365 \times .794 + P \times .190 = T_R \times .999$

$7280 + 1880 + P \times .190 = T_R$

(1.) $9160 + .190 P = T_R$

Projection of forces on line P_1 perpendicular to MN must equal zero.

$T_m \times \sin 2°26' - T_c \times \sin 37°28' + P \times \cos 10°57' + T_R \sin 2°26' = P_1$

$7280 \times .0425 - 2365 \times .608 + P \times .982 + T_R \times .0425 = 3710$

$309 - 1438 + .982 P + .0425 T_R = 3710$

(2.) $-4839 + .982 P + .0425 T_R = 0$

Substituting the value of T_R from equation (1.) in equation (2.)

$-4839 + .982 P + 9160 \times .0425 + .190 \times .0425 P = 0$

$-4839 + .982 P + 389 + .008 P = 0$

$-4450 + .990 P = 0$

$P = 4500\# > 3220\#$

(Indicating that reactive pressure is developed.)

Substituting this value in equation (1.)

$9160 + 855 = T_R$ or $T_R = 10{,}005$ #/lin. in.

If we had assumed P = 3220#, we would have found that P_1 = 2468# < 3710# (which is the assumed minimum value during filling.)

The stress in the pile on the centerline of the cofferdam, disregarding friction between soil and piling is

2967 × 35.25 ÷ 12 = 8720#

Dividing this value by the cosine of 29°-11.5' checks the value of T_R as found from above computations.

$T_R = \frac{8720}{.873} = 9990$#

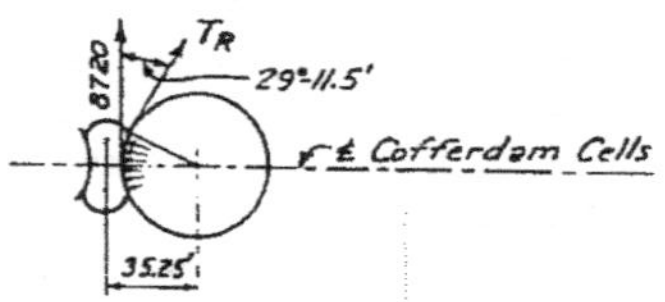

Figure 34
Design Analysis - Sheet 2 - Interlock Stresses in Circular Cells - Kentucky Dam

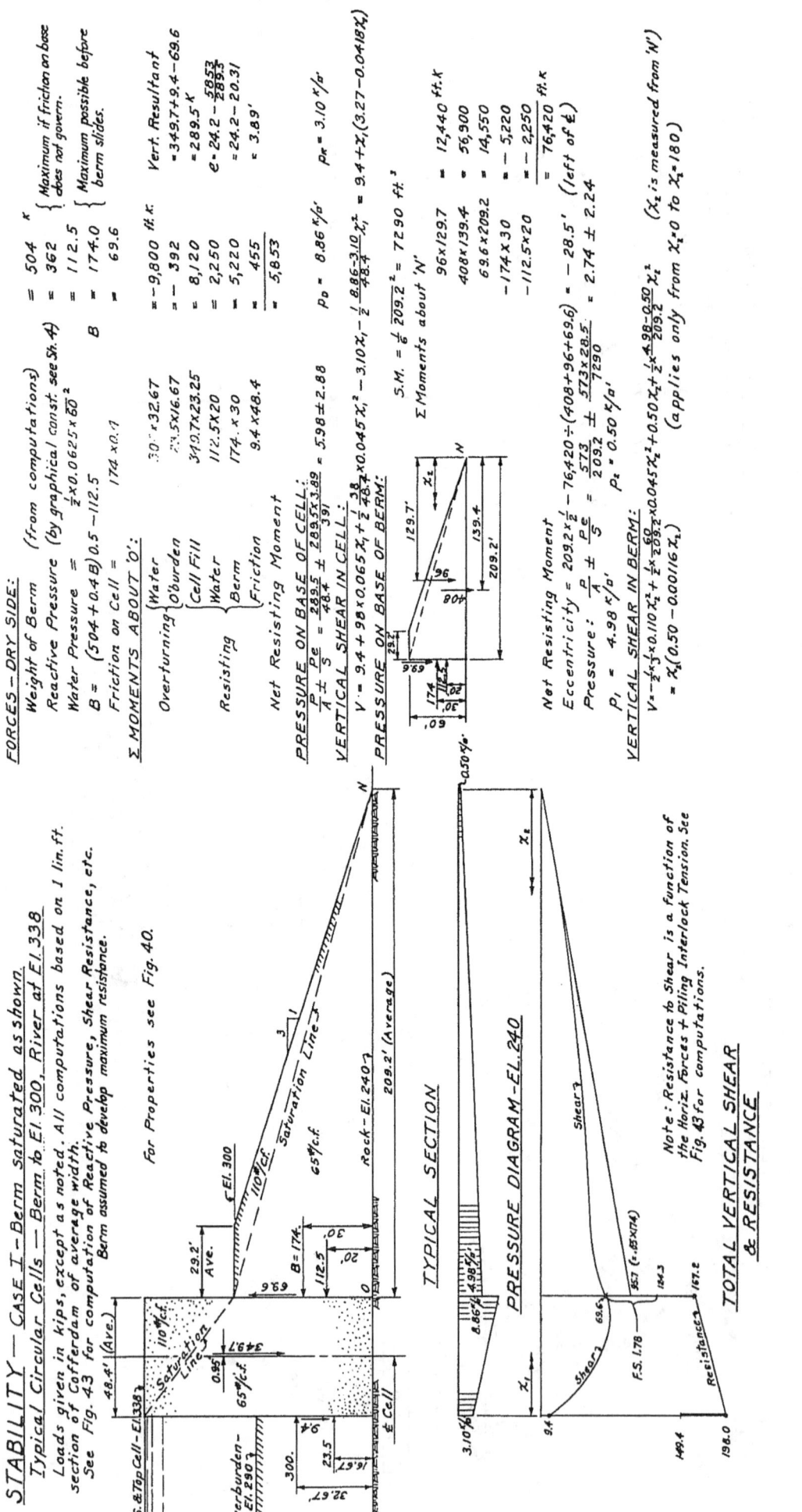

Figure 35
Design Analysis - Sheet 3 - Stability in Circular Cells With Berm at Elevation 300 - Kentucky Dam

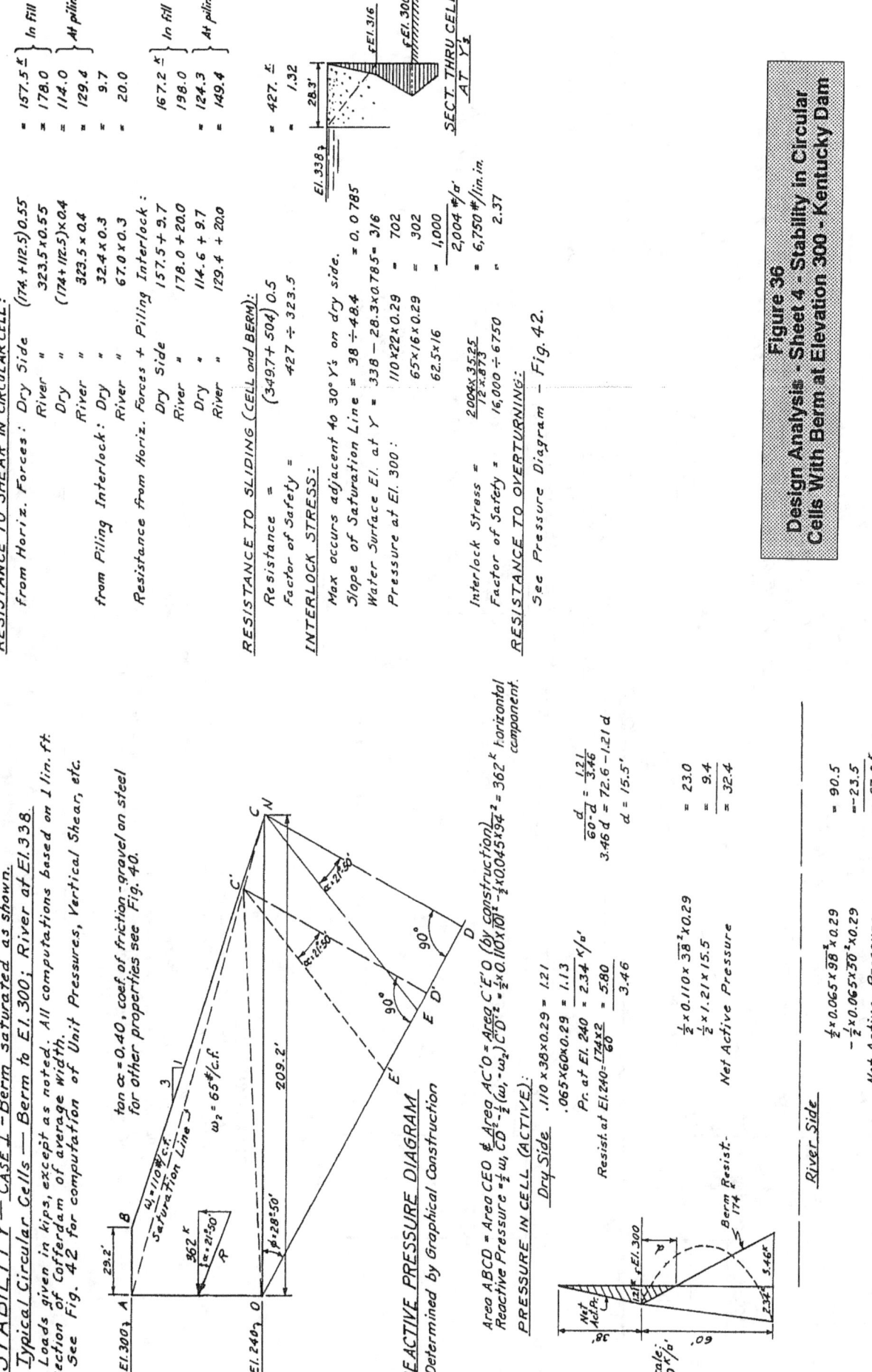

Figure 36
Design Analysis - Sheet 4 - Stability in Circular
Cells With Berm at Elevation 300 - Kentucky Dam

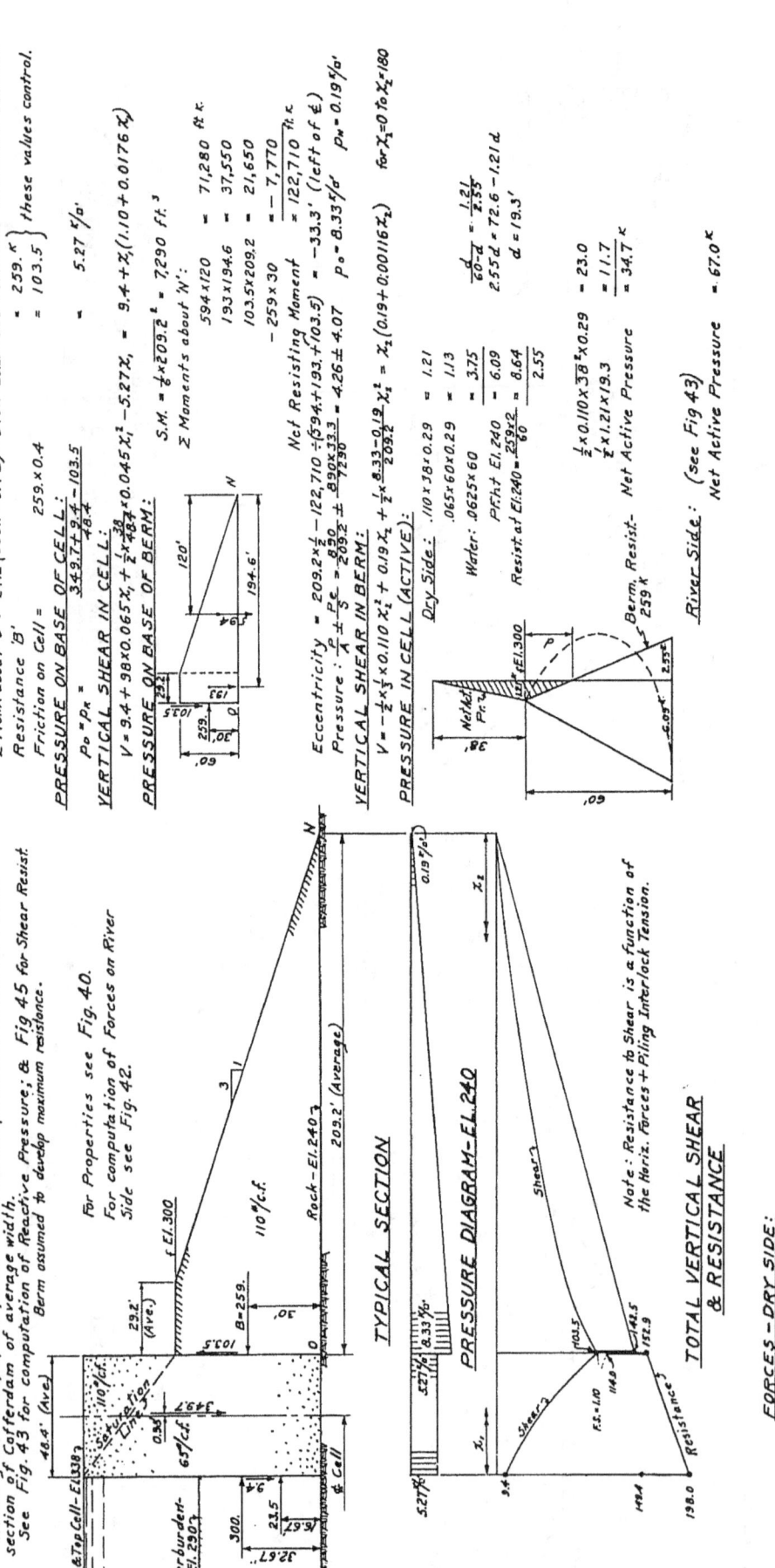

Figure 37
Design Analysis - Sheet 5 - Stability in Circular
Cells With Berm at Elevation 300 - Kentucky Dam

STABILITY — CASE II - Berm completely drained.
Typical Circular Cells   Berm to El. 300; River at El. 338.

Loads given in kips, except as noted. All computations based on 1 lin. ft. section of Cofferdam of average width.
See Fig. 44 for computation of Unit Pressures, Vertical Shear, etc.
Berm assumed to develop maximum resistance.

RESISTANCE TO SHEAR (CIRCULAR CELL):

from Horiz. Forces:	Dry Side	259 × 0.55	= 142.5 K	} In fill				
	River "	323.5 × 0.55	= 178.0					
	Dry "	259 × .4	= 103.5	} At piling				
	River "	323.5 × .4	= 129.4					
from Piling Interlock:	Dry "	34.7 × 0.3	= 10.4					
	River "	67.0 × 0.3	= 20.0					
Resistance from Horiz. Forces + Piling Interlock:								
	Dry Side	142.5 + 10.4	= 152.9	} In fill	Dry Side	103.6 + 10.4 = 114.0 K	} At piling	
	River "	178.0 + 20.0	= 198.0		River "	129.4 + 20.0 = 149.4		
Factor of Safety:	in Cell	114.0 ÷ 103.5	= 1.10					

RESISTANCE TO SLIDING:

Resistance = (349.7 + 787.) 0.5 = 568. K
Factor of Safety = 568 ÷ 323.5 = 1.75

RESISTANCE TO OVERTURNING:
See Pressure Diagram — Fig. 44.

INTERLOCK STRESS:
Interlock Stress (see Fig. 43)   6,750 #/lin. in.
Factor of Safety (" " 43)   2.37

Figure 38
Design Analysis - Sheet 6 - Stability in Circular Cells With Berm at Elevation 300 - Kentucky Dam

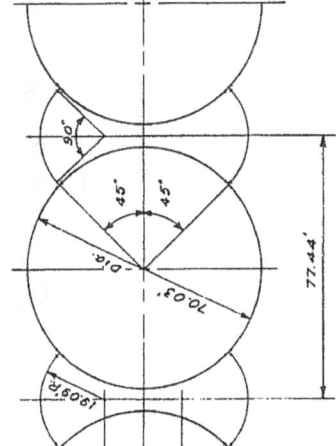

Properties:

Area: Main Cell = 3,852 s.f.
       Conn. Cell = 908 s.f.
       Total = 4,760 s.f.

Ave. Width = 4,760 ÷ 77.44 = 61.5'

Sect. Modulus based on Ave. Width:
$S = 1/6 \times \overline{61.5}^2 = 630$ ft.^{3}

ASSUMPTIONS AND DATA:

Cell fill to be river run gravel with all minus 1/4" material screened out.
  Weight of drained fill $\gamma = 110$ #/c.f.
  Weight of submerged fill $\gamma_s = 65$ #/c.f.
  Angle of internal friction
    (for dry or saturated fill) $\phi = 28°-50'$; Tan $\phi = 0.55$
  Coeff. of friction - gravel on rock (sliding)   Tan $\phi = 0.55$
  Coeff. of friction - gravel on steel ($\alpha = 21°-50'$)   Tan $\alpha = 0.40$
  Coeff. of friction - in piling interlocks $f = 0.3$

Horizontal resultant of fill pressure computed according to Coulomb's theory--Used in computing piling tension and shear resistance:

Active, $P_a = 1/2 \gamma h^2 \times \dfrac{\cos^2 \phi}{\left(1 + \sqrt{\dfrac{\sin(\phi + \alpha) \sin \phi}{\cos \alpha}}\right)^2}$ $C = 0.29$
(level surface)

Ratio of horizontal to vertical pressures on a vertical plane just prior to failure in shear-- Used in computing shear resistance of cell fill:

$K = \dfrac{\cos^2 \phi}{2 - \cos^2 \phi} = 0.623$

Allowable stresses in piling:
  Tension in web   24,000 #/sq. in. (for 70,000 #/sq. in. ultimate)
  Interlock tension   12,000 #/lin. in. (for 16,000 #/lin. in. min. ultimate)

Factor of safety 1.25 minimum--for sliding, shear, and interlock.

Min. Pool Elev. 632
Normal Pool Elev. 634--Assumed W.S. during cell filling.
Top of Cofferdam Elev. 638--Used for max. design head.
Rock Elevation Ave. El. 580--Used for design.

Saturation line--Assumed to be at 1/2 the depth of water outside cofferdam. Field to maintain drainage through weep holes so that area below saturation line (measured on ₵ of cells) will be not more than 1,900 sq. ft. Foundation rock to be grouted to insure tight base.

Overburden--None assumed for design.

This design procedure follows that outlined in this Chapter on Steel Sheet Pile Cellular Cofferdam Design.

All computations based on 1 lin. ft. section of cofferdam of average width. The weight of the piling and the concrete slab and crane runway have not been included in the design.

Figure 39
Design Analysis - Sheet 1 - Upstream Cofferdam for Units 15 and 16 - Hales Bar Dam

WEIGHT OF CELL FILL:
61.5 × 29.0 × 0.110 = 196.2^k
61.5 × 29.0 × 0.065 = 115.8^k (no piling incl.)
Total   312. ^{k}

FORCES--RIVER SIDE:
Water Pressure = 1/2 × 0.625 × $\overline{58}^2$ = 105.0^k
Friction-fill on piling = 105 × 0.4 = 42.0^k

RESISTANCE TO SLIDING:
Resistance = 31.2 × 0.55 = 171.6^k
Factor of Safety = 171.6 ÷ 105 = 1.64

RESISTANCE OF RIVER SIDE PILING TO RISE:
Summation of Moments about "O" (weight of fill does not affect the result as it is balanced by upward pressure on the base):
M_o = 105 × 19.33 = 2030 ft. k.
M_r = 42.0 × 61.5 = 2580 ft. k.
Factor of Safety = 2580 ÷ 2030 = 1.27

OVERTURNING:
Summation of Moments about "O":
M_o = 105 × 19.33 = 2030 ft. k.
M_r = 312 × 1/2 × 61.5 = 9600 ft. k.
Factor of Safety = 9600 ÷ 2030 = 4.72

PRESSURE ON BASE OF CELL:
Summation of Moments about "C":
M_c = 105 × 19.33 = 2030 ft. k.
$\frac{W}{A} \pm \frac{M_c}{S} = \frac{312}{61.5} \pm \frac{2030}{630} = 5.08 \pm 3.22$
P_o = 8.30 k/$\square$' P_r = 1.86 k/$\square$'

INTERLOCK TENSION:
The max. tension will occur at the point of max. bulge of the cell. This point has been assumed at 1/4 H above the base.
0.29 × 0.110 × 29 = 0.925
0.29 × 0.065 × 14.5 = 0.273
0.0625 × 14.5 = 0.907
p = 2.105 k/$\square$'

$t_{max.} = \frac{1}{12} p \times \frac{77.44}{2} \times \sec 45° = 9.62/\text{lin. in.}$

Factor of Safety = 12 ÷ 9.62 = 1.25

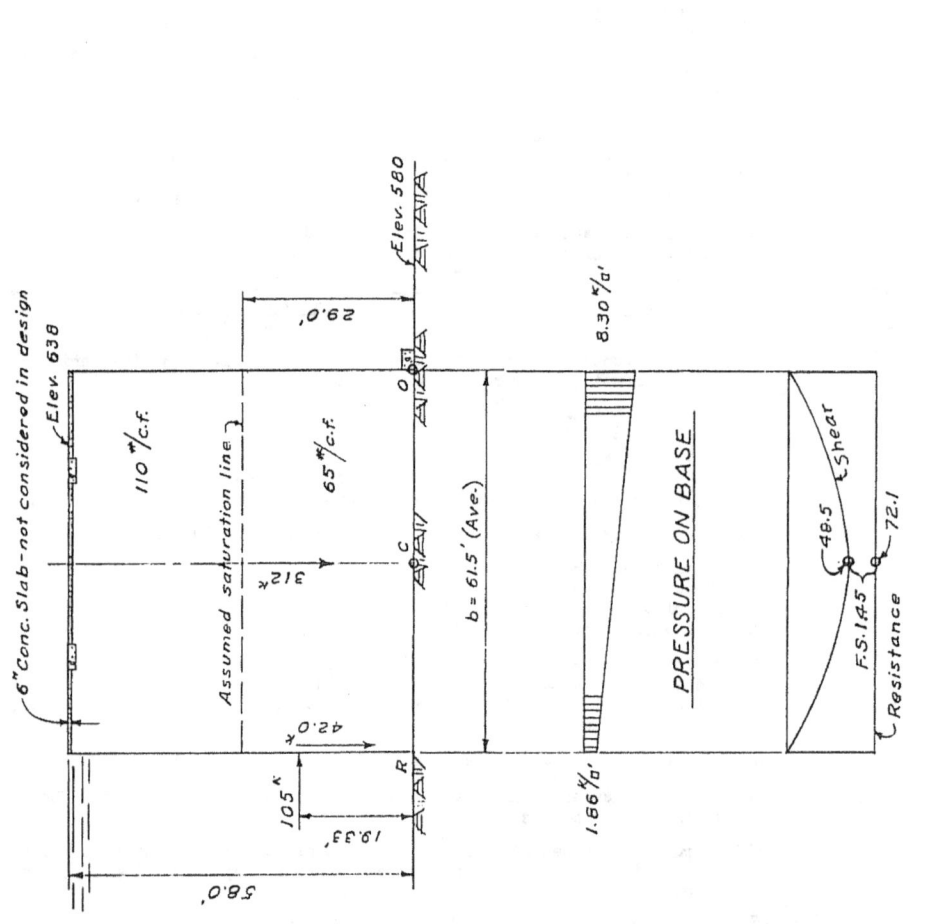

Figure 40
Design Analysis - Sheet 2 - Upstream Cofferdam
for Units 15 and 16 - Hales Bar Dam

SHEAR AND SHEAR RESISTANCE:

The shear in the cells is considered to act on vertical planes similar to the horizontal shear in a beam. The tendency for the cells to fail in shear is resisted by the cell fill plus the resistance of the piling to slip in the interlocks.

Investigation of the shearing forces involves several phases, as follows:

(1) The cells are filled with the W.S. at El. 634, and a shear resistance, which is a function of the pressure in the submerged fill, is developed.

(2) Now, the cofferdam is unwatered with W.S. at El. 634 (or lower) and the cells are subjected to a shearing force. The shearing resistance developed during filling should be great enough to resist this or there may be large deflections before the full resistance is mobilized.

(3) As the cofferdam is unwatered, the pressure within the cells will increase and consequently the shear resistance will increase to about a maximum when it is fully unwatered.

(4) Now, if the W.S. rises to El. 638 (top of cells) the resistance mobilized in (3) may be considered to resist the increased shear.

(1) SHEAR RESISTANCE DEVELOPED DURING FILLING:

Fill:

$1/2 \, K \, \sigma \, x \, H^2 = 1/2 \times .623 \times 0.110 \times \overline{58}^2 =$ 115.1^k

$-1/2 \, K \, (\sigma - \sigma_s) \times d^2 = -1/2 \times .623 \times 0.045 \times \overline{54}^2 = \underline{-40.8^k}$

Area abe in fig. $= 74.3^k$

Shear Resist. in fill = 74.3 × 0.55 $- 40.8^k$

Piling:

The pressure contributing to frictional resistance in piling interlocks is represented by area abc.

$0.110 \times 4 \times .29 = .440 \times .29 = 0.128$

$0.065 \times 39.5 \times .29 = 2.57 \times .29 = \underline{0.745}$

     p $= 3.01 \times .29 = 0.873$

$.128 \times 4 \times 1/2 = 0.256$
$.128 \times 39.5 = 5.06$
$.745 \times 39.5 \times 1/2 = 14.70$
$.873 \times 14.5 \times 1/2 = \underline{6.33}$

     Area abc $= 26.35^k$

Piling Resistance = 26.35 × 0.3 = 7.9^k

Total Shear Resistance = 40.8 + 7.9 = 48.7^k

(2) SHEAR IN CELLS DURING UNWATERING OF COFFERDAM:

Consider cofferdam to be unwatered with W.S. at El. 634.

Water Pressure = $1/2 \times 0.0625 \times \overline{54}^2 = 91.2^k$

$M_c = 91.2 \times \dfrac{54}{3} = 1640$ ft. k.

Shear on ℄, $Q = \dfrac{3}{2} \times \dfrac{1640}{61.5} = 40.0^k$

Factor of Safety during unwatering = 48.7 ÷ 40.0 = 1.22

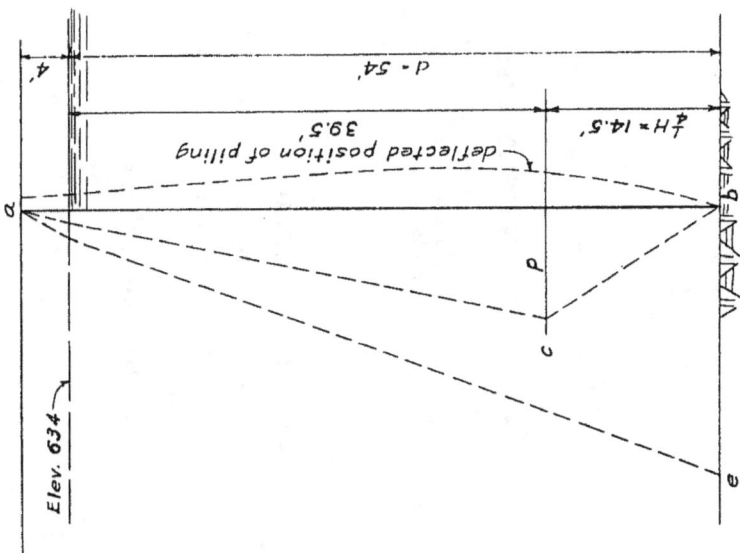

Figure 41
Design Analysis - Sheet 3 - Upstream Cofferdam
for Units 15 and 16 - Hales Bar Dam

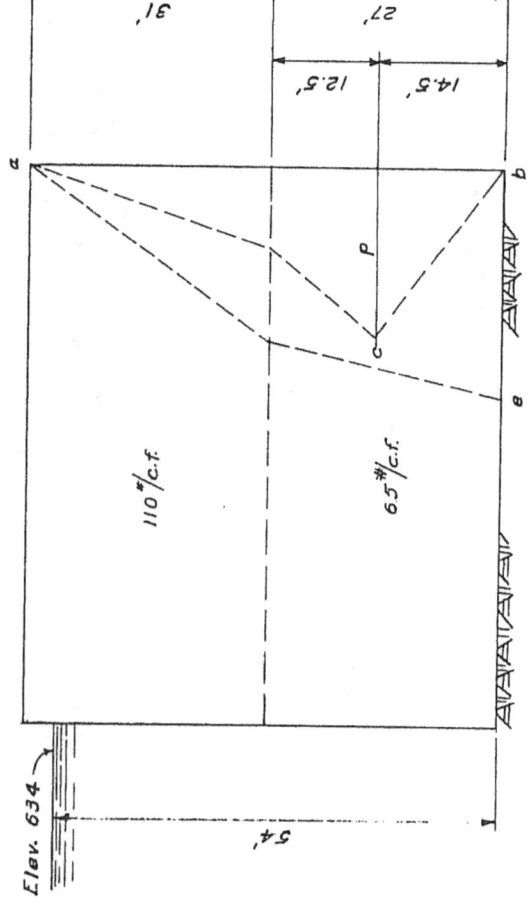

Elev. 634

110 #/c.f.

65 #/c.f.

54'

31'

27'

12.5'

14.5'

(3) SHEAR RESISTANCE AFTER UNWATERING:

$0.29 \times 0.110 \times 31 = 0.990$
$0.29 \times 0.065 \times 12.5 = 0.235$
$0.0625 \times 12.5 = \underline{0.782}$
$p = 2.007$

Shear Resistance of Fill:

$1/2 \times \sigma \times \overline{58}^2 = 1/2 \times .623 \times 0.110 \times \overline{58}^2 = 115.1^k$
$-1/2 \times (\sigma - \sigma_s) \times \overline{27}^2 = -1/2 \times .623 \times 0.045 \times \overline{27}^2 = \underline{-10.2^k}$
Area abe $= 104.9^k$

Fill Shear Resist. $= 104.9 \times 0.55 = 57.6^k$

Shear Resistance of Piling:

$1/2 \times 0.29 \times 0.110 \times \overline{31}^2 = 15.30$
$0.29 \times 0.110 \times 31 \times 12.5 = 12.35$
$1/2 \times 0.29 \times 0.065 \times \overline{12.5}^2 = 1.47$
$1/2 \times 0.0625 \times \overline{12.5}^2 = 4.88$
$1/2 \times 2.007 \times 14.5 = \underline{14.55}$
Area abc $= 48.55^k$

Piling Shear Resistance $= 48.55 \times 0.3 = 14.6^k$
Total Shear Resistance $= 57.6 + 14.6 = 72.2^k$

(4) SHEAR IN CELLS WITH W.S. AT EL. 638:

Consider that the river now rises to El. 638; then:

Shear on ℄, $Q = \frac{3 M_c}{2 b} = \frac{3 \times 2030}{2 \times 61.5} = 49.5^k$

Factor of Safety $= 72.2 \div 49.5 = 1.46$

GENERAL NOTES:

The W.S. should not exceed El. 634 during cell filling and cofferdam unwatering. A lower elevation would be preferable.

Allowable interlock stresses may be exceeded unless the piling is securely anchored to prevent expansion of the cells at the base. The piling should be driven sufficiently for it to "bite" into the rock. As an added precaution a small concrete berm should be placed around the inside of the cofferdam immediately after unwatering.

Figure 42
Design Analysis - Sheet 4 - Upstream Cofferdam
for Units 15 and 16 - Hales Bar Dam

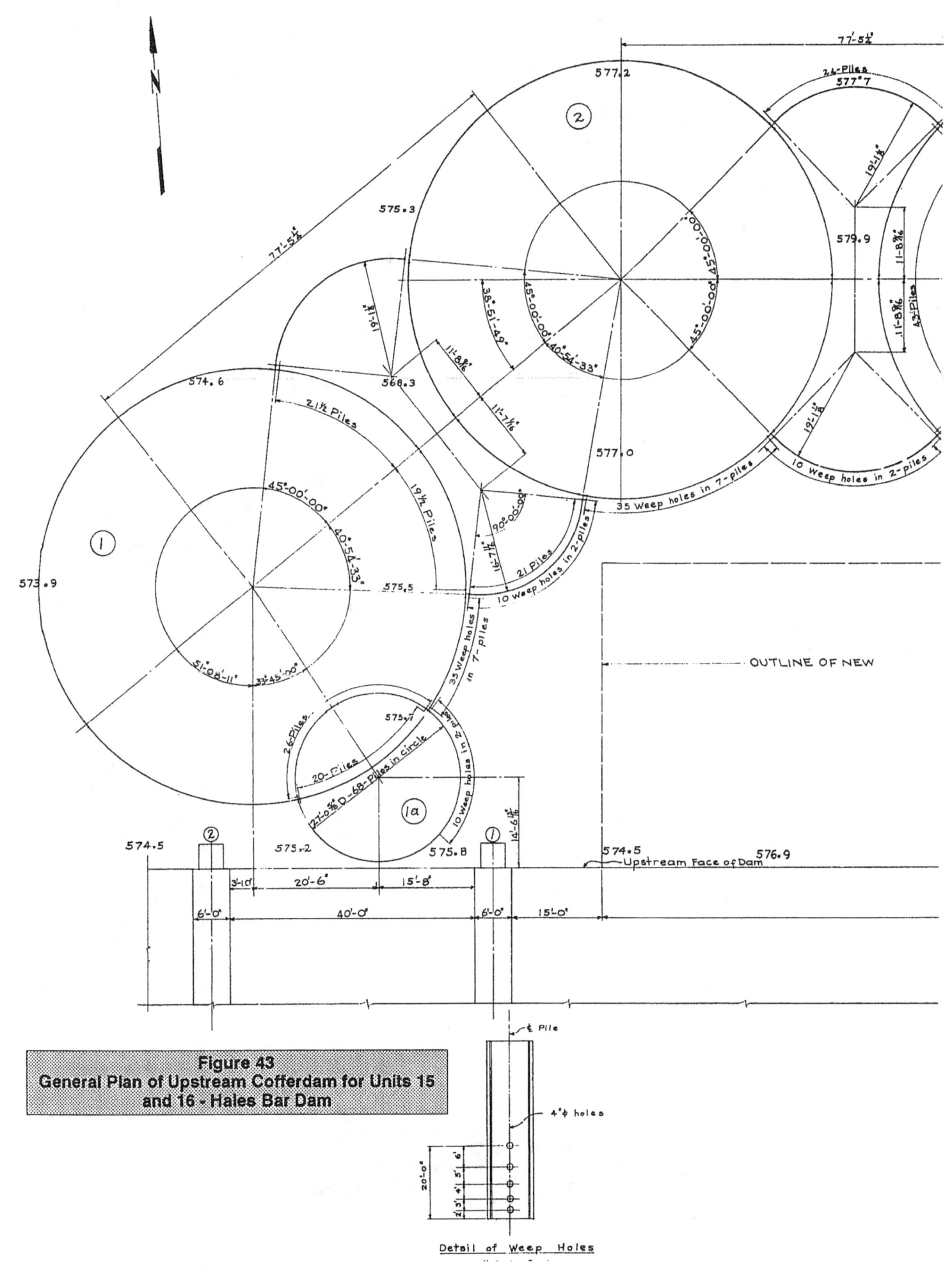

Figure 43
General Plan of Upstream Cofferdam for Units 15 and 16 - Hales Bar Dam

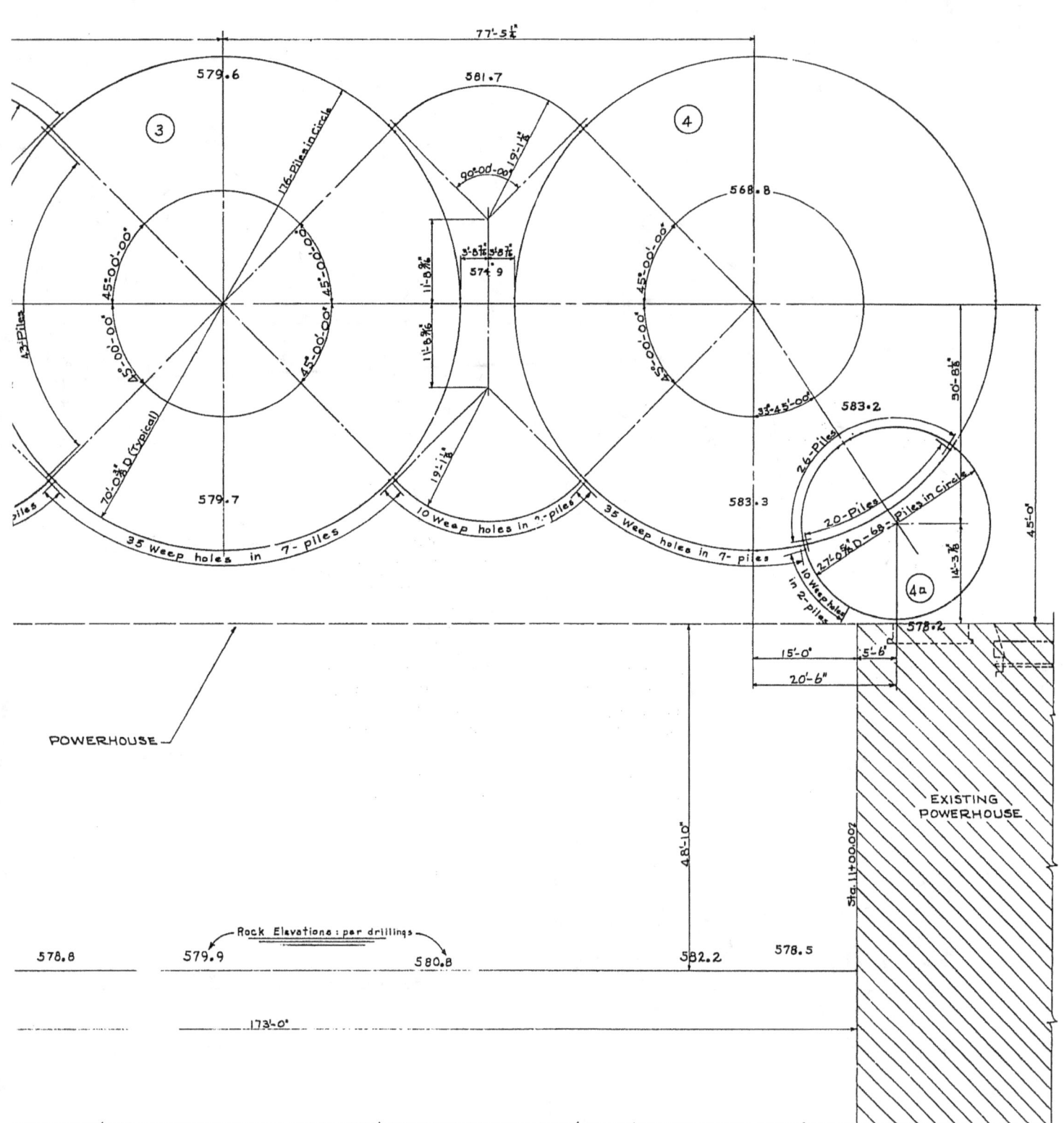

**Figure 43 (Continued)
General Plan of Upstream Cofferdam for Units 15 and 16 - Hales Bar Dam**

CHAPTER 6

FINAL LAYOUT

Before the final layout of the cofferdam can be completed the number of construction stages and the portion of the construction project to be included in each stage should be known. Then, the layout, alignment, and detailing of the cofferdam system are ready to be put in final form and considerations involved at this point are:

1. Provisions for an adequate channel to accomodate stream flow and navigation where required, during each construction stage.
2. The inclusion of a sufficient area inside of each cofferdam enclosure to insure enough room to carry on the necessary construction operations.
3. Construction features to be included inside the cofferdam, such as berms, ramps, cutoff walls, cofferdam drainage system, sumps, etc.
4. Use of earth dikes or other supplementary means to reduce the number of steel sheet pile cells otherwise required.
5. Details of the design of the steel sheet pile cells to be used (see chapter 5, "Design").
6. Alignment of the cells, and of the cofferdam as a whole.
7. Tie-in requirements and details.

Size of Cofferdam Enclosure

The size of each enclosure should be held to a minimum for economical reasons unless the area can be made larger without added expense. Advanced planning and a knowledge of the details of the final design of the cofferdam are essential in order that the final size of the enclosure will be sufficient to provide room enough for ramps, construction bridges, berms, working area for construction equipment, etc. (See figures 1, 3, 43 and 44.)

Auxiliary Cofferdam Structures

Earth dikes, or earth dikes with steel sheet pile cutoff walls to seal against leakage, are frequently used for inshore enclosures where water is shallow, or to facilitate early construction operations while the sheet pile cells are being built. Short earth dikes are also used at the shore end of cell walls where part of the adjoining flood plain is to be included, or between groups of cells where the alignment passes over high ground. (See figures 1 and 3. Also see "Tie-Ins," page 102). Earth dikes should be anticipated, investigated, and included (where feasible) in the original layout.

Cell Alignment

Considerations involved in cell alignment are:
1. Selection of a standard cell size, or sizes, which will give the minimum number of required piles.
2. Cell and connecting cell layout and detail.
3. Controlling factors governing the deflection angles in both cell and piling alignment.
4. Alignment and details of tie-ins between stages, to riverbanks, and to the permanent structures.

Selection of Cell Size

The final cell design will indicate the required cell size or sizes. It is desirable, however, to have only one size cell for each entire cofferdam stage. This is not practical in every case, due to the condition and elevation of the bedrock. It may be possible, by compromising between a number of indicated sizes, to attain one size that will meet all design requirements. It is important that this decision be made before proceeding further with the final layout. The additional construction costs resulting from having to build and use more than one size template, and complications in construction procedure in the case of more than one size cell, may well exceed the added cost resulting from an increase in the cell size to attain standardization.

In selecting the number of piling to meet the adopted cell size, the least even number of piling approaching this size should be taken, as an even number is necessary in every case to ensure proper interlock relations for the closure pile. Also a saving of a few piling per cell may mean a material saving in handling, driving, and extracting costs when accumulated over the entire project.

Cell and Connecting Cell Layout and Details

The layout details (figure 45) are considered as standard, although not binding. The number of piling and other pertinent data for this general arrangement, covering a wide range in cell sizes (in multiples of six piles), is shown in table 2. The total piling in the cell is increased in multiples of six so that the angle $\theta = 30°$ may be kept constant. This condition holds true since $2\theta = 60° = 1/6$ of $360°$. As the number of piles, center to center of the tees, is increased by one then the total number in the cell as a whole must be increased in the same proportion, or in other words in multiples of six. By the same reasoning it may be shown that (for $\theta = 45°$) the cells will contain multiples of four piles. For estimating purposes, the data for intermediate cell sizes may be had by interpolation. The interlock stresses in the upper range in cell sizes, however, may approach the critical limit and should be carefully checked.

The angle θ may be varied where a greater width of connecting cell or more piling in the connecting cell is desired. In doing so, however, for circular cells it can be varied only in multiples of the angle per pile, based on the number of piles in the cell. In no case is it advisable to make it appreciably less than $30°$, since this would tend toward reducing the width and watertightness of the connecting cell and the stability of the cofferdam as a whole. However, at times it may be desirable to make the angle θ greater than $30°$. By increasing this angle, the average width of the cofferdam, b, will be increased without increasing the diameter of the cell D (see figure 45). This may be used to advantage in tight places, as was the case of the upstream cofferdam at Hales Bar Dam (figure 43) previously discussed on page 49.

To have used larger cells at Hales Bar Dam would have offered more obstruction to the discharge of water through the ole generating units and over the spillway. Larger cells would also have required the use of a larger template. As previously stated, had $60°$ between tees been used, a cell diameter of 78 instead of 70 feet would have been required.

However, there are two disadvantages in increasing the angle between tees. One of these is that interlock stress at the tees will be increased, as it is a function of secant θ (see discussion on "Interlock Tension," page 36). This is offset to some extent, however, since the dimension L will be decreased. Thus, for cofferdams of average widths varying from 18 to 48 feet, the net effect will be an increase in the interlock stresses in the tee piles of 9 percent for the cell using an angle of $\theta = 45°$ as compared to the cells using an angle of $30°$. In most cases this will not be serious, but it should always be considered. The other disadvantage is that the amount of steel sheet piling per lineal foot of cofferdam will be increased for cofferdams of the same average widths as above (18 to 48 feet) the number of piles per lineal foot will average about 10 percent more for the cells using $\theta = 45°$ as compared to those using an angle of $30°$. However, this objection may not be

PILE

Whether you need heavy duty steel sheet piling, H-bearing pile, lightweight steel sheet piling, pipe piling, or custom produced spiralweld pipe piling, L.B. Foster has it.

Get it at a competitive price, or rent it for even bigger savings.

Coated piling? Sure. And specialty items including Y's, T's, fabricated sections, redeb and soffit panels. And the most complete line of equipment for driving pile including vibro, diesel, and hydraulic pile drivers, pile threaders, and ground release shackles.

Whatever you need. Whenever you need it. Call us. 412-928-3400.

DRIVERS

Whatever kinds of piling you're driving...for whatever purpose...and whatever the jobsite conditions...L.B. Foster Company has the pile drivers you need.

We'll sell or rent you pile drivers and give you the best sales and technical support in the business to keep your job on schedule.

- The revolutionary IHC Hydrohammer is much faster, more efficient with total control of driving operations. Designed for both onshore and offshore use, it delivers 90-95% of the available energy to the pile.
- L. B. Foster Vibro Driver®/Extractors are built for long-term performance and include a state of the art hydraulic power system. Our technically advanced transmission (exciter) gives longer gear, shaft and eccentric life.
- The Foster/Kobelco single-acting diesel hammer features a water cooling system with a 9 & 13 position rack-style fuel pump for easy starting and energy control.

Pile drivers, piling, accessories—whatever you need, whenever you need it. Call us. 412-928-3400.

L.B. FOSTER COMPANY

P.O. Box 2806, Pittsburgh, PA 15230

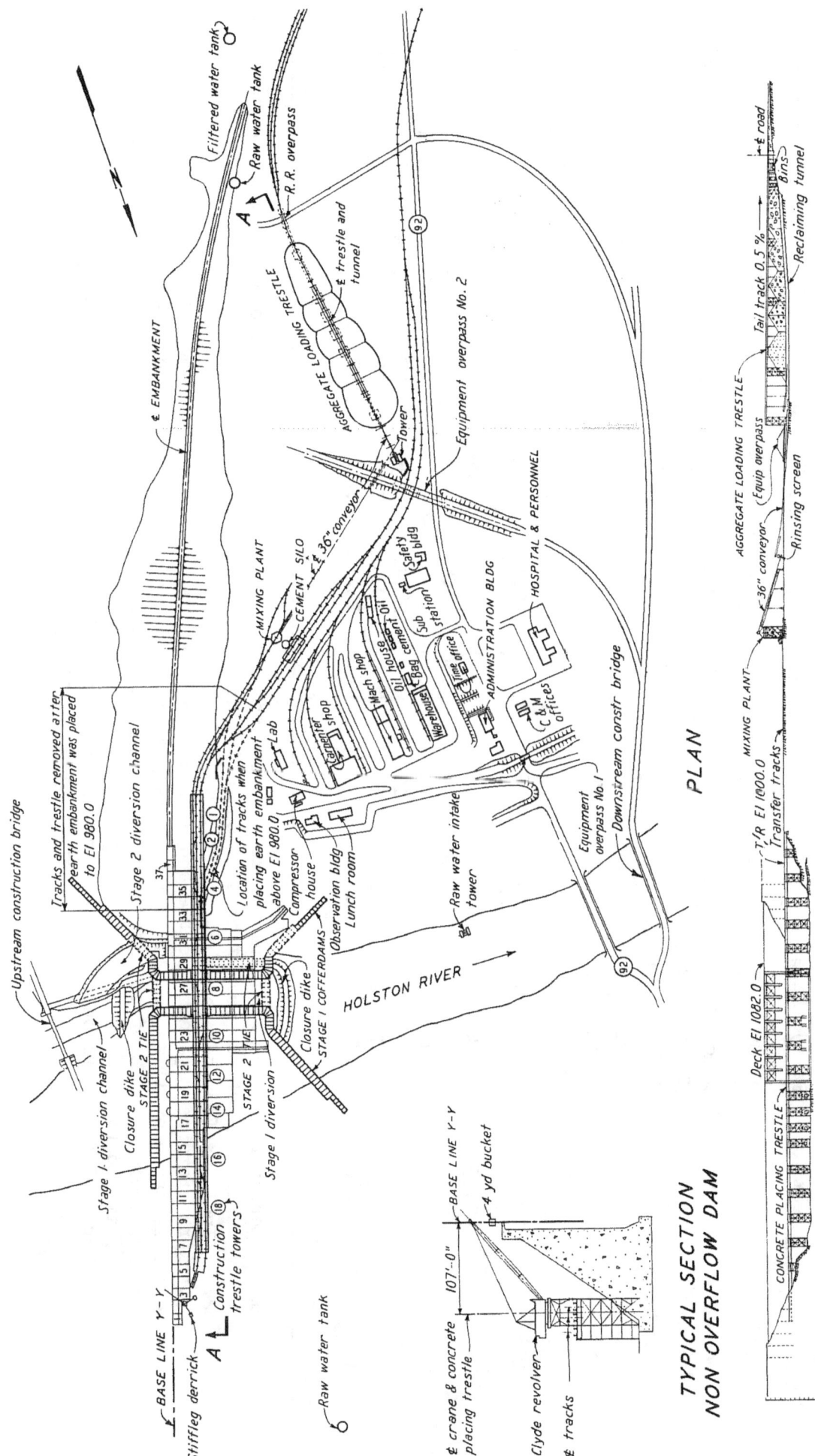

Figure 44
General Plan of Construction Plant and Cofferdam Arrangement - Cherokee Dam

Table 2
Design Data for Cells of Varying Diameters, Based on Layout Details Shown in Figure 45

No. Piles in Cells*	Theoretical Cell Diam. "D" (Feet)	2L (Feet)	m No. of Piles	Radius O1P (Feet)	n No. of Piles	x (Feet)	A (Feet)	B (Feet)	C (Feet)	Av. Width b (Feet)	Area (Sq Ft) 1 Cir. Cell	Area (Sq Ft) 1 Conn. Cell
48	19.10	24.86	7	7.16	6	-1.09	10.22	8.85	3.58	14.64	287	77
54	21.49	26.92	8	7.16	6	-0.50	11.41	9.88	3.58	16.56	363	83
60	23.87	28.99	9	7.16	6	0.10	12.60	10.91	3.58	18.48	448	88
66	26.26	32.26	10	8.35	7	-0.34	13.80	11.95	4.18	20.27	542	112
72	28.65	34.32	11	8.35	7	0.26	14.99	12.98	4.18	22.23	645	118
78	31.04	36.40	12	8.35	7	0.86	16.19	14.02	4.18	24.17	757	123
84	33.42	38.46	13	8.35	7	1.46	17.38	15.05	4.18	26.11	877	127
90	35.81	41.70	14	9.55	8	1.01	18.57	16.08	4.77	27.91	1007	157
96	38.20	43.78	15	9.55	8	1.61	19.77	17.12	4.77	29.88	1146	162
102	40.58	45.84	16	9.55	8	2.21	20.96	18.15	4.77	31.83	1293	166
108	42.97	47.90	17	9.55	8	2.80	22.15	19.18	4.77	33.82	1450	170
114	45.36	51.18	18	10.74	9	2.37	23.35	20.22	5.37	35.58	1616	205
120	47.75	53.24	19	10.74	9	2.97	24.54	21.25	5.37	37.56	1791	209
126	50.13	55.32	20	10.74	9	3.56	25.73	22.29	5.37	39.53	1974	213
132	52.52	57.38	21	10.74	9	4.16	26.93	23.32	5.37	41.53	2166	217
138	54.91	60.64	22	11.94	10	3.72	28.12	24.35	5.97	43.29	2368	257
144	57.30	62.72	23	11.94	10	4.32	29.32	25.39	5.97	45.27	2579	261
150	59.68	64.78	24	11.94	10	4.92	30.51	26.42	5.97	47.27	2797	264

Notes:
1. Dimensions in the "x" column marked (-) indicate that the center lies on the opposite side of the centerline of the cell from the connecting arc.
2. * Includes four tee piles.
3. For reference regarding source and application of above data, see figure 54.
4. This table applies to all 15-inch straight piling, with the angle θ (figure 54) = 30°.
5. The average width of the cofferdam b equals the area of one circular cell and one connecting cell divided by 2L.
6. For tables of data on cofferdams with angle θ (figure 54) = 45°, see Steel Sheet Piling (Handbook) published by the U. S. Steel Corporation, pages 38 and 39.

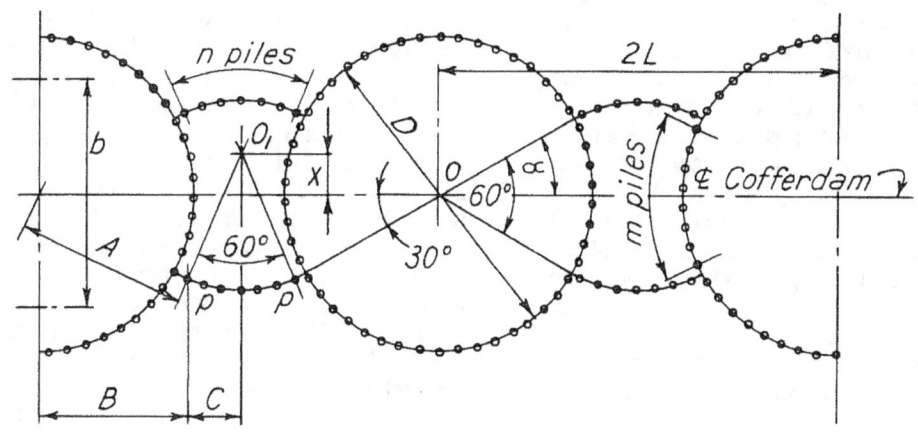

**Figure 45
Typical Layout Arrangements - Circular Cell Steel Sheet Pile Cofferdam**

**Table 3
Method of Listing Steel Sheet Piling Required for Guntersville Cofferdam**

Number of Cell	Lengths of Piling - Typical Cells					
	River Side			Lock Side		
	Circular Cell		Connecting Cell	Circular Cell		Connecting Cell
	35-SP6	2-FT61	8-SP6	35-SP6	2-FT61	8-SP6
1-4, 7-12	51' 0"	51' 0"	51' 0"	46' 0"	46' 0"	46' 0"
13	33' 0"	38' 0"	51' 0"	33' 0"	38' 0"	46' 0"
17-20	46' 0"	46' 0"	46' 0"	38' 0"	33' 0"	38' 0"
23, 24	46' 0"	43' 0"	46' 0"	38' 0"	33' 0"	38' 0"
25-28	43' 0"	43' 0"	43' 0"	38' 0"	33' 0"	38' 0"
29, 31-34	43' 0"	38' 0"	43' 0"	38' 0"	33' 0"	38' 0"
36	38' 0"	38' 0"	51' 0"	38' 0"	33' 0"	38' 0"
37	51' 0"	51' 0"	None	38' 0"	43' 0"	None

serious as in the case of short cofferdams or cofferdams built under restricted conditions such as encountered at Hales Bar Dam. (See "Sectional All-Welded Pipe Template--Hales Bar Dam," page 78).

The values of the terms in table 2 (for $\theta = 30°$) have been determined as follows:

A = one-half of the theoretical diameter of the cell, D, plus 8 inches. (The 8-inch dimension is standard for riveted tees used with the 15-inch-wide piles shown in figure 11)

The use of the point P (figure 45) as the point of tangency for the connecting cell arc causes the arc to pass through the working point of each pile as it should, with an equal deflection angle for each pile in the arc. To move the point of tangency to the point of intersection between the arc and the theoretical driving line of the cell (as shown in the manufacturers' catalogs), however, would mean that the piles in the arc would be tangent to the working line at the center of each pile, or that the deflection angle for the first pile would be greater than the rest. The latter condition would cause an unbalanced deflection stress in the leg of the tee pile.

$\theta = 30°$ (for $\theta = 45°$, see Steel Sheet Piling (Handbook) [formerly] published by U. S. Steel Corporation).

Angle $PO_1P = 60°$.

Arc PP: Determined by the number of piles used in the connecting arc. A different number of piles from that shown in table 2 may be used to suit special cases.

O_1P: Based on the length of the arc PP (O_1P is limited by the permissible deflection in the piling interlocks) and where an even shorter radius is required, as in the connecting cell arcs for cloverleaf cells at Kentucky cofferdam, bent piling will have to be used. (See bent pile detail B-1, figure 10)

B and C: Computed from data above.

x: Depends on the length of the radius O_1P.

m: Depends on the theoretical diameter of the cell and angle.

2L = 2 (B + C).

b = average width of the cofferdam. This is equal to the area of one circular cell and one connecting cell divided by the distance center to center of

In detailing the cells and connecting cells, right- and left-hand tees and the number of intervening piling must be shown. Tees are shown correctly for position in an adjoining and enlarged sketch (as in figures 6, 7, and 10. The intervening piling may be indicated either by length or final top elevation. Where bedrock is irregular and soundings are available, a table may be placed in the same drawing with the cells, showing lengths of piling required in each cell. (See table 3.)

Deflection Angle in Cell Alignment--Deflection angles in the alignment of cells can be made only in multiples of the angle per pile based on the number of piles in the cell at which the deflection occurs. This is to ensure proper interlock relations between tees and cell arcs.

Where deflection in alignment is dependent on deflection in the piling interlocks, such as for single sheet pile cutoff walls, the maximum deflection in any one interlock will be controlled by permissible deflections between individual piling.

Tie-In Details--For a variety of tie-in arrangements with other structures, see figures 1, 59 and 61. Details for tie-ins must be worked out in conjunction with cell alignment to ensure feasibility of tie-in arrangement and design and to avoid costly alterations which might otherwise occur in making tie-in connections fit cell location and alignment. (See later discussion on "Tie-Ins," page 102).

CHAPTER 7

CONSTRUCTION

The construction of cofferdams as discussed in this chapter covers the various features, methods, and equipment used by TVA throughout the construction of its steel sheet pile cellular cofferdams. The discussions appear in the following order: template design and use, handling and setting sheet steel piling, timber trestle, steel sheet pile driving and driving equipment, load bearing piling, pile hammers and their selection, tie-ins, closure cells, cell filling, floodgates and sluiceways, foundation treatment, and unwatering and maintenance.

TEMPLATE DESIGN AND USE

The selection of type of template to be used during construction of steel sheet pile cells can materially affect the construction time as well as the construction costs of a cofferdam. The selection of a prefabricated, all-welded metal template similar to one of the templates used by TVA is recommended. This recommendation is based on results such as minimum materials involved in template construction; maximum reuse of template (experience showed that the life of the template was usually much greater than the life of the cofferdam); increased speed in cofferdam construction due to the relatively short time required in moving and resetting template; and speed and ease in racking piling, as at Hales Bar project where guides for the T-piles were built into the template. At Pickwick Dam, where the 1-piece, all-welded pipe template was used, the lock cofferdam cells were built in less than half the time estimated as necessary based on the use of built-in-place templates. Timber templates, either built in place or prefabricated, may be required however for special or off-size cells where the number of such cells to be built is small.

Welded 1-Piece Pipe Template

The 1-piece, all-welded template, constructed from standard black pipe (shown in figures 46 and 47), is typical of the design used in construction of the first steel sheet pile cellular cofferdams by TVA. A pipe hub and additional pipe struts were added to the original design to give additional stiffness to the frame. At Guntersville Dam the pipe hub was enlarged to 24 inches in size so that it could be used for passing the hydraulic materials to the bottom of the cell during cell filling operations. This process was necessary where the template was left in place as a support for the piling until sufficient material was in place as a support for the piling until sufficient material was in place to make cell stable and permit its removal. Structural members of the template might otherwise have been damaged by the falling material.

Wooden walkways were bolted to the top and bottom rings of the circular templates. These served as working platforms for the men and at the same time as the actual tempplate for setting the piling. The overhang protected the frame during pile driving. At Pickwick cofferdam the outside diameter of the timber deck was made 4 inches less than the theoretical diameter of the driving line of the piling. On later projects the diameter of the template was increased to reduce this difference to 3 inches, 2-1/4 inches, and finally to 2 inches for the Kentucky cofferdam. It was found that the increase in diameter of the template took up much of the slack in the piling interlocks, thereby reducing the tendency of the cells to bulge when filled. The bulge in the cell was considered a distinct advantage in that its tendency is to prevent sliding of the piling in interlock, thereby increasing its value in resisting shear in the cell fill. The bowed piling resulting from the bulge, however, increases the placing problem due to the difficulty in meshing the interlocks of the bowed piling with those of the straight piling in the connecting cells and should be avoided during the period of setting the piling. In building a template for the smaller diameter cells this difference between the outside diameter of the template and diameter of the driving line of the piling should be greater since the total amount of slack in the piling to be absorbed is less.

Suggested cell and template diameters are as follows:

Cell Driving Diameter	Number of Piles in Cell	Welded 1-Piece Pipe Template Diameter	Difference
42 ft 11-3/4 in.	108	42 ft 9-1/8 in	2-5/8 in.
47 ft 9 in.	120	47 ft 6-5/8 in.	2-3/8 in.
54 ft 1-1/4 in.	136	53 ft 11-1/8 in.	2-1/8 in.
58 ft 10-5/8 in.	148	58 ft 8-5/8 in.	2 in.

To determine template diameter for any size cell see figure 47.

Alignment and Setting of 1-Piece Pipe Template--The 1-piece pipe template was lifted, moved, set, and aligned in one operation. It was handled with a floating crane, using four cables attached to the lower panel points. Alignment was established by means of transit set up at a convenient location. At Chickamauga the template beams for the connecting arcs were attached to the main template and to the last cell constructed, as a means of guiding the main template during setting operations. At Guntersville cofferdam, however, the main template was set and aligned before attaching the template beams for the connecting arcs.

The eight spuds shown in figure 46 were necessary only where the stream was swift, making it difficult to hold the template in place.

The eight spuds shown in figure 46 were necessary only where the stream was swift, making it difficult to hold the template in place. Only four spuds were used, for the most part, at Chickamauga and at Fort Loudoun cofferdams. In the use of eight spuds it was found that the template was disturbed

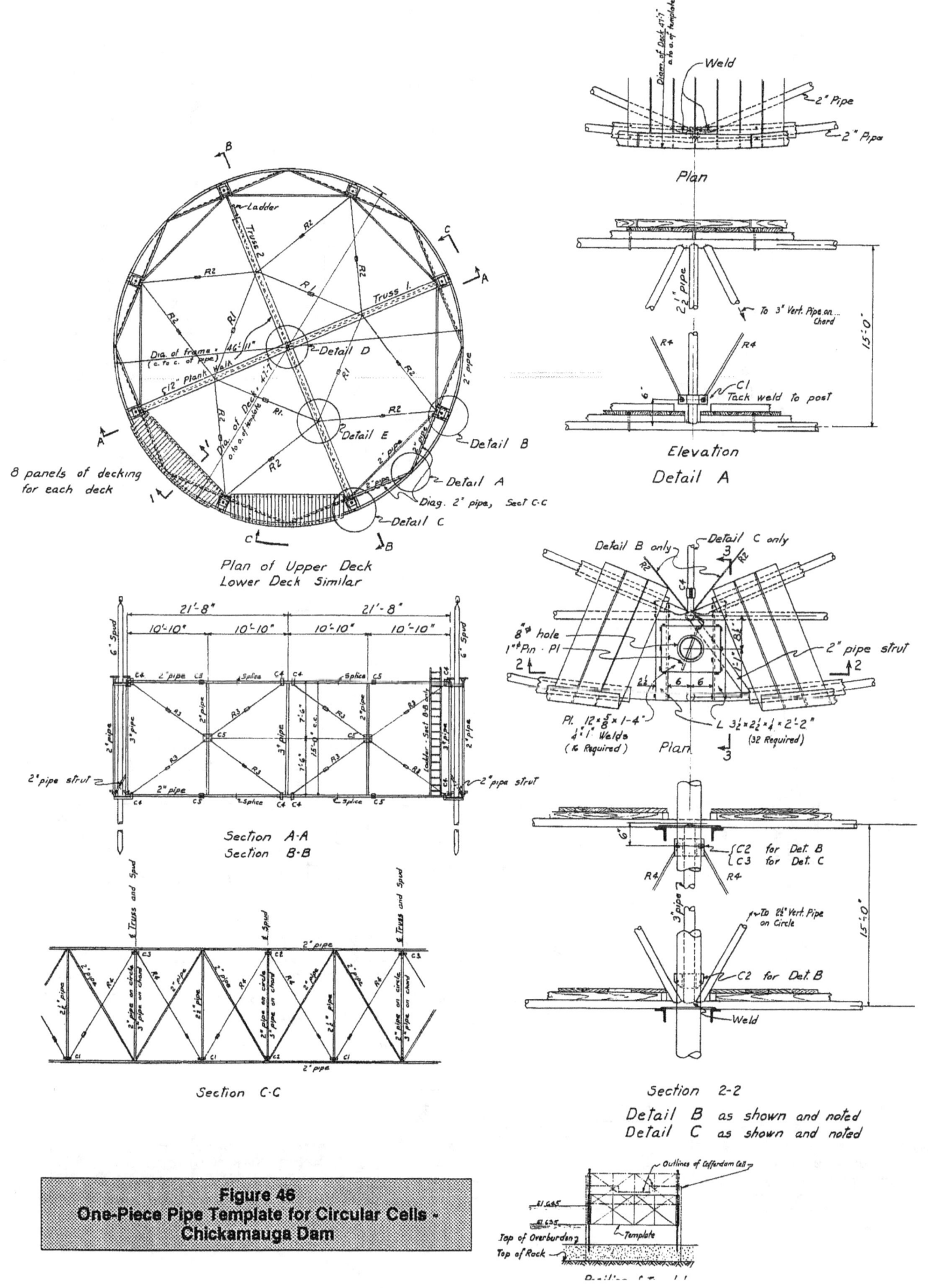

Figure 46
One-Piece Pipe Template for Circular Cells - Chickamauga Dam

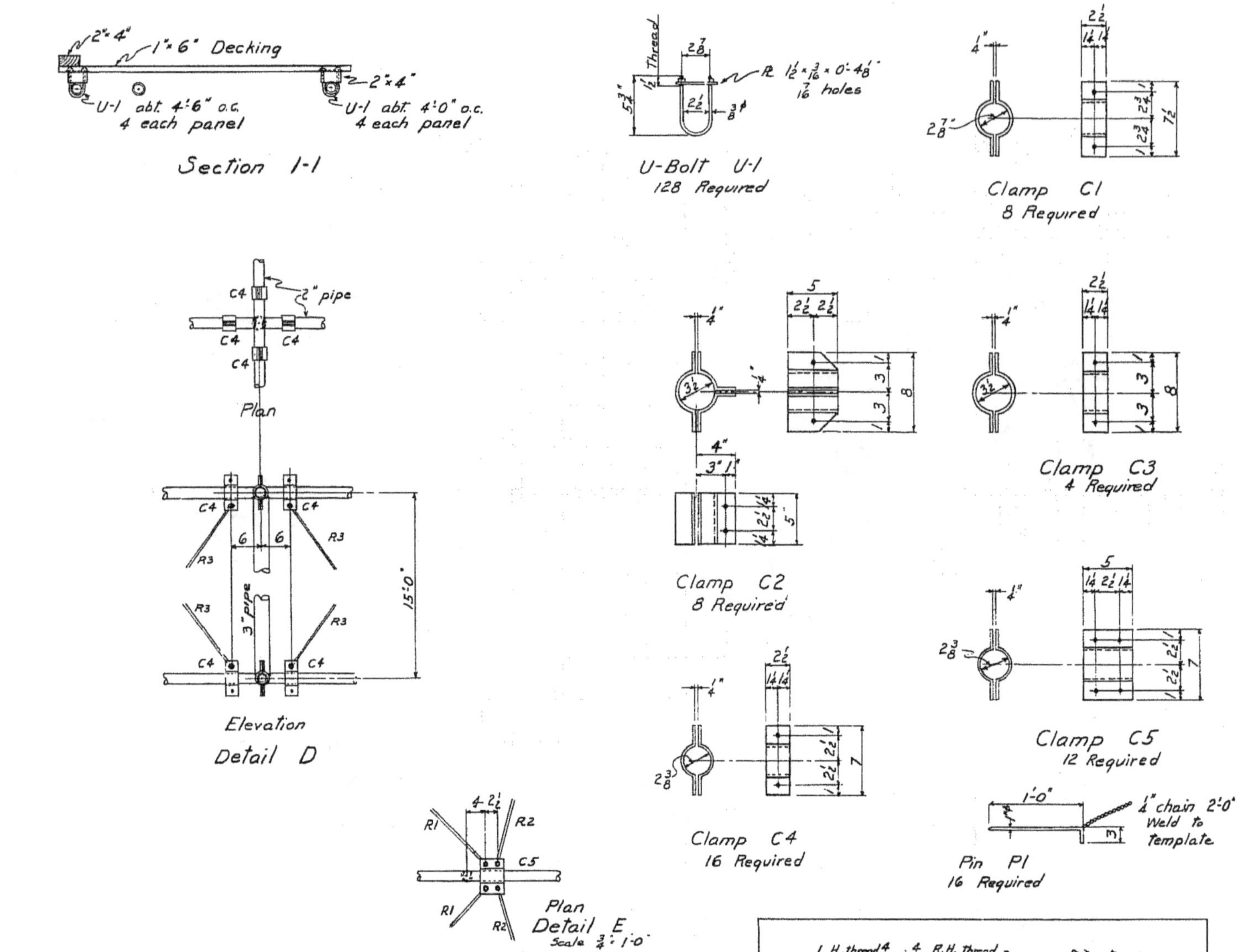

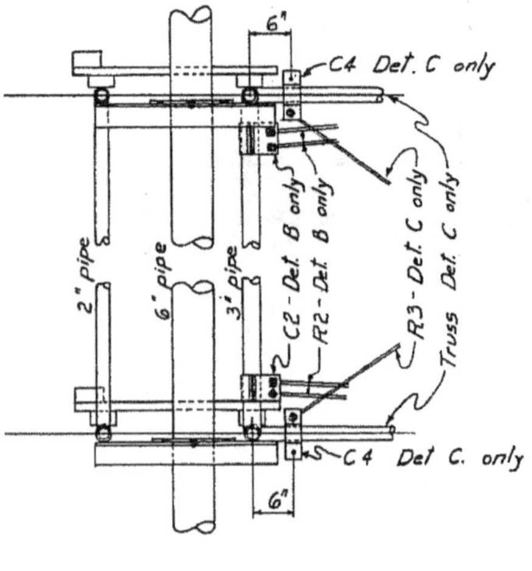

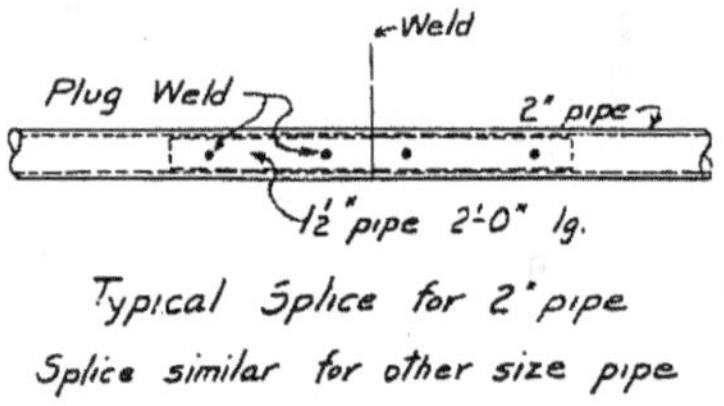

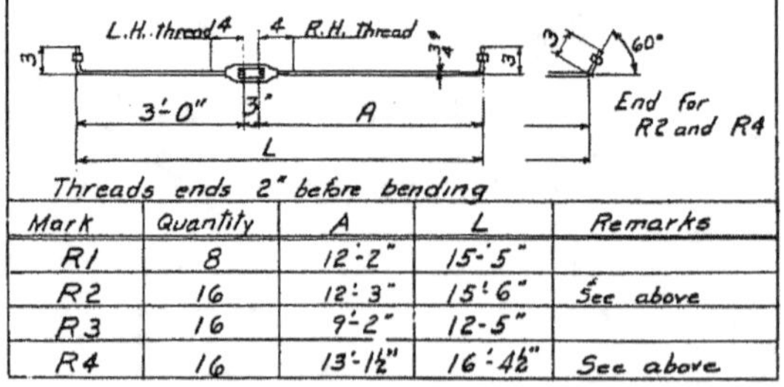

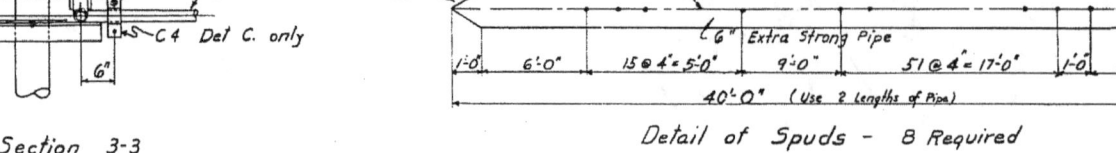

Figure 46 (Continued)
One-Piece Pipe Template for
Circular Cells - Chickamauga Dam

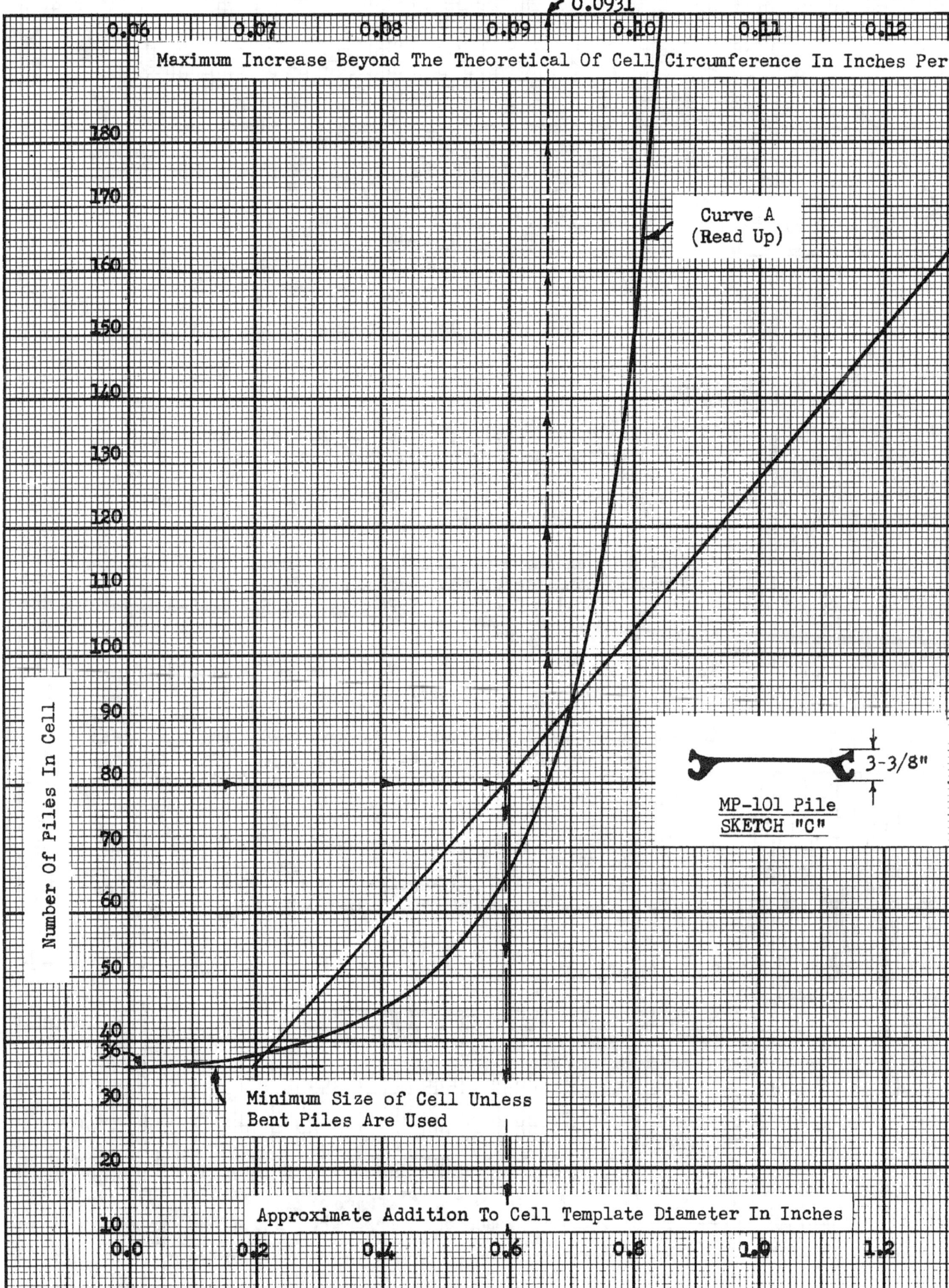

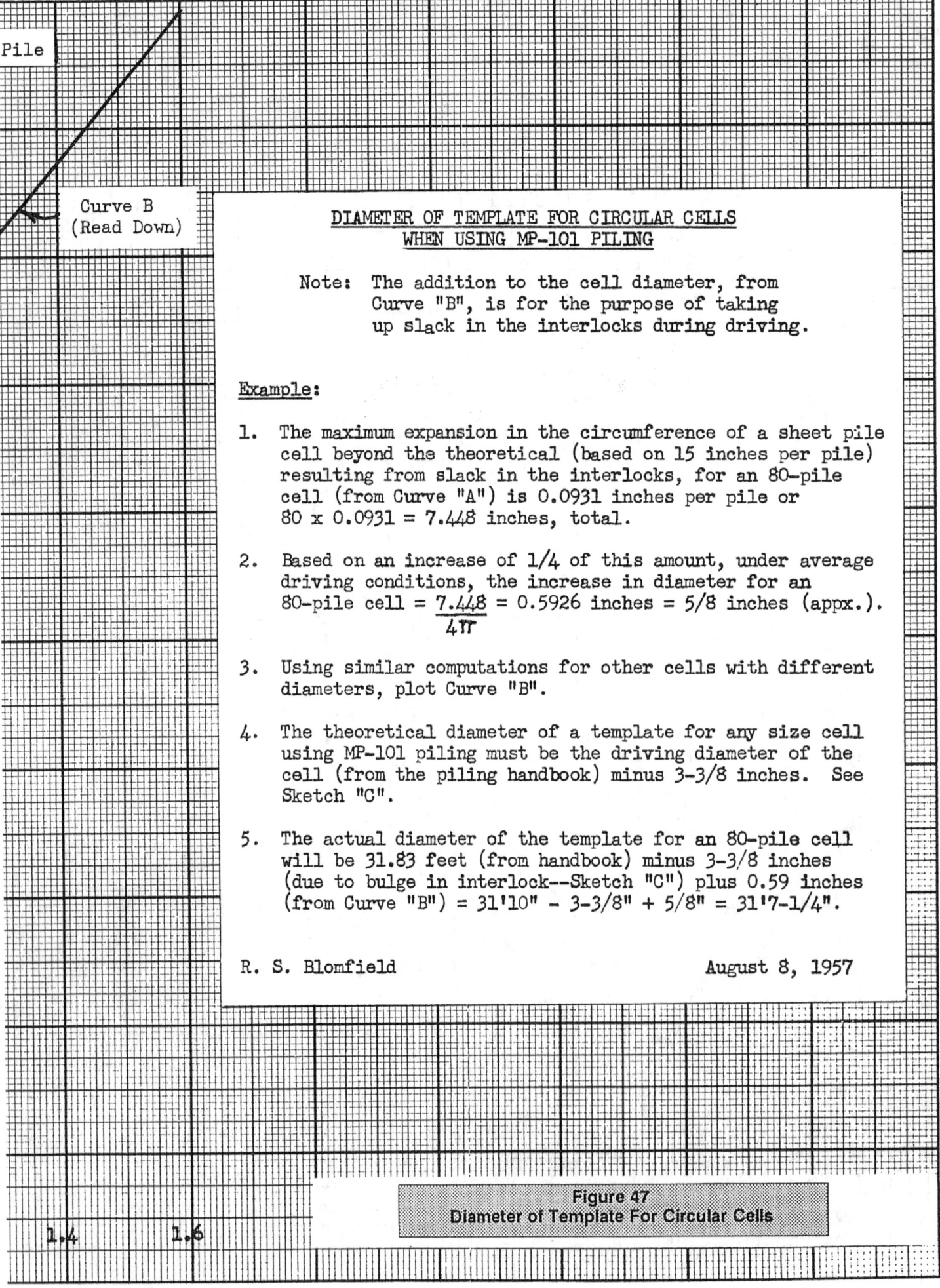

DIAMETER OF TEMPLATE FOR CIRCULAR CELLS
WHEN USING MP-101 PILING

Note: The addition to the cell diameter, from Curve "B", is for the purpose of taking up slack in the interlocks during driving.

Example:

1. The maximum expansion in the circumference of a sheet pile cell beyond the theoretical (based on 15 inches per pile) resulting from slack in the interlocks, for an 80-pile cell (from Curve "A") is 0.0931 inches per pile or 80 x 0.0931 = 7.448 inches, total.

2. Based on an increase of 1/4 of this amount, under average driving conditions, the increase in diameter for an 80-pile cell = $\frac{7.448}{4\pi}$ = 0.5926 inches = 5/8 inches (appx.).

3. Using similar computations for other cells with different diameters, plot Curve "B".

4. The theoretical diameter of a template for any size cell using MP-101 piling must be the driving diameter of the cell (from the piling handbook) minus 3-3/8 inches. See Sketch "C".

5. The actual diameter of the template for an 80-pile cell will be 31.83 feet (from handbook) minus 3-3/8 inches (due to bulge in interlock--Sketch "C") plus 0.59 inches (from Curve "B") = 31'10" - 3-3/8" + 5/8" = 31'7-1/4".

R. S. Blomfield   August 8, 1957

Figure 47
Diameter of Template For Circular Cells

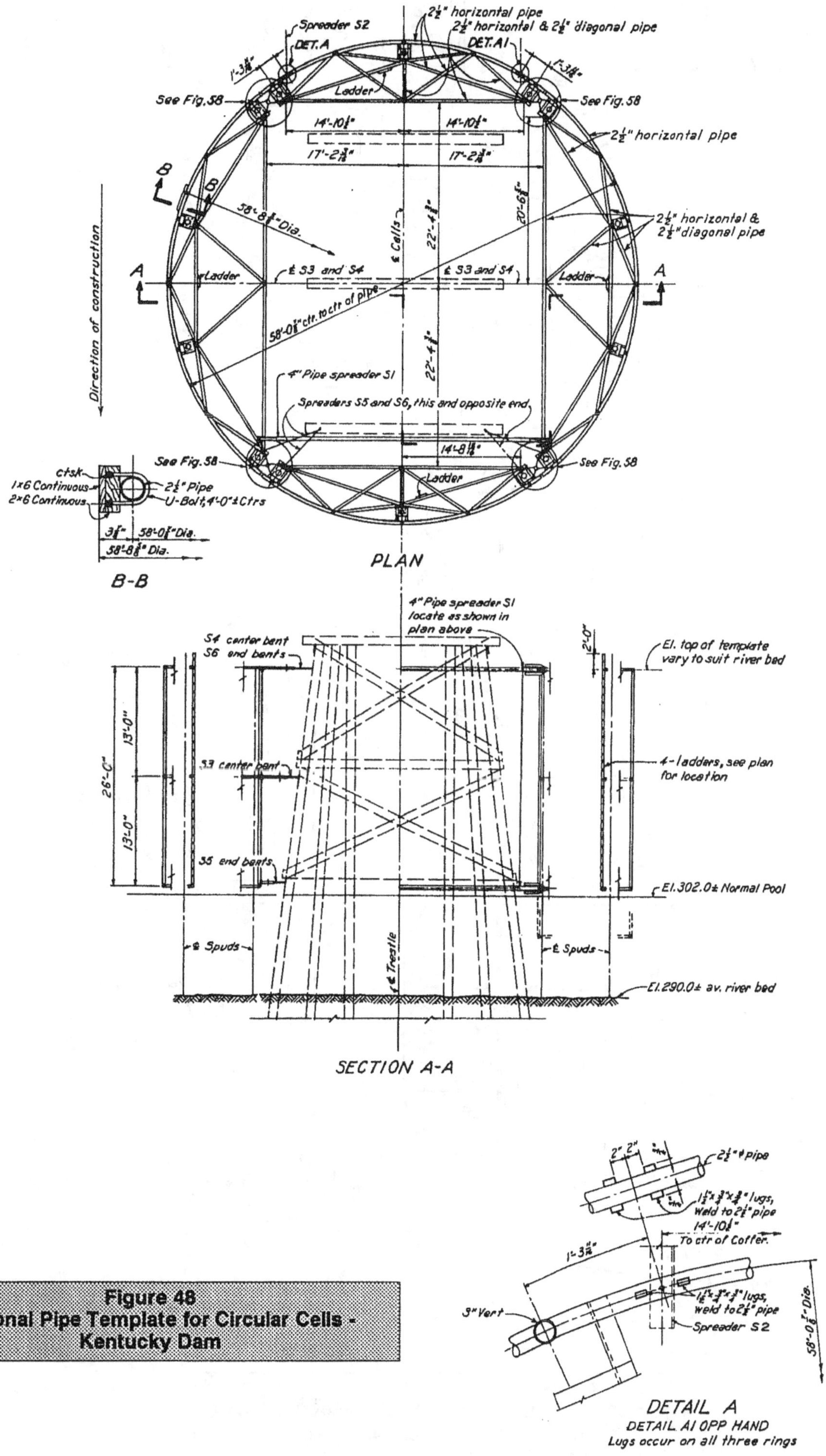

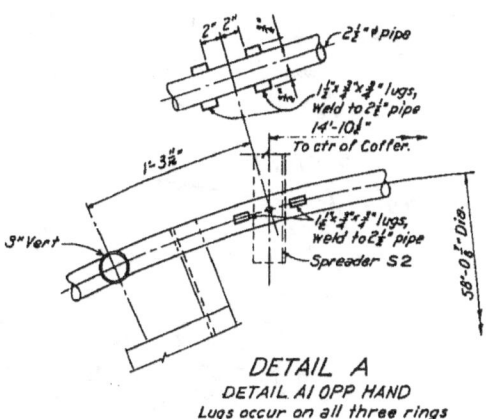

**Figure 48
Sectional Pipe Template for Circular Cells -
Kentucky Dam**

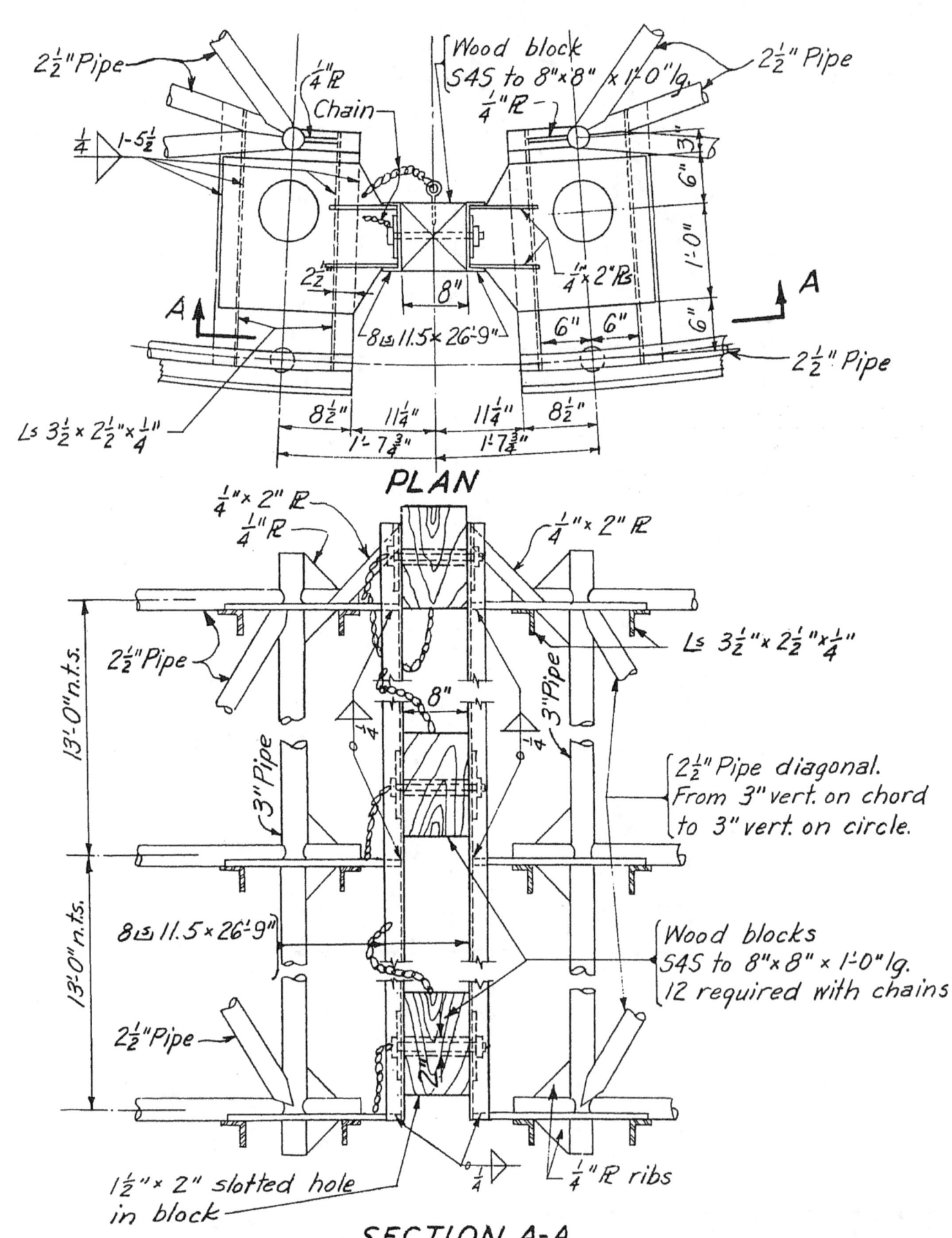

Figure 49
Details of Connection for Sections of Sectional
Pipe Template - Kentucky Dam

less for alignment and level if four spuds were dropped at a time. After the first four were dropped and adjustments made, then the other four were dropped and final adjustments made. It was found advisable to drive at least three of the spuds firmly into the riverbed in every case to assure that misalignment did not occur during pile setting and driving operations. The spuds were usually set by driving lightly with the pile hammer. Spuds were made of 6-inch extra strong pipe and were pointed at the bottom end to facilitate driving. Holes 1-1/16 inches in diameter and spaced 4 inches center to center were drilled in each spud to receive a 1-inch-diameter pin. The pin supported the template frame and served as a means of adjusting the fram for height and level.

Where little or not overburden exists for driving spuds, and where the water is swift, it may be necessary to anchor the template in position through the use of guy cables. This is especially true when setting a template for closure (see "Closures," page 109).

Connecting Cell Template--Two arcs made of 6-inch H-beams, weighing 20 pounds per lineal foot, were pin connected to each side of the template near the top and bottom rings to serve as templates for the connecting cell piling. The other ends of the beams were bolted to the tee piles in the last circular cell set and during setting operations were used to hold the template at the proper distance from the previously constructed circular cell.

Pipe Template Data--The following two tabulations give statistical data covering the 1-piece pipe templates used at Pickwick, Chickamauga, and Guntersville.

1-PIECE PIPE TEMPLATE HANDLING DATA

Project	Basis for Study	Moving & Setting Time per Cell
Pickwick	26 cells	3 hr 50 min.
Chickamauga	107 cells	6 hr 59 min.
Guntersville	150-hr time study	4 hr 12 min. per 100 tons piling

1-PIECE PIPE TEMPLATE DIMENSION AND RELATED DATA

	Pickwick Cofferdam	Guntersville Cofferdam	Chickamauga Cofferdam
Template frame diameter	58'0"	42' 1-3/4"	46' 11"
Template outisde diameter	58' 6-5/8"	42' 8-3/8"	47' 6"
Cell theoretical diameter	58' 10-5/8"	42' 11-3/4"	47' 9"
Difference in cell and template diameter	4"	3-3/8"	3"
Template weight (tons)	10	11.5	12.35; 12.20
Number of spuds	8	8	8 4

Wood Ring Template

At Guntersville Dam the overburden was thin, making the piling penetration insufficient to permit pipe template removal until after some of the cell fill had been placed. In this case the template was left in the place until part of the fill was placed, after which the template was removed. At the same time a prefabricated wood ring was placed inside of the cell near the top of the piling to give support to the cell until the rest of the fill could be placed.

Sectional All-Welded Pipe Template--Kentucky Cofferdam

Three all-welded pipe templates were used in construction of the Kentucky cofferdam. A timber trestle for supporting the pile driving equipment was located along the centerline of the cofferdam with bents inside the cells, thus preventing the use of the 1-piece pipe template. As a result each pipe template was fabricated in four sections, so designed as to fit around the trestle bents, and when assembled formed a full circle having a diameter of 58 feet 8-5/8 inches. (See figures 47 and 48). The outside diameter of the template was only 2 inches less than the theoretical diameter of the cell. This produced the effect of taking up the slack in the piling interlocks, thereby eliminating the tendency of the cells to bulge excessively when filled. Template sections were pin connected and the assembled template was supported on pipe spuds and the timber trestle.

Spreaders--Spreaders were used to connecte the template to the trestle bent. These were attached to the template frame near the upper and lower rings and were constructed by slipping a 2-inch pipe over the threads and welding it to the runner nut of a 10-ton pushpull jack. The spreaders were pin connected to the template so that when they were not in use they could hang free. This arrangement afforded a rigid method for aligning and fixing the template for position. Four spuds were required for each of the two large sections and three spuds for each of the two small sections. The spuds functioned in the same manner as for the 1-piece template.

Template for Connecting Cells--Templates for the connecting cell arcs at Kentucky were independent timber segments fabricated on a 9-foot 5-1/4-inch radius. These segments were shop fabricated from 2-inch lumber and were supported on 3- by 10-inch knee braces attached to the end piling of the timber bents.

Template Removal--These templates were removed in sections, loaded on a barge, and moved to the next position.

Sectional, All-Welded, Outside-Type Pipe Template--Hales Bar Cofferdam

A radically different type of template for racking and driving the steel sheet piling was designed by TVA's Construction Plant Branch for the upstream cofferdam at Hales Bar. The 1-piece pipe templates previously used by TVA had a maximum diameter of approximately 58 feet and depended on pipe spuds both for supporting the weight of the template and for holding it in position. The sectional template used at Kentucky Dam, also with a maximum diameter of approximately 58 feet, was supported on spuds and on the timber piling of the construction trestle. Both of these were inside templates. The cells for Hales Bar cofferdam, however, were 70 feet in diameter and the piling had to be driven in 60 feet of water into relatively shallow overburden. It would have been very difficult to hold such a template in true position by means of the pipe spuds alone, as not lateral restraint was possible along the length of the spuds except at the top. This further aggravated the problem of keeping the piling in a true vertical position during racking and driving.

Another difficulty precluding the use of standard 1-piece pipe template was the problem of removing a template of this size from the completed cell and setting it in position for the next cell. It would have been too heavy and unwieldy for practical use.

The above considerations led to the conception, adoption, and design of the outside type of template--consisting of outside and inside sections--discussed below. both sections were made up of welded pipe trusses, using 2-1/2-inch standard black steel piping throughout with but few exceptions. (See figures 50, 51 and 52).

Outer Template Section--The outer template, fabricated in four quarter sections, was octagonal in shape, measuring 83 feet from outside to outside, with the innter edge forming a circle apprixmately 72 feet in diameter. It was 25 feet in height and consisted of three decks spaced 12 feet 6 inches apart. The upper deck carried a timber platform 2 inches thick with innter edge cut on a circle of the proper diameter to permit setting the piling for a cell having a 70-feet diameter. A skirt guide set in the plane of the lower deck, together with the timber deck at the top, served as two of the three points of contact needed to keep the piling in a true vertical position during setting and driving. The skirt guide was made from a 13- by 1/4-inch steel plate and was attached at regular intervals to brackets welded to the inner chord of the outer template. It was set at a 45° angle downward so that as the piling was lowered it would be deflected toward the proper position. A 1-1/4-inch reinforcing bar was welded to each edge of the plate to give it stiffness. This gave a skirt plate that was sufficiently rugged to absorb the shock resulting from the piling striking it as it was lowered into position.

The template quarter sections each weighed approximately 8 tons. This permitted the template to be disconnected and floated in sections on pontoons to the position of the next cell. The quarter sections were joined near the inner and outer edges of the pipe fram. The pipe key for joining the sections was made from 1-1/2-inch pipe filled with grout to give it additional stiffness. A 3 inch hole through a plate at the upper deck and short sections of pipe near the middle and lower deck served as guides for the 25-foot-long pipe key. A stop at the bottom prevented the pipe key from passing on through. The sections wer disconnected simply by lifting out the pipe key that held them together. After piling had been racked and the cell fill partially placed, the pipe pins were removed and the template sections floated to the next position. Where there was enough room to operate, the template was removed in half sections instead of quarter sections.

Each quarter section was supported on eight pontoons spaced so that it would float as a single unit. The pontoons were 5 feet long by 3 feet 6 inches in diameter and were fabricated from 14-gage well casings. The desired buoyancy (1 foot 2 inches to 2 feet 6 inches) was obtained by adding or removing water through the valve provided in each pontoon. (See figures 50 and 52 for position of pontoons.) It should be noted here, however, that the vertical position of the template was maintained through the use of the pontoons, with the pipe spuds being used to support the pipe template in the level position.

Four spuds of 6-inch-diameter extra heavy pipe were used to anchor the outer template in lateral position. After the spuds were dropped the template was again checked for alignment and then leveled up and the sections joined together. Special guides and winches (described later) were installed for positioning and drawing the sections together. The winches were also used to take the pressure off the pipe pins so that they could be removed easily.

The spuds were located at the midpoint of each quarter section, just inside the outer edge of the frame. A 7-1/8-inch-diameter hole through the gusset plate at that point, at each of the three deck levels, served as a guide to hold the spuds in a true vertical position. In some instances, when working in turbulent water, etc., it was necessary to anchor the template in position laterally with wire cabes and floating timber struts.

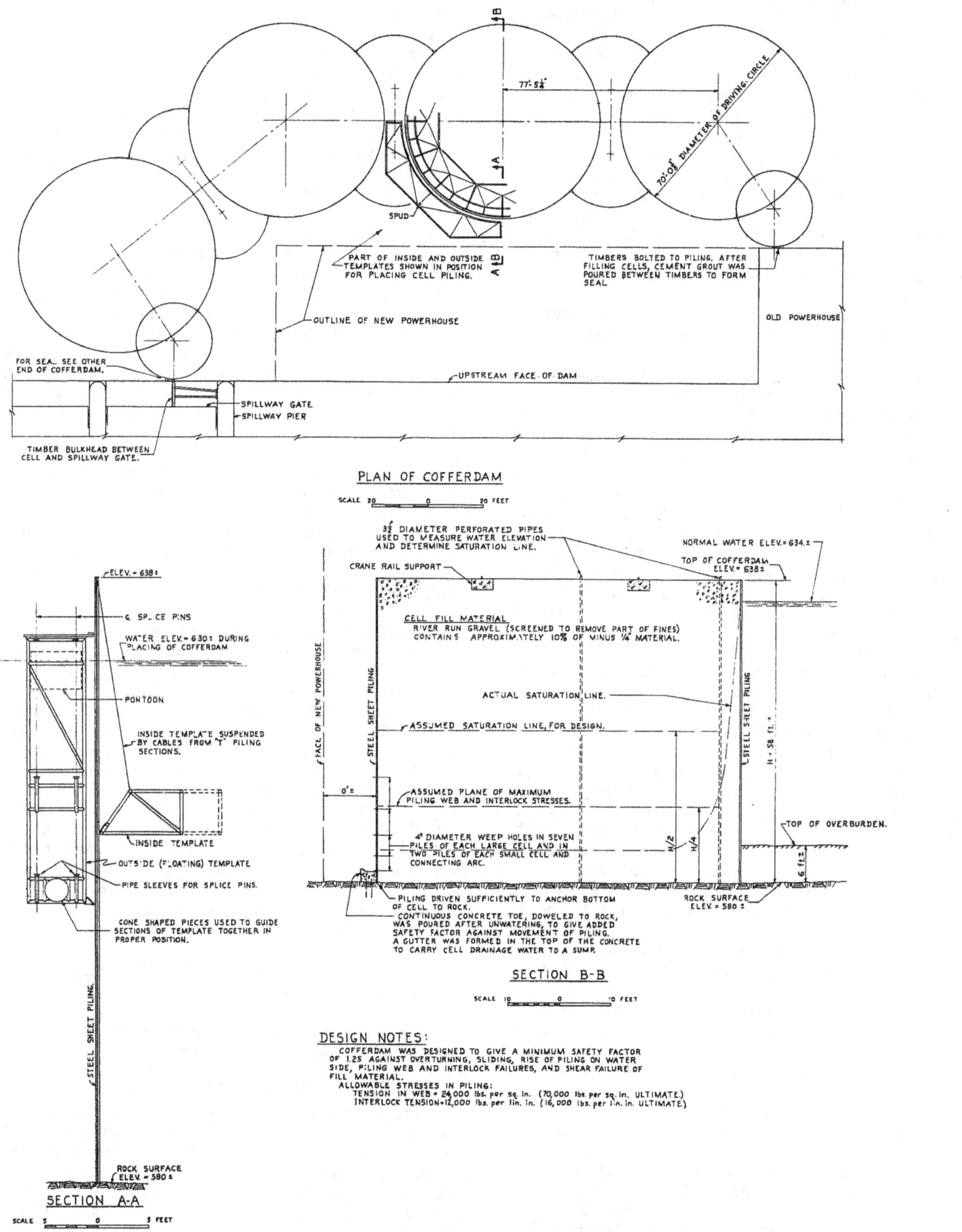

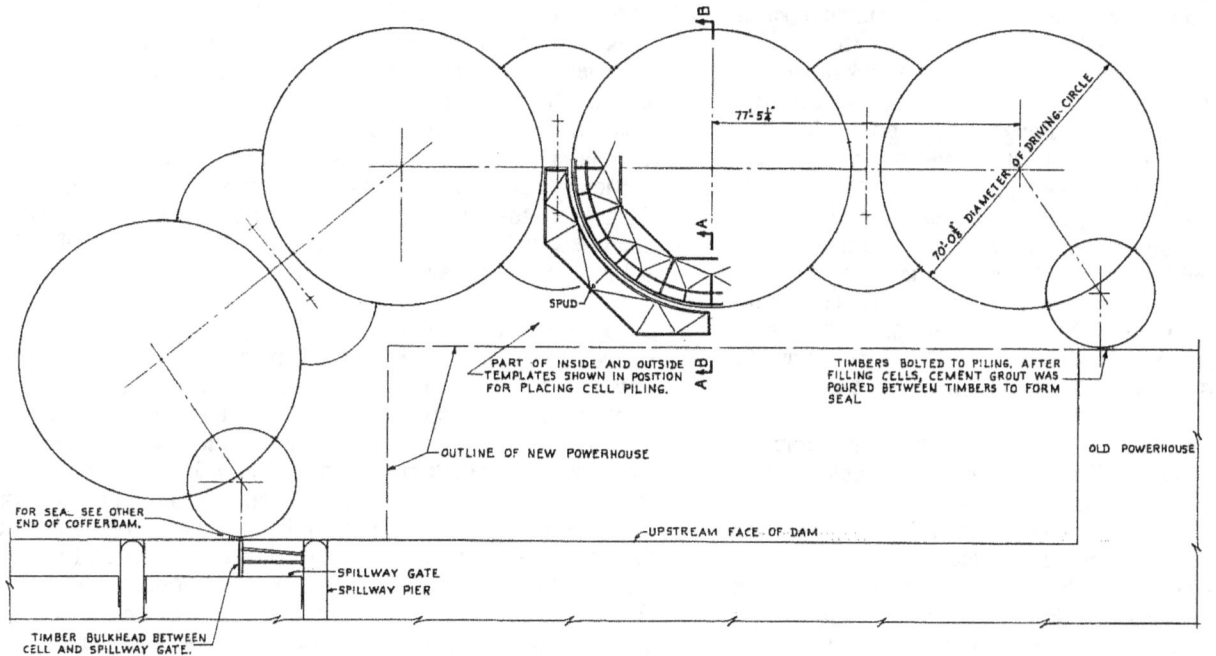

**Figure 50
General Arrangement of Inside-and-Outside-Type
Pipe Template - Hales Bar Dam**

Inner Template Section--The inner template was made up in one complete unit. It was made up of two horizontal decks spaced 4 feet apart. In plan the two decks were perfect circles along the outer edge, with the circle for the top deck having an outer diameter of approximately 64 feet and the circle for the outer edge of the bottom deck having a diameter of slightly less than 69 feet 6 inches. The inner template was an 8-sided figure with each side about 19 feet in length (See figures 51 and 52).

Short lengths of 12- by 1/4 inch plates were welded completely around the outer diameter of the inner template to form a skirt or fender that would deflect the piling into the space between the two templates. The outer perimeter along the bottom plane served as the third contact point to support the piling in the true vertical position. This inner template weighed only 5-1/4 tons and little difficulty was experienced in handling it.

The large open space in the middle of the inner template facilitated placing the fill. The inner and outer templates were both removed as soon as sufficient fill had been placed to render the cell stable. The balance of the fill was then placed while the next cell was being constructed.

Special Features--One of the unique features of this outside Type template was the tee guides attached along the inner face of the outer template section. (See detail "B," figures 51 and 52.) Several of these were required since the tee piles did not come in the same place for all of the cells. By using these guides the tee piles for a cell could be set prior to racking the piling. And, since the guides held the tee piling perfectly plumb, the only thing left to be done was to drop the piling into place. The tee guides were spaced so that the intervening space between tees was correct for receiving the proper number of piling; and since the tee were plumb, there was no wedging of piling or damage to interlocks due to misalignment. The time saving element due to this one feature alone was sufficient to warrant a considerable expenditure for template construction.

Another special feature of this template was the take-up for the outer template frame. As is normal in this type of cell construction, the cells stretch due to the play in the interlocks when the cell is filled. This would normally have pressed the walls outward against the outer template and caused the pipe keys joining the quarter sections to jam, thereby making removal impossible. The take-up was a simple hand-operated winch located on the tope deck at the junction point of each two quarter sections. Two take-ups were used at each location, one near the outer edge of the outer template and the other near the inner edge. Cables running diagonally downward from the winches to the bottom deck passed through a system of shackles so that by merely turning the crank on the winch the sections could be drawn evenly and snugly together. This was done before any piling was set. Then, after the piling had been set, driven, and the cells partly filled, and after the resulting expansion had taken place in the cell due to the slack to the interlocks, the winch take-ups were released. This allowed the outer template to move away from the cell, thereby relieving any binding in the pipe keys at the quarter section junction points.

Template Setting--The sections of the outer template were floated into position, connected, spotted for position, and anchored in place by means of the pipe spuds. The tee piles were then set and driven to rock. Extreme care was taken to see that the template was accurately located for each cell. In some cases the template had to be secured firmly in position through the use of wire cable sand floating timber struts. The inner template was then lowered into place, leveled, and anchored in position by means of wire cables attached to the top of the tee piles already driven. The rest of the piling was set and driven, the cell filled to the bottom of the inside template, and the inner template removed. The cell was then completely filled. The outer template was removed just as soon as the cell was sufficiently stable to permit its removal.

Cloverleaf Template--Kentucky Cofferdam

The cloverleaf cells in the Kentucky cofferdam were constructed by using built-in-place timber templates in combination with sections of the standard circular cell templates. These standard sections were used without difficulty, as the radius of the cloverleaf arcs was just 3-1/4 inches less than the radius of the standard cell. Templates for the cross walls consisted of two guides, each made of 2- by 12-inch timbers spaced at 26 feet vertically.

Timber Templates

Normally the prefabricated pipe template will serve satisfactorily for the bulk of the cell construction, but built-in-place and prefabricated timber templates will also be required. These will be used for tie-ins, closure, and other odd size cells. Timber templates may be used where the number of cells of a given size is too small to warrant the fabrication of a pipe template. The use of built-in-place templates should be held to a minimum since more time is required for their construction than is the case for the shopbuilt template. Templates left in place to support the cell until the fill material can be placed will be buried and lost so far as salvage goes. For built-in-place templates which may be removed before the fill is placed, shop-fabricated segments or sections may be used to save time and duplication in construction.

Timber Templates--Kentucky Cofferdam--At Kentucky cofferdam three prefabricated timber templates were used in addition to the three all-welded pipe templates. the first timber template was constructed for use until the first pipe template could be completed. A second template, having the same radius as the standard pipe template, was built in the field around the first timber pile trestle bent in stage 2 and consisted of shop-built segments fabricated from 2-inch lumber and supported on 3- by 10-inch knee braces fastened to the trestle pile. The third timber template built was a completely assembled floating circular template used in the construction of the closure cell in stage 2 cofferdam. This template was built on the riverbank and floated down river to position. The base of the template was a large circular platform built on 12- by 12-inch timbers covered with 2-inch boards. The radius was the same as that of the pipe template and carried a horizontal ring of equal diameter supported on timber bents approximately 11 feet high. The upper ring was made up of segments cut from 2-inch-thick lumber and carried a handrail for security of the workmen. The timber bents were cross braced for rigidity.

HANDLING AND SETTING STEEL SHEET PILING

TVA has found it very helpful on most jobs to bill the piling according to the lengths required in each cell. This facilitates ordering, scheduling shipments, and placing of piling in the cells. (See table 3, for typical billing and piling).

Piling is generally brought to the point of construction on barges. Considerable time may be saved if barges are loaded with the proper lengths of piling to accomplish the job. Prepared charts showing the length of piling required for each cell and the combination of available lengths to meet these needs should also be prepared for cells to be build from reused piling.

Reused piling should be subjected to rigid inspection and servicing for length, alignment, damaged interlocks, and for interlocks filled with sand, gravel, and dried mud. Gradation for length, accomplished during this servicing operation, will help to minimize multiple handling and splice welding later.

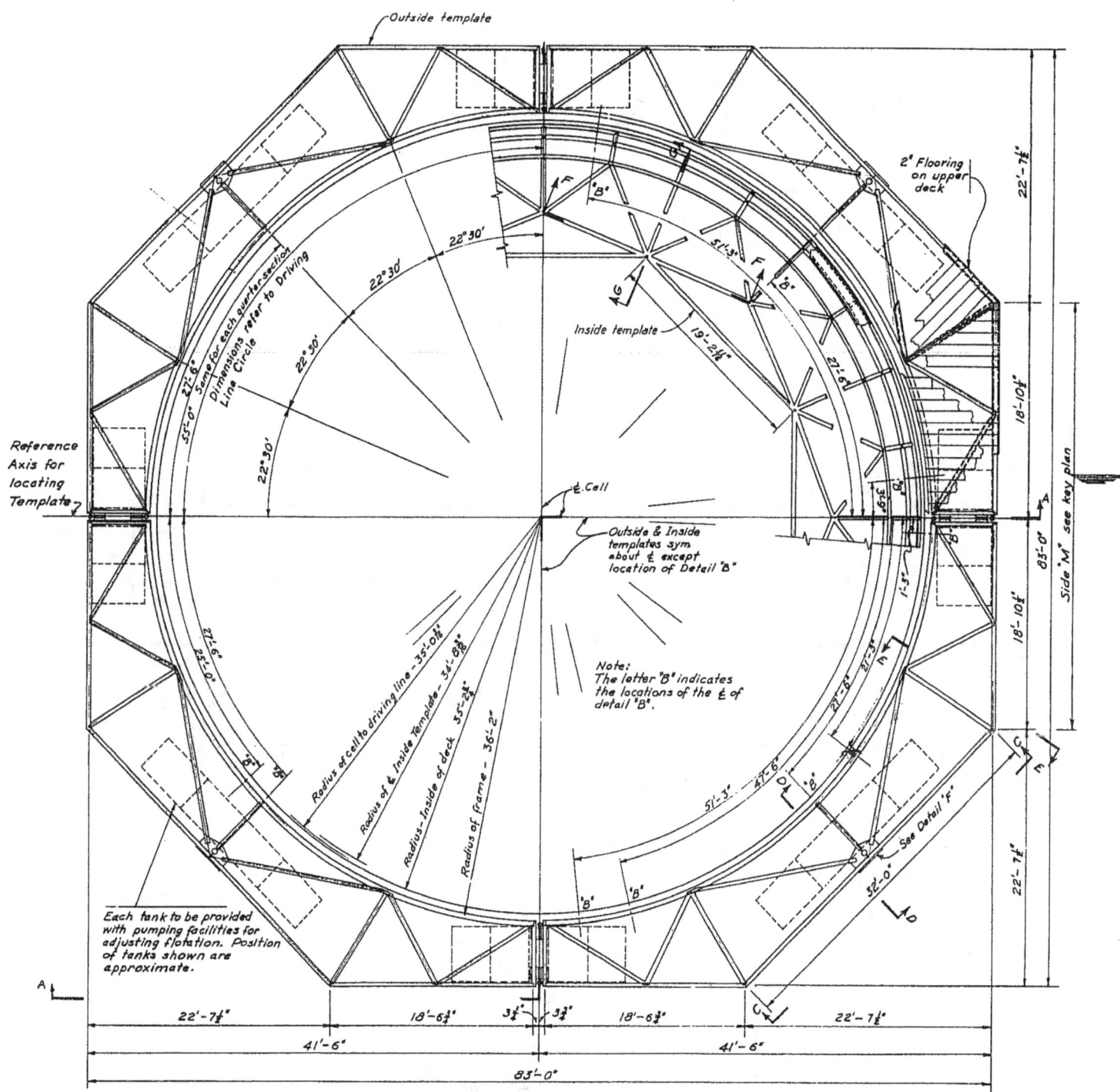

PLAN

**Figure 51
Details of Floating, Sectional, Outside-Type Pipe
Template - Hales Bar Dam**

KEY PLAN

SECTION A-A

ESTIMATE OF WEIGHT

Outside template (Quarter section):

Steel -	5.61 T.
Timber -	1.4
Min. lifting wt.	7.01 T.
Spuds	.74
Tee guides (℄)	.5
Total	8.25 T. (Total for ¼ section)

Inside template:—

Total	5.3 T.

GENERAL NOTES

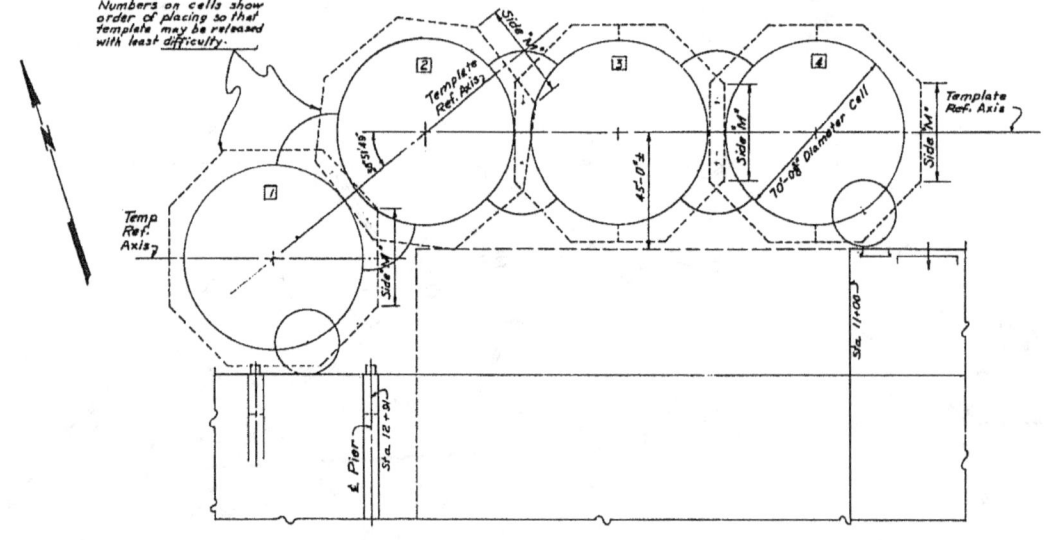

Figure 51 (Continued)
Details of Floating, Sectional, Outside-Type Pipe Template - Hales Bar Dam

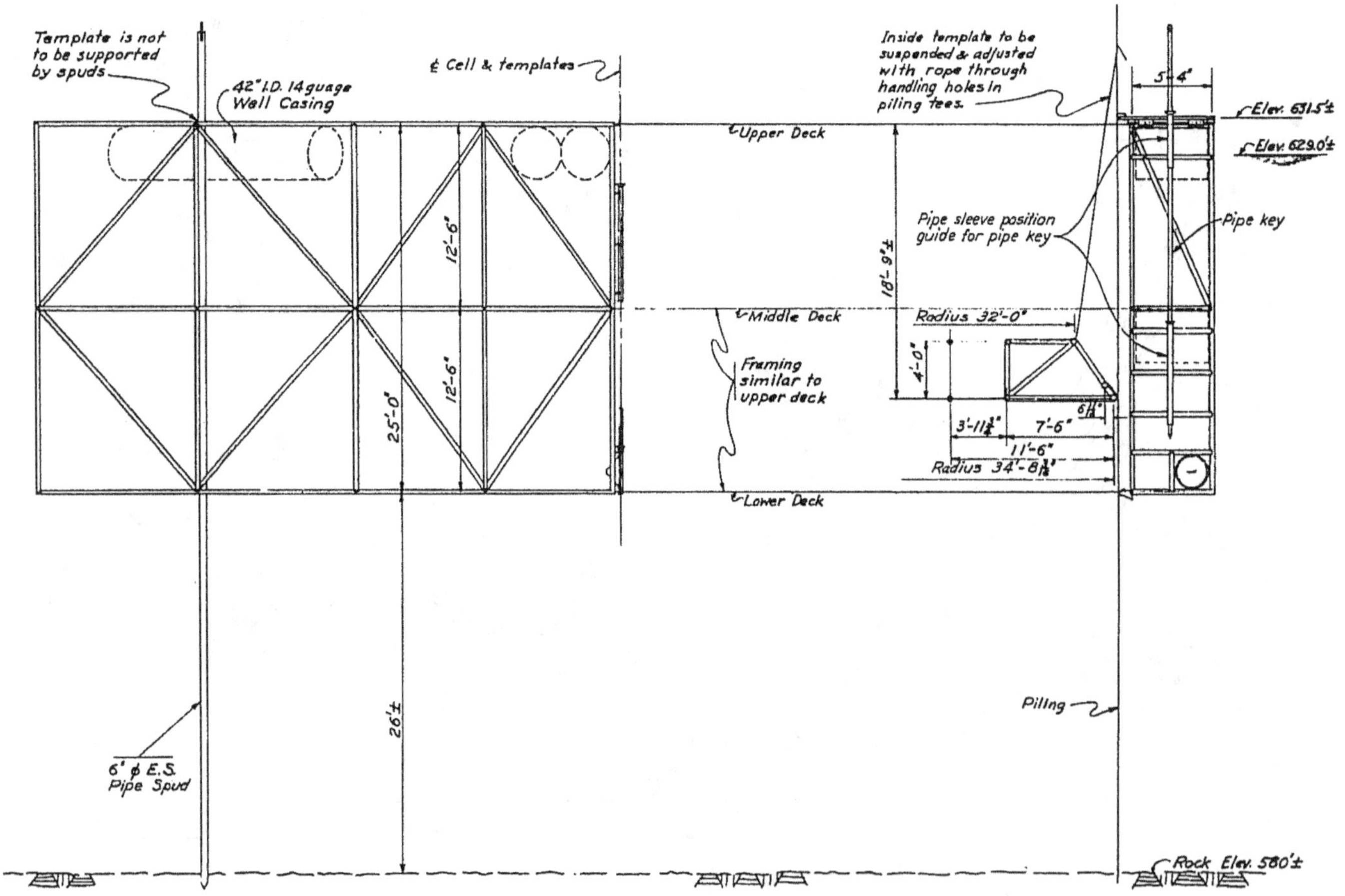

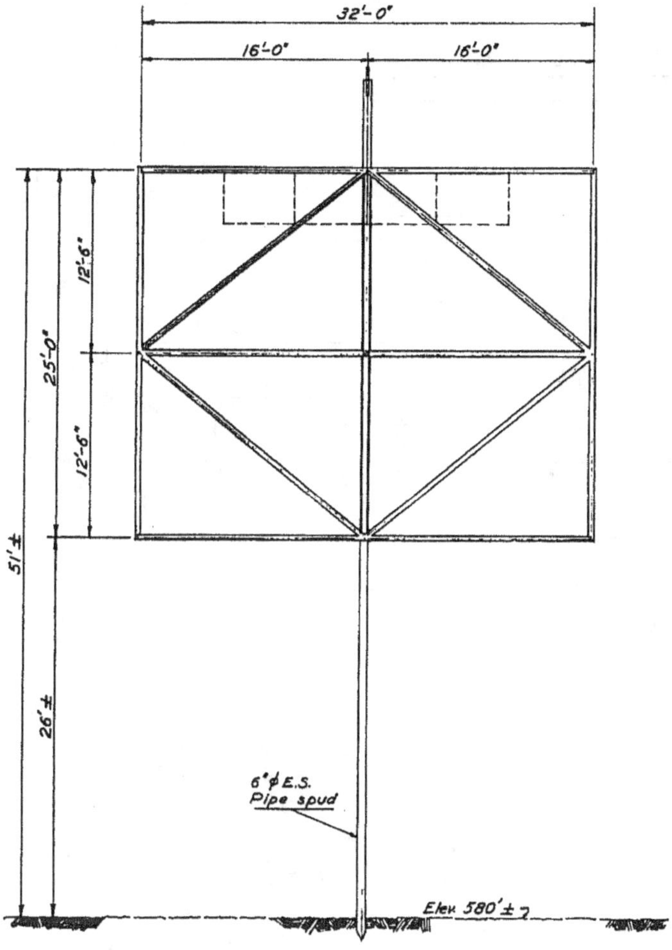

SECTION C-C

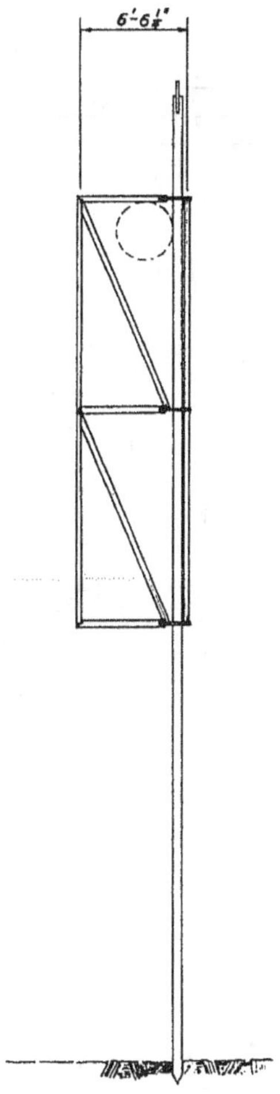

SECTION D-D

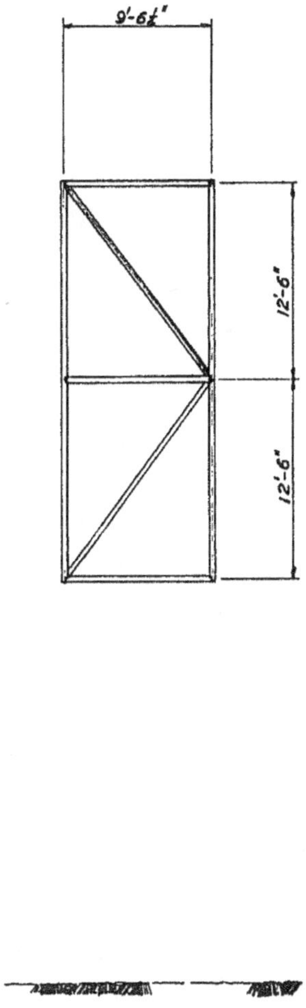

SECTION E-E

PLAN

ELEVATION

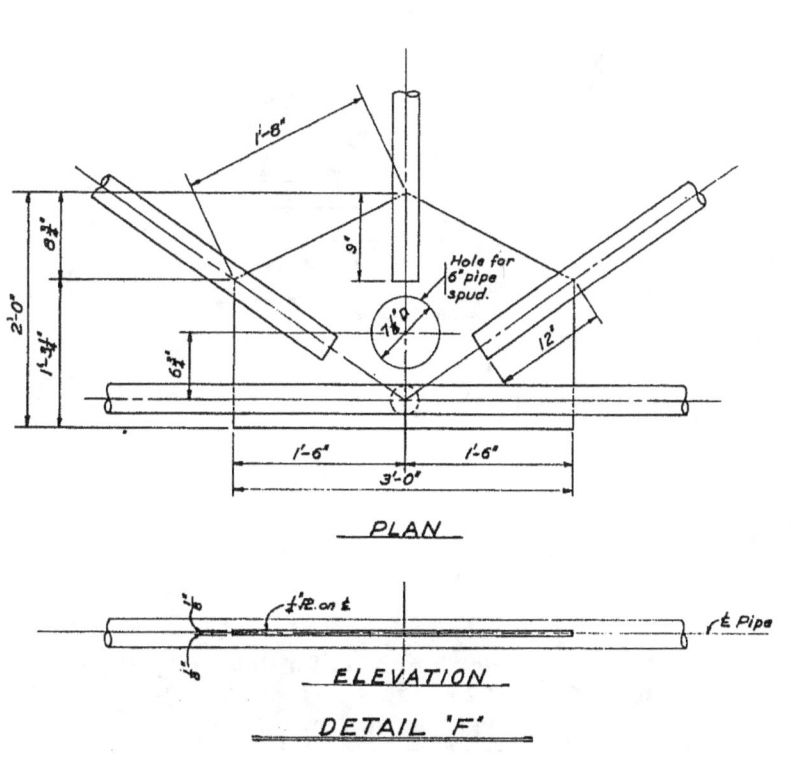

DETAIL "F"

Req'd. on all decks.

ELEVATION

PLAN

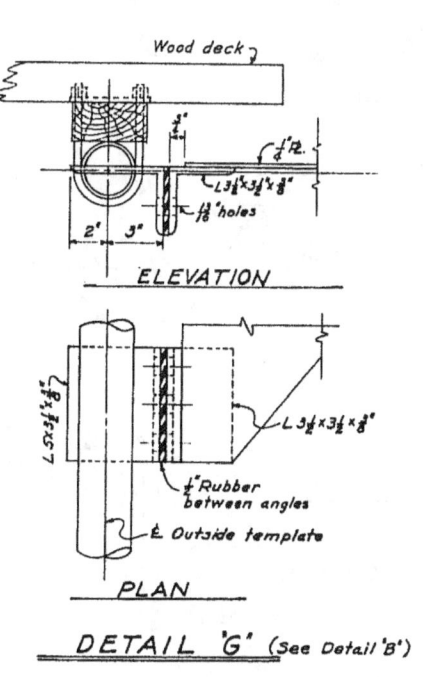

DETAIL "G" (See Detail "B")

2- Req'd at each
deck level.

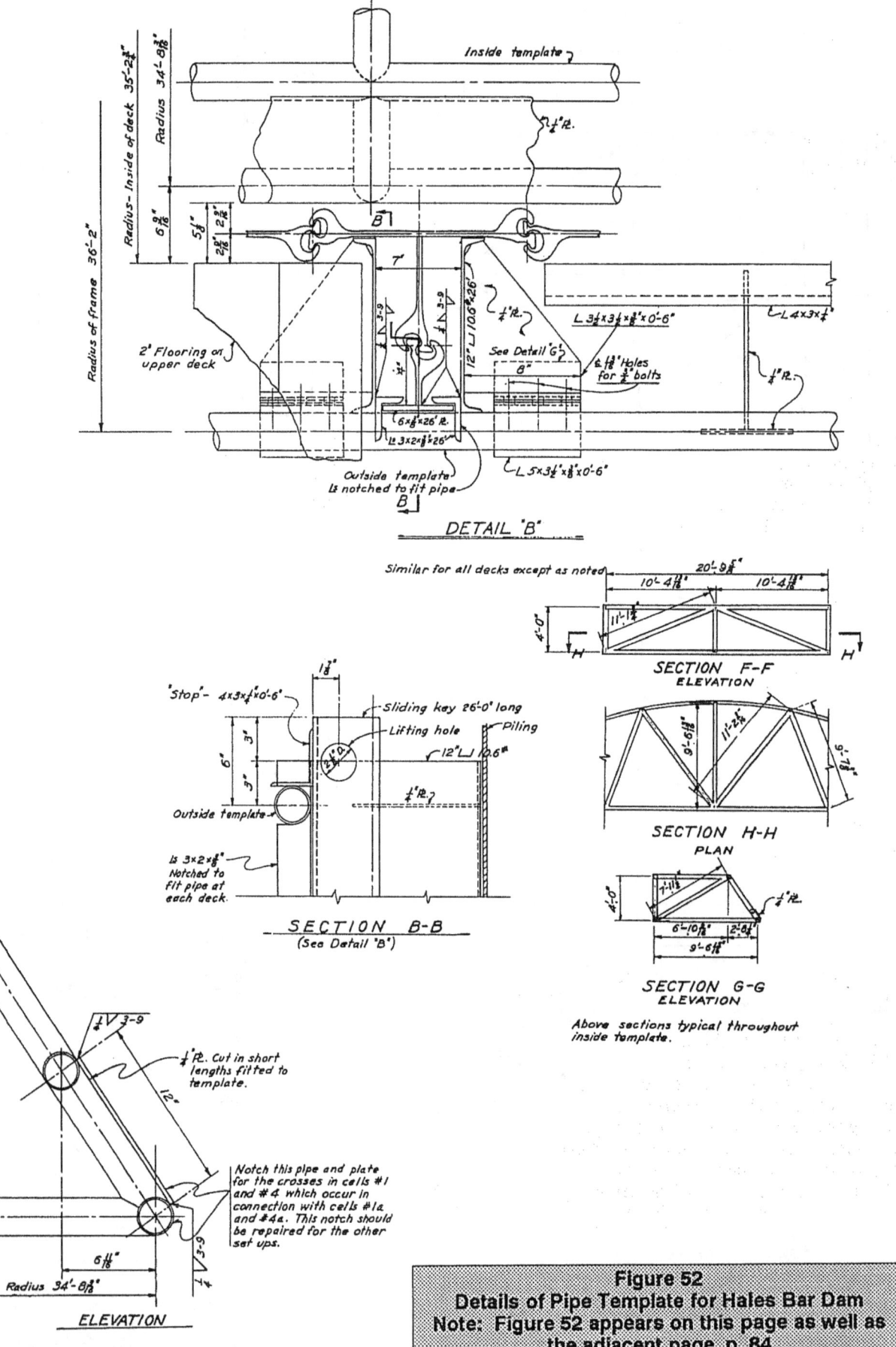

Figure 52
Details of Pipe Template for Hales Bar Dam
Note: Figure 52 appears on this page as well as the adjacent page, p. 84

Warped piling makes threading of interlocks difficult and will slow up the setting operation.

Nail or paint marks along the outer edge of the upper and lower walkways of the template, as markers for the position of each pile, will aid in a quick check of the piling for position and plumb.

Pile Setting

The majority of the piling set on TVA projects was handled with floating rigs. Where the trestle is used, as was the case at Kentucky cofferdam, a revolving crane operating from the trestle may be used. Where the cells are constructed and filled before proceeding with the next cell, the skid rig may be used by operating from the top of the completed cell. As piling is set by any of the above pieces of equipment, it is guided into position and the interlocks engaged by a workman standing in iron stirrups atop the last pile set.

Order of Setting Piling

The setting, or racking, of piling in a cell must be completed before driving is started. The tee or wye piles are the key piles in the cell, and must be located accurately for psition and plumb to avoid difficulty with tie-in to connecting walls. Racking piling alternately between the two sides to bring them to completion simultaneously is preferable since racking in one direction around the entire cell will tend to magnify any accumulation in lean that may develop. The process to be followed, otherwise, must be the one that will satisfy the job conditions and permit efficient operation of the driving equipment.

Pile Setting Procedure on TVA Projects

The procedure followed at Pickwick, where the 1-piece template was used, was to start with the connecting cell walls at the tee connection in the last cell constructed. Racking proceeded along the two connecting cell walls to the tees in the cell under construction, then along each side of the cell in the direction of construction, to closure on the far side. Next the connecting cell guides were removed and the piling between the tees was racked (see figure 53-a). Where two setting rigs were available the operation progressed along the two sides of the cell simultaneously. Where only one rig was available the operation progressed in the same manner by racking alternately along the two sides of the cell. In order to hold the piling in place until cell closure could be completed, approximately every tenth pile was tied to the center of the template with a block and tackle guy.

At Fort Loudoun cofferdam, where many of the cells were driven on dry land or without too much interference from stream flow, the method of racking in one direction around the entire cell was followed without encountering difficulty.

At Guntersville cofferdam, on the other hand, the lock cofferdam cells were built from upstream to downstream with the racking in each cell starting at the upstream side. As racking progressed, however, the piling had a tendency to lean downstream due to the current of the stream. To correct this the procedure was changed, with the piling in the connecting cell and the upstream half of the cell being racked first, after which racking was starting with the two downstream tee piles. Racking proceeded thereafter in each direction from these two tees until cell closure was completed. The connecting cell template beams were removed next and the segment of the cell between the two tees set last. This method (figure 53-b) was used very little, however.

At Watts Bar cofferdam the piling was racked in the order shown in figure 53-c.

At Kentucky cofferdam, however, construction was accomplished largely from a construction trestle. Here trestle construction, template erection, and pile setting and driving were accomplished more or less on the production line basis with the equipment working in tandem (see figure 44).

The trestle construction was normally kept several cell lengths ahead of cell construction. Pile and template setting was then accomplished by a trestle mounted Clyde whirley electric revolving gantry crane working from a position ahead of construction. This crane was assisted in this operation by floating equipment. For a complete description of the operation cycle at Kentucky cofferdam see "Sheet Steel Pile Driving and Driving Equipment," page 95, and figure 53-c.

In constructing the upstream cofferdam for the Hales Bar powerhouse addition, the floating outside template provided an effective means of combating the tendency for the piles to lean. All piling between tees slipped into position without binding, thereby eliminating closure difficulties so frequently encountered with piling that is not plumb. The template had built-in guides for all tees and crosses to hold in exact position and plumb. All of the tees and/or crosses were set first and the straight piling was then racked to fill in the gaps between. No specific order of racking was required. the piling in the connecting cell was racked up last. Difficulty in racking the connecting cells was experienced in only one instance, and this was no doubt due to deflection in piling alignment caused by difficult driving through the overburden. The piling in this instance had to be pulled and redriven. A considerable amount of construction trash was present in places, causing the piling in the cells and arcs to drift during driving. Under conditions of this kind the plan of making the arcs a little longer than the theoretical to provide sufficient slack in the piling to accommodate some drifting should be considered. The allowance for drift should not be construed to mean, however, that laxity is permissible in template location. The greatest care should be taken to locate the main template accurately.

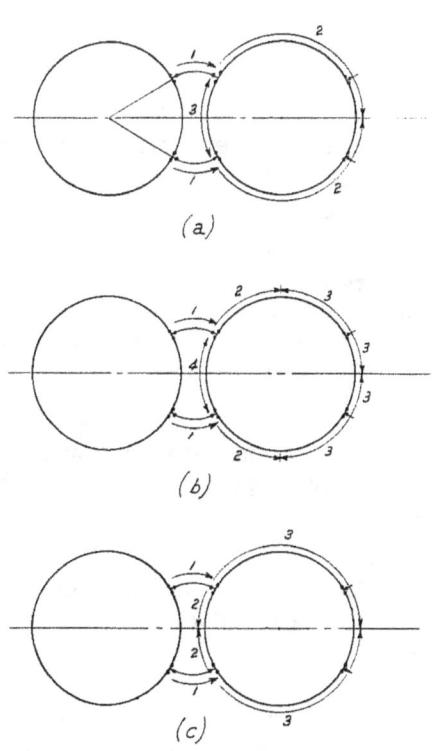

**Figure 53
Procedure for Racking Piling for Inside-Type Template Shown in Figures 55 and 57**

Table 4
Summary for Time Study for Cofferdam Construction at Chickamauga, Guntersville, and Pickwick Landing Dams

General Data	Chickamauga No.	Chickamauga Diameter	Guntersville No.	Guntersville Diameter	Pickwick No.	Pickwick Diameter
Number of cells in time study	38	47' 09"	37	42' 11-3/4"	16	58' 10-5/8"
Tie-in walls	2		2		-	
Tie-in cells	-		-		1	
Total	40		39		17	

	Chickamauga	Guntersville	Pickwick
No. of piles in time study	5152	4740	2856
Average length of piles	42.7	40.2	51.0
Total tons (piles used)	4637	3753	2830 (approx)
Average depth of penetration (feet)	13.5	5.8	11.0

Time Study	Chickamauga Hours per Cell	Guntersville Hours per Cell	Pickwick Hours per Cell
Moving and setting template	7.55	4.04	3.84
Setting piles	13.74	8.50	10.00
Driving piles	13.83	5.87	4.50
Splicing piles	11.40		
Moving rigs	3.77		
Delays: Rig	3.86		
Hammer	0.32		
Miscellaneous	3.93		
Idle	0.40		2.72
Total	58.80*	18.41**	21.06

*Extremely difficult foundation and driving conditions resulting in a pile penetration in excess of the expected, requiring splicing of piles and a longer construction period.
**Includes delays.

Pile Setting Crews

Normally the same crew assigned to a rig for setting piling also remained with that rig throughout driving operations. the number of persons to be assigned to these crews was determined independently for each job and they were selected to best fit job requirements. One pile setting crew, exclusive of welders and personnel required to operate and maintain the setting rig and other floating equipment (for four TVA jobs), is shown below:

	Kentucky (One Rig)	Pickwick (Two Rigs)	Guntersville (Two Rigs)	Chickamauga (Two Rigs)
Forman	1	1	1	1
Carpenters		8*	-	-
Laborers	1		-	-
Riggers	-		8	10	10
Rigger Helpers	-		5	4	4
Inspectors		1	1	1	1
Total	11	15	16	16

*Used instead of riggers because of jurisdictional agreement between unions.

TIMBER TRESTLE

Floating and land based equipment was sufficient to meet construction needs on all TVA steel sheet pile cofferdam construction except at Kentucky Dam. Here the depth of water, pile penetration, and magnitude of the cofferdam made it obvious from the start that ppile driving was going to be the bottleneck in cell construction. To gain a height that would permit greater ease in handling the 75- and 80-foot lengths of piling and provide working space for enough equipment to expedite cell construction, a plan was adopted whereby land equipment, mounted on a high trestle, would be used for handling the bulk of the pile setting and driving (see figure 54).

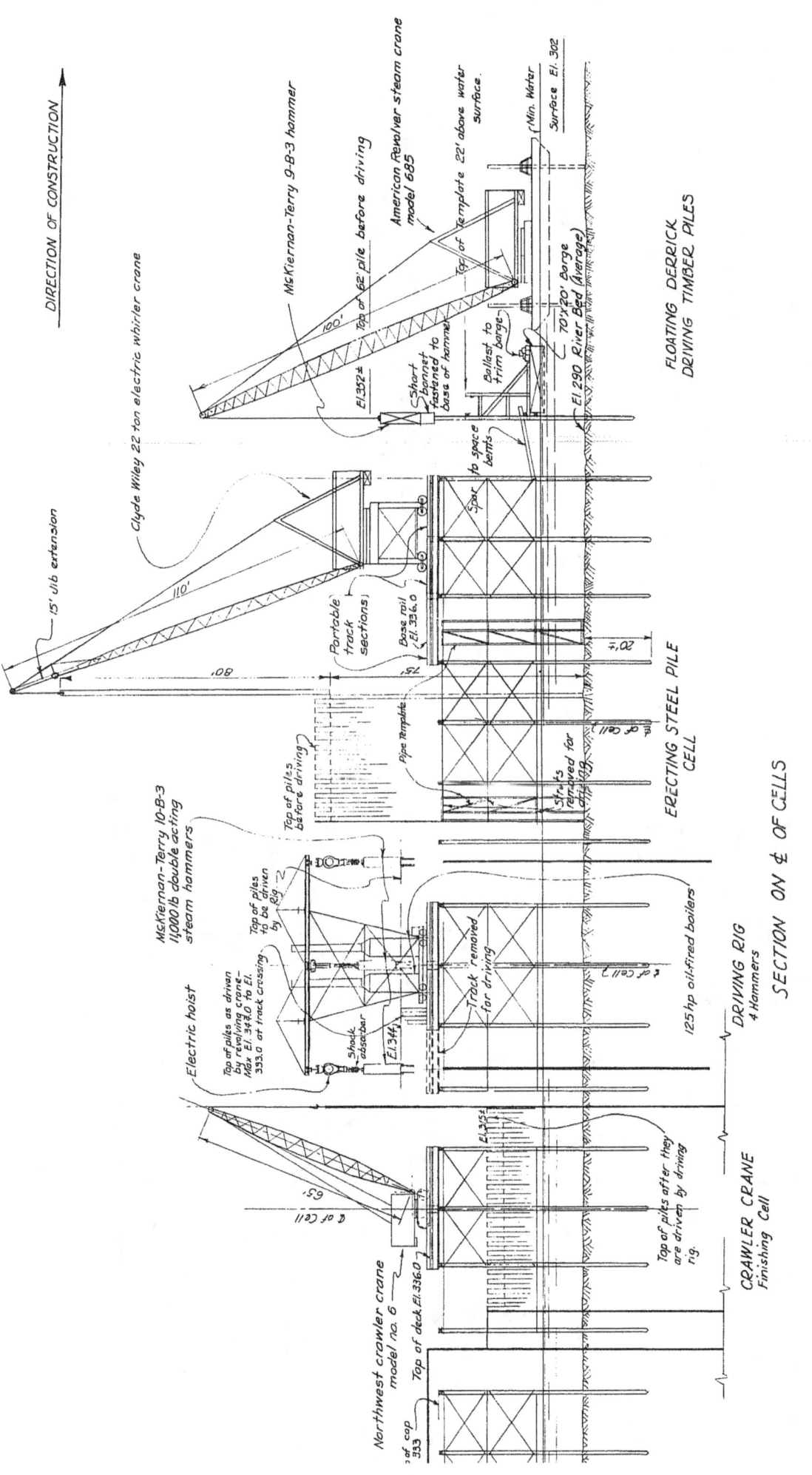

Figure 54
Methods Used in Setting and Driving Steel Sheet Piling - Kentucky Dam

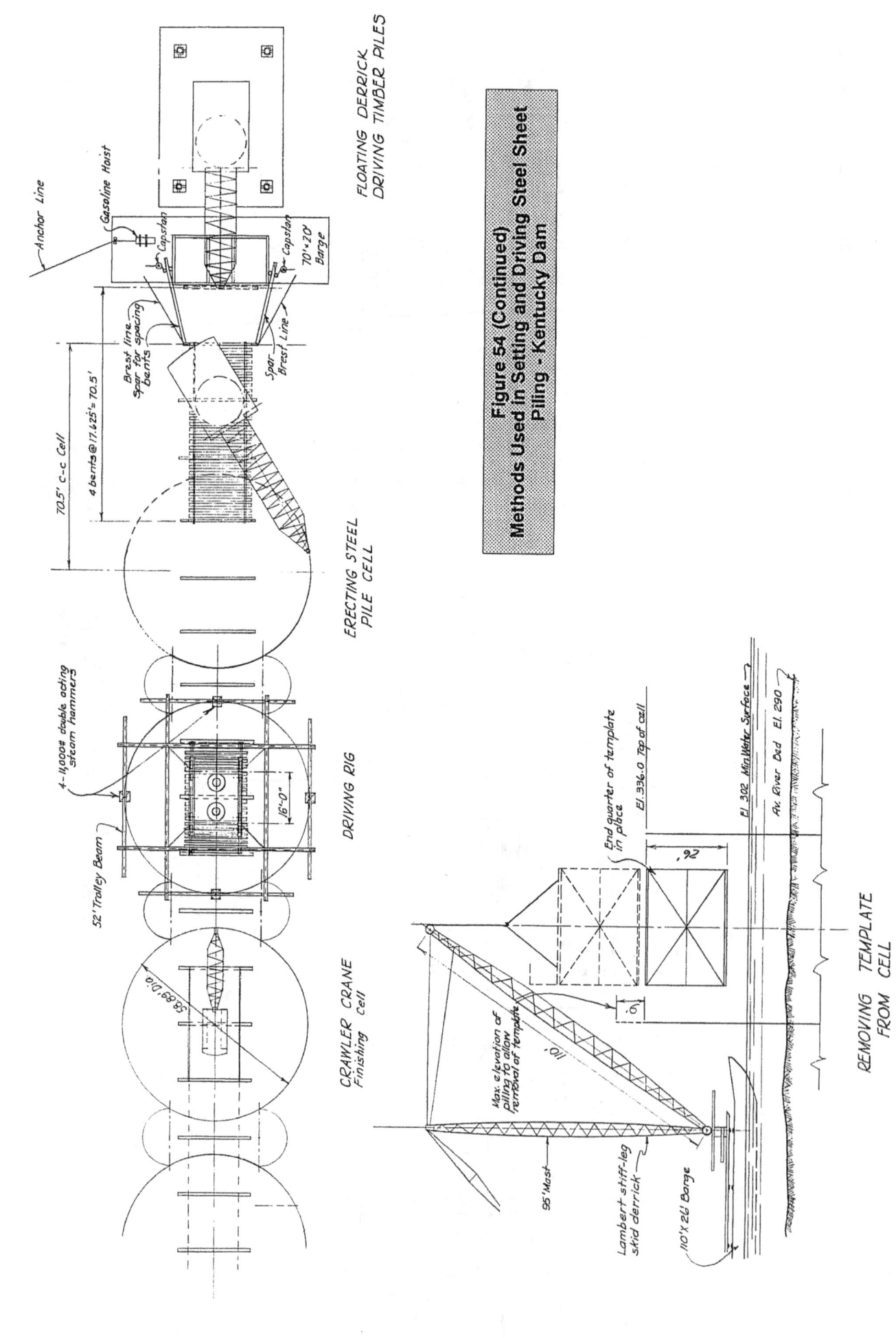

NOTES:
All bolts to be 1"φ, 3" thread both ends, length to suit field conditions. Provide two MI Washers for each bolt when bearing on wood.
All holes for 1"φ bolts to be 1 1/16" diameter.
Length of piles as shown on this drawing is approximate. True length to be determined from soundings.

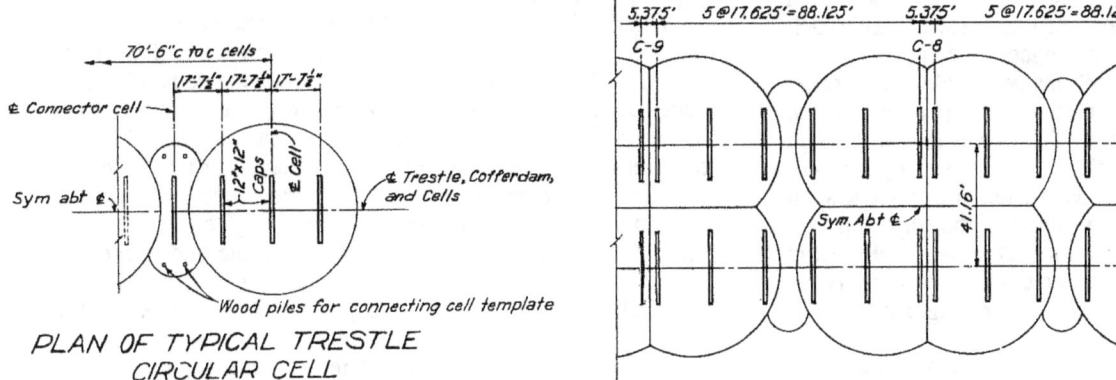

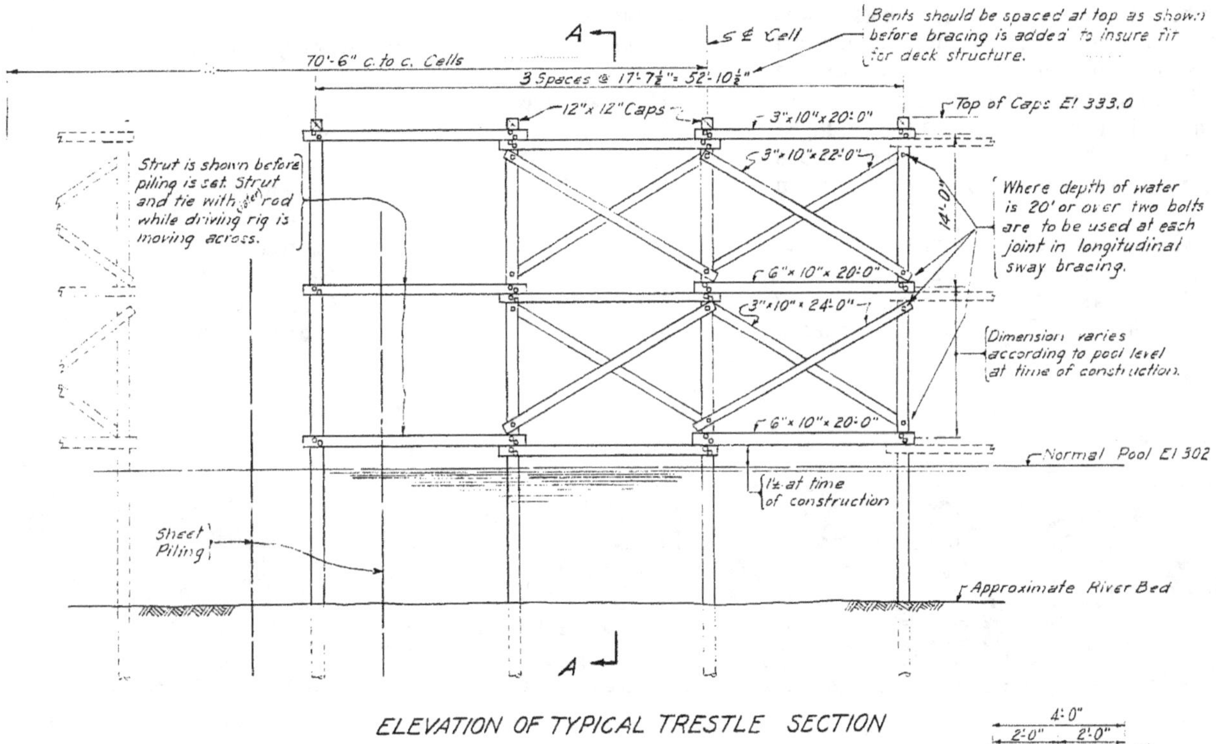

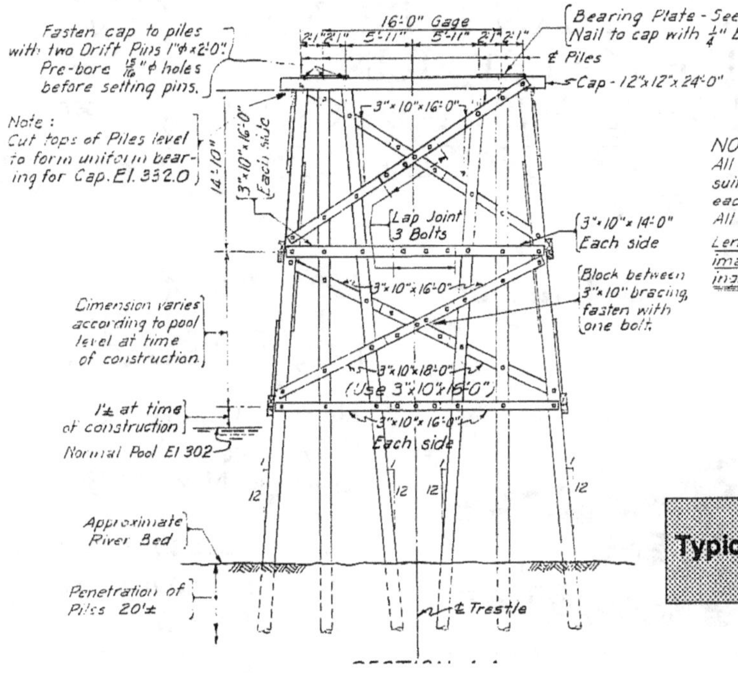

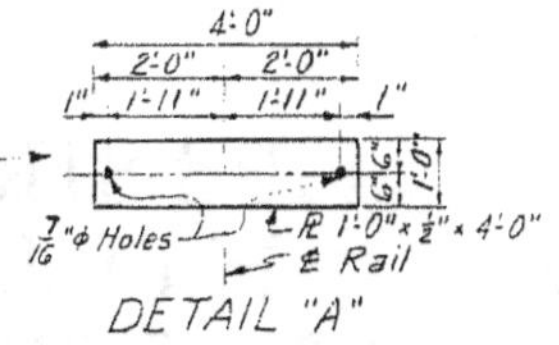

Figure 55
Typical Details of Cofferdam Construction Trestle - Kentucky Dam

Kentucky Cofferdam Trestle

For the construction of the circular cells a trestle was built consisting of a 3-bent, self-supporting trestle tower inside each cell and a single bent inside each connecting cell.

Two parallel trestles were built inside the cloverleaf cells, each being located 20.58 feet from and parallel to the longitudinal centerline of the cells. This placed one bent inside each quadrant of each cloverleaf cell. (See figure 55). Movable deck units with crane rails attached were used to span between bents (see figure 56). Material in the stage 1 trestle involved 83,900 lineal feet of untreated Douglas fir piling and 991,500 fbm of creosoted timber bracing. The piling had a 9-inch minimum tip diameter and varied in length from 58 to 75 feet.

Timber Pile Template--At Kentucky Dam a 22-foot-high template composed of vertical and sloping timbers, placed in pairs to form slots in correct relation to each pile, was attached to overhand the sides of a 20- by 70-foot steel barge and used for setting the timber trestle piles. The barge was counterweighted with short pieces of steel piling to counteract the weight of the template. Working platforms were provided at two levels and used during driving and capping operations. A system of spars, hand-winches, and capstans was arranged in a convenient manner on the barge deck to provide a means for spotting and holding the template barge in proper position during pile driving.

Timber Pile Driving--Piles were set and driven by an American revolver, oil-fired, steam-operated, floating crane using a McKiernan-Terry model 9B3 steam hammer. Piles were driven to give a minimum of 50,000 pounds of bearing capacity as figured by the "Engineering News formula." See page 177 and Cofferdams[1] and Pile-Driving Handbook.[2]

Where piles were driven to below grade before attaining the required bearing capacity, they were cut off level with the horizontal bracing system. They were then capped and a pony bent added to bring the bent to required grade. In some cases it was necessary to add piles to the bents to meet bearing requirements. The piles were held in place during driving by means of ropes in the hands of workmen stationed on the working platforms of the template barge. The two vertical piles were driven first to anchor the template and to prevent the template being forced out of line during driving of the battered piles. Piles were capped before removing the template. Bracing was added later.

For cloverleaf cells the template used in driving piling for the trestle in the circular cells was moved to one end of the cloverleaf cells of two bents in line from the same location. Otherwise the procedure was the same as that used for the circular cells.

The timber pile driving crew was organized as follows:

Number of Men	Duty
1	Foreman
2	On raft
4	On template
1	On steam hammer throttle
1	Helper - odd jobs
1	Crane operator
1	Crane fireman

IF IT'S STEEL, AND IT GOES INTO THE CONSTRUCTION OF A CELLULAR COFFERDAM, THEN... WE'VE GOT IT!

We also purchase surplus inventories

PIPE STRUTS ▪ WIDE FLANGE BEAMS
H-PILING ▪ STEEL SHEET PILING
HIGH STRENGTH SECTIONS (A-572 Gr. 50)
ANGLES ▪ PLATE ▪ CHANNEL ▪ PIPE
& MORE!

New or used

Call: Nestor

PH) 516-546-7900
PH) 516-542-9100
FAX) 516-542-9087

Nestor Palahnuk
P.O. Box 95
Merrick, N.Y. 11566

The only good thing about new steel is it becomes used . . .
And We Have It!

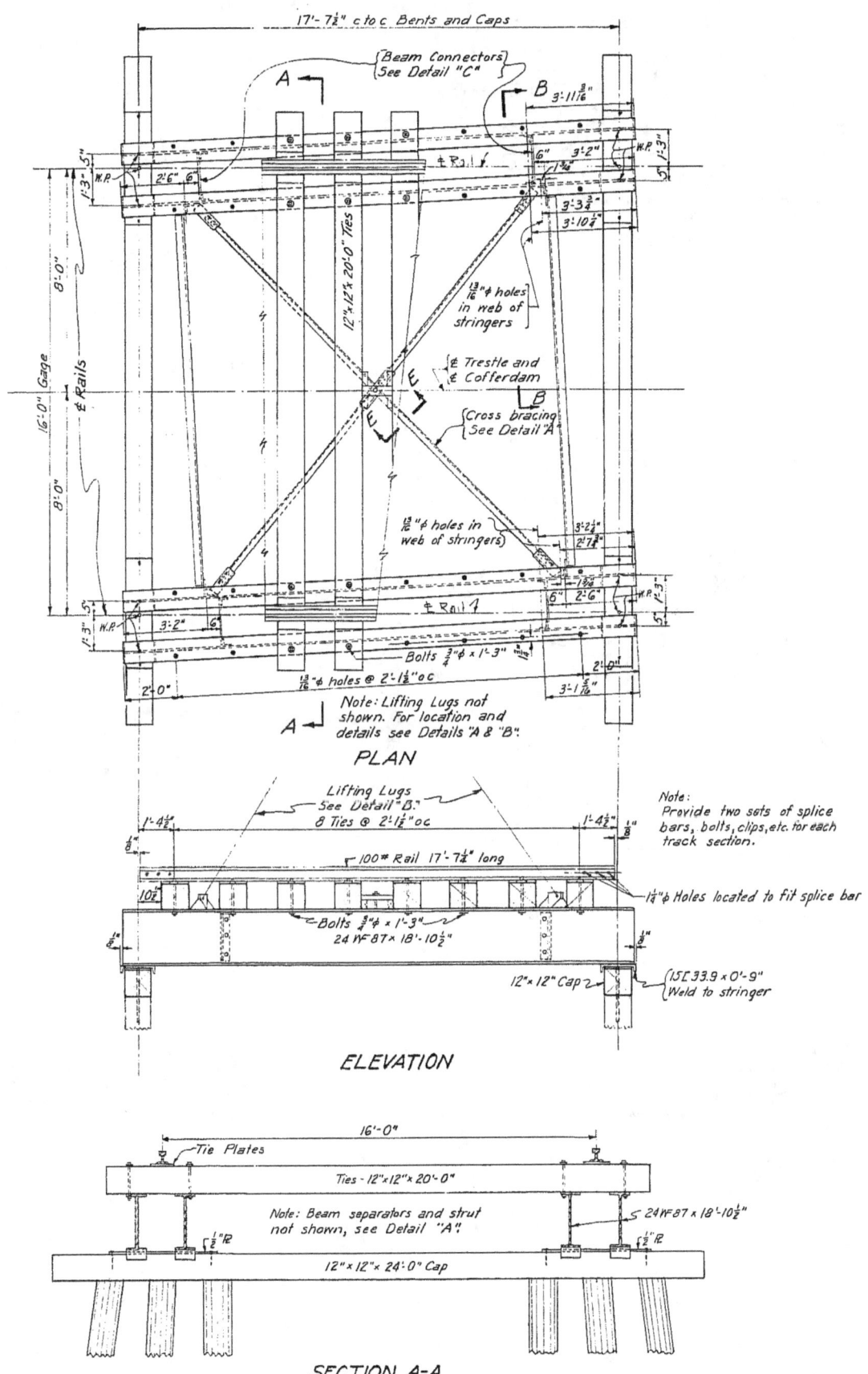

Figure 56
Cofferdam Construction Trestle Deck Details -
Kentucky Dam

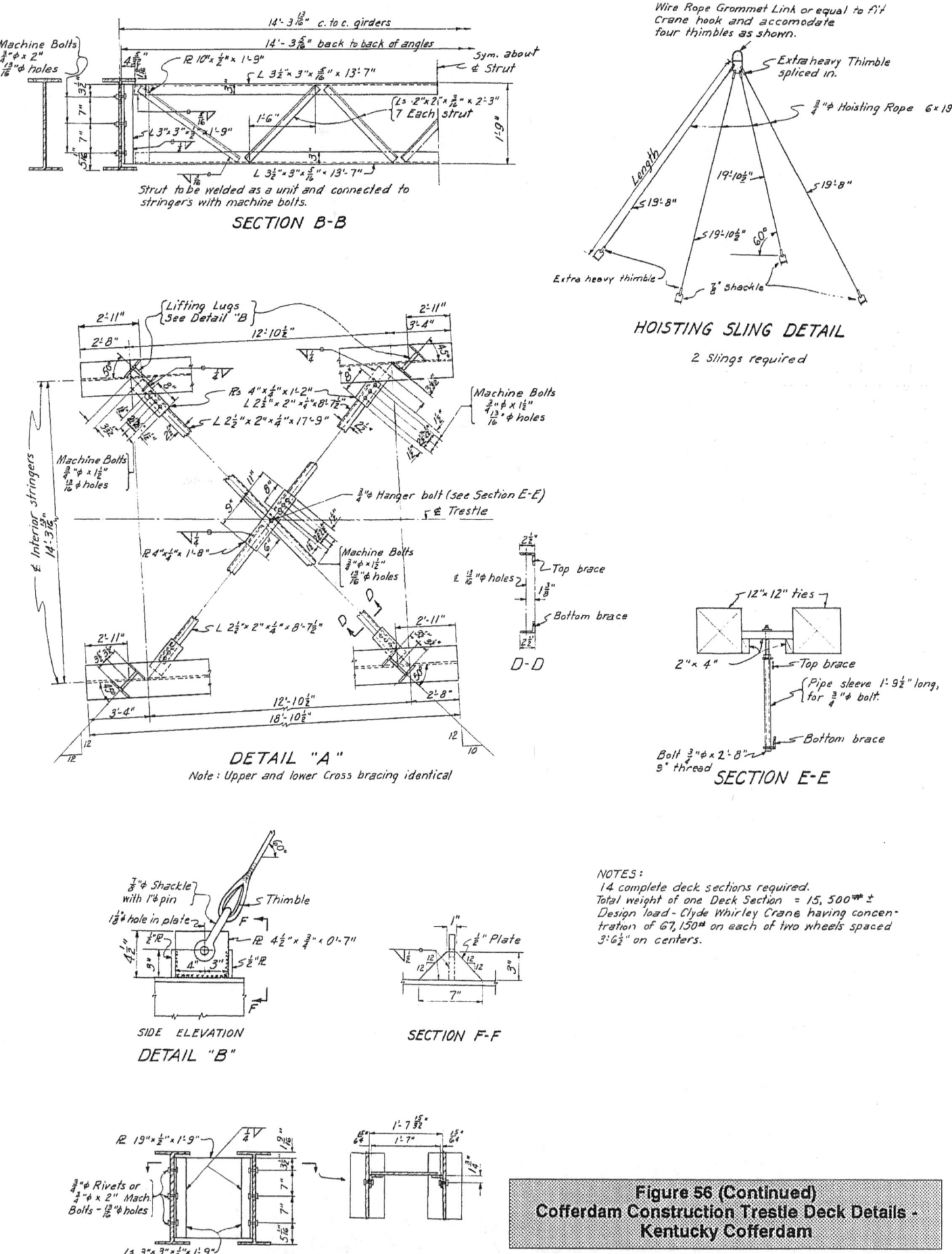

Figure 56 (Continued)
Cofferdam Construction Trestle Deck Details - Kentucky Cofferdam

Trestle Deck Section--For details of the deck section see figure 56. The 24-inch wide flange beams in adjacent deck sections were placed on a skew with the crane rails to permit overlapping at the bent support. The 15-inch channel bearing shoe attached to the beams rested on a 1/2-inch steel bearing plate attached to the bent caps and required no further connection between the bent and the deck section. The lateral bracing system in the deck section gave rigidity to the section and at the same time served as the top horizontal bracing for the trestle. Fourteen complete deck sections were built. These were placed in service in the trestle as required, being moved ahead in each case by the crane as work advanced.

STEEL SHEET PILE DRIVING AND DRIVING EQUIPMENT

Marking Piling--Before starting the driving, paint marks should be placed a few feet from the top of each pile in such manner as to provide a check on the amount burned off the top of each pile due to battering and on the amount of lead at the bottom between adjacent piles.

Guys for Sheet Piling--Unless steps are taken to control the piling during driving the cell will no retain its true circular shape. A block and tackle attached to approximately each fifty pile, and anchored to the template or trestle, as at Kentucky Dam, will afford a satisfactory means of holding the cell to its true shape. Similar block and tackle assemblies were used in holding piling for plumb and position in preparation for driving. With the outside and inside combination template used at Hales Bar Dam, little guying was necessary. The three points of contact provided by the templates seemed to hold the piling in position satisfactorily.

Order of Driving--The order in which the piling in the cell is driven after they have all been set or racked is relatively unimportant. It is important, however, that the wye and tee piles be located accurately for position and plumb. The pile racking program should therefore be arranged accomplish this (see "Order of Setting Piling," page 86).

Where piling penetration is slight or the stream swift, difficulty will be experienced in holding the piling to true position and alignment. Deep piling penetration in overburden filled with boulders, and a faulty rock foundation, may result in a similar problem. No solution for this type of problem can be offered in advance but rather it is a problem to be solved by the construction personnel.

Experience on TVA Projects

Guntersville Cofferdam--At Guntersville Dam the piling penetration was slight (averaging only 5.8 feet for the lock cofferdam). In driving the piling in each cell the tees, with a few adjoining piles for support, were driven first. This set up a control for the rest of the piles in the cell and they were driven in pairs to best suit the efficient operation of the driving equipment. Where the tops of the piles indicated that the bottoms of two adjacent piles were not at the same elevation after driving in pairs, they were again driven singly.

In stage 2 of the same project, difficulty was experienced with gain in the over-all length of the upstream arm of the cofferdam. This was thought to result from a flattening of the cells caused by pressure on the sides of the cells from the swift current. As a result a more rigid method for controlling the location of each cell was devised. Driving started by driving the two tee piles on the side of the cell toward which construction was progressing. This was followed with the driving of each tenth pile and the remaining two tee piles. The rest of the piling was then driven in the order best suited to the driving equipment.

Chickamauga Cofferdam--Driving conditions here were very difficult due to the irregular rock surface which contained many deep solution channels and seams. This required that driving be done in two steps. All piling was driven until the hammer leads approached the top walkway on the template. The template was then removed and such piling that had not already been completely driven was driven to full penetration. As a result of this condition, considerable splicing of piling was necessary (see "Splicing Piling," page 24). The usual order of driving was to drive the connecting cell first, the tees for the connection to the next cell, and then to continue driving around the cell to closure (see figure 53-c).

Kentucky Cofferdam--At Kentucky cofferdam the trestle was built first and served as a support for the cell template. Setting of the piling was then carefully controlled, thereby making it unnecessary to further control the position of the cell. Since driving was conducted in stages, with the Clyde whirley crane driving the first stage and the special driving rigs, completing the operation, a specific order of driving would have been difficult to follow. The piles, however, were driven in pairs around the cell, always holding the lead of each pair of piles to a maximum of 2 feet above or below the top of adjoining pairs. This helped to keep the piling plumb and in proper position.

Piling was driven on the modified production line basis as indicated in figure 54. Equipment used in setting and driving the major portion of the piling was mounted on the trestle and consisted of two electrically operated Clyde whirley cranes with 95-foot booms equipped with 15-foot jib extension, two special job-built driving rigs, and two small Northwest diesel-powered crawler cranes with 65-foot booms. One each of the above, in the order listed, was required for a single pile driving operation (see figure 54).

The Clyde whirley cranes were designed to operate on a 16-foot-gage track and had an operating capacity of 22 tons at a 25-foot radius. Each crane was supported on a 20.5-foot gantry and was propelled on swivel-type, motor-driven, 2-wheel trucks. The base of the crane rails was set at elevation 336, or 34 feet above minimum water level. Evan at this height the cranes required the 15-foot-long jib extension on the boom in order to provide the reach necessary to handle the 75- and 80-foot-long sections of steel sheet piling.

The crawler cranes were too light to do heavy pile driving, but in following up the main driving operation they were idean for racking the short top course of 6-foot piles and for testing all the cell piles singly for proper seat on rock foundation.

Two barge-mounted, oil-fired, steam-operated, 8-ton skid derricks equipped with 110-foot booms and two barge-mounted, oil-burning, steam-operated, 22-ton cranes with 100-foot booms were available to assist the trestle-mounted equipment in setting and driving piling.

The normal operating procedure, after the trestle had been constructed, was to set the template for the next cell to be constructed. The template was handled by the trestle-mounted Clyde whirley gentry crane operating from a position slightly behind the cell construction operations, aided when required by the floating equipment. After the template was set the gantry crane moved out onto the trestle surrounded by the template. It then removed the portable deck section from the rear side of the cell and placed it onto the trestle section ahead, thus clearing the back section for cell construction. The piles were then set by the same gantry crane, including the 6-foot pile followers for the 75-foot piles. The piles were then driven by the gantry crane, in pairs as previously described, to elevation 344. A gap at the back of the cell was driven to elevation 333 to permit entrance of the special driving rig into the cell. Driving by the gentry crane was then discontinued,

The piles were driven from elevation 344 to bedrock in pairs by the special driving rig. Since one pile in every pair was 80 feet long and the other 81 feet long (a spliced pile made up of one 75- and one 6-foot section), the latter always reached rock first. This was true as the top of each pair of piling was always kept as nearly at the same elevation, during driving, as possible. Driving was stopped at this point by the special driving rig and was completed later by the crawler crane or one of the floating cranes. the amount of driving for the follow-up cranes was materially reduced by the above procedure since only the 80-foot piles had to be driven the remaining distance to rock. Otherwise every pile would have had to be tested for seating on rock. The follow-up crane removed the 6-foot length of piling and drove the adjoining 80-foot length of piling until it was seated on rock. The top course of 6-foot lengths of the 81-foot piling was then replaced to bring the top of the cell to required grade.

Driving Specification for Steel Sheet Piling

In driving the piling to refusal, the word "refusal" is defined as the number of blows required to drive the pile 1 inch. This may vary at various elevations depending upon what average elevation has been established as bedrock. The specifications set up for the number of blows will also depend upon the condition of the foundation for the particular project. Requirements set up on various TVA projects were as follows:

Project	Number of Blows
Fort Loudoun	25
Watts Bar	40
Chickamauga	50*
Pickwick	300**
Kentucky	***

*Much piling entered the numerous cracks, crevices, and solution channels at this project and therefore penetrated to a much greater depth than was generally considered bedrock. The number of blows required was therefore reduced progressively to as low as five blows for piling reaching a depth of as much as 30 feet below the riverbed, or bedrock.

**At Pickwick cofferdam the bedrock was relatively smooth. It was not known to what extent this surface would resist the tendency of the cells, as a whole, to slide. Attempts were made, therefore, to get as much penetration of the piling into the rock as possible in order to increase the factor of safety against sliding. A penetration of from 1/4 inch to 1 inch was obtained. There seems to be no instance on record either in TVA or on other projects, however, of failure through sliding. The 300 blows specified seems to have been unwarranted in the light of records and of the procedure followed on other TVA jobs. Excessive driving of this nature is certain to reduce the life of the piling for reuse.

***No requirements were set up at Kentucky. Driving was continued until it was certain that piling was firmly seated on rock. (See procedures outlined under "Interlock Failure," page 24, and "Experience on TVA Projects--Kentucky Cofferdam," page 91.)

Special Job-Built Pile Driving Rigs for Kentucky Cofferdam

Two of these rigs (one is shown in Figure 54) were designed and built by TVA for pile driving at the Kentucky cofferdam. Each rig was a self-contained unit designed to handle four steam-driven, 11,000-pound, double-acting McKiernan-Terry model 1033 hammers. Steam supplied by this rig was also used for operating the hammer on the follow-up crawler crane.

The essential parts of the rig included a 17-foot-square car body mounted on four 2-wheel, swivel-type trucks spaced 16 feet on centers both ways. The car body supported two 125-horsepower, oil-fired, vertical-type boilers, oil storage tanks, water tanks, and transformer equipment for operating the cranes and hoists, and a mushroom-shaped steel tower structure for supporting the driving equipment. Each rig carried two sets of geared trolleys, from which were suspended two standard 20-inch I-beams. Each 20-inch I-beam served as a runway for an 8-ton electric motor-driven hoist supporting in turn an 11,000-pound hammer. A compression-type coil spring shock absorber, equipped with safety-type hook, served as a connecting link between hammer and hoist. The overhead trolley had a 13-foot travel from a point 18 feet to a point 31 feet from the centerline of the car body. The trolley hoists traveled at right angles to the direction of travel of the overhead trolleys, and their travel was limited to 22 feet either side of the centerline of the car body. This 2-way motion enabled the four hammers to cover a 62-foot-diameter circular cell except for a 36-foot square in the center. The original hand chain and pendent rope operation of trolley hoists, although satisfactory, was later replaced by more economical motor-driven, pushbutton controls. This change was actually completed only on one rig.

One platform at car body level and one 16 feet higher provided for convenience and safety of operators. The upper platform served as working space for the throttle operators (one for each two hammers). Controls were grouped, as far as possible, for operation from a central point, and located as to give the operator an unobstructed view of the equipment under his control. Oil, water, and power lines leading to the rig, and the steam lines leading out, were brought in over the top of the tower to clear operating equipment. Connections to the source of supply were made of flexible hose or cable.

Equipment Summaries

Tables 5, 6, and 7 give the principal items of construction equipment used by the TVA in driving the steel sheet piling for the Pickwick, Guntersville, Chickamauga, Watts Bar, For Loudoun, and Kentucky cofferdams.

LOAD BEARING PILING

In that the use of the timber trestle as an aid in building large sheet pile cofferdams can be a major construction consideration, the following design information has been included in this chapter on construction.

The basic design of load bearing piling rests upon three equally important considerations. These are:

1. Study of the depths, characteristics, and load carrying capacities of the soil by means of adequate boring and soil sampling, and the driving of test piles.
2. Study of pile types and pile driving equipment by means of an adequate dynamic pile driving formula.
3. Study of pile carrying capacities by means of the staticc formula for frictional resistance.

The extent to which these studies are conducted will depend largely upon the extent to which the load bearing pile will be used and the duration and magnitude of the loads to be supported.[3]

Test Piles

The driving of several test piles on the site in order that

Table 5
Summary of Pile Driving Equipment - TVA Projects

	Crane Capacity (Tons)	Crane Mount	Boom Length (Feet)	Hammers (Each)	Pick-wick Dam	Gunters-ville Dam	Chicka-mauga Dam	Watts Bar Dam	Fort Loudoun Dam	Kentucky Dam
Job-built rigs	--	Trestle	--	4	--	--	--	--	--	2
Clyde whirley revolving elec. crane (16' gage, 20' 6" gantry)	22	Trestle	95 (plus 15' ext.)	1	--	1	--	2*	--	2
American revolver elec. crane	--				--	--	--	1*	--	--
American revolver crane (steam)	22	Barge	75 - 95	1	1	2	--	--	--	2
McKiernan-Terry, Lambert-National skid rig (steam)	8	Barge & skid	110	1	2	2	2	2	2	2
American locomotive crane	--	Track			--	1**	--	--	--	--
Northwest crawler crane (diesel)	--	Trestle	50 - 65	1	--	--	1**	--	1***	2
Pile driving hammers:										
a. McKiernan-Terry model 9-B3					2	2	3	3	3	7
b. McKiernan-Terry model 10-B3					--	--	--	--	--	12
c. Union model No. 2					--	--	--	1	--	--
Electric welding machines (pile splicing)					--	--	2	--	--	--
Portable steam boilers:										
	50 hp				--	--	--	--	1	--
	60 hp				--	--	--	1	--	--
	100 hp				--	--	--	1	--	--

*On low gantry. **Used for unloading piling. ***Used for construction of shore cells.

Table 6
Summary of Barges Used in Cofferdam Construction - TVA Projects

	Size	Fort Loudoun Dam	Watts Bar Dam	Chicka- mauga Dam	Gunters- ville Dam	Pick- wick Dam	Kentucky Dam
Derrick barge (for American revolver crane)	62 x 38 x 6' 6"	-	-	-	1	1	1
Derrick barge (for American revolver crane)	56 x 40 x 5' 6"	-	-	-	1	-	1
Barge (for skid rigs)	90 x 40 x 7'	1	1	-	-	-	2
Barge (for skid rigs)	100 x 26 x 6' 6"	-	-	-	2	2	-
Barge (for skid rigs)	110 x 26 x 6' 6"	2	2*	4*	-	-	1
Barge (supply and misc. construction)	70 x 26'	-	1	-	-	-	-
Barge (supply and misc. construction)	70 x 20'	-	-	-	-	-	3
Barge (supply and misc. construction)	100 x 26 x 6' 6"	-	2	-	2	3	-
Barge (supply and misc. construction)	110 x 24 x 5'	-	-	4	-	-	-
Barge (supply and misc. construction)	110 x 26 x 6' 6"	2	6	-	2	-	2
Barge (supply and misc. construction)	120 x 28 x 7'	-	-	-	2	-	2
Barge (template--pile driving)	70 x 20'	-	-	-	1	-	1
Barge (for electric welding machine), shop-built (wood)	30 x 16 x 3' 6"	-	-	1	-	-	-
Barge (for electric welding machine [wood])	36 x 12 x 3' 4"	-	-	-	2	-	-

*One skid rig mounted on two barges.
All barges constructed of steel except as noted.

Table 7
Summary of Towboats and Motor Launches Used in Cofferdam Construction - TVA Projects

Crafts	Owner	Horsepower	Type Hull	Size	Fort Loudoun Dam	Watts Bar Dam	Chicka- mauga Dam	Gunters- ville Dam	Pick- wick Dam	Kentucky Dam
Towboats:										
Duck	TVA	75	Wood	30' x 9'	-	-	-	-	-	-
Buffalo	TVA	112	Wood	35' x 9'	-	-	-	x	-	x
Joe Wheeler	TVA	180	Steel	50' x 12' 6"	-	-	-	-	-	x
Paint Rock	TVA	85	Steel	42' 5" x 10' 3"	-	-	-	x	-	x
Mary Lou	TVA	180	Steel	65' x 16'	-	-	-	x	x	x
Bobby Bill	TVA	170	Steel	35' x 9'	-	-	-	-	x	x
Mary Bond	TVA	45 - 50	Wood	57' x 11'	-	-	x	-	-	-
Robert	TVA	165	Wood	40' x 11'	x	-	-	-	x	-
Betty Lou	TVA	85	Wood	33' x 9' 5"	-	x	-	-	-	-
Elk	TVA	100	Wood	40' x 10'	-	-	-	-	-	x
HP #2	TVA	45	Wood	50' x 16'	x	x	-	-	-	-
Sea Lion	Private	---	Steel	64' x 12'	-	-	x	-	-	-
Howard Powell	Private	---	---	---	-	-	Third Stage	-	-	-
Motor launches:										
Sir Patrick	TVA	45		26' x 8'	-	x	-	-	-	-
Betty Jane	TVA	--		20' x 6'	x	-	-	-	-	-
Big Sandy	TVA	115		---	-	-	-	-	x	x
Ellen "A"	TVA	95		25' x 6' 8"	-	-	-	x	-	-
Iuka	TVA	35		24' x 8'	-	x	x	-	-	-
Dodge Passenger	TVA				-	-	-	x	-	-

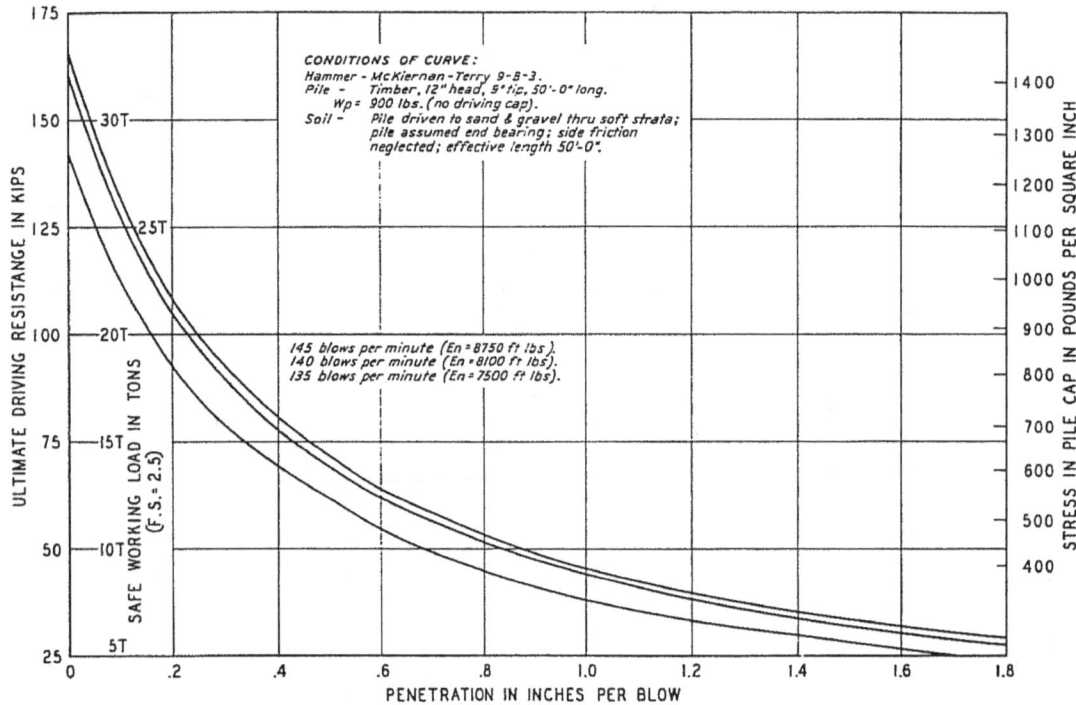

**Figure 57
Timber Pile Penetration and Bearing Value Curves**

driving records may be studied is most desirable. These test piles should be driven to a deeper tip grade than is expected for actual conditions so as to determine the characteristics of the soil carrying capacity below the expected pile tip grade. Redriving of several of the test piles after the ground around the piling has had time to settle will provide valuable information as to the increase or decrease in frictional resistance. This resistance usually increases in impervious strata and changes but little in cohesionless soil, although a considerable decrease has been noted in some coarse grained pervious strata. One of the most important factors causing misleading results from test loading of a single pile is the disregard for the time element between time of driving and time of testing.[4] Test loading carried to failure, if possible, will provide frictional values including end resistance.

Pile Driving Formulae

Kentucky cofferdam was the only TVA cofferdam where timber piling was used to any appreciable extent. Load capacities for these piles were determined by means of the widely used Engineering News formula. A number of these dynamic formulae have been proposed and used by leading engineers in the past, and the assumption upon which they all are based is that the ultimate carrying capacity of the pile is equal to the dynamic driving force. The basic principle behind them all is that the weight of the ram multiplied by the stroke may be equated to the driving resistance multiplied by the net penetration of the tip of the pile.[5] Such a formula can only apply in the case of a cohesionless strata such as sand, gravel, or permeable fill, however, in which case the resistance acting while the pile is being driven bears a reasonably close relationship to that acting on a pile carrying a static load. In the case of driving piling in plastic materials, such as soft clay or fine grained silt, there is no relation between the temporary resistance to driving and the permanent resistance to the applied load on the pile. However, the stress in the pile during driving in cohesive soils and the selection of pile hammers may be made on the basis of the dynamic formulae, but the results must not be made on the basis of the dynamic formulae, but the results must not be used to determine permanent load carrying capacities. In the case of mixed soils composed of both classes, good judgment in the use of a formula must be relied on.

The Engineering News Formula--In the face of up-to-date methods for designing piling, and available data, this formula although widely used in the past is being used less frequently. The factors of safety actually obtained are frequently either much greater or much less than advisable due to the attempt to provide a short formula that will fit all conditions. Its use is not recommended.

Comprehensive Pile Driving Formula for Practical Use--With the accumulation of data available in Pile-Driving Handbook,[6] and in view of ease of handling and accuracy which may be obtained through use of the following comprehensive dynamic formula, it is recommended in preference to other methods. The formula for use with double acting and differential acting steam hammers is as follows:

$$R = \frac{12 \, e_f E_n}{s + (1/2)(C_1 + C_2 + C_3)} \times \frac{W_r + e^2 W_p}{W_r + W_p}$$

See Pile Driving Handbook, page 191, formula 1b, where:

R_u = the ultimate carrying capacity of the pile (pounds)
W_r = the weight of the falling mass in pounds
e_f = manufacturer's rated efficiency of the hammer being used
E_n = the rated energy of the hammer per blow in ft-lb
e = coefficient of resitution[7]
W_p = weight of the pile in pounds, including the shoe

and driving cap for drop hammers and single-acting hammers and including the weight of shoe and anvil for double-acting and differential-acting type hammers.

In the case of steel sheet piling where the piles are driven in pairs, the weight of both piles should be included.[8]

- s = the final penetration in inches for one blow of hammer (using average of last 5 blows for drop hammers and last 20 blows for other types)
- C_1 = the temporary compression allowed for pile head and cap
- C_2 = the temporary compression allowed for pile
- C_3 = the temporary compression allowed for soil

A practical example follows (for reference see Pile-Driving Handbook,[9] page 190, and tables I-IV, pages 164-169):

Assumed conditions:
Hammer: McKiernan-Terry, No. 9-B3
Blows per minute = 145
E_n (energy) = 8,750 foot-pounds
Pile = timber, Douglas fir, 50 feet long (12-inch cap, 9-inch tip)
W_p (pile weight) = 900 pounds
W_r (hammer weight) = 1,600 pounds
e_f (for McKiernan-Terry double-acting hammer) = 0.85
e = 0.4 (see footnote 7, page 178)
R = working load for single pile = 50,000 pounds (specified)
Factor of safety = 2.5
R_u = 50,000 x 2.5 = 125,000 pounds
P_1 = stress per square inch in pile head = R_u/area pile head = 125,000/113 = 1,100 pounds per square inch
C_1 = 0.10[10]
P_2 = R_u/Av. area of pile = 125,000/86.6 = 1,440 pounds per square inch
C_2 = 1440/1000 x 0.012 x 50 = 0.58 inch[11]
P_3 = R_u/Tip area = 125,000/63.6 = 1,965 pounds per square inch
C_3 = 0.05 inch[12]
Substituting in formula 1b:[13]
125,000 = 12 x 0.85 x 8750/(s + (1/2)(0.10 + 0.58 + 0.05)) x (1600 + 0.4^2 x 900/1600 + 900)
s = 0.134 inch to give working load of 50,000 pounds

By completing a number of similar computations covering a range of possibilities which will be consistent with the job in question and equipment in use, a graph similar to that shown in figure 57 may be prepared for field use by plotting final penetrations against ultimate driving resistances or against working loads for any selected factor of safety. This graph is very convenient for the inspector because he can pick the pile bearing value for any penetration directly from the curve.

Static Formula--After the pile lengths have been tentatively selected, but before ordering the piles, it is necessary to consider the total lengths of embedment in load resisting friction strata. It is also necessary to consider the resulting carrying capacities strictly from a friction point of view after, however, deducting such percentage for point resistance. This will provide a rough check on designed lengths. Values of friction from actual tests are given in table VII of Pile-Driving Handbook[14] as an aid to judgment because definite test information as to the particular case under consideration is lacking.

This static formula may be expressed in the form:
f_u = R_u - R_t/A_s in which f_u = ultimate friction value, in pounds per square foot
R_t = amount of ultimate load assumed ccarried on tip, in pounds
A_s = surface area of portion of pile acting in friction, in square feet

Tests have indicated that piles displacing all of the earth within their perimeter develop much greater friction values than those which do not, and this fact should be borne in mind when considering friction values. It is generally advisable to obtain a perimeter as large as possible in the friction load carrying strata and not to use reduced perimeters at these levels.

Formulae for Analysis of Energy Losses and Check of Computations--The following formulae will serve to determine the amount of the total applied energy lost due to various causes. They give a check on the efficiency of the equipment being used and serve as a check on the compuations by the dynamic formula.

Net effective energy available- for driving ultimate- resistance to driving final penetration under last blow	Total kinetic energy applied by hammer	Loss in energy due to impact	Loss in energy due to elastic compression of		
			Pile head and cap	Pile	Soil
*R_us =	$e_f W_r h$ -	$e_f W_r h$ W_p (1-e^2) / (W_r+W_p)	- $R_u C_1$/2	- $R_u^2 1$/2AE	- $R_u C_3$/2
**R_us =	12 $e_f E_n$ -	12$e_f E_n$ W_p (1-e^2) / (W_r+W_p)	- $R_u C_1$/2	- $R_u^2 1$/2AE	- $R_u C_3$/2

*For use with drop hammer and single-acting steam hammers.
**For use with double-acting and differential-acting steam hammers.

Where
A = Average of cross-sectional area of piling at butt and at center of resistance to driving, in square inches.[15]
E = Modulus of elasticity for pile material
h = Height of free fall of ram, in inches, for drop hammer; the normal (shortest) stroke of ram, in inches, for single-acting steam hammers; and the normal stroke of ram for double-acting and differential-acting steam hammers.[16]
1 = Length of pile in feet, measured from head to center of resistance to driving.

Note: Other symbols in above formulae are the same as shown in the paragraph under heading "Comprehensive Pile Driving Formula for Practical Use," page 99.

These formulae are of assistance in selecting the proper weight of hammer to use. For economical and efficient driving, a reasonable portion of the applied energy should remain available for driving. Where such is not the case and the remaining useful energy is small, slight uncertainties in assumptions may be as great numerically as the value of the net energy remaining for driving. This indicates that the formula may be too sensitive and that perhaps too much dependence should not be put into the results obtained. It is desirable, then, that an efficient size of hammer be used. It is better to select a hammer that is on the heavy side than one that is too light. In general, the hammer should be as large as can safely be used without damaging the pile. the computed stresses in the pile should also be compared with the yield point of the pile material, reduced by a factor of safety, to prevent exceeding such a value.

It should be noted here that where the steel sheet piling is driven to rock to form cofferdam cells the load bearing value of the piling is unimportant. However, the formulae previously discussed may be found useful in estimating driving resistance and in selecting the proper hammer size.

PILE HAMMERS AND THEIR SELECTION

The size of the hammer affects the time required to drive the pile and is a reason for selecting as heavy a hammer as possible without overstressing the pile. The heavier hammer will permit a low-velocity blow resulting in the delivery of a higher ratio of effective energy to the pile. The light hammer requires high-velocity blows to develop the required energy. A blow that strikes the pile sharply is largely dissipated in heat and elastic pile deformation at the point of impact and tends to damage the pile head by brooming or crushing. The heavier hammer will be more efficient as it will produce a higher energy blow with a lower velocity, producing a "follow through" pushing blow rather than the sharp striking blow.

> *Editors Note*
> Modern vibratory hammers were not available to the TVA constructors at the time for driving or pulling. See Pile Buck® Specifications for Pile Hammers, page 230.

The proper selection of hammer, then, will reduce waste energy and thereby make a greater percentage of the energy of the blow available for moving the pile. Studies conducted in England indicate the average efficiency of a hammer blow on a wood pile or concrete pile (where a wood mat is used between the metal driving head and the top of the pile) to be as follows:

Ratio of Ram Weight to Pile Weight	Efficiency of Blow (Percent)
1:4	25
1:2	37
1:1	54
2:1	70
4:1	85

Since all items of waste energy are destructive to the pile, their magnitude must be controlled or the effect distributed throughout the pile. Driving heads or mats reduce the intensity of the stresses at the end of the pile by distributing the blow locally.

Limitations Due to Type of Pile

The type of pile to be driven will have considerable influence on the selection of the hammer to be used. Steel sheet pile has a high modulus of elasticity, thereby reducing elastic deformation losses. This pile also has a low driving resistance due to its small cross section area. This makes it possible for a fairly light hammer to operate efficiently. Concrete piles with a much lower modulus of elasticity (resulting in a higher elastic deformation loss) and with a larger cross section area (causing higher driving resistance) will usually require a larger and heavier hammer of low-impact velocity. A driving mat will help to reduce the tendency to crush the driving head of the pile. Timber piles are similar to concrete piles in their hammer requirements The effective energy delivered to the pile is low compared to the gross energy in a high-velocity blow, and under such conditions the timber pile tends to broom. To avoid damage to the pile, then, the movement or value of penetration (s) must be of sufficient magnitude to keep the stress in the pile within the allowable limits unless the elasticity of the pile is sufficient to absorb all of the applied energy when the value of (s) is zero. The stress at the head of the pile is usually the governing stress.

Types and Sizes of Hammers

No particular type of hammer has advantages which make it pre-eminent for all classes of work. Generally the selection for size is of much greater importance than selection for type: Factors affecting the selection of the hammer for size are:

1. Availability and cost of purchase or rental.
2. Available steam or air pressure (some hammers require greater pressures than others).
3. Headroom clearance, accessibility of piles, battered piles, etc.

Where the hammers are being selected for the construction of a steel sheet pile cofferdam, requirements for driving the steel sheet piling will normally prevail since it will involve the major portion of the driving. Considerations involving the driving of timber piling must be included, however, where timber trestles, timber piling for built-in-place templates, etc., are to be included as part of the structure.

Where the hammers to be selected are considered for type, several factors are involved. These include:

1. Double versus single-acting hammers.
2. Driving requirements of the piling.
3. Limitations set up based on the type of pile to be used.
4. Extractor requirements (some hammers may be used as extractors)

Drop Hammers--the drop hammer consists of a heavy ram which is raised in leads by a hoist line, then allowed to drop on the pile. This type is of little use today due to the slow driving rate when compared with driving rates attained through the use of steam, or air, hammers. Where only a few piles are to be driven, the use of the drop hammer may be warranted.

Single-Acting Hammers--The single-acting hammer meets requirements in every particular except that it lacks a high frequency of blows. The hammer consists of a heavy ram, raised by steam (or air), but falling only by force of gravity. Advantage in using this type of hammer in comparison with the double-acting type is the low velocity of the ram when striking the pile, thereby resulting in reduced impact losses and damage to the pile. Disadvantages are: (1) greater head room required, (2) requires guides for the ram in every case, and (3) has a slower driving speed. Where this type of hammer is employed, a sufficiently heavy ram to offset the low velocity must be used.

In using the single-acting type of hammer, the actual stroke should be measured at a time close to the final penetration. Strokes are frequently less than the normal figures. Due to the sensitivity of the pile driving formula at the point where the net force at the tip is small compared with the applied force, a reduction of a few inches in the stroke with consequent reduction of applied energy may make a large percentage reduction in the net energy, that is, after deducting losses occurring during driving.

Rebuilt hammers sometimes have a shorter stroke than the normal, and such hammers should be checked for this condition so that computations may be corrected accordingly.

Double-Acting Hammers--The double-acting hammer

employs steam (or air) both for raising the ram and for increasing the velocity on the downward stroke. In comparison with the single-acting hammer, the double-acting hammer (although operating with a rapid succession of blows and weighing no more than the single-acting hammer) has insufficient weight for the energy developed. This results in the delivery of an injurious and less effective blow. The double-acting hammer is more satisfactory for pile driving than the single-acting type, however, under most driving conditions.

Advantages of this type are:

1. Requires less headroom clearance due to shorter stroke.
2. Reduced stress in the pile due to reduced driving resistance.
3. Increased driving speed.
4. Can be used for driving without guides in certain instances.
5. Can be used for subaqueous driving.

The rapidity with which the blows follow one another with this type of hammer keeps the pile in motion thereby reducing inertia, friction, and point resistance. this may offset the destructive effect of greater impact in this hammer over that in the single-acting type.

The double-acting and differential-acting (next paragraph) types must be run at full listed speed as the net allowable energy at the pile tip falls off rapidly at less speeds. This drop off is particularly sharp in the case of undersize hammers. It should also be noted here that the number of strokes per minute should always be noted on pile driving reports, especially if for any unavoidable reason the speed is less than the maximum specified. If the speed has fallen off and is not noted, the energy is reduced, and the smaller penetration obtained will indicate falsely high bearing values. Strokes must be counted while driving the pile and not with the hammer running free.

Differential-Acting Hammers--The differential-acting hammer employs an entirely different steam cycle from the other types, and its effectiveness is based on a proper ratio between the piston areas for raising and driving the ram. The hammer combines the advantages of the single- and double-acting types, and in addition a 30 percent saving in steam consumption over other types is claimed. In comparison with the single-acting type (both having equal energy), the differential-acting type with heavier ram and shorter stroke will have the least velocity upon striking the pile. This lessens both the impact losses and the possibility of damaging the packing and pile head. The tendency to damage is directly proportional to the square of the velocity.

Data for Selection of Proper Size Hammer

Tables 8 through 13 give data from which the proper size or type of hammer may be selected for certain conditions. Additional hammer data not included in these tables may be found in table IV, pages 165-169, of Pile-Driving Handbook.[17]

Tables 8 and 9 give the weight, stroke, energy delivered, and power required for the particular light and heavy hammers listed. Table 10 gives the approximate energy of blow in foot-pounds per blow required in driving steel sheet piling, timber piling, or concrete piling through both low and high resistance soils. Table 11 gives the proper size Warrington-Vulcan single-acting hammer to fit each condition set up in table 10. Table 12 gives the proper size McKiernan-Terry double-acting hammer to satisfy each condition set up in table 10, and table 13 gives the proper size Super-Vulcan differential-acting hammer to satisfy each condition set up in table 10. Tables 8 and 9 do not include hammer data for the Super-Vulcan differential-acting hammer, but this data may be found in table IV, page 166, of Pile-Driving Handbook.[18]

Jetting

In some cases it may be necessary to resort to jetting to reach the necessary tip grade without overstressing.[19] The skin friction and point resistance for a pile being driven through sand, silt, or close to existing structures are generally very high and may be materially reduced through the use of the jet, particularly in the case of concrete piling in sand, where the use of the jet is almost indispensable. Jetting should be stopped several feet above final pile point elevation and the pile driven the remaining distance to final grade. For piles which are jetted down the static test load may be low, however, with the skin friction and point resistance being only a small percent of the ultimate values attained after the ground has had time to settle in around the pile. Final driving may even be deferred for a specified period to give the soil time to set, and to attain higher bearing values. (See specifications for driving precast concrete piling.[20]) This condition rarely exists, however, when using steel sheet piling.

Hammer and Driving Data for TVA Projects

The steel sheet piling on TVA projects was normally driven in pairs, using a double-acting pile hammer driven either by steam or air while supported from a standard crane or driving rig. A McKiernan-Terry hammer, model 9B3, was sufficient to meet all TVA needs except at Kentucky Dam where, due to depth of piling penetration, a McKiernan-Terry hammer, model 10B3, was used. Although air was used for driving the hammers in a number of cases, steam was preferred since the correct air pressure necessary to give smooth, efficient operation is hard to maintain (see table 8).

Fishtail Guides--Pile hammers of the double-acting type, such as those used by TVA, were designed to swing free from the bottom of the pile driving rig and operate without guides. Guides of some type were found to be necessary to hold the hammer in line, however, particularly in the case where two of the steel sheet pilings were being driven simultaneously. The guides shown in detail in figure 58 were therefore fabricated and attached to the hammer so that when it was in position the guides extended past the top of the piling to rest on either side and guide the hammer for accurate driving. Heavier guides of similar design were used on the McKiernan-Terry hammer, model 10B3.

TIE-INS

The cofferdam, in general, must be tied in to the riverbank and to completed portions of the structure. For the first stage of construction the cofferdam will be tied in only to the riverbank, as a rule, but during subsequent stages both types may be required.

Tie-Ins to Riverbank

Where the cofferdam joins the steep sloping shoreline, the first cell is usually located at a point where the top of the cell intersects the sloping bank. Variations of this scheme include the addition of a single wall of steel sheet piling, connected to the cell at one end and extending shoreward to form a cutoff wall to aid in sealing against leakage. In the case of a rocky shoreline prohibiting the driving of piling, a timber crib may be built immediately adjoining the shore and connection between the crib and cell accomplished by means of tee piling bolted to the timber structure.

The most commonly used type of shore tie-in on TVA projects was one immediately adjoining a wide flood plain. In such cases, where the flood plain was lower than the tope of the cofferdam cells, protection from floodwaters along the shore side was accomplished by constructing an earth dike to a height of approximately 4 feet above the top of the cells. This dike extended from the upstream to the downstream arms of the cofferdam and enclosed as large an area of the flood plain as required for construction purposes. The extra height was necessary in order to provide adequate protection from the floodwaters. A cutoff wall, usually consisting of a single row of steel sheet piling, was frequently driven to rock along the dike with ends connected to the cells to aid in sealing against leakage from the flood plain. (See figure 1, for cutoff walls used at Kentucky Dam.) Whether the cutoff wall is to join the upstream and downstream arms of the cofferdam to shield the entire cofferdam enclosure from flow from the flood plain (as along the right bank at Kentucky Dam, figure 1) or extends from the end of the cofferdam straight in toward the shoreline (as along the left bank at Kentucky Dam, figure 1) will depend, on of course, upon the type of overburden, location of rock, and extent of the flood plain. Consolidation of the overburden through grouting may also prove effective in stopping the flow of water. Final judgment will usually be left to the field construction forces.

Tie-Ins to completed Dam Structure

As a rule the cofferdam for a succeeding stage will overlap the one already in place, thereby permitting the tie-in for the succeeding stage to be built in the dry. In the case of repairs or additions to existing structures, however, tie-ins between cofferdam and structure may have to be made under water.

Provision for a tie-in to the upstream face of the dam is relatively simple, since the face of the dam is either vertical or can be made so by adding a concrete pier at the point of tie-in. The pier must extend from rock foundation to the full height of the proposed cofferdam. The connection between the dam face or pier and the cofferdam is accomplished by embedding a section of steel sheet piling in the concrete with interlock extended. (See figure 59 for typical details of this type of tie-in).

A tie-in at the downstream face of the dam spillway is difficult because of the curvature of the sloping face. The method used in accomplishing the tie-in, in this case, will vary depending upon existing conditions. The most commonly used method, however, is to construct a timber pier between adjacent spillway piers with the crib tailored to fit the sloping face. Tee piles bolted to the downstream face of the crib may then be connected to the sheet pile cell. At Guntersville project the connection between the sheet pile cells and the sloping face of the lock wall consisted of two rows of standard steel sheet piling cut to fit the sloping face and spaced 26 feet 7-3/4 inches apart. These were tied together with wales and tie rods and sealed against leakage along the bottom of the piling by setting piling into a recess filled with a bituminous material. (See figure 60.) A connection to a vertical face was used at Chickamauga cofferdam and is shown in figure 61. Other methods can be developed to fit local conditions.

Piling in the cofferdam cells, resting upon the concrete spillway apron, must be carefully fitted to prevent passage of water beneath the cells. Timber recesses were encountered which were filled with mastic to receive the ends of the sheet piling. At Fort Loudoun Dam sandbags were used instead of mastic. At Kentucky cofferdam recesses were formed in the apron concrete and filled with mastic for receiving the ends of the piling and sealing against leakage under the cells. In every case where the cells are connected to the dam structure; close collaboration between dam design forces and construction forces is necessary since embedded piling, extra piers, and recesses, to satisfy the requirements, may affect the design of the dam structure.

Tie-Ins at Kentucky Cofferdam

A special arrangement of cells was devised for Kentucky cofferdam to provide for the overlap between stages 1-B and 2 (see figure 1). The cloverleaf cells built along the river side of stage 1-B provided stability for this cofferdam without need for a supporting berm. This permitted construction of the dam to be brought close to the inside of the cells and allowed the tie-in for stage 2 to be built in the dry without much overlap. The triangular group of cells C-1 to C-5 inclusive and C-10 to C-14 inclusive (figure 1) were built integrally with stage 1-B and remained in place as a part of stage 2. To complete stage 2, then, it was only necessary to place the tie-in cells C-15 to C-17 inside of stage 1-B, the upstream and downstream arms, and remove cells C-6 to C-9 inclusive.

The upstream tie-in presented no difficulty and was made in the usual manner by embedding vertical piles in the face of the dam and connecting them to a cloverleaf cell by small arcs.

The downstream tie-in presented a major problem, however, largely because of the great size of the cells. A final

Table 8
Data on Heavy Duty Pile Drivers

Model	Weight in Pounds Unit	Ram	Strokes Per Min. (Max.)	Length (In.)	Energy (Ft-Lb Per Blow)	Steam Reqd (B.HP)	Compr. Air Reqd* (Cu Ft)	Equiv. Fall (Feet)
McKiernan-Terry--Double Acting								
9B3	7,000	1,600	145	17	8,750	45	600	5.5
10B3	10,850	3,000	105	19	13,100	50	750	4.4
11B3	14,000	5,000	95	19	19,150	60	900	3.8
Union--Double-Acting								
1-1/2A	9,200	1,500	135	18	8,680	35	450	5.5
1A	10,500	1,600	130	18	10,020	40	500	6.3
1	10,500	1,850	125	21	12,725	40	600	7.9
0	14,500	3,000	110	24	19,850	50	750	6.6
0A	17,000	5,000	95	21	22,050	60	800	4.4
00	21,000	6,000	85	36	54,900	125	--	9.1
Super-Vulcan--Differential-Acting (closed type)								
3000	7,250	3,000	133	12-1/2	7,260	40	488	2.4
5000	12,140	5,000	120	15-1/2	15,100	60	880	3.0
8000	18,480	8,000	111	16-1/2	24,450	80	1,245	3.0
14000	27,980	14,000	103	15-1/2	36,000	100	1,425	2.6
20000	39,050	20,000	98	15-1/2	50,200	120	1,745	2.5
Warrington-Vulcan--Single-Acting								
2	6,700	3,000	70	29	7,260	25	580	2.4
1	9,600	5,000	60	36	15,000	40	975	3.0
0	16,250	7,500	50	39	24,375	60	1,450	3.25

*100 pounds pressure for McKiernan-Terry, Union, and Super-Vulcan pile drivers;
  80 pounds pressure for Warrington-Vulcan pile drivers.

design approved by the TVA Design personnel provided two concrete walls. Each of these was 7 feet thick and built to an elevation that was 2 feet below minimum low water elevation. These walls extended downstream from piers 7-8 and 9-10 with the downstream face vertical and contained an embedded section of steel sheet piling with interlock extended to tie-in with arcs from the nearby steel sheet pile cell. The concrete walls were heavily reinforced and tied together at the downstream face with 10-inch wide flange beams at 49 pounds per foot (spaced as shown in figure 62.) to resist the outward thrust of the sand and gravel fill material placed between the walls. This arrangement formed the base for additional tie-in cells that were required to briing the cofferdam at this point to above high water. Upon removal of the cofferdam, the 10-inch wide flange beams were burned off and the concrete walls left in place since they were below water and had no ill effect on the flow of water over the spillway. Cell 16 (the tie-in cell) and half of cell 15 rested on the spillway apron concrete and were set in asphalt filled recesses (see figure 63), to seal against leakage along the bottom of the cells as previously described. Concrete pads resting on the rock foundation were constructed and recessed in a similar manner to seat the piling for the other half of cell 15 extending beyond the spillway apron. Piling was racked with the Clyde whirley crane, but no driving was required since the weight of the piling was sufficient to seat itself in the asphalt filled recess. Cell 16-A was constructed between the two concrete tie-in walls and was tied in to piers 7-8 by means of a small timber crib. The space between the cantilever walls, the spillway structure, and cell 16 was filled with gravel upon completion of the cantilever walls. A small cell, 16-A, was then driven part way into this fill and connected to cell 16 and to pier 8 with connecting sheet pile arcs.

CLOSURE

The construction of the last cell in a cofferdam, usually in the upstream arm, often meets with difficulty because of the restricted opening and the swift flow of water through the opening. This condition normally occurs in the last stage of construction, along with maximum restriction of the river channel, and unless pile setting is accurately controlled the opening may vary so much in size that the standard cell template will not fit. This then may require the construction of a special template and an odd size cell. The method and procedure adopted to facilitate closure will depend on local conditions and should be anticipated and planned for in advance.

Table 9
Data on Light Duty Pile Drivers

Model	Weight in Pounds Unit	Ram	Strokes Per Min.	Length (Inches)	Energy Deliv. (Ft-Lb Per Blow)	Steam Reqd (B.Hp)	Compr. Air Reqd (Cu Ft)
McKiernan-Terry--Double-Acting							
5	1,500	200	300	7	1,000	20	250
6	2,900	400	275	8-3/4	2,500	25	400
7	5,000	800	225	9-1/2	4,150	35	450
Union--Double-Acting							
5	1,625	210	250	9	1,010	10	100
4	2,800	370	200	12	2,100	12	150
3	4,700	700	160	14	3,660	20	300
3-A	5,200	820	150	13-1/2	4,390	25	350
2	6,600	1,025	145	16	5,755	25	400
Super-Vulcan--Differential-Acting (closed type)							
600	1,887	600	225	7-1/2	1,125	16	214
1100	3,040	1,100	181	9	2,180	20	242
1800	4,274	1,800	150	10-1/2	3,600	25	308
Warrington-Vulcan--Single-Acting							
4	1,400	550	80	21	825	8	62
3	3,700	1,800	80	24	3,600	18	380

Table 10
Data for Selection of Pile Hammers for Driving Concrete, Timber, and Steel Sheet Piling Under Average and Heavy Driving Conditions

Length of Pile (Feet)	Depth of Penetration (Percent)	Sheet Pile*			Foot-Pounds Per Blow Timber Pile		Concrete Pile	
		Light	Medium	Heavy	Light	Heavy	Light	Heavy

Table A--Driving through earth, sand, loose gravel--normal frictional resistance

25	50	1000-1800	1000-1800	1800-2500	3600-4200	3600-7250	7250-8750	8750-15000
	100	1000-3600	1800-3600	1800-3600	3600-7250	3600-8750	7250-8750	13000-15000
50	50	1800-3600	1800-3600	3600-4200	3600-8750	7250-8750	8750-15000	13000-25000
	100	3600-4200	3600-4200	3600-7500	7250-8750	7250-15000	13000-15000	15000-25000
75	50		3600-7500	3600-8750		13000-15000		19000-36000
	100			3600-8750		15000-19000		19000-36000

Table B--Driving through stiff clay, compacted gravel--very resistant

25	50	1800-2500	1800-2500	1800-4200	7250-8750	7250-8750	7250-8750	8750-15000
	100	1800-3600	1800-3600	1800-4200	7250-8750	7250-8750	7250-15000	13000-15000
50	50	1800-4200	3600-4200	3600-8750	7250-15000	7250-15000	13000-15000	13000-25000
	100		3600-8750	3600-13000		13000-15000		19000-36000
75	50		3600-8750	3600-13000		13000-15000		19000-36000
	100			7500-19000		15000-25000		19000-36000

Weight per lin ft	20 pounds	30 pounds	40 pounds	30 pounds	60 pounds	150 pounds	400 pounds
Approx pile size	15 inches	15 inches	15 inches	13" dia.	18" dia.	12" square	20" square

*Energy required in driving single sheet pile. Double these when driving two piles at a time.

Table 11
Recommended Sizes of Warrington-Vulcan Pile Hammers for Various Types of Piling Under Average and Hard Driving Conditions

Length of Pile (Feet)	Depth of Penetration (Percent)	Sheet Pile*			Timber Pile		Concrete Pile**	
		Light	Medium	Heavy	Light	Heavy	Light	Heavy

Table A--For piling to be driven through earth, sand, or gravel with normal frictional resistance

Length of Pile (Feet)	Depth of Penetration (Percent)	Light	Medium	Heavy	Light	Heavy	Light	Heavy
25	50	3	3	3-2	3	3-2	2	1
25	100	2	2	2	3-2	2	2	1
50	50	2	2	2	2	2	1	1-0
50	100	2	2	2-1	2	2-1	1	1-0
75	50		2-1	1		1		0-OR
75	100			1		1		0-OR

Table B--For piling to be driven through stiff clay and compact gravel with high frictional resistance

Length of Pile (Feet)	Depth of Penetration (Percent)	Light	Medium	Heavy	Light	Heavy	Light	Heavy
25	50	3	3	3-2	2	2	2	1
25	100	2	2	2	2	2	2-1	1
50	50	2	2	2	2-1	2-1	1	0
50	100		2	2		1		0-OR
75	50		2-1	1		1		OR
75	100			1		1-0		OR
Weight per lin ft (lb)		20	30	40	30	60	150	400

*Sizes of hammers listed are based upon driving two sheet piles simultaneously.
**A driving head is required for all sizes of concrete piles.

Table 12
Recommended Sizes of McKiernan-Terry Pile Hammers for Various Types of Piling Under Average and Hard Driving Conditions

Length of Pile (Feet)	Depth of Penetration (Percent)	Sheet Pile			Timber Pile		Concrete Pile	
		Light	Medium	Heavy	Light	Heavy	Light	Heavy

Table A--For piling to be driven through earth, sand, or gravel with normal frictional resistance

Length of Pile (Feet)	Depth of Penetration (Percent)	Light	Medium	Heavy	Light	Heavy	Light	Heavy
25	50	5	5	6	7	7	9B3	9B3
25	100	5	6	6	7	9B3	9B3	10B3
50	50	6	6	7	9B3	9B3	9B3	10B3
50	100	7	7	7	9B3	10B3	10B3	11B3
75	50			7	9B3		10B3		11B3
75	100			9B3		11B3		11B3

Table B--For piling to be driven through stiff clay or compact gravel with high frictional resistance

Length of Pile (Feet)	Depth of Penetration (Percent)	Light	Medium	Heavy	Light	Heavy	Light	Heavy
25	50	6	6	7	9B3	9B3	9B3	9B3
25	100	6	7	7	9B3	9B3	10B3	10B3
50	50	7	7	9B3	9B3	10B3	10B3	10B3
50	100		9B3	10B3		10B3		11B3
75	50		9B3	10B3		10B3		11B3
75	100			11B3		11B3		11B3
Weight per lin ft (lb)		20	30	40	30	60	150	400

Table 13
Recommended Sizes of Super Vulcan Differential-Acting Pile Hammers for Various Types of Piling - Under Average and Hard Driving Conditions

Length of Pile (Feet)	Depth of Penetration (Percent)	Sheet Pile*			Timber Pile				Concrete Pile	
		Light	Medium	Heavy	Light		Heavy		Light	Heavy

Foot-Pounds Per Blow

Table A — For piling to be driven through earth, sand, or gravel with normal frictional resistance

Length	Depth	Light	Medium	Heavy	Light		Heavy		Light	Heavy
25	50	1100	1100	1800	1800		1800-3000		3000	5000
	100	1800	1300	1800-3000	1800		1800-3000		3000	5000
50	50	1800-3000	1800-3000	3000	1800-3000		3000		5000	5000- 8000
	100	3000	3000	3000	3000		3000-5000		5000	5000- 8000
75	50		3000	3000			5000			8000-14000
	100			3000-5000			5000			8000-14000

Table B — For piling to be driven through stiff clay or compact gravel with high frictional resistance

25	50	1800	1800	1800	3000		3000		3000	5000
	100	1800	1800-3000	1800-3000	3000		3000		3000-5000	5000
50	50	1800-3000	3000	3000	3000-5000		3000-5000		5000	8000
	100		3000	3000			5000			8000-14000
75	50		3000	3000-5000			5000			14000
	100			5000			5000-8000			14000

| Weight per lin ft | 20 pounds | 30 pounds | 40 pounds | 30 pounds | | 60 pounds | | 150 pounds | 400 pounds |

*Sizes of hammers listed are based on driving two sheet piles simultaneously.

Closure Procedures

Chickamauga Cofferdam, Stage 3--Before cofferdam construction had reached the closure stage, the river current was so swift that most of the overburden had been scoured from the bed of the river leaving very little anchorage either for the cells or the template. To facilitate closure, a pontoon bridge consisting of four barges lashed together end to end was extended out from the shore. Over this 12-cubic-yard trucks hauled spoil material to build a cutoff dike reaching from the main cell structure to the tie-in cell at the dam. The water was thereby diverted through the low spillway bays, permitting closure of the cofferdam to be completed with little difficulty.

Guntersville Cofferdam, Stage 3--During the closure of the cofferdam for this stage the water passing through the closure opening was swift, having a drop in head of approximately 2.7 feet. The template for the closure cell was therefore difficult to hold, and it became necessary to build four mooring cribs approximately 60 feet upstream. The template was then secured in position through the use of anchor cables attached to the mooring cribs.

Piling along the upstream side was racked first to break the flow of the water, thereby aiding in setting the piling along the downstream side of the template. As pile racking reached a mooring cable, a second cable was installed by passing it through a hole burned in the pile last set. The first cable was then removed and racking of the piling continued. Pressure from the water was so great against this cell, before construction was completed, that the pipe template was badly buckled along the bottom ring, making removal impossible. The frame did not collapse, however, and was left in place in the cell fill until the cofferdam was removed. This experience indicates that the template is weak for such procedure, and that it should be reinforced along the upstream or pressure side whenever similar conditions are expected.

Kentucky Cofferdam, Stage 2, Upstream Arm--Construction of the closure cell in the upstream arm of this cofferdam was difficult due to the swiftness of the water passing through the opening. Construction was further complicated by a rise in streamflow of 30,000 to 75,000 cubic feet per second just as closure was started. The plan of procedure (as indicated in figure 64) was to construct a closure wall of sheet piling across the opening to hold the water back until the cell could be built. The closure wall was built by working from each end toward the middle, racking alternately with panels made up of one tee pile and five standard piles. Some difficulty was experienced in anchoring the frame or template for the closure wall to the cells on either side of the opening. This was due to weak piling interlocks in the cell, and it became necessary to do some repair work to the cells. In one instance the anchor cable was extended to the barge located parallel with the cofferdam immediately upstream.

The two barges moored in the gap upstream from the barrier wall served as working platforms, one of them carrying a double-drum hoisting engine which was used to tighten the cable guys from the wall piling.

To facilitate setting of the piling for the closure wall a rigid fram template to accomodate five standard piles was made up from two standard piles 25 feet long braced rigidly with horizontal pieces of similar material. This template was racked into position with the interlock at one end attached to the tee pile in the abutment cell, then secured in position with guy cables. Next a 75-foot-long tee pile was racked into position at the other end of the template. As soon as this tee was set and driven the intervening template was removed and a panel, consisting of five standard piles bolted securely together with 1-inch bolts passing through heavy timbers at top and bottom of piles, was set into position in its place. This process was repeated until three panels were placed at each end of the barrier wall. A pile handling rig working at each end of the wall performed the above operation.

At this point the intensity of the river flow had scoured out most of the overburden along the riverbed, making it impossible to hold the long tee pile in position with cables as had been done up to this time. A single template section to span the remaining opening was then fabricated from standard piling, with the 75-foot lengths of piling from which it was made being carefully spaced to receive panels made up to five piles each. After fabricating the template in one piece it was separated into two sections, with each section being handled separately by one of the cranes. Two two sections were then fastened together to form one template. This template was racked in position to fill the remaining gap in the wall, and driven until the piles were seated on rock . The five panels were then set and attempts made to drive them to rock. Pressure from the watter was so great, however, that driving had to be stopped, leaving an opening of 2 to 4 feet in height between the bottoms of the piles and the bedrock. The opening was finally sealed, first by dumping large rock along the downstream face of the wall to break the force of the flow and then by pumping sand and gravel into the area along the upstream face of the wall. The complete operation of installing the wall required approximately two weeks, after which the wood template was floated into position for setting the piling for the closure cell. Upon completion of the closure cell the closure wall was removed.

Kentucky Cofferdam, Stage 2, Downstream Arm--Construction of the closure cell in the downstream arm was made easy by construction of the upstream closure cell and elimination of the flow of the river through the cofferdam area. This downstream closure consisted of one full circular cell and two connecting cells. Piling for the cells was first set and those for the connecting cells driven to final position. The upstream and downstream arcs of the circular cell were then removed to allow passage of the dredge out of the cofferdam area, after which the piling was reset and driven to final position.

CELL FILLING

The cofferdam fill material may be classified under two general heads--Wet and dry. Under the first the fill is placed hydraulically, normally by suction dredge, pumping the material directly from source of supply to cell. Where a suitable material is available this type of cell filling is highly satisfactory. However, all the fill requirements for the cofferdam will not be met throught this one method alone. The tie-in cribs and cells, resting for the most part of dry land, and other special conditions may either require or be adapted to the use of a dry fill. For handling this class of fill the dump truck or clamshell is normally used (see Chapter 4, "Fill Materials").

Hydraulic Dredging

This means of filling the steel sheet pile cells was adopted by TVA because of the presence of sand and gravel deposits overlying bedrock along the Tennessee River. The adoption of this method on any other project for cell filling, however, will be limited by the presence of a fill material suitable for hydraulic handling within the immediate vicinity of the cofferdam. This should fall within a radius of approximately 2000 to 2500 feet. Suitable materials outside the immediate limits of the project may be reclaimed by dredging, however, and moved to the cells by other means. This was done in several instances by TVA. In such cases the material may be either loaded onto barges and moved directly to the cells or stockpiled on shore and moved later, by truck or other conveyance, to the cofferdam. The method used is generally influenced by the equipment available to do the hauling, the

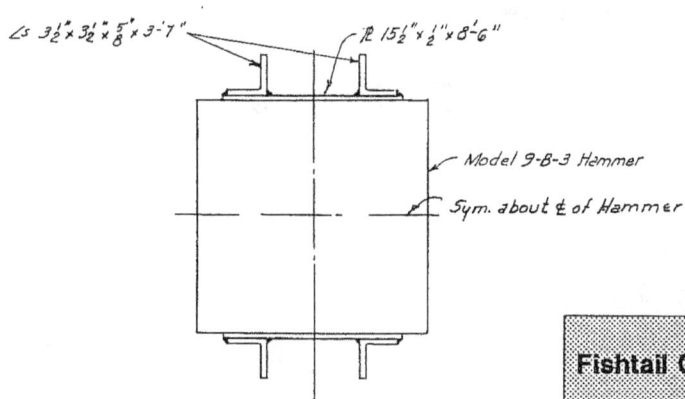
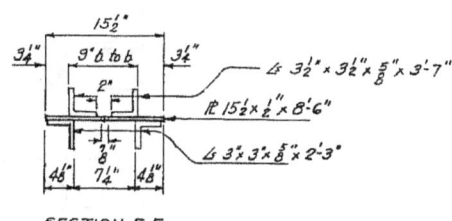
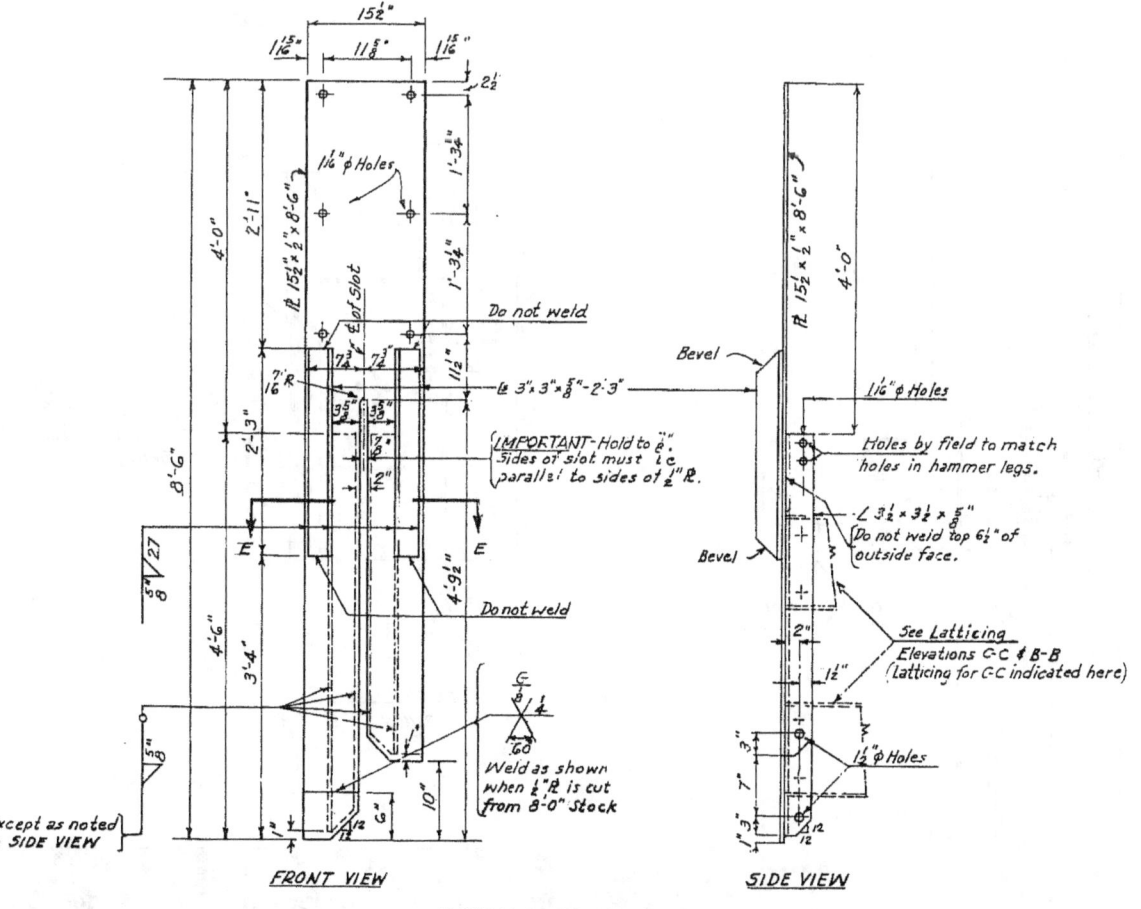

**Figure 58
Fishtail Guide for McKiernan-Terry Model 9B3 Pile Hammer - Kentucky Dam**

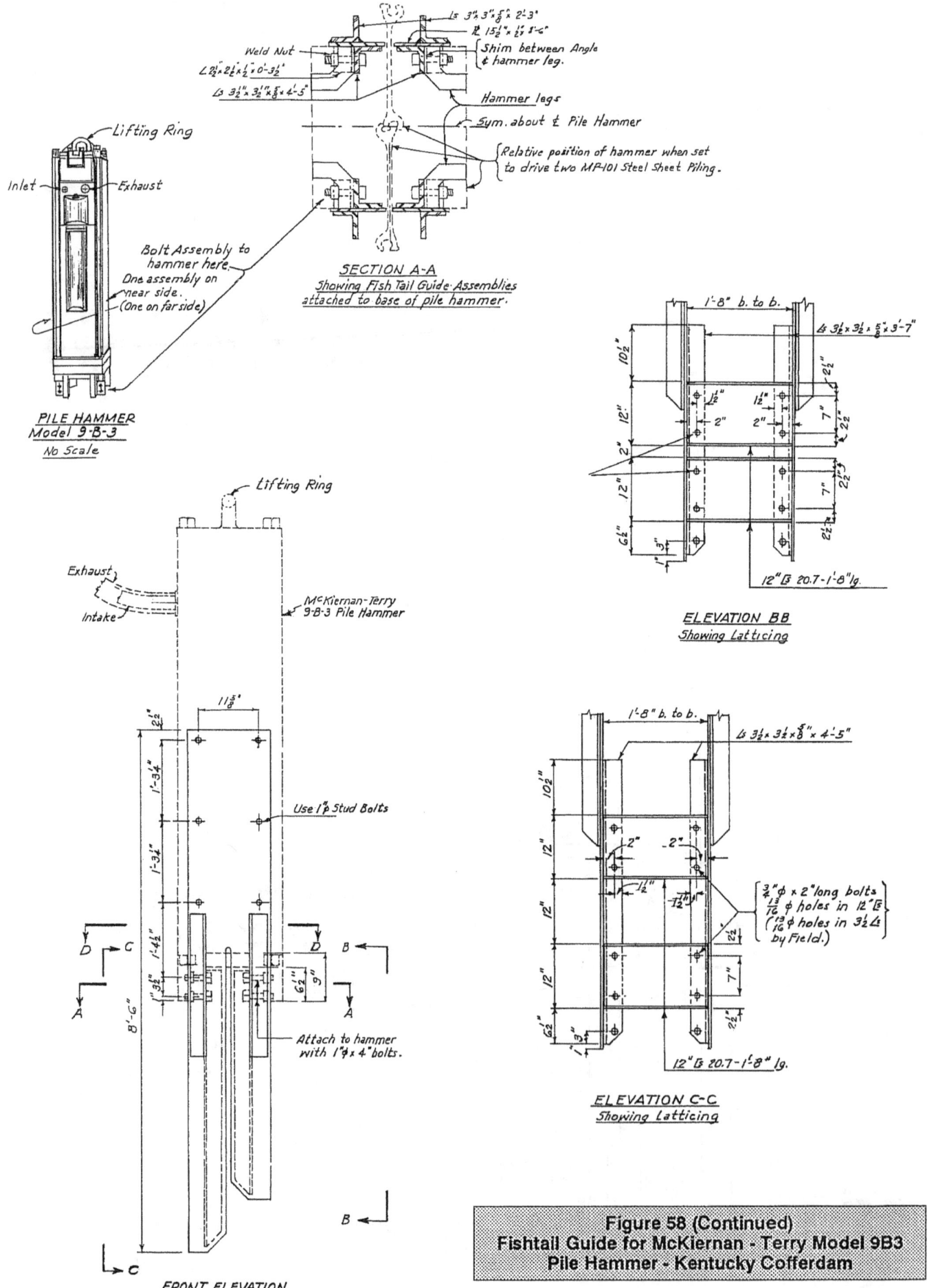

Figure 58 (Continued)
Fishtail Guide for McKiernan - Terry Model 9B3
Pile Hammer - Kentucky Cofferdam

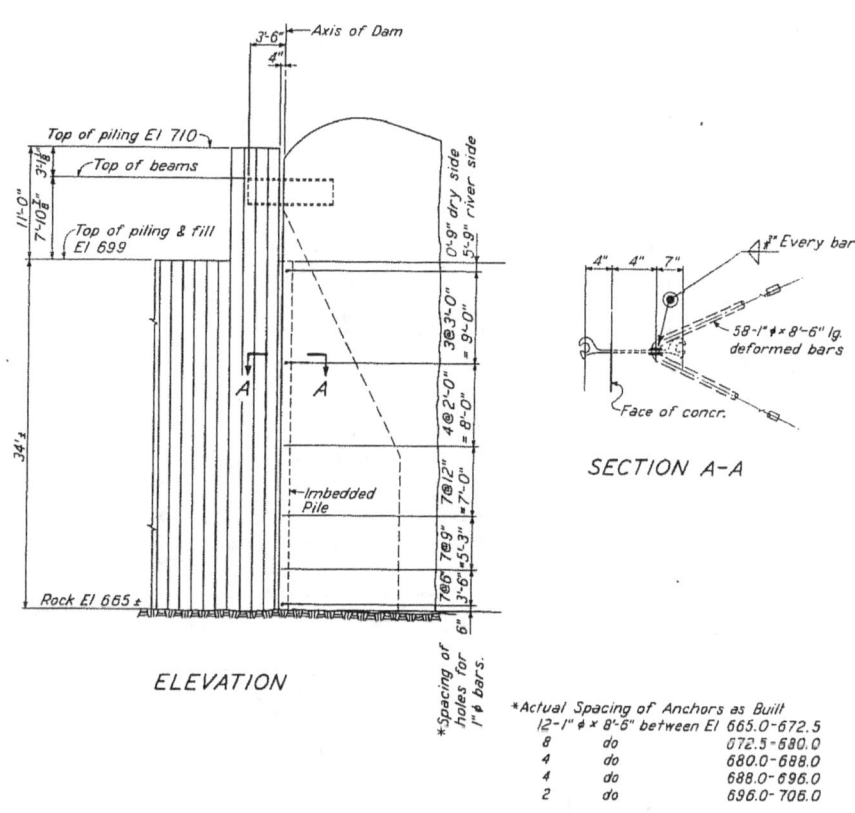

**Figure 59
Steel Sheet Pile Tie-In to Upstream Face of Watts Bar Dam**

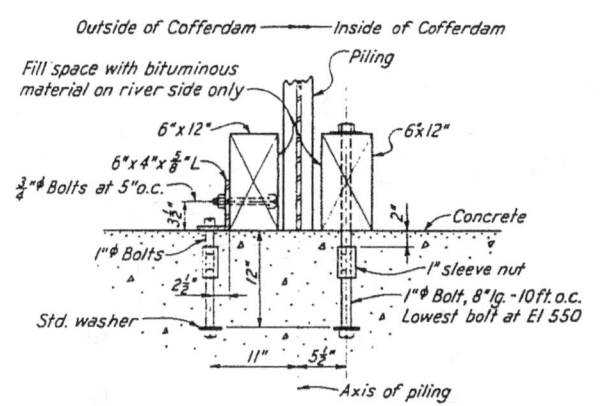

**Figure 60
Method of Sealing Against Leakage for Steel Sheet Pile Resting on Concrete - Guntersville Dam**

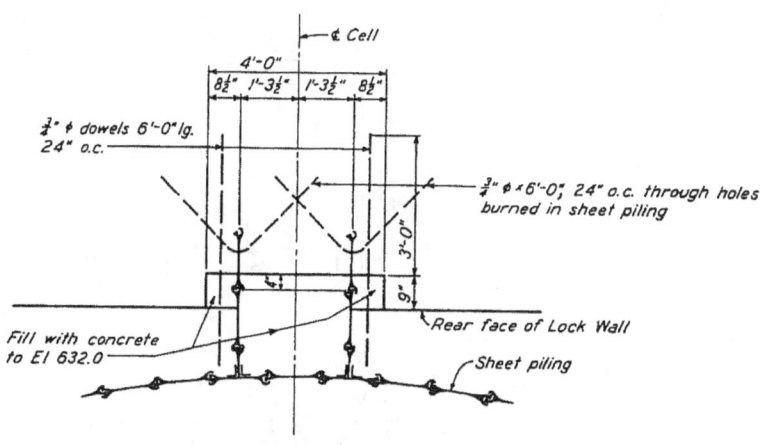

**Figure 61
Seal Between Steel Sheet Pile Cell and Vertical Face of Concrete Lock Wall - Chickamauga Dam**

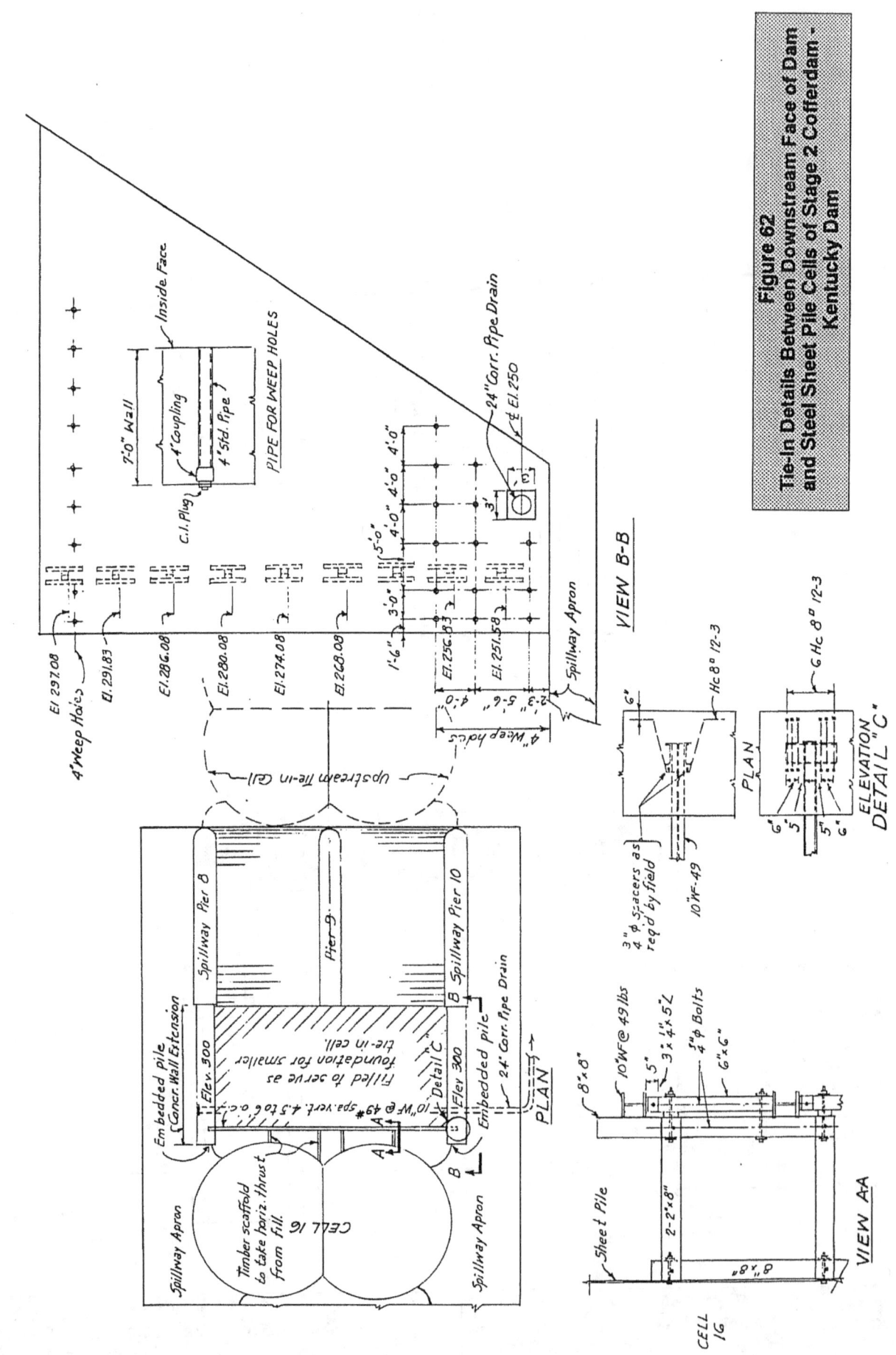

Figure 62
Tie-In Details Between Downstream Face of Dam and Steel Sheet Pile Cells of Stage 2 Cofferdam - Kentucky Dam

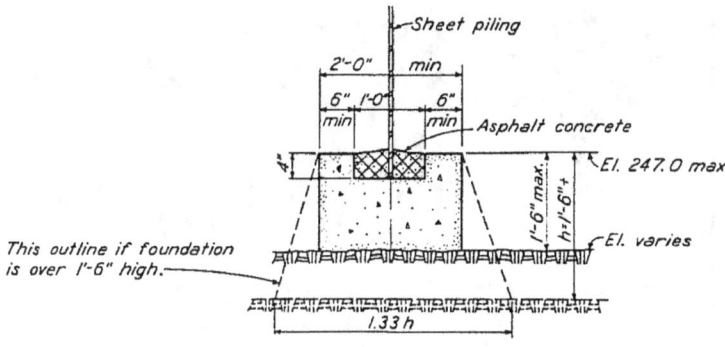

**Figure 63
Asphaltic Concrete Filled Recess for Sealing Against Leakage Between Steel Sheet Piling and Spillway Apron Concrete - Kentucky Dam**

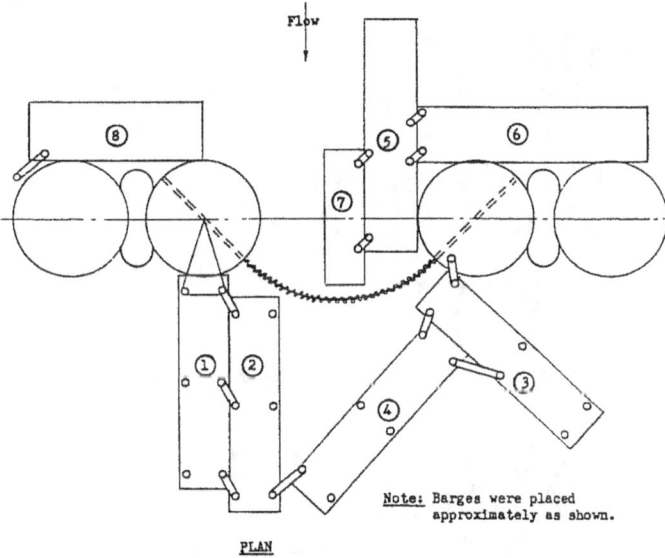

LEGEND
1. Lambert Skid-Derrick
2. Storage barge for 5-pile sections.
3. Lambert Skid-Derrick
4. Storage Barge for Tees.
5. Barge for Storage and Work Area.
6. Barge.
7. Barge.
8. Barge.

**Figure 64
Closure in Upstream Arm of Stage 2 Cofferdam - Kentucky Dam**

means by which it can be moved the cheapest, and the time available in which cells must be filled.

Where suitable fill material overlies the dam site it can be removed by the dredge and placed directly into the cells. This material must be removed in any event, and where it can be used for fill a considerable saving may be effected. Removal of the dredge from inside the cofferdam should be planned for just before closure of the downstream arm of the cofferdam.

In filling the cells the fill is pumped from dredge to cofferdam through a sectional pipeline which may be extended, as required, to reach each cell. Where the overburden is light, affording little support for the cell piling, the template is usually left in place until sufficient fill has been placed to render the cell stable. Some templates used by TVA were constructed with a large pipe as the hub, into which the dredge line discharged. This conveyed the fill material to below the template as a means of protecting the template from damage caused by falling material. Where cells are sufficiently stable to remove the template before filling the cell, then filling may be scheduled in whatever manner best suits job conditions. In any event, however, the filling of the connecting cell should be kept behind the filling of the circular cell on either side of it to ensure that the full circular shape of each circular cell is maintained.

Weep holes must be provided in the cell piling to permit drainage of the water from the fill material. These holes are normally along the inside of the cofferdam and should be of such number and size as to prevent excessive interlock stresses caused by high head inside the cell, and at the same time prevent the loss of an excessive amount of the fill material through the holes. The holes can be burned through the piling after the cells are in place, but if the river level is high during construction the holes should be made before setting the piling to ensure early drainage.

The efficiency of the dredging operation will depend on the amount of suitable material available in the deposits at hand. An adequate survey should therefore be made in advance to determine the extend of the deposits before the method of filling the cell is adopted. See tables 14 and 15 for a summary of cell filling operations on TVA projects and a record of the operations of TVA-owned dredge "Dallas" while working in Pickwick lock cofferdam. It will be noted that (except at Pickwick and Kentucky cofferdams) hydraulic filling of the cells was done under contract by private dredging firms.

Dredging Specification--The following specification covered the filling of the second-stage cofferdam at Guntersville and the furnishing of river-run gravel for the dam.

Section 1--Location and description of work. The Guntersville Dam, now under construction by the TVA, is located on the Tennessee River in Marshall County, Alabama, about nine miles downstream from Guntersville, Alabama, and approximately at Mile 349 (U. S. Engineer's mileage of 1909).

The dam will consist of earth embankment across the flood plain and a concrete navigation lock, spillway and powerhouse structure in the river channel.

The first stage of construction within the limits of the river, comprising the construction of the lock within a cofferdam consisting of earth-filled steel sheet piling cells, is approaching completion.

The TVA will pull the steel sheet piling from the lock cofferdam and will re-drive it to form a cellular cofferdam for the second construction stage which is expected to include 15-1/2 sections of concrete masonry spillway immediately adjacent to the lock. The number, however, is subject to change to a minimum of 2-1/2 and a maximum of 17-1/2.

Under Item 1, the Contractor shall fill the cells of the second-stage cofferdam with the fill material from the first-stage cofferdam as it is released, supplemented if necessary by material dredged from the bed and bar or bank of the river at such points as directed.

Any fill material released from the first-stage cofferdam which is in excess of that required for filling the second-stage cofferdam shall be removed and deposited at locations as directed.

The Contractor shall not dredge closer to the second-stage cofferdam than (50) feet, except as specifically permitted or directed by the Engineer.

Under Item 2, the Contractor shall furnish river-run gravel for the riprap blanket of the south dam and deposit it as directed in one (1) and in two (2) stockpiles on the south bank of the river within 1500 feet upstream or downstream from the axis of the dam and not less than 500 feet from the riverbank.

Section 2--Contract drawing. The accompanying drawing, entitled "Guntersville Dam, Second-Stage Cofferdam, General Plan and Sections N-6182.1" and dated January 8, 1937, shows the approximate location and construction of the cofferdam. Certain controlling dimensions and elevations are as follows.

Approximate length of area enclosed of cofferdam	600-800 ft.
Approximate width of area enclosed by cofferdam	350 ft.
Top of cofferdam--river side upstream and south end	El. 575
Top of cofferdam--river side downstream	El. 572
Top of cofferdam--encl. side upstream and south end	El. 570
Top of cofferdam--enclosed side downstream	El. 567
Top of fill upstream and south end	El. 570
Top of fill downstream	El. 567
Normal low water	El. 550-555

Section 3-- Contractor's plant. The Contractor shall provide a dredge having a capacity of not less than one hundred fifty (150) tons of solid material per hour and shall provide all boats, barges, screens, and other equipment and accesories for the proper and expeditious performance of the work, and shall at all times so man, maintain, and operate his plant as to do the work in the manner, and at, and within the time required.

Section 4--Construction of cofferdam cells. The second-stage cofferdam will be of the steel sheet pile circular cell type and the successive cells tied together by "connecting cells" of steel sheet pile diaphragms in short arcs joined to the cells by fabricated T-pieces of sheet piling.

The construction of the cofferdam will probably be commenced at the lock end of the upstream side, and will proceed toward the south bank of the river by the construction of successive cells and connecting diaphragms, then along the river side, and finally along the downstream side toward the lock until closure is made. The TVA, however, reserves the right to revise this procedure at any time and construct the cells in such other order as deemed most desirable by it. Cells No. 2-2 on the upstream side and No. 2-36 on the downstream side of the second-stage cofferdam will be left in place from the first-stage, and the new cells between these and the lock walls will be filled by the TVA.

The TVA proposes to construct five (5) to ten (10) cells per week, working 24 hours per day, less lunch periods, for five days per week and approximately 10 hours Saturday, with no work on Sundays or holidays. However, the TVA reserves the right to revise its working sechedule and working hours at any time deemed by it advisable. Such changes may include working on Sundays and holidays whenever require by an emergency.

Section 5--Filling the cofferdam cells. To avoid collapse or deformation of completed cells, it is necessary that each be filled without delay after completion of the pile driving operation. The Contractor shall, therefore, conduct his operations as follows:

(a) Start filling each circular cell not later than the time of completion of the cell immediately succceeding or sooner if the Engineer so directs, and promptly carry the fill to not less than elevation 555 nor more than elevation 564.

SEABOARD STEEL CORPORATION

STEEL SHEETING AND PILE HAMMERS FOR CELLULAR COFFERDAMS

- **VIBRATORY PILE DRIVERS/EXTRACTORS**
- **FLAT WEB STEEL SHEET PILING**
- **H-PILING**
- **DIESEL PILE HAMMERS**
- **AIR/STEAM PILE HAMMERS**
- **SWINGING AND FIXED LEAD SYSTEMS**
- **VERTICAL EARTH AUGERS & FLIGHTING**
- **POINTS, SPLICERS & ACCESSORIES**

Seaboard Steel Corporation
P.O. Drawer 3408
Sarasota, Florida 32430

PH) 813-355-9773
FAX) 813-351-7064

Table 14
Data on Filling and Filling Equipment for Steel Sheet Pile Cofferdam Cells - TVA Projects

Project	Stage	Type	No. Cells	Cell Diameter Feet	Cell Diameter Inches	Cubic Yards of Fill Dredge	Cubic Yards of Fill Clam	Cubic Yards of Fill Truck	Cofferdam Started	Filling Ended	Dredge Name and Owner	Dredge Type	Dredge Size	Pump Motor Volts	Pump Motor Horsepower	Cutter Head Motor Hp	Cutter Head Number Blades	Capacity Tons per Hour
Pickwick	Lock	Circular	2	40	10-1/2	83,000	10,200	--	Aug. '35	Oct. '35	Dallas (TVA)	Elec.	16"	2300	800	200	6	250-280
	Lock	Circular	24	58	10-5/8	93,800	--	--	7/20/35	8/15/35	Dallas (TVA)							
	2	Circular	22	58	10-5/8	129,000	9,600	--	6/4/37	8/18/37	Dallas (TVA)							
	3	Circular	26	58	10-5/8													
Guntersville	Lock	Circular	37	42	11-3/4	61,100	--	--	4/30/35	6/8/36	Dallas (TVA)	Diesel-elec.	16"	2300	500	75	8	--
	2	Circular	30-1/2	42	11-3/4	72,400	6,900	--			Columbus[1]	Columbus[1]	15"		600	--	7	300
	3	Circular	6	54	1-1/4						Greenville[1]	Elec.						
	3	Circular	11	54	1-1/4	31,300	32,300	--			Greenville[1]							
	3	Circular	13	42	11-3/4													
Chickamauga	Lock	Circular	38	47	9	75,300	--	--	2/20/36	7/22/36	Columbus[1]	Diesel-elec.	16"	2300	500	75	8	--
	2	Circular	34	47	9	67,100	2,000	--	5/8/37	7/20/37	Greenville[1,2]	Elec.	15"		600	--	7	300
	3	Circular	33	47	9	58,500	17,500	12,200	2/5/38	5/14/38	Greenville[1]							
	3	Circular	1	58	10-5/8													
Watts Bar	Lock	Circular	33	42	11-3/4	31,900	44,400	--	8/16/39	11/1/39	--	Diesel-elec.	15"			--	--	150-400
	1-A	Circular	17	42	11-3/4		--	--	5/21/40	6/18/40	Minneapolis[3]							
	2	Circular	33-1/2	42	11-3/4	46,200	--	21,200	10/11/40	11/17/40	Minneapolis[3]							
Fort Loudoun	2	Circular	30	47	9	18,400	--	55,830			Chickasaw[4] and one other	--	--	--	--	--	--	--
Kentucky	1-B	Circular	35	58	10-5/8	304,500					Dallas (TVA)							
	1-B	Circular	2	39	9-1/4													
	1-B	Cloverleaf	8	(See figure 3)														
	2	Circular	24	58	10-5/8	206,200		40,600			Dallas (TVA)							
	2	Circular	1	41	4-1/2						Dallas (TVA)							
	2	Cloverleaf	2-1/2	(See figure 3)							Dallas (TVA)							

1. American Aggregate Corporation.
2. 85,600 cubic yards pumped to storage piles, then cells filled from these piles.
3. Minneapolis Dredging Company.
4. Birmingham Slag Company.

Table 15
Operating Data on Hydraulic Dredge "Dallas" Filling Pickwick Lock Cofferdam

Period	Gross Operating Time Hr.	Gross Operating Time Min.	Delays Mech. Hr.	Delays Mech. Min.	Delays Pump Hr.	Delays Pump Min.	Delays Misc. Hr.	Delays Misc. Min.	Delays Ashore Hr.	Delays Ashore Min.	Delays Total Hr.	Delays Total Min.	Net Operating Time Hr.	Net Operating Time Min.	Cu Yd Delivered	Power (Kwh Used)	Cu Yd per Net Operating Hour	Power (Kwh per Cu Yd)	Percent Total Dredge Time	Average No. Feet of Pipeline
9/1-9/15 1935	101	00	2	00	0	00	4	35	12	00	18	35	82	25	23,800	65,000	289	2.73	38	1,809
9/15-9/30 1935	312	00	8	30	10	20	48	55	13	00	80	45	213	15	61,022	187,000	264	3.06	100	1,072
10/1-10/15 1935	44	30	0	00	0	00	18	55	2	35	21	30	23	00	8,415	19,000	366	2.26	15	800
Total	457	30	10	30	10	20	72	25	27	35	120	50	336	40	93,237	271,000	277	2.91	--	1,234
Clamshell															-10,237					
Revised figures for dredge													336	40	83,000	271,000	247	3.27	--	1,234

(b) At no time shall more than two circular cells be allowed to stand with the fill lower than elevation 555.
(c) Each connecting cell shall be filled to elevation 555-564 upon, but not before, the completion of the corresponding fill in the two adjacent circular cells. This fill shall in no case be carried higher than that in the adjacent circular cell.
(d) Completion of the filling of each circular and each connecting cell shall be made when the circular and connecting cells next succeeding have been filled to elevation 555-564, except as deferred with the approval or at the direction of the Engineer.
(e) All circular cells and the connecting diaphragm cells shall finally be filled to the elevations given in Section 3, except as another level be specifically ordered by the Engineer in writing.

Section 6--Material for item 2. The gravel shall be river-run material with the fines removed so that not more than 10 percent by weight of the representative sample of any one hour's production shall pass an A.S.T.M. standard square mesh No. 4 laboratory sieve.

Section 7--Source of material. River-run gravel shall be secured from the area outside that bounded by lines 800 feet upstream and downstream from the axis of the dam, except as the Engineer may permit procurement from within the prohibited area at such points and under such conditions as he may direct. Waste material from the production of gravel shall be disposed of as directed.

Section 8--Measurement and pay. The total quantity of cofferdam fill to be paid for under Item 1 shall be the number of cubic yards dredged by the Contractor from the lock cofferdam and from the river and deposited in the circular and connecting cells of the second-stage cofferdam together with the number of cubic yards dredged by the Contractor from the lock cofferdam and disposed of at locations as directed.

The quantity deposited in the cofferdam shall be measured above the actual level of the river bottom established by cross sections taken prior to the filling operations, with horizontal areas computed from design dimensions as shown on drawings.

The quantity removed from the lock cofferdam and otherwise disposed of shall be computed by measurement of "cut" before dredging, using the method of cross section and average end areas.

No payment will be made to the Contrator for any material placed by the TVA or left in place from the first-stage cofferdam.

The total quantity shall be computed from measurements of the stockpile, or piles, by methods of cross section and average end area.

Special Hydraulic Placing Method--Chickamauga Cofferdam, Stage 3--During this stage the sand and gravel fill material was reclaimed by dredge from the riverbed and stockpiled on shore, from which it was hauled with dump trucks to the cofferdam site to be placed hydraulically by combining it with water at a land-based pumping station. (See figures 65 and 66). The fill material was dumped directly into the receiving hopper from the trucks mounted on a timber ramp 10 feet high. Dishcarge from the receiving hopper into a second hopper, or sluicing chamber, was controlled with a pneumatically operated gate . Water for the sluicing chamber entered at a 45-degree angle from a single nozzle fed by two 12-inch lines, fed in turn by two 14-inch vertical, centrifugal pumps located offshore near one of the cofferdam cells. The velocity of water entering the sluicing chamber was sufficient to force the fill material along a 16-inch feeder line to the cells.

This method gave a placing capacity ranging from 100 cubic yards per hour for the first cell to 40 cubic yards per hour for a cell 470 feet distant. An Allis-Chalmers horizontal centrifugal booster pump rated at 1740 gallons per minute at 200-foot head, powered by a 150-horsepower General Electric motor, was added at this point, increasing the placing capacity to 80 cubic yards per hour. This gradually decreased at the rate of 7 cubic yards per hour per 100 feet over a distance of an additional 560 feet to the last cell. Water was fed from the booster pump into the 16-inch feed line through a wye connection.

Special Hydraulic Placing Method--Hales Bar Dam--The hydraulic filling of the upstream cofferdam at hales Bar Dam during the installation of generating units 15 and 16 was performed in a somewhat different manner from that normally used by TVA during earlier cofferdam construction. This was necessary since the size of the cells made it essential that the cell fill be free draining to ensure stability.

The material which came from the riverbed at a point downstream of the dam passed through a barge-mounted booster pump and thence over the existing dam structure to the cofferdam. Here it was passed through a revolving screen with only that material retained on a No. 4 screen being used as cell fill. The dredge line discharged into a hopper that fed the screen. Material passing the screen was collected in a waste pan and discharged back into the river outside the cells. Material retained on the screen was directed by means of a chute to the approximate center of the cell as cell fill.

The above procedure was essentially satisfactory and produced a fill material that was free draining. It is suggested, however, that a grizzly might have been placed ahead of the screen to bypass the large boulders, thereby improving screen operation and reducing wear on the screen and discharge chute.

Truck and Clamshell Filling

This type of cell filling was used by TVA principally in topping out cells that had been partly filled by the hydraulic dredge and for supplementing the dredging operations to speed up the work. An effort was made in every instance to fill hydraulically those cells located in the bed of the stream. This was not always possible, however. For instance, at Fort Loudound Dam where the dredge fell behind schedule, it became necessary to fill some of the cells by other means. Crib tie-ins and cells built on dry land at both Fort Loudoun and Watts Bar Dams were filled with dry material, using trucks and crane-operated clamshell. (See table 14 for tabulation of quantities placed by the various means.) Where cells cannot be filled by the hydraulic method the fill must of necessity be placed dry. In using trucks for filling circular sheet pile cells it should be kept in mind that the trucks will have to be backed in for dumping. It should also be remembered that the cell on either side of the connecting cell must be filled first, in order that the true circular shape of the cell be maintained. This will mean that a bridge will have to be built over the connecting cell to reach the next cell. The connecting cell will then be filled last. Where fill material can be hauled directly to the cells by barge, cranes equipped with clamshell buckets may be used with satisfactory results.

In filling the crib-type cofferdam, high cribs should not be filled by dumping from the top. TVA had one failure from this method of filling, which was probably caused by the impact of falling material against the crib timbers. High cribs should either be filled progressively with crib construction or they should be filled carefully to minimize the effect of impact.

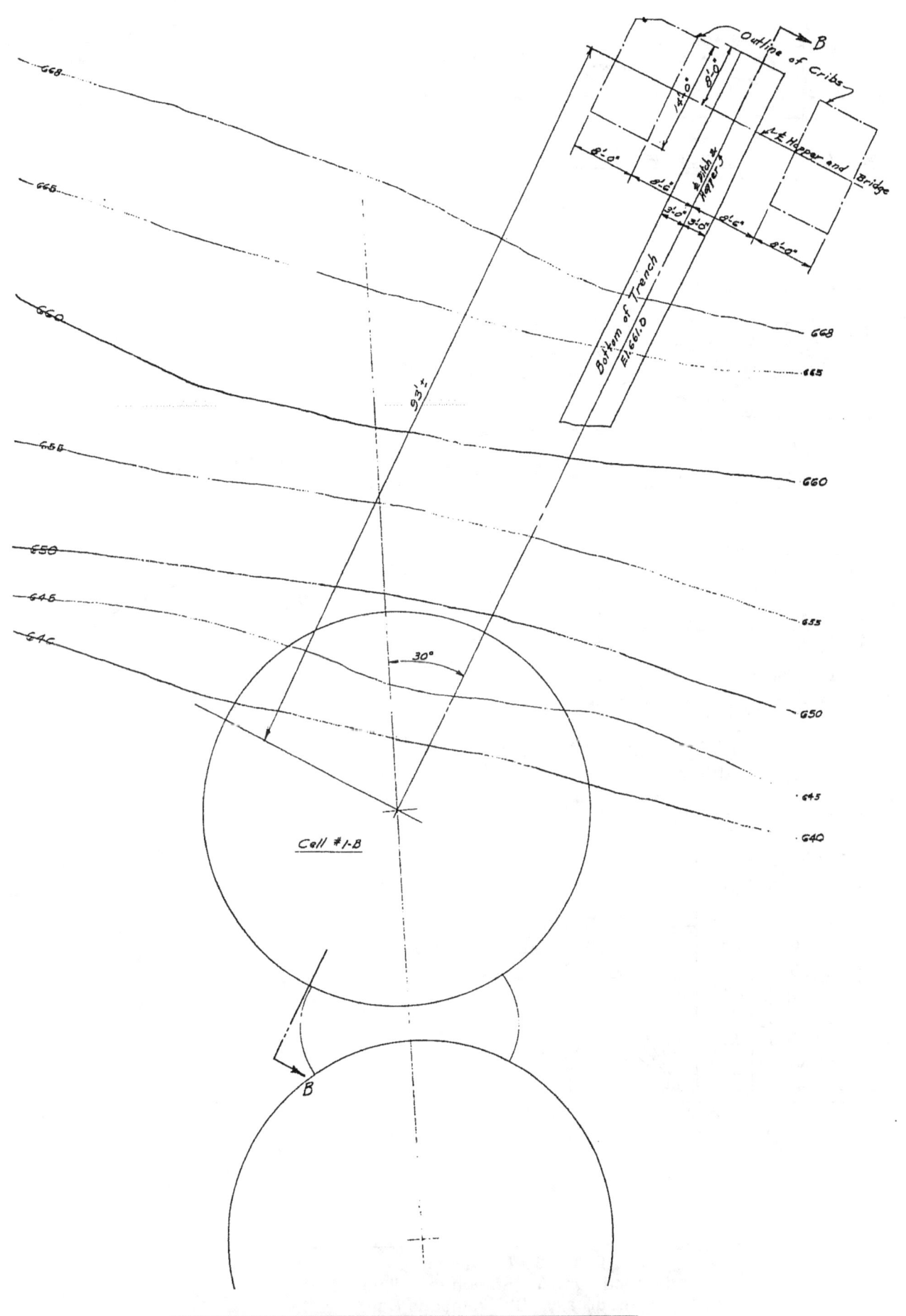

**Figure 65
Land-Based Pumping Station for Filling Cells
Hydraulically - Chickamauga Dam**

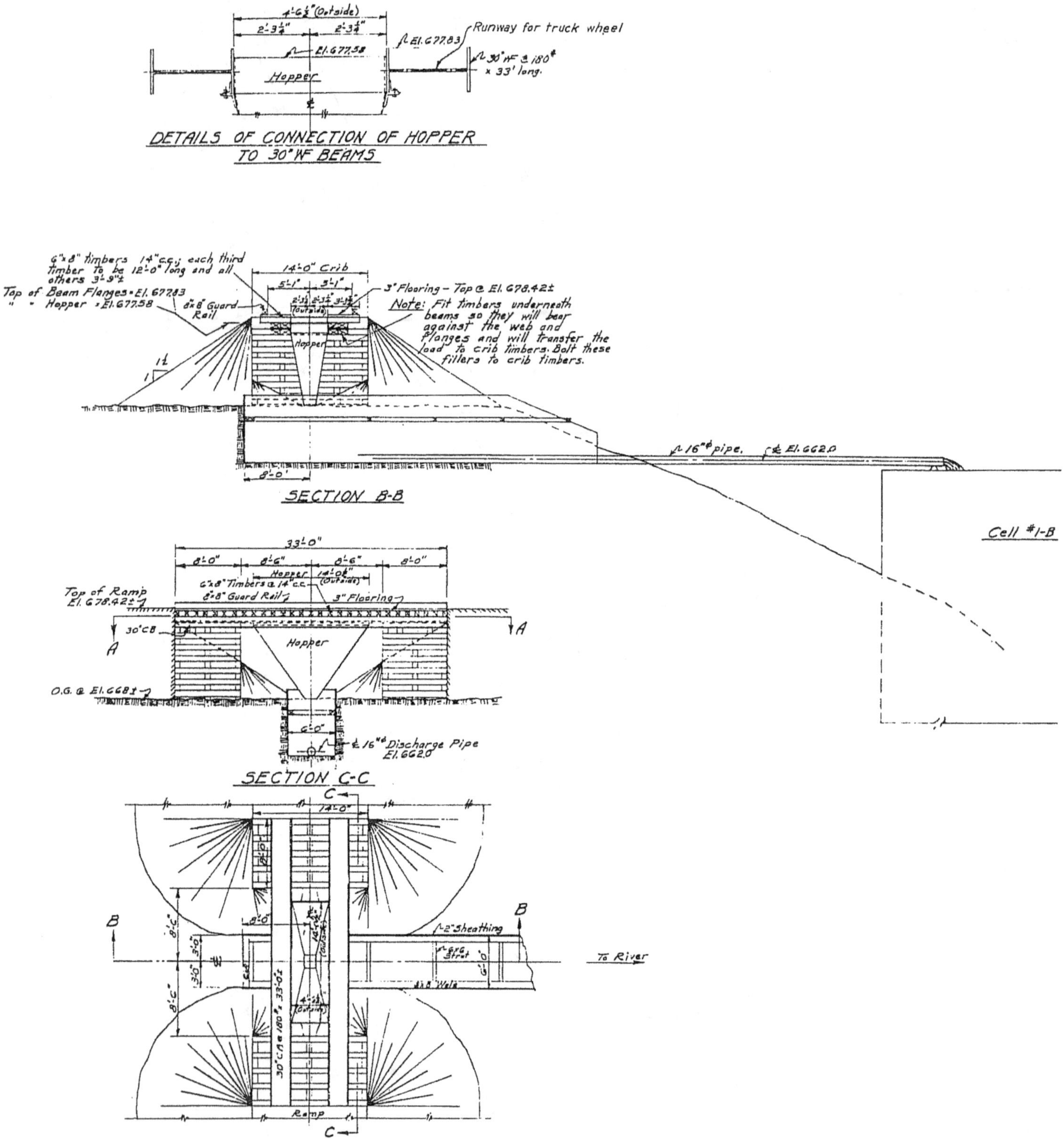

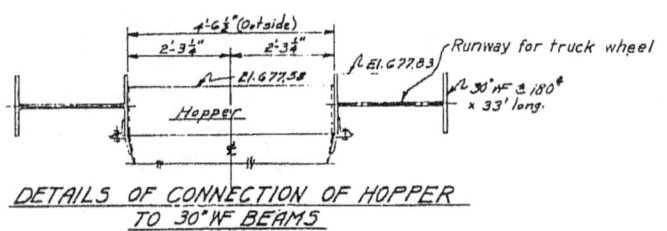

**Figure 65 (Continued)
Land-Based Pumping Station for Filling Cells
Hydraulically - Chickamauga Dam**

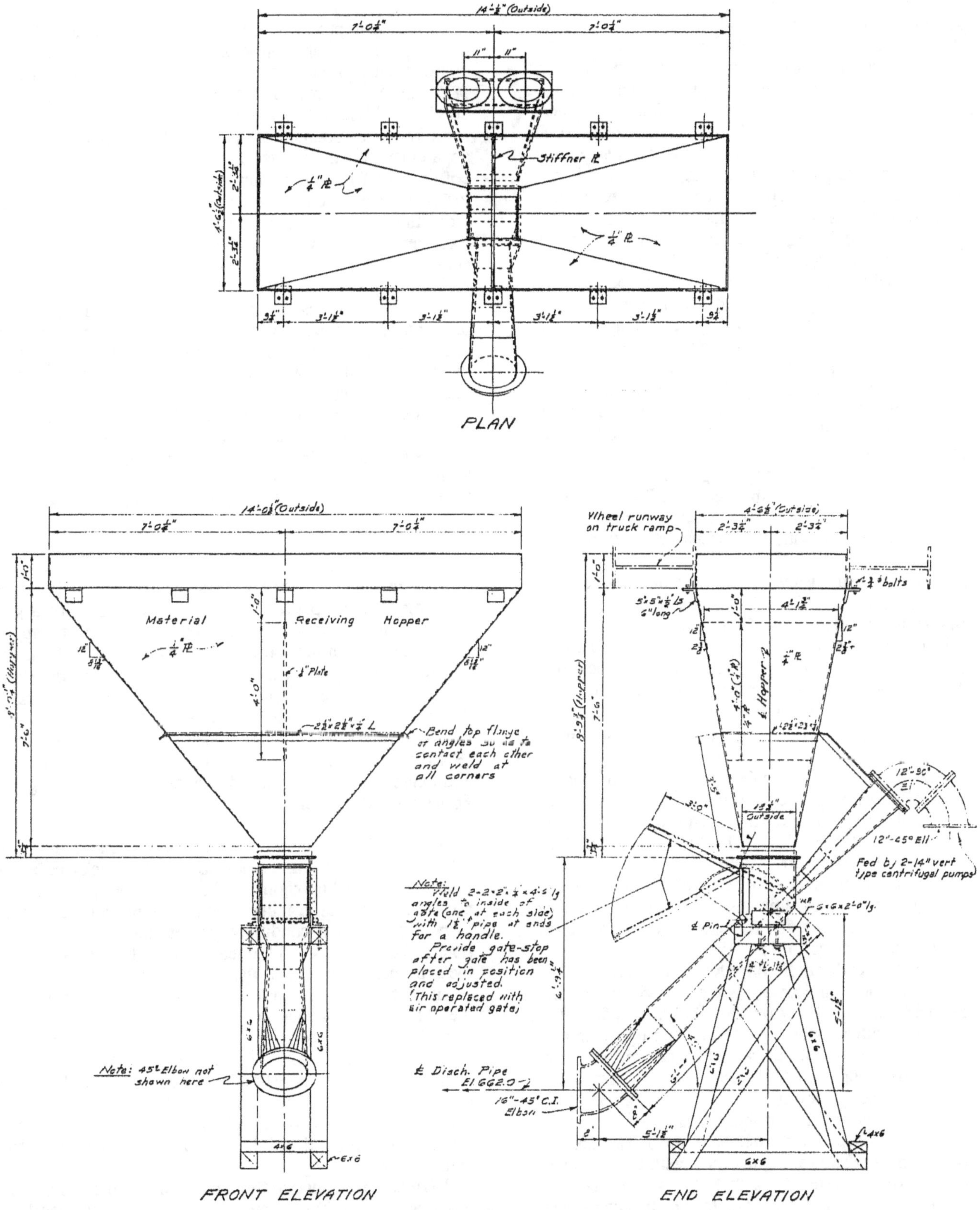

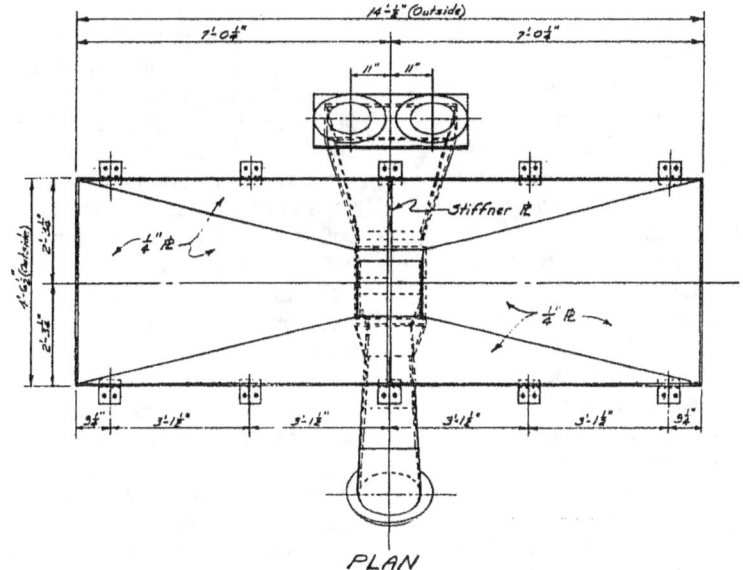

Figure 66
Details of Sluicing Hopper Used in filling Cells Hydraulically - Chickamauga Dam

FLOODGATES AND SLUICEWAYS

Flooding of the cofferdam area by permitting floodwaters to overtop the cells may cause serious damage to the cofferdam. Fill material washed from the cells and berm can materially weaken and even cause collapse of a portion of the structure. To control this flooding operation, floodgates, sluiceways, or both, should be installed in the cofferdam. Such facilities should be a part of all large cellular steel sheet pile cofferdams. Preflooding of the cofferdam, however, will not necessarily entirely prevent scour of the cell fill, where the cells are completely submerged. It will reduce scouring materially and, in general, will ensure against collapse of the structure. A careful check of flood stage predictions must be kept as a basis for determining when equipment should be removed from the cofferdam area and the gates opened for flooding. Where serious leaks occur due to the flood stage, it may become necessary to flood the cofferdam to equalize pressures and prevent serious damage to the cofferdam, even if the predictions do not indicate possible overtopping by floodwaters. Floodgates and sluiceways are also used for flooding the area upon completion of the work and just prior to cofferdam removal.

The size and number of flood openings must be determined by calculations based on size of the area to be flooded and the possible rate of rise of the river. Results should provide gates which will flood the area within a reasonable length of time, say 2 to 4 hours.

Sluiceways are usually installed by placing a steel pipe through a hole cut in the piling of one of the connecting cells. The piping is then welded in place for security and to prevent leakage around the pipe. Flow through the intake end is easily controlled by means of a sliding gate operated from the top of the cell. However, berms at the discharge end must be protected. This can be accomplished by means of a pipe extension, timber flume, concrete flume, or heavy riprap protection for the fill.

Floodgates may also be installed in one or more of the connecting cells, in which case the piling may be cut off at the proper level and the cell capped with concrete to provide a flow line for the water. Gates installed on TVA projects for this type of opening usually consisted of vertical timbers or needle beams. These could be raised for removal by jacking or by lifting with a crane hook. Beams raised in this manner will be swept into the cofferdam by the force of the water, however, unless provisions are made to retain them.

Kentucky Cofferdam Floodgates and Sluices--Two floodgates and one sluiceway were provided in the downstream arm of stage 1-b cofferdam at Kentucky Dam, but in stage 2 one floodgate was provided for the upstream arm and one floodgate and one sluiceway for the downstream arm. (See figure 1 for general location.) An 8- by 8-foot concrete box was used for the sluiceway. (See figure 67 for details of the gate and box. For details of the floodgate see figure 68. Note the brackets attached to the needle beams for jacking the needle beams in pairs and for raising them with a crane hook. The beams were covered with canvas to increase watertightness, as shown. Note, also, the needle beams used to close the opening above the sluice gate. These beams were bolted securely in place, however, and were not removable for flooding the cofferdam.

Floodgates for Other TVA Projects--Floodgates or sluiceways, sometimes both, were used on all TVA steel sheet pile cellular cofferdams. Flooding facilities on these projects were similar in design to those used at Kentucky but were generally smaller and less elaborate in detail because of the small size of the area enclosed.

FOUNDATION TREATMENT

A thorough program of foundation consolidation through grouting to seal against leakage should be conducted in conjunction with cell construction and should be completed prior to cofferdam unwatering. Solution channels and crevices, although temporarily sealed by natural materials, form potential sources of leakage that will become more apparent with the increase in head of the water outside the cofferdam. Thorough preliminary core drill investigations of the site to be occupied by the cofferdam will give some indication of the amount of grouting to be expected. Experience on TVA projects, however, has been that considerably more grouting is usually required than originally anticipated.

Interlock ruptures present another threat of leakage. Cell failure may be minimized by consolidation of the overburden or portions of the cell fill in the vicinity of the rupture, provided the location of the rupture can be established. It must be borne in mind, however, that the cell fill should be grouted except in special cases where other means of sealing are ineffective. Grout from the fill will leak into the piling interlocks, making pile extraction difficult if not impossible. (See "Interlock Failure--Kentucky Cofferdam," page 24, for methods of investigation conducted during pile driving at Kentucky cofferdam.)

Grouting Equipment

A complete grouting unit (the same as used for foundation grouting on other parts of the project) consisted of a grout mixer, agitator, water meter, and two grout pumps all completely assembled on a single skid-mounted fram with piping so arranged that it was necessary only to connect air, water, and grout supply lines to put equipment into operation. Several types of grout stops were used to confine the grout to specific areas and to prevent its leaking into the overburden. (For these and a detailed discussion of grouting methods and equipment see TVA Technical Report No. 22, Geology and Foundation Treatment.)

Foundation Treatment at Kentucky Cofferdam

Grouting of the foundation for the Kentucky cofferdam represented the maximum effort on the part of TVA construction forces in foundation treatment, due not so much to poor foundation conditions on this job but rather to the potential danger of small foundation faults being magnified by the extremely high head of water behind the cells. Drilling and grouting to consolidate the rock foundation and portions of the 50 to 75 feet of sand and gravel overburden were started prior to and continued for some time after the cell construction was completed. Details of grouting procedure followed at Kentucky cofferdam may be accepted as a guide for foundation treatment procedure to be followed on future projects of a similar nature. (See figure 69 and 70 for typical method of recording drilling and grouting data obtained during cofferdam construction.)

Before Cell Construction--Diamond core drills mounted on barges and working through 3-1/2-inch pipe casings drilled holes at a 45-degree angle to a vertical depth of approximately 40 feet. This brought the holes from 5 to 10 feet below existing seams in the rock foundation (see figure 70). The 45-degree angle at which the holes were drilled was for the purpose of intercepting both vertical and horizontal seams. These holes were spaced on 60-foot centers. Vertical holes in figure 70 were drilled after the construction of the cells was completed.

To grout the inclined holes, a packer was set in the hole at the top of the rock to prevent the escape of grout into the

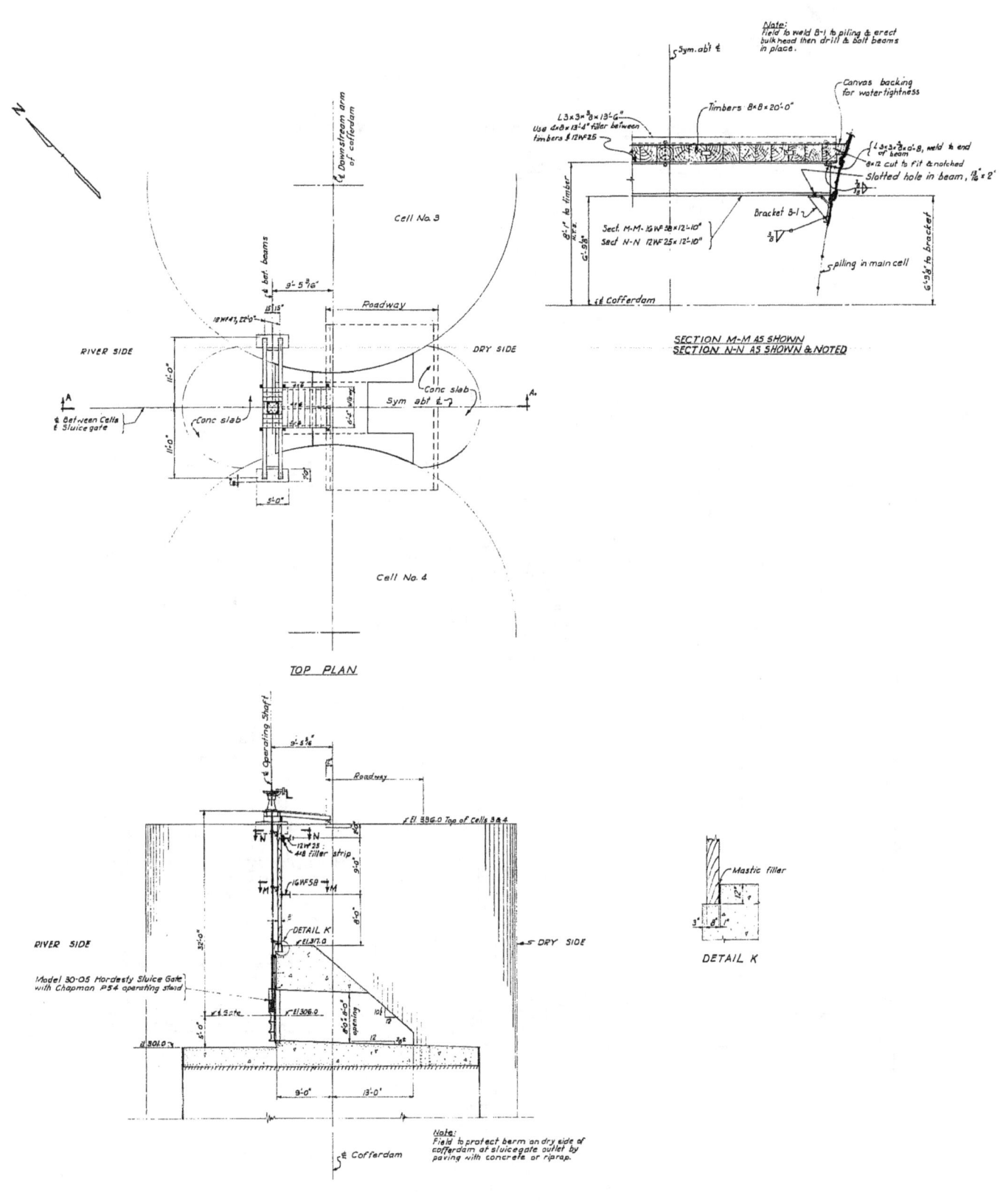

**Figure 67
Cofferdam Floodgate Showing Sluice Gate Below
and Bulkhead Above - Kentucky Dam**

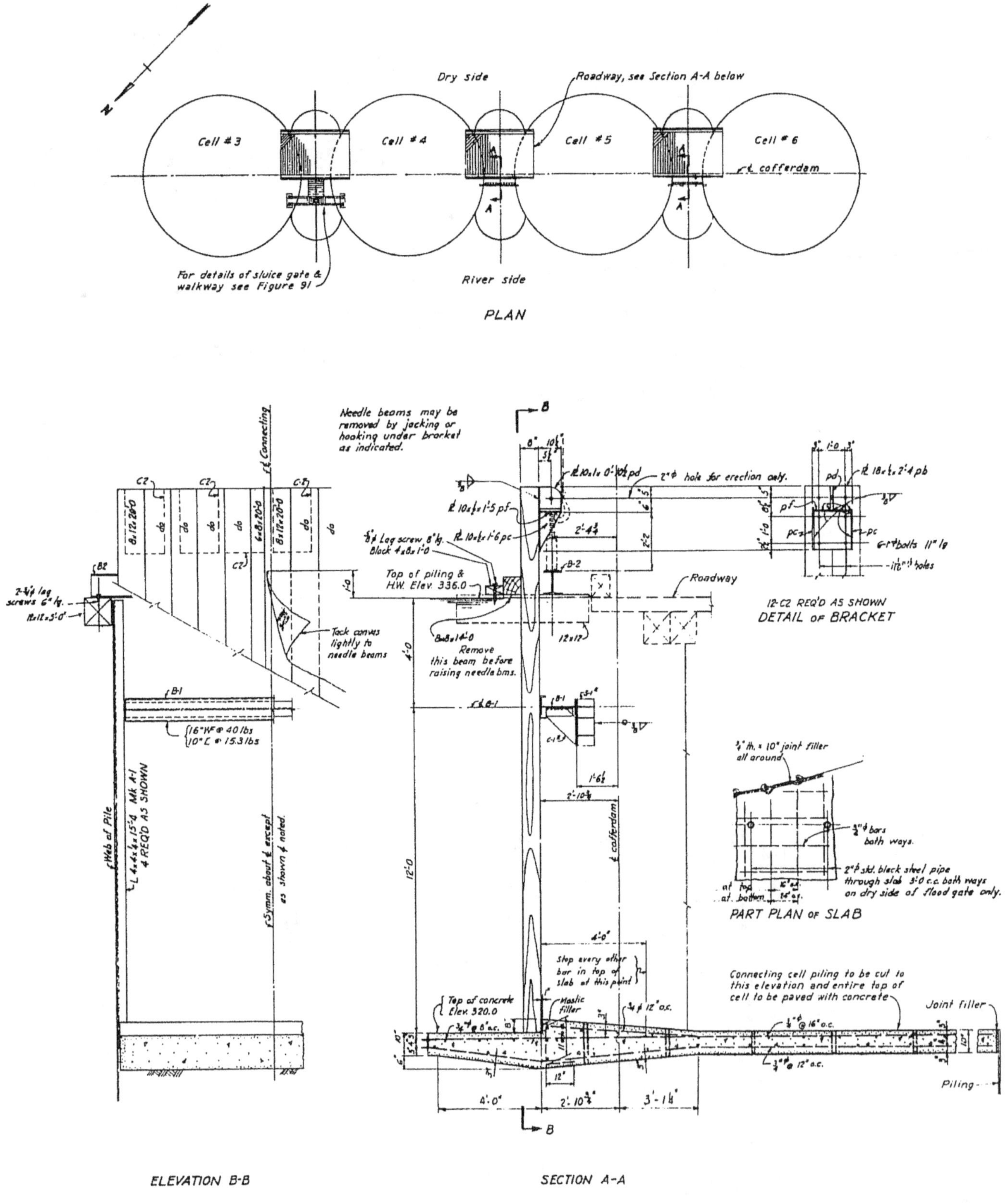

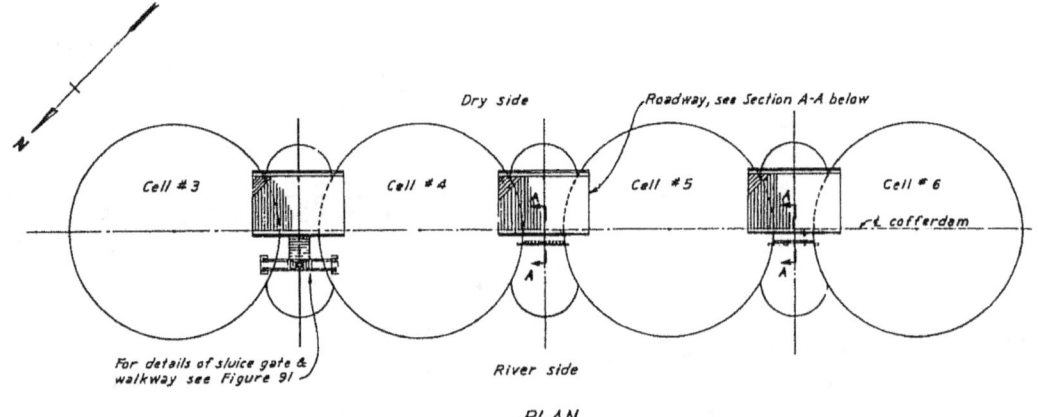

Figure 68
Bulkhead-Type Floodgate - Kentucky Dam

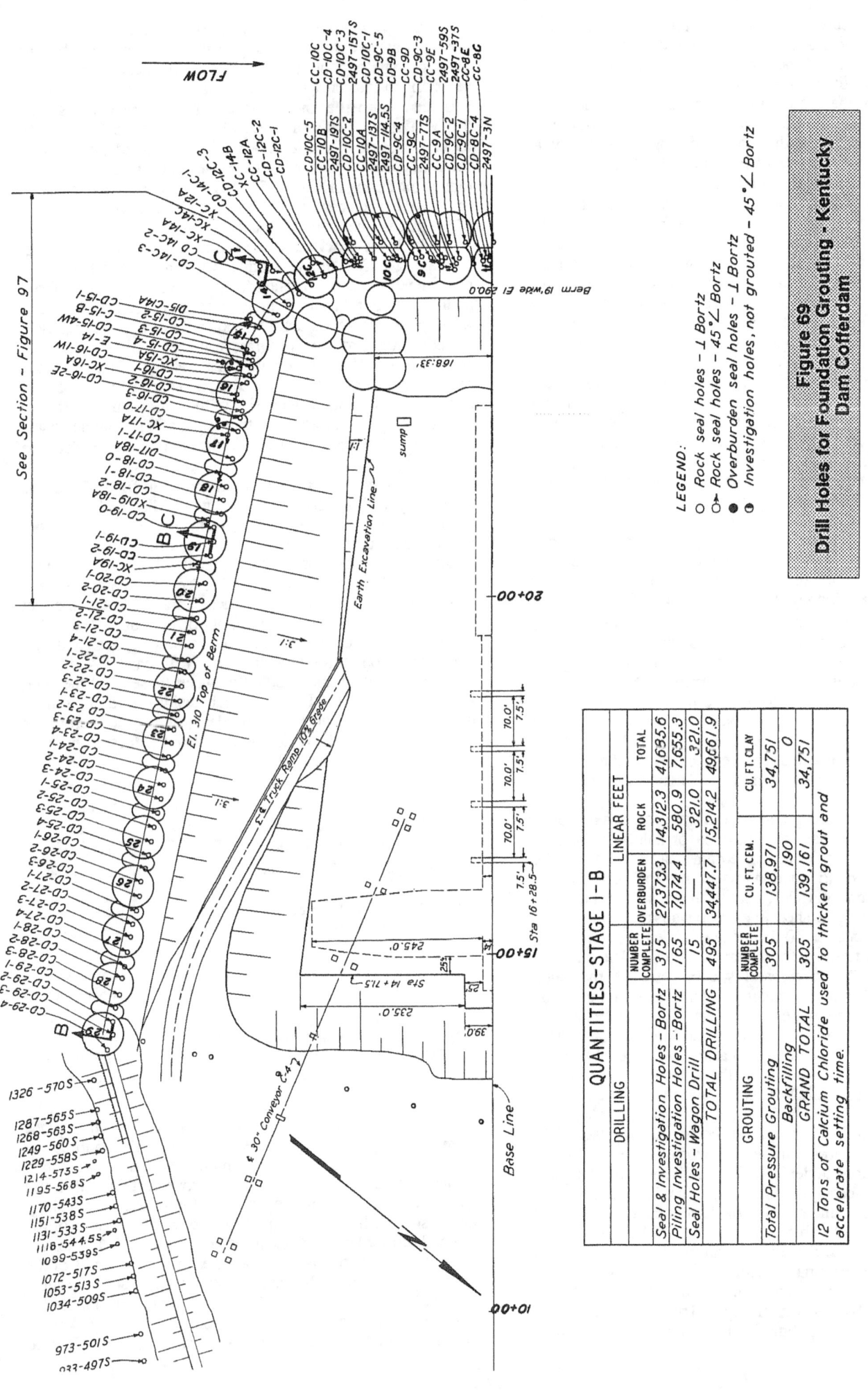

Figure 69
Drill Holes for Foundation Grouting - Kentucky Dam Cofferdam

overburden. Grout was injected to refusal into the rock foundation under pressures up to 20 pounds per square inch. Where the condition of the rock was so broken as to prevent proper seating of the pack and of sealing against leakage into the overburden, the grout was injected directly through the casing. This procedure was never very satisfactory, however, as it was difficult to seat the inclined casing so as to prevent the escape of grout. As a result, pressure at the top of the casing was held to 20 pounds per square inch.

After Cell Construction, Stage 1-B--After the cofferdam was constructed grouting was continued from the top of the completed cells through vertical holes until a continuous cutoff curtain, with holes 20 feet on centers, was completed for the entire length of the cofferdam. These holes were located on a line approximately 12 to 15 feet inside the centerline of the cells (see figure 69) and were drilled through 3-1/2-inch-diameter casings to a depth of 40 feet into bedrock or to below existing deep seams.

Where possible the packer was set in the grout holes near the top of the rock. If the condition of the rock was such that the packer could not be set near the top of the hole, however, then it was lowered and set just above the lower seams, after which the bottom part of the hole was grouted. The packer was then removed and the rest of the hole grouted through the casing. Before the holes were grouted, water was injected at grouting pressure into the holes to determine approximately the amount of grout that would be required. Grout was then injected, starting with a grout having a water-cement ratio of 2.0, which was gradually thickened until a pressure of 20 pounds per square inch was reached. As soon as refusal was reached the mix was again thinned and pumped continuously for 15 minutes to check refusal. The grout mix varied from 2.0 to 0.6 water-cement ratio, the latter being used where the grout was injected through the casings. A limit of 100 cubic feet of cement per hole was set when grouting through the casing. The amount of grout taken by each hole varied from a very small amount to as much as 1200 cubic feet. The additional holes added along the river arm to ensure positive cutoff resulted in an ultimate spacing of holes in this location of 10 feet center to center.

Clay-cement grouting of stage 1-B was used as a matter of economy to consolidate a large area of solution channels existing along the edge of the cofferdam area and adjoining the lock. Only the rock foundation was grouted for the area inside the limits of the cell and sheet pile cutoff walls, but for the area outside the limits of the sheet pile cutoff walls the grouting was extended to include the 40 feet of overburden above the rock foundation (see figure 1 for general location). The mix was 1-6-5 (cement-clay-water) and was injected in the same manner as the standard cement grouts. Injection was made almost entirely through the pipe casing. This was because of the broken state of the rock foundation. Grouting was carried to refusal in stages by raising the casing 2 to 3 feet at a time.

During Stage 2--In this stage the same general procedure was followed as for stage 1-B, except that all holes were drilled vertically. A grout cutoff curtain for the cofferdam arm parallel with the stream was effected during stage 1-B. Grouting in the upstream and downstream arms of the cofferdam was done under 30 to 50 pounds pressure per square inch, using a low cost grout made in batches in the proportions of 2 cubic feet of cement, 1/3 cubic foot of bentonite, 4 cubic feet of sand, and 13 cubic feet of water. As little as 6 cubic feet of water was used in some instances. Holes were drilled and grouted to give a final spacing of 20 feet on centers for the full length of each cofferdam arm. Drilling and grouting were conducted from barges until interrupted by construction, to be completed later by grouting from the top of the completed cells.

River Arm Sheet Pile Seal, Stages 1-B and 2--After driving was completed, a study of the position of the bottom of the piling along the river arm indicated that some piling had been driven into crevices and solution channels while others had been stopped above bedrock. In such cases holes were drilled along the side of these piles to at least 5 feet in good rock. Grouting of the rock was then completed first through the drill casings, after which the overburden was grouted by raising the casing in stages of from 1 to 2 feet at a time, grouting to refusal in each case, until a height of 3 feet above the bottom of the pile in question was reached.

Left Bank Rock and Overburden Treatment, Stage 2--Stage 2 cofferdam was flanked along the left bank by a broad level flood plain rising to approximately 80 feet above bedrock. Ground water drainage into the cofferdam from this flood plain was considerable and, although not completely effective, the grouting program together with the drainage control was sufficient to handle the leakage. The sheet pile protection wall along the west bank, and the cellular sheet pile retaining wall along the west side of the cofferdam area (figure 1) backed up by the grout curtain (figure 71), were sufficient to control the embankment during operations inside the cofferdam.

In grouting the rock foundation shown in figure 71, a line of holes spaced on 120-foot centers was drilled and grouted to refusal under pressures ranging from 30 to 50 pounds per square inch using a neat cement mixture of 1.0 to 0.6 water-cement ratio for holes inside the limits of the west embankment and a mixture of 2 cubic feet of cement, 1/3 cubic foot of bentonite, 4 cubic feet of sand, and 13 to 6 cubic feet of water for holes outside the limits of the dam. The initial holes were widely spaced to prevent grout loss by interconnections through seams known to exist. Additional holes were drilled and grouted between these until the final spacing was 30 feet. The 3-1/2 inch casings used in grouting the rock were left in place to be used in grouting the overburden later. Grouting of the rock was completed before grouting of the overburden was started.

Elevation	Mix by Volume	Pressures, Psi	Limit per Foot of of Depth
237 (bedrock) to 250	1 bentonite - 15 to 9 water	30	8 cubic feet of bentonite
250 to 265	1 bentonite - 15 to 9 water	25	8 cubic feet of bentonite
265 to 280	1 bentonite - 15 to 9 water	20	8 cubic feet of bentonite
280 to clay blanket	1 cement, 1/3 bentonite - 10 to 6 water	15	16 cu ft cement, 5-1/3 cu ft of bentonite

The holes at 30 feet on centers with casings left in place from grouting the rock were used first in grouting the overburden shown in figure 71. Intermediate holes were then drilled to rock and grouted until a grout curtain with holes spaced at 15 feet on centers extended for the full length along the west bank from upstream arm to downstream arm of the stage 2 cofferdam. A neat cement grout of 1.5 to 0.6 water-cement ratio was used for holes inside the limits of the west embankment, with a limit on cement being set at 50 cubic feet per linear foot of hole. Mixes prescribed in the tabulation below were used outside the limits of the west embankment. Grout was injected under a pressure ranging from 15 to 30

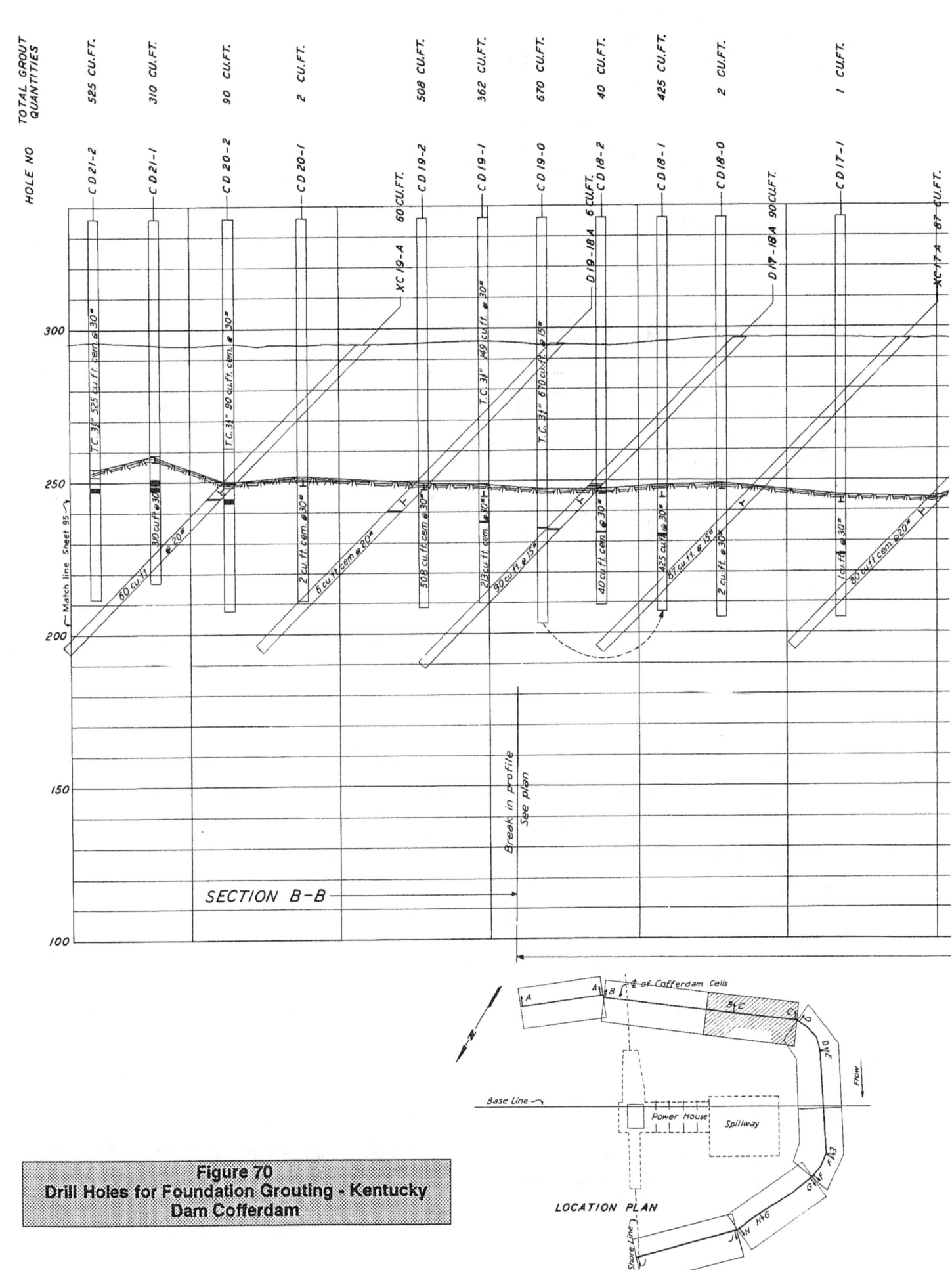

Figure 70
Drill Holes for Foundation Grouting - Kentucky Dam Cofferdam

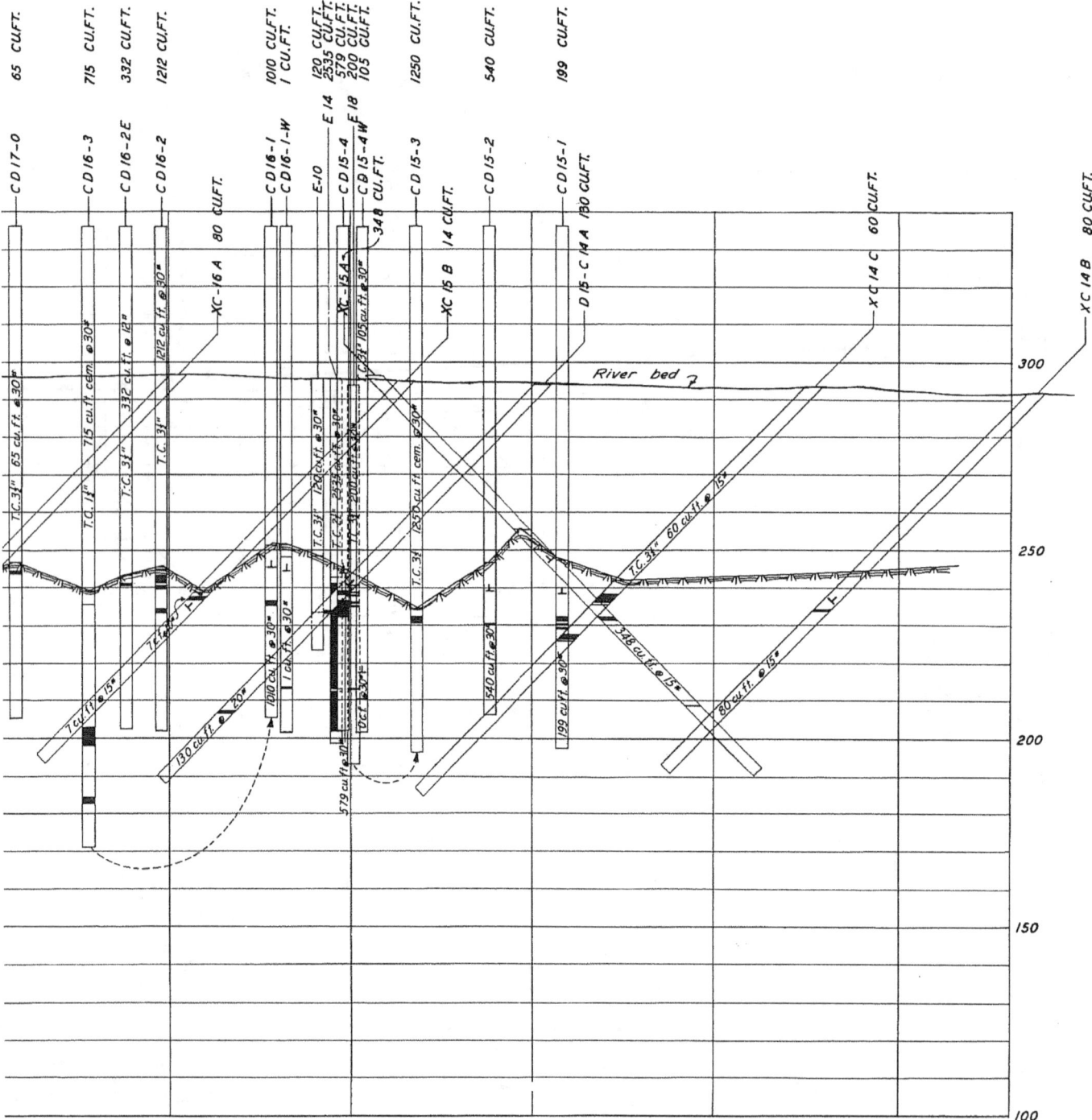

SECTION C-C – LOOKING UPSTREAM

**Figure 70 (Continued)
Drill Holes for Foundation Grouting - Kentucky Dam Cofferdam**

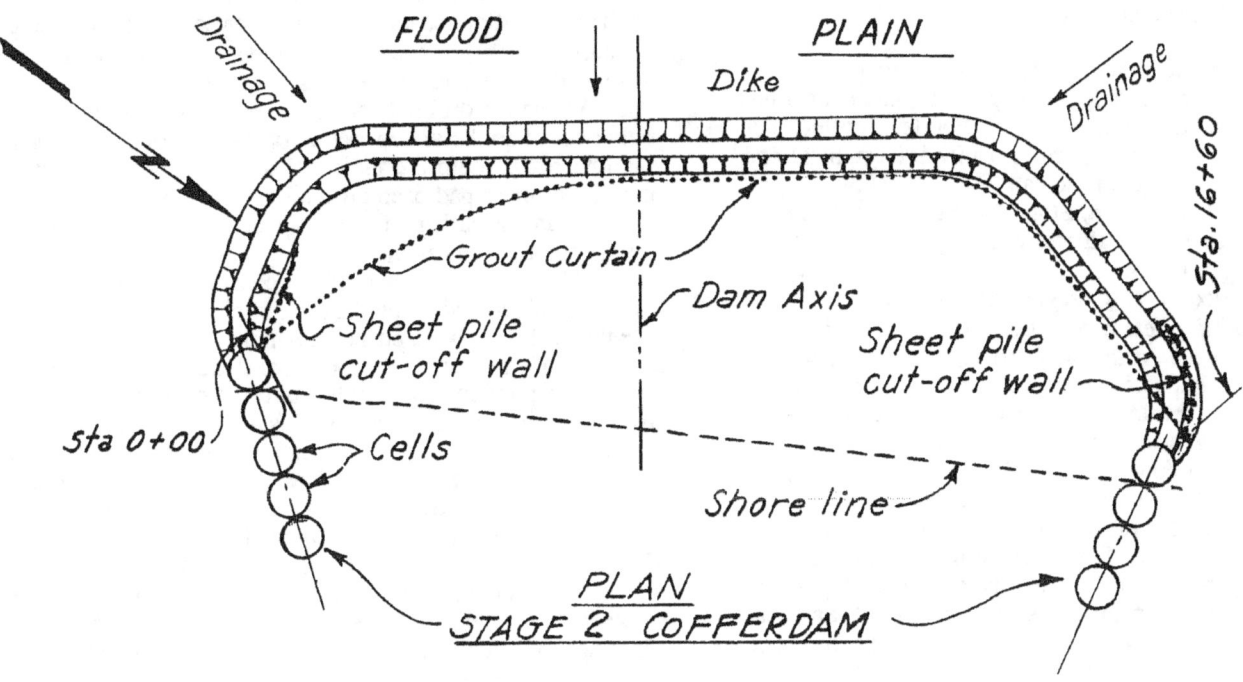

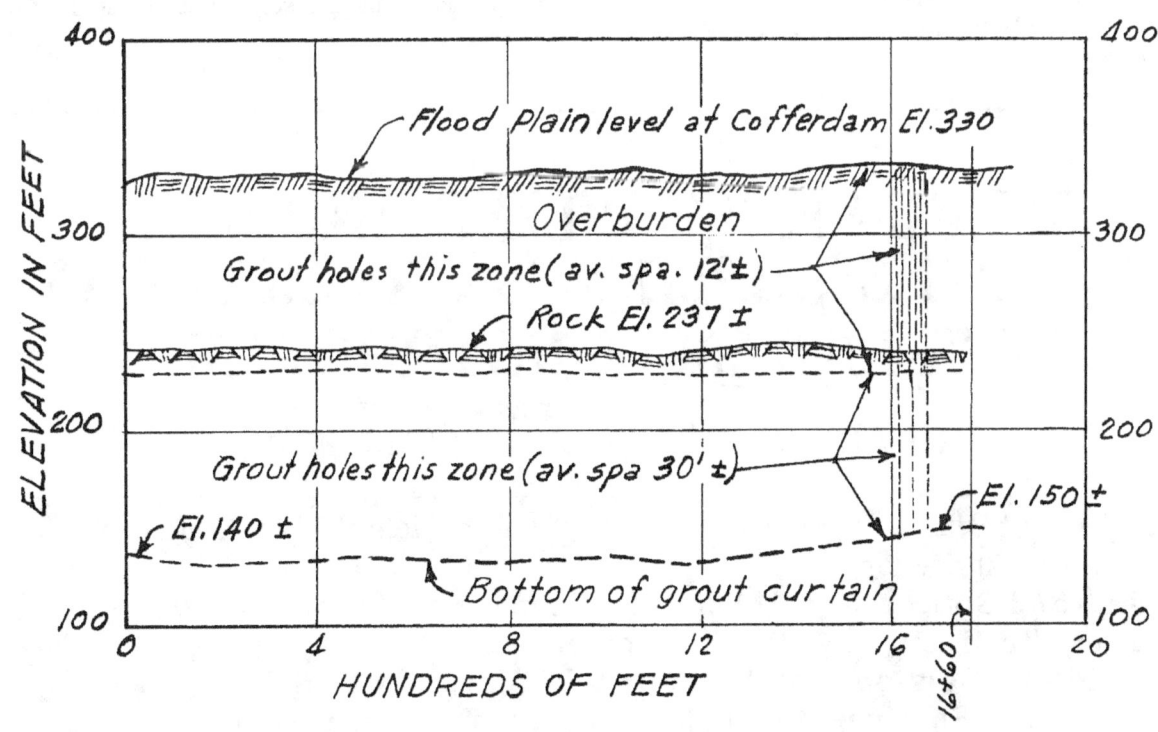

Figure 71
West Bank Protection Against Flood Plain Leakage at Kentucky Dam, as shown in Figure 98

pounds per square inch. The fill was grouted to refusal until the full depth of the overburden below the overlying clay blanket was consolidated. This was to a depth of approximately 63 feet above bedrock.

Foundation Treatment on Other TVA Projects

Since foundation conditions and grouting procedures for Guntersville, Chickamauga, and Fort Loudoun cofferdams were unlike those of Kentucky cofferdam, a brief summary of procedures on each project follows.

Guntersville Cofferdam, Stage 2--The foundation under this cofferdam ranged in structure from good rock free of overburden to good rock covered with a stratum of residual boulders embedded in sand, gravel, and silt.

No foundation treatment was attempted until after construction of the cells was completed. Four- and six-inch-diameter pipe casings were then sunk to rock on 5-, 10-, or 20-foot centers along the centerline of the cells. Sinking of the grout casings was accomplished by alternately driving and washing. The drop weight used for driving was suspended from, and operated by, a drill rig. Drills 3-1/2 and 5 inches in diameter were used for drilling purposes, and the size of the drills determined the use of the 4- and 6-inch casings.

In grouting the foundation a mix of standard portlant cement and water, of varying proportions, was used. Sawdust and calcium chloride were added to the mix for large water-carrying seams that could not be sealed with the cement and water grout. The foundation condition varied so widely from hole to hole that each hole became an individual problem. No standard grout mix with a fixed water-cement ratio was ever developed. However, the methods of procedure can be grouped into four general classifications, A, B, C, and D, which are described below.

A. Where the cells rested on a good rock foundation, free of overburden, the casings were driven through the cell fill to rock. Holes were then drilled in the rock to approximately 2 feet below existing seams and faults, a packer set in each hole at the top of rick, and the holes grouted to refusal--similar to the procedure followed at Kentucky cofferdam.

B. Where the drill hole entered a seam or crevice and the packer could not be seated in the top of the hole, the packer was lowered beneath the fault and the lower seams grouted first. After these had been grouted to refusal the packer was removed and grouting of the hole completed by grouting directly through the casing.

C. This method was used to consolidate the layer of residual boulders overlying the solid rock foundation. In this case the drill holes were pushed down through the boulders to a distance of 2 feet below the top of solid rock. Grout was then injected by pumping directly through the casing, raising the casing in stages and grouting to refusal at each stage.

D. Where the drill holes could not be pushed down through the boulders because of the inflow of sand and gravel, the drilling was stopped, the pocket of sand and gravel consolidated through grouting, then the drilling continued. The procedure, after drilling was stopped, was to wash a 1-1/2- or 2-inch pipe down through the pocket to rock, then grout the pocket by raising the pipe in stages of 1 to 2 feet at a time and grouting to refusal at each stage until the pocket was consolidated. Drilling was then resumed until solid rock was reached. Where another pocket was encountered, the above procedure was repeated. Grouting through the hole to consolidate the overburden and rock for its full depth was accomplished by using a suitable combination of the above methods.

For...

STEEL SHEET PILING AND PILE DRIVING EQUIPMENT FOR CELLULAR COFFERDAM CONSTRUCTION

CALL... (516) 579-PILE

- **Steel Sheet Piling** - Foreign or Domestic, Sales or Rentals; Light, Medium and Heavy Weight Sections
- **Points and Splicers** - For H-Piling, Pipe, Timber or Steel Sheet Piling (Call or Write for our new Point and Splicer Catalog)
- **Pipe Piling, Caisson Pipe, Dredge Pipe** - All Sizes & Lengths, New & Used
- **Pile Driving Equipment** - Vibratory, Diesel, Air/Steam, Sales or Rentals
- **H-Piling and Structural Shapes** - Foreign or Domestic
- **Augers** - Sales or Rentals

EAST COAST PILE

87 Barnyard Lane, Levittown, NY 11756
Dial: (516) 579-PILE

Table 16
Grouting and Foundation Treatment To Stop Leakage in Cofferdam - Chickamauga Dam

Hole No.	Cement	Bentonite	Sand	Oats	Sawdust	Cinders	Remarks
11	400						Refusal 75#
21	852			47	46	6	Pressure 0# leaks
31	554					187	Refusal 75#
41	888						Refusal 50#
42	14						Connected with 41
51	89	17	36		6	63	Pressure 40# leaks
52	72	24	96				Pressure 0# leaks
1 in cell 6							
61	821	671	454				Refusal 40#
	499	87	348				Refusal 50#
71	1,859	197	690	263	322	209	Refusal 40#
81	826	301	1,097	29		88	Refusal 50#
82	346	102	404		6	8	Refusal 75#
83	114	38	152				Refusal 40#
84	736	289	1,189	Wheat bran	2	2	Pressure 10# leaks
91	662	503	236	82	677	690	Pressure 0#
92	655	45	31		181	282	Pressure 0# leaks
93	97	35	92		12	12	Refusal 75#
101	330	110	357				Refusal 75#
102	183	24	96		100	113	Pressure 0# leaks
111	428	131	518		8		Refusal 50#
112	177	53	212				Refusal 40#
1 in cell 12							
	88	16	64				Refusal 40#
151	264	250	140		183	198	Pressure 0#
161	39	15	58				Refusal 75#
114	123	41	164				Pressure 40# leaks
73	2	2	2				Refusal 55#
39	38		76				Refusal 50#
39A	12		19				Refusal 50#
38	4		4				Refusal 50#
Pipe in seam for spillway excavation	518						
Totals	11,688	2,949	6,752	338	1,540	1,855	

Note: All cement and prescription grout was injected through the tops of casings.

Grouting pressures ranged from 5 to 50 pounds per square inch, depending upon the condition of the foundation.

Casing removal followed completion of the grouting operations. The pipe casings were removed with a standard pile extractor suspended from a crane. The pile extractor was air operated. Casings that could not be removed in this manner were left and were removed with the cell fill during cofferdam dismantling.

Drilling and grouting equipment consisted of four 3-1/2 by 5-1/2 inch electric Ingersoll-Rand, calyx core drills and two grout machine assemblies similar to those used at Kentucky cofferdam except that only one grout pump was used in each case instead of two as at Kentucky.

Chickamauga Cofferdam, Stage 1--The foundation under the Chickamauga cofferdam was honeycombed with seams and solution channels filled with clay. Preliminary foundation investigations, however, were made with wash borings and did not reveal the true foundation condition. (Core drilling should be used.) As a result, inadequate provisions were made for consolidating the foundation for the first stage, or lock cofferdam. To complicate the situation further, grouting of the foundation was delayed until after the construction of the cofferdam was completed and unwatering operations started. Unwatering the cofferdam relieved pressure from within the cells, thereby activating the pressure due to water on the outside of the cells. This forced the saturated material from many of the clay-filled solution chanels and developed a number of serious leaks. In fact, the leakage which existed throughout the construction of this cofferdam was of such magnitude that the foundation treatment which followed was little more than a program of the "stopgap" variety. A large number of pumps were kept in constant service throughout the entire construction period.

A number of methods and materials were used in attempting to seal off the solution channels and stop foundation leakage into the cofferdam. These included cutoff dikes, auxiliary cofferdams, sandbags, blankets of impervious materials spread over the bed of the river in an attempt to intercept the openings to the solution channels, and grout mixtures of various types. Grout mixtures were used that contained cement, sawdust, cinders, oat chaff, and bentonite. A log of the grouting operations is given in table 16.

Detailed reference is made to grouting operations at Chickamauga Dam during cofferdam construction for two reasons. The first is for the purpose of emphasizing the difficulties that may very likely occur where insufficient attention is given to foundation investigation and consolidation. The second is for the purpose of describing the various methods of grouting that were used, particularly in grouting with asphalt.

The cement grout mixtures were adequate in stopping leaks in seams and small crevices but were ineffective in sealing off the large water-filled solution channels. Asphalt grout was used to seal off these but for economical reasons was later replaced with roofing pitch. The roofing pitch proved to be more satisfactory than asphalt because of its low shrinkage factor and because it was not subject to deterioration.

For experimental asphalt grouting, hot wire method, asphalt was injected through a 3-inch core drill hole, using a standard asphalt heating kettle and piston pump. The asphalt was kept in a liquid state in the hole by means of the American Grout Company's patented "Hot Electric Wire" process. The wire which lay along the length of the hole was abandoned upon completion of the grouting operation. This process was not satisfactory because the asphalt was not kept in a sufficiently liquid state to fill the seams and crevices completely.

The asphalt grouting with the live steam method was developed by TVA, using live steam to keep the asphalt in a liquid state during grouting operations. The piping used for this method made it necessary to drill a 6-inch-diameter hole rather than the 3-inch-diameter hole used in the method above. The holes were drilled with 6-inch-diameter calyx core drills. Since the cofferdam had been unwatered before the grouting was started, the holes were drilled inside the cofferdam close to the cells. This saved having to drill down through the cell fill. Six-inch steel pipe casings were sunk to rock and were sealed to the rock at the bottom with cement grout or clay to prevent escape of the grout and steam. The holes were then drilled. The steam was fed under pressure through a 2-inch pipe inside the 6-inch casing, and the exhaust steam was returned through a 1-inch pipe inside the 2-inch steam pipe. The 6-inch casing was capped at the top entrance and the escape of the steam was controlled by a valve on the 1-inch pipe extending through the cap. The asphalt for this operation was melted in kettles and pumped directly down inside the 6-inch casing under pressures varying from 0 to 75 pounds per square inch depending upon the nature of the requirements. Piston-type force pumps proved more satisfactory than gear pumps since the piston pumps required little or no maintenance.

Other methods to control leakage were also used. The clay blanket spread over portions of the riverbed to fill and seal entrances to solution channels was placed along the upstream arm by bulldozing the material out from shore. Loose clay dumped along the river arm was soon swept away by the current, however, and its use was found to be much more effective when in bags. These were dropped along the outside of the cells or hauled on a small wooden barge to fill openings appearing along the bed of the river. The entrances to these openings were located by watching the action of small tin can buoys or by attaching a bag of clay to the end of a rope for sounding. The entrance to a channel was indicated by the pull or suction on the bag as it was drawn into the opening. Other expedients included old mattresses and tarpaulins, weighted down with rocks. In one instance a semicircular sheet pile cell was driven on the outside of the cofferdam in an attempt to seal off the large leak which occurred under the connecting cell between cells 7 and 8. This arrangement was only partially effective, however.

Chickamauga Cofferdam, Stages 2 and 3--The foundation condition and the grouting requirements for stages 2 and 3 were anticipated and the necessary preparations made. Six-inch drill casings were set in each cell ahead of cell filling to eliminate the necessity of drilling through the cell fill. Drill and grouting operations were similar to the procedure followed on other projects and no further trouble was encountered. No asphalt grout was used. Grout mixes varied from neat cement for tight holes to a mix composed of bentonite, sand, and sawdust. Cement and calcium chloride were added to the latter mix for some holes.

None of the maintenance difficulties of stage 1 developed during these stages, demonstrating the great advantage of adequate foundation grouting before unwatering the cofferdam, a fact which had already been indicated.

For Loudoun Cofferdam, Stage 1--Foundation grouting and consolidation along the earth dike used in stage 1 was extensive. Here again, no grouting was done prior to cofferdam unwatering. Leakage did not start until foundation excavation had been extended to a depth of approximately 20 feet below the riverbed. It appeared at that time in the form of boils, with the water reaching the cofferdam area through a network of cavities along the almost vertical strike joints intersecting the bedding seams. This condition was almost continuous from the river to the cofferdam area and it was obvious at this time that grouting would be required before excavation could be completed. After a number of attempts by grouting short areas, grouting was extended along the full length of the north short areas, grouting was extended along the full length of the north sheet pile wall with holes approximately 40 feet deep spaced on 2-foot 6-inch centers. This was later extended along the upstream arm with holes spaced 10 feet on centers until the flow of the underground water was intercepted. As excavation progressed, leakage from sinkholes near the downstream end of the lock wall became apparent. This was stopped by means of a grout curtain cutoff which extended from the downstream end of the sheet pile wall to the lower guard wall. Holes for this cutoff were spaced on 30-foot centers. In one instance an auxiliary earth dike was built and the area next to the main cofferdam flooded to equalize the pressure until grouting could be completed. All holes except a few wagon drill holes, at first, were drilled with diamond drills. Some cement grout was used at first but, due to the volume required, clay-cement grout was adopted. Asphalt and pitch were also used where the solution channels and flow of water required it. Grouting materials used during stage 1 were as follows:

Cement - 184,244 bags
Clay - 484,696 cubic feet
Asphalt - 9,408 cubic feet

The effectiveness of cofferdam grouting was quite evident as revealed during foundation excavation. It was also evident that this grouting aided materially in final dam foundation consolidation and cutoff operations.

Fort Loudoun Cofferdam, Stage 2--One exploratory drill hole at the center of each cell, drilled prior to driving the sheet piles, revealed a better rock foundation than in stage 1 but indicated some badly weathered joints and bedding planes similar to those found in stage 1.

Upon completion of cell filling and before unwatering the cofferdam, two holes in each main cell and one hole in each connecting cell were drilled with diamond drills along the river side of the cells. These holes, together with the exploratory holes, were then grouted using clay-cement grout except a few holes that were tight under water testing. These were grouted with a neat cement grout. Additional closely spaced holes were later drilled and grouted in four cells immediately downstream of the spillway. No leakage occurred through the rock foundation but considerable leakage appeared in the cells immediately upstream and downstream from the dam spillway upon flooding stage 1 cofferdam. The gravel fill in these cells was then grouted with clay by jetting pipes to rock and raising the pipe in stages as the clay was pumped into the gravel fill. Some asphalt was used in the cells at the spillway.

Although the above grouting program was highly successful, it was felt that at least some of the watertightness was due to the large amount of grout materials used in stage 1 cofferdam area.

Quantities for grouting in stage 2 were:
26,169 linear feet diamond drill holes
55,662 bags cement
101,428 cubic feet clay
125 barrels asphalt

Effect of Grouting on Pile Removal--At Fort Loudoun Dam, consolidation of the cell fill was used in some instances as a means of stopping leakage. The cement grout found its way into the piling interlocks, however, making removal by pulling impossible. No satisfactory means was found for overcoming this difficulty. Extreme care should be exercised, therefore, in grouting near piling to be removed through pulling.

UNWATERING AND MAINTENANCE

A considerable amount of maintenance and additional construction may be required after the cofferdam has been constructed and unwatered. This may extend throughout the life of the cofferdam.

Such maintenance will usually involve:
1. Maintenance pumping to remove seepage water. (The amount of seepage water to be handled will depend to a large extent on the original condition of the foundation rock--as at Chickamauga cofferdam, stage 1, discussed on page 131; the thoroughness and effectiveness of the grouting program prior to unwatering operations; the existence of flood plains adjoining the cofferdam area as shown in figure 1, the ineffectiveness of piling interlocks caused by damage during driving; and the type of cell fill used and the method employed in placing the fill to attain watertightness.)
2. Repairs to eliminate seepage. (This may include additional foundation grouting or consolidation of cell fill through grouting, as discussed under "Foundation Treatment at Kentucky Cofferdam," page 121; the construction of sheet pile cutoff walls; the construction of drains to handle the water; and the construction of short auxiliary cofferdams.)
3. Repairs to piling damaged during driving (see "Interlock Failure--Kentucky Cofferdam," page 24). The extend of such damage will be revealed upon unwatering of cofferdam, or as foundation excavation progresses in the area close to the cells.
4. Repairs of damages caused by floods. These damages may be held to a minimum through the use of adequate floodgates, and through the removal of equipment from the cofferdam area prior to flooding. Damages to the cell fill as the result of scour will depend to a large extent on the depth to which the cells are submerged during flooding, on the swiftness of flow, and on the resistance of the cell fill to scour.

CHAPTER 8

REMOVAL

Following a description of pile extraction procedures and pile extractors this chapter is devoted mainly to the problems of cofferdam removal encountered on TVA projects.

> *Editors Note*
> Modern vibratory hammers were not available to the TVA constructors at the time for driving or pulling. See our chapter on Vibratory Driver/Extractors - page 215.

PILE EXTRACTION

Pile pulling is expedited through the use of special extractors which transmit an upward pulling blow to the pile. The Union hammer is inverted for use as an extractor, while the McKiernan-Terry and Vulcan manufacturers make a special extractor. All types are designed for use with a crane transmitting a direct pull on the extractor while the extractor jerks the pile upward. The jerking effect breaks the skin friction and allows the crane to lift the piling out. None of the extractors are designed to operate under extreme crane pulls. Where the piling is particularly hard to extract it may become necessary to provide a special crane or derrick hoist in addition to the one handling the extractor.

Table 17
Physical Data on Pile Extractors

Make	Model	Weight Unit (Lb)	Weight Ram (Lb)	Stroke Per Minute	Stroke Length (Inch)	Energy (Ft-Lb per Blow)	Steam (B.Hp)	Compressed Air Free Air (Cfm)	Compressed Air Pressure Delivery (Psi)	Max Crane Pull (Tons)
Vulcan	200	1500	200	550	2	250	18	312	150	25
	400	2850	400	550	2	500	25	615	150	25
	800	5400	800	550	2	1000	40	1330	150	25
McKiernan-Terry	E-2	2600	200	450	3	700	35	450	100-125	50
	E-4	4400	400	400	3	1000	35	450	100-125	100
Union	5	1625	210	190	9	910	10	100	100	
	4	2800	370	120	12	1390	12	150	100	
	3	4700	700	120	14	1780	20	300	100	
	3A	5200	820	120	13-1/2	2320	25	350	100	
	2	6600	1025	115	16	2375	25	400	100	
	1-1/2A	9200	1500	100	18	2500	35	450	100	
	1A	10300	1600	95	18	3520	40	500	100	
	1	10000	1600	95	21	4080	40	600	100	
	0	14500	3000	80	24	3930	50	750	100	

Also, see Pile-Driving Handbook,[1] table IV A, pages 170-171.

---

1. Robert Dunning Chellis, Pile-Driving Handbook (New York: Pitman Publishing Corp.), 1944.

Provisions for selecting the extractor equipment should be made well in advance of pile pulling operations through the use of test piles driven on the job to duplicate actual pulling conditions. The principal factors affecting the difficulties to be expected are (a) the depth of penetration of the piling and (b) the damage to the piling resulting from driving conditions. Other difficulties in extraction, arising on TVA projects, were due to piling being wedged into position during driving operations and piling locked in place by grout used in consolidating the fill against leakage.

Extraction Procedure

Before proceeding with pile extraction the fill inside the cells should be removed to water level, berms should be removed and, in the case of single sheet pile cutoff walls, the fill along the piling should be removed to as great a depth as practical. This will reduce the skin friction in each case and facilitate pile pulling.

Extractor Size

The Vulcan models 800 and 800A and the McKiernan-Terry model E-4 extractors, operating from cranes ranging up to 22 tons in capacity, served adequately to meet TVA pile extraction needs except in extreme cases. Extreme difficulty in pile extraction was experienced at Guntersville, Pickwick, and Kentucky Dams (see following discussions). See, also, table 17 for manufacturers' data on pile extractors.

REMOVAL PROBLEMS ON TVA PROJECTS

The following discussions illustrate both the normal and special cofferdam removal and pile extraction problems encountered by TVA on its main-river projects.

Guntersville Cofferdam

Pile removal on this project was ideal and met with little resistance except for piling in the cutoff walls where the special rig described below was used. The overburden was light and ranged from none at midstream to a maximum of 35 feet in depth at the shoreline.

The river cells were removed with a 22-ton, gantry-mounted whirley crane operating from the top of the cells; an 8-ton, barge-mounted skid rig; and a 22-ton, barge-mounted revolving crane operating from the river. All three were used both for clamming fill material from the cells and for extracting piling, as the occasion required. Two Vulcan pile extractors, models 800 and 800A, operating on steam or air, were used to extract the piling.

The first step was to remove approximately one-fourth of the fill from the cell with a clamshell bucket, piling it in the water around the outside of the cell. Since there was very little overburden present, this helped to equalize the pressure on the piling and make removal easier. It also acted as a support for the piling to prevent distortion which would have otherwise occurred when the cell spread, after removal of the first pile. The first pile pulled was the most difficult and was therefore selected with care by the crew foreman. This pile was attached to the extractor by means of one or two of the 2-inch pins provided. Removal of the other piling in the cell was usually accomplished by attaching the pile to the extractor with a wire rope sling and shackle. The 2-inch pin connection between the extractor and the pile was used only on piling that was hard to remove.

A typical crew consisted of a foreman and seven riggers for each crane in operation. A special track crew, attached to the gantry mounted whirley crane, moved the track ahead as required.

A small amount of the fill clammed from the cells was loaded on barges and moved to the next stage. The material left, or dumped into the river during cell removal, was dredged from the riverbed by means of the hydraulic dredge and either pumped to the next stage for filling cells or was pumped to the shore and disposed of.

Special Pile Puller for Guntersville--Piling in the cutoff walls was difficult to loosen with the standard pile extractor. A special rig was therefore built, consisting of a 50-foot gin pole mounted on two 10- by 10-inch by 6-foot 0-inch long skids. The gin pole was guyed with four lines, each rigged with a block and tackle to facilitate adjustment. Extracting equipment consisted of a Vulcan model 800 extractor, a 13-part tackle with a 7500-pound single line pull, and an electric hoist powered by a 60-horsepower motor, giving a maximum pulling capacity of 46,000 pounds. This rig was cumbersome and is not recommended for extensive operations.

Pickwick Cofferdam

Pile extraction was accomplished with a Vulcan model 800 pile extractor operating from a 22-ton American revolving crane, except in the case of the cutoff wall driven along the earth embankment. In this case the embankment was not to be distrubed and no material could be excavated along the face of the wall to relieve skin friction. Instead, a special pile pulling device (figure 72) with a maximum pulling capacity of 175 tons was built. The frame was constructed with 16-inch posts, supporting beams made up of three 8- by 16-inch timbers. This device was cumbersome and time consuming but represents a means by which pulling power may be increased without undue expense for equipment. It may be used successfully where only a small amount of piling has to be pulled.

Kentucky Cofferdam

Cofferdam removal at Kentucky Dam was a major operation. It was therefore necessary to set up a detailed plan of procedure to be followed. The step-by-step procedure prescribed at previous projects for removal of circular cells did not work at Kentucky because of limited working space and fouling of the cutter head by scrap steel in the cell fill being removed. The system finally adopted was much the same for both circular and cloverleaf cells, and is shown in detail in figure 73.

Removal Equipment--Equipment used in the removal operation was as follows:

1. Two Clyde electric gantry cranes (for short time).
2. Two 22-ton American revolver barge-mounted cranes.
3. Two McKiernan-Terry, Lambert-National, 8-ton barge-mounted skid rigs.
4. Three barge-mounted extractor units.
5. One skid-mounted extractor unit.
6. One Northwest shovel (used for cell removal).
7. Sluice barge (Figure 73).
8. Hydraulic dredge "Dallas" (TVA-owned).
9. Repair barge
10. Ten Vulcan model 800A extractors and two McKiernan-Terry model E-4 extractors.

Equipment listed in items 4, 5, and 7 above was developed for pile removal at Kentucky cofferdam and is described in detail below.

Sluice Barge--The sluicing barge consisted of a sluice gun mounted on top of a 35-foot-high tower on one end of a 70- by 20- by 4-foot barge. This was used to wash the fill from the cells through the "window" created by burning off eight piles at water level and raising them so that the dredge cutter head could be inserted for further removal of the fill material. River water was pumped to the sluice gun through a 12-inch line by a gasoline-engine-driven, 5500-gallon-per-minute at 50-foot-head Dayton-Dowd pump (see figure 73).

Extractor Units--Since 100-ton-capacity cranes were not available for purchase for pile removal at Kentucky cofferdam, three barge-mounted and one skid-mounted land-based extractor units were built on the job to accomplish this operation.

Pile extractor unit 1 was mounted on a 90- by 40- by 7-foot barge and was powered by a Lambert-National, 8-ton, skid rig

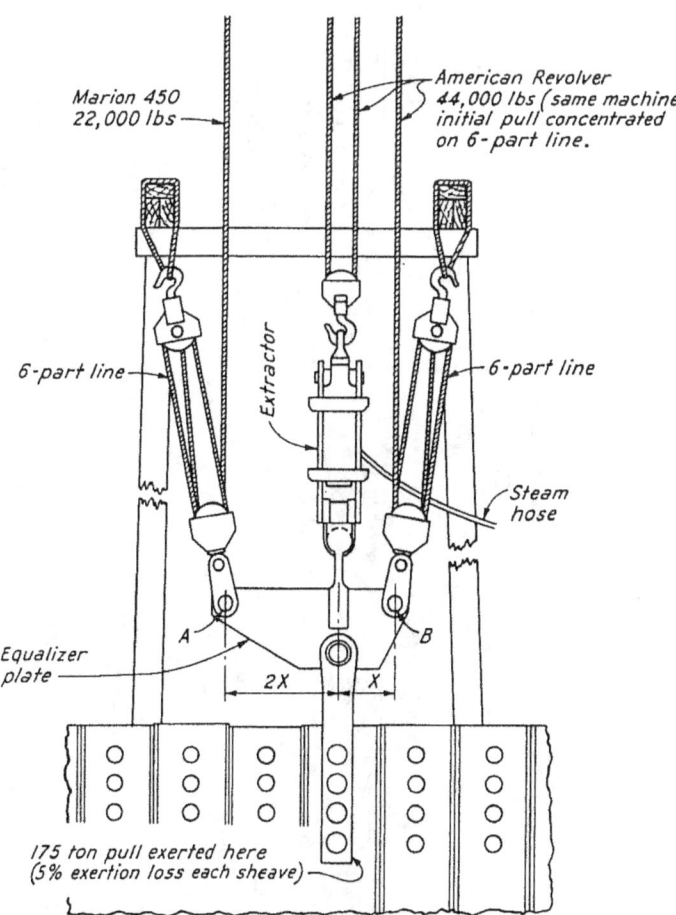

**Figure 72
Specially Designed Piling Extractor - Pickwick Landing Dam**

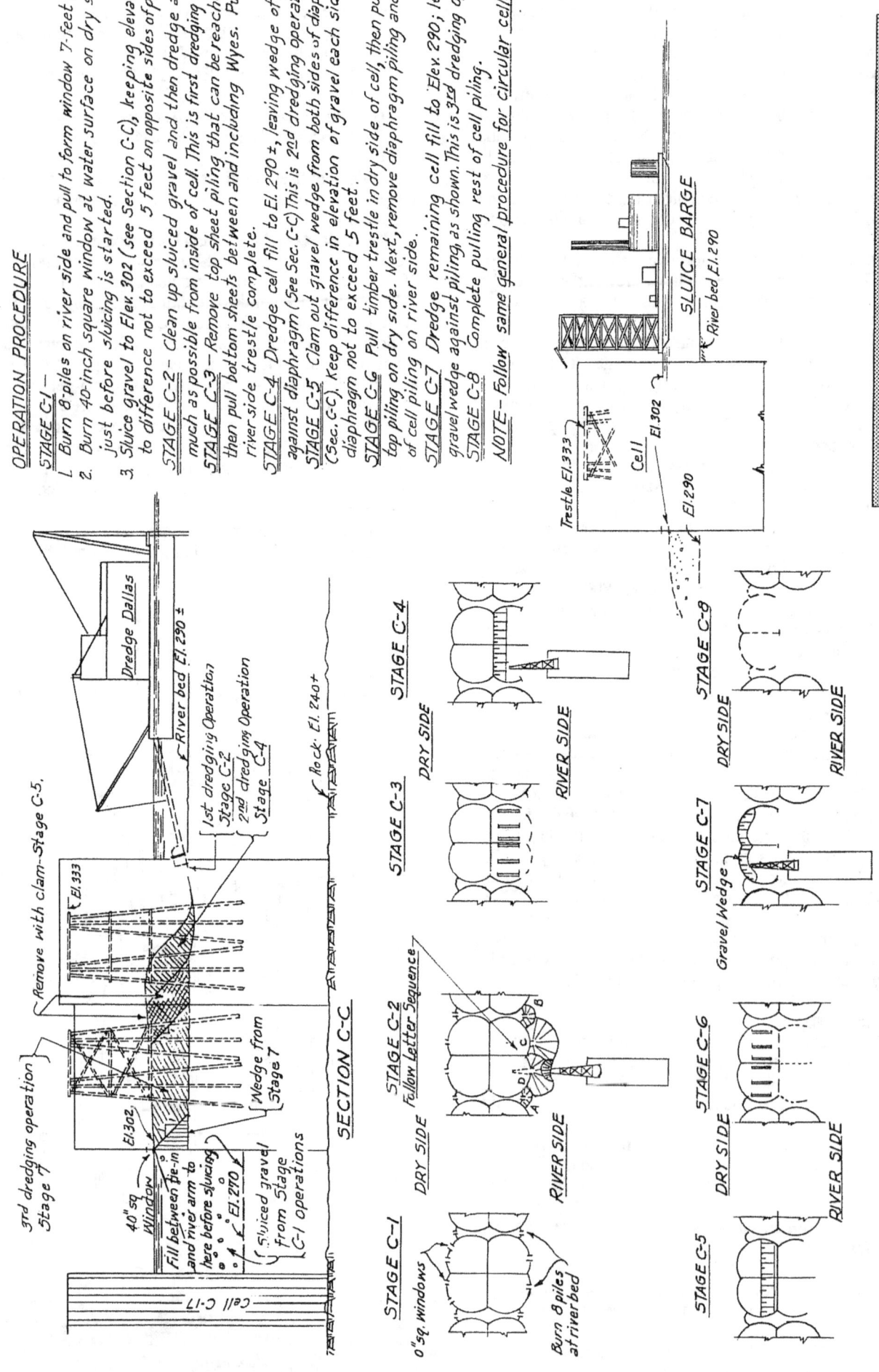

Figure 73
Cofferdam Removal Procedure - Kentucky Dam

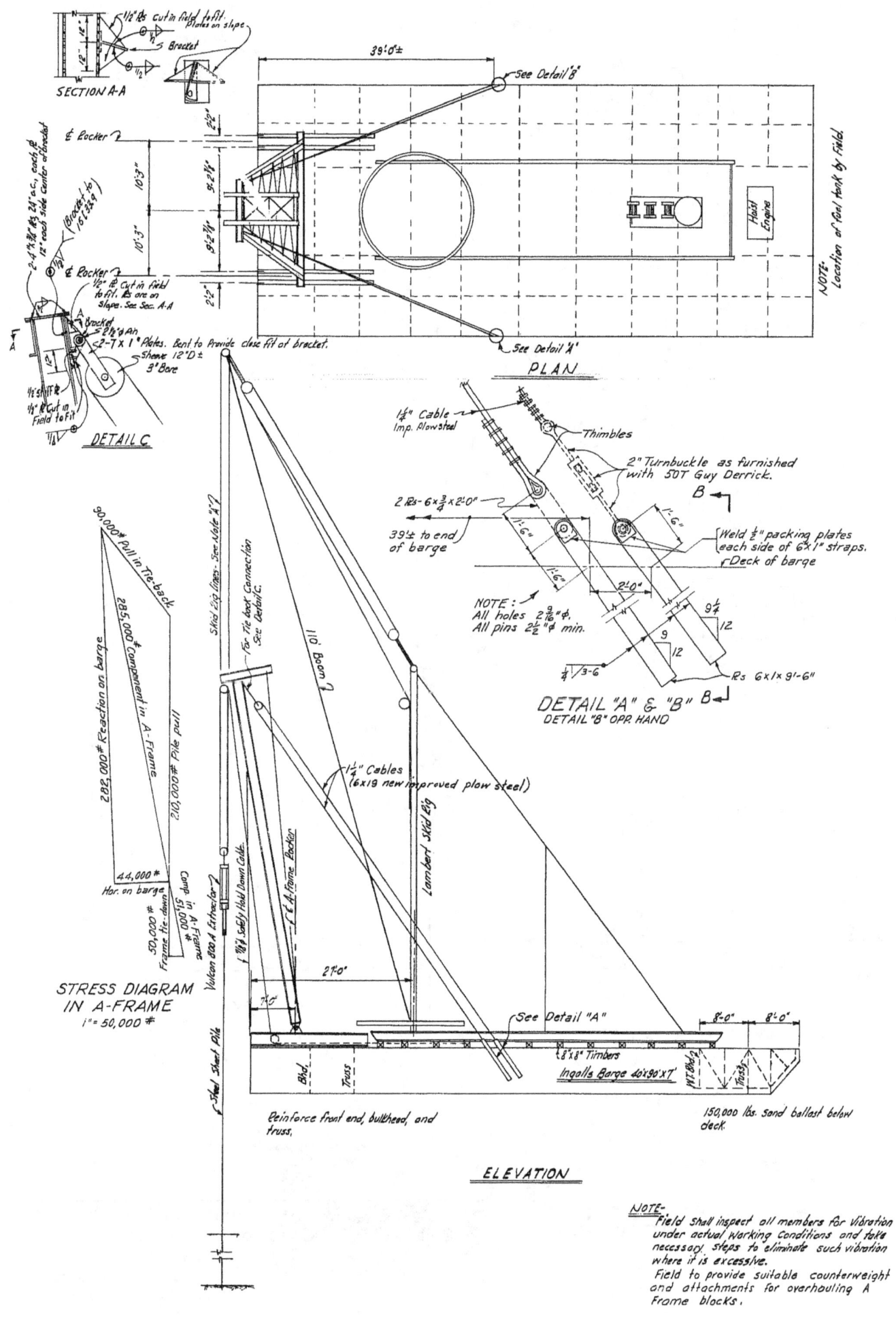

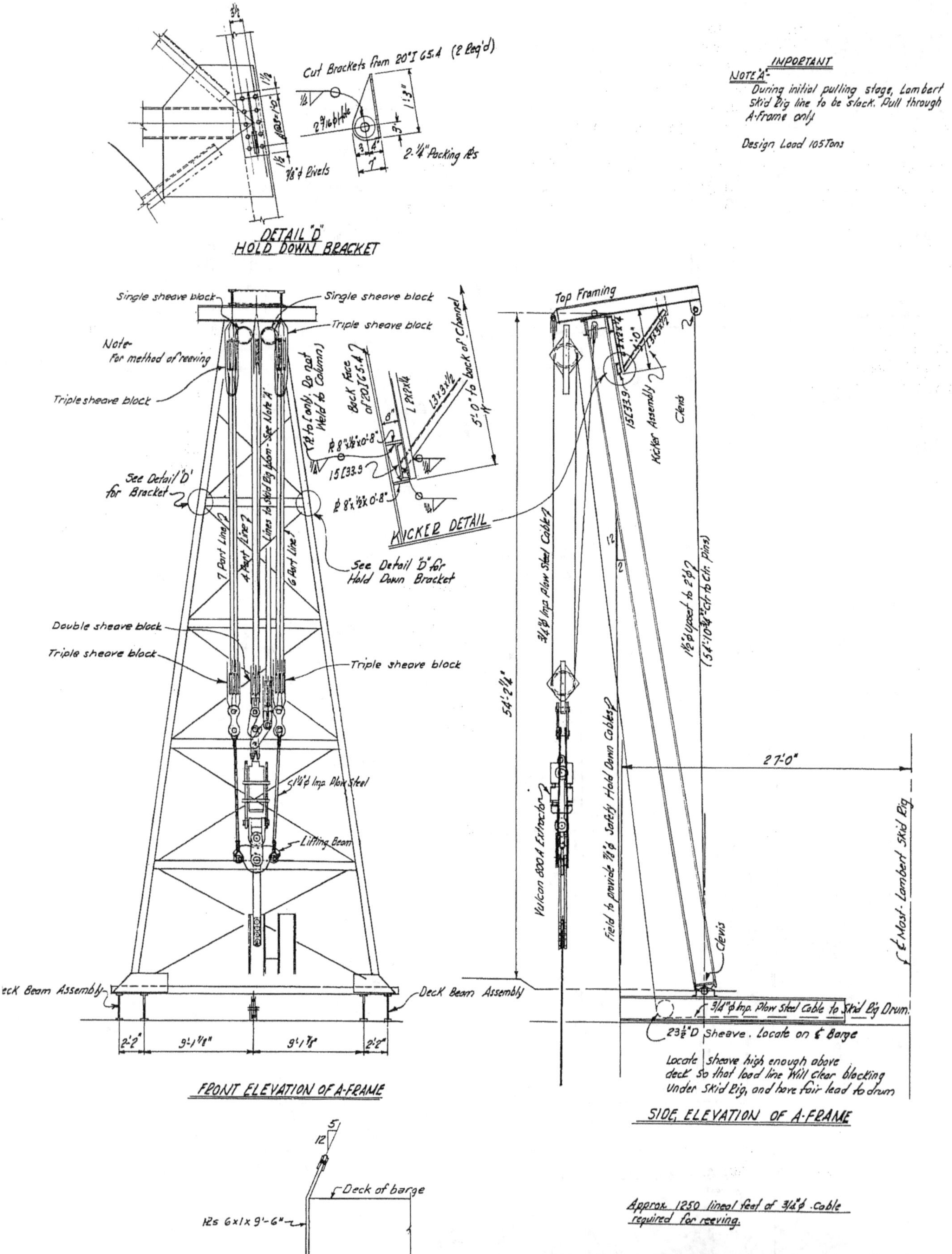

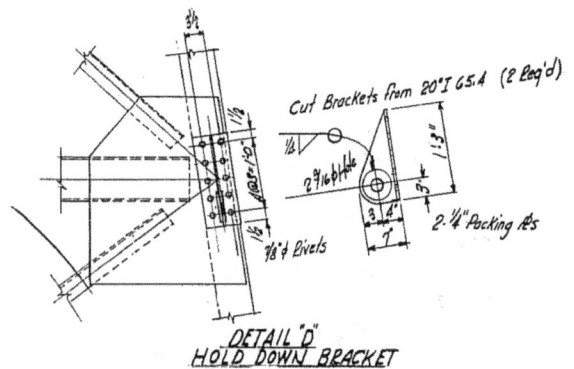

Figure 74 (Continued)
Details of Piling Extractor No. 1 for Cofferdam Removal - Kentucky Dam
Note: Figure 74 appears on this page as well as the adjacent page, p. 136

derrick (see figure 74). The barge was reinforced transversely and longitudinally with alternate watertight bulkheads and built-up truss sections. The truss and bulkhead under the A-frame were further reinforced by adding vertical and diagonal braces and plates to resist the A-frame reaction transmitted to the barge. Watertight compartments at the rear corners of the barge, filled with 150,000 pounds of sand, served as a counterweight to balance the pull on the A-frame. The all-welded steel A-frame, 55 feet high, was pin connected to the deck beam assembly with 5-inch-diameter pins. The frame normally operated on a forward batter of 2 in 12 and was held in position with hold-down and tie-back cables. The head frame, also pin connected to the A-frame with 5-inch-diameter pins, held the sheave assembly and extractor free of the A-frame. the 3-sheave assembly and resulting 17-art line were reeved from extractor to skid rig auxiliary drum. The load line, used to support the extractor while preparations were being made to pull, was attached to the extractor by a second lift link and could be slacked off during initial hard pulling without disengagement. (See figure 74 for the sling assembly hook-up used.) The pulling procedure for pile extractor unit 1 was as follows:

Pulling Procedure

With extractor attached to a pile, and the sheaves tightened up until the barge deck was level with the water, a sustained pull of approximately 100 tons was exerted on the pile. The extractor was then started, with the 100-ton pull being maintained during the operation until the pile came free. The skid rig load line, which was slack during the operation, was now tightened up to take the load of the extractor and pile so that the pulling sling and hook could be disengaged and hauled down in preparation for pulling the pile (see figure 74). As an aid in starting the hard-to-pull piles (the first ones in the cell to be pulled, in particular) a regular driving hammer was seated on the adjoining pile or in some cases on both adjoining piles and operated simultaneously with the extractor. The up-and-down action in the interlock aided in the loosening process.

Pile extractor unit 2 had a 60-foot timber tower, due to the shortage of steel. The skid rig used to power the extractor and the barge and sheave assemblies was the same as that used for extractor unit 1. The pulling procedure was the same as that used for extractor unit 1.

Pile extractor unit 3 had a 50-foot high timber tower similar to the one used on extractor unit 2 and was mounted on a 110- by 30- foot steel barge, counterweighted at the rear end with 50 tons of steel sheet piling stacked on the barge deck. The unit was powered by a 10- by 12-inch, double-drum, compound-geared, oil-fired steam hoist, using a 160-horsepower boiler which developed a 200-pound-per-square-inch working pressure. The average line pull was 42,000 pounds with a maximum first layer pull of 60,000 pounds. The load line was reeved to six parts, using one triple sheave block at the tower head frame and one triple sheave block at the bottom for supporting the extractor. Pulling operations were, in general, the same as described for the extractor units 1 and 2. Extractor unit 3 was used largely for pulling timber trestle piling and the first steel piles in the cells.

Pile extractor unit 4 was a land-operated, skid-mounted, 60-foot timber tower pile pulling rig used in extracting steel sheet piles along the west bank protection wall. Tower members were dapped at the joints and the joints secured with 1-inch bolts and wide steel straps bent to fit the rounded timber members. The head frame was constructed of heavy oak timbers reinforced with steel channels bolted to two sides. The

assembly. The 50-foot horizontal bottom members extended 20 feet beyond the timber back stays to support a platform and hoisting engine. The platform and hoisting engine acted as a counterweight. The unit was powered by a Lambert, double-drum, oil-fired, 40-horsepower steam hoist, with a single line pull of 9000 pounds at a speed of 175 feet per minute. Steam was supplied by a 100-horsepower boiler, with a working pressure of 200 pounds per square inch. The hoist was reeved to a 13-part line, using a 4-sheave block at the head frame and three 2-sheave blocks at the extractor. To maintain a sustained pull on the pile during extracting operations, lines were tightened up until the counterweighted rear end was raised free of the ground. The unit was moved from place to place by sliding over greased logs.

Special Methods of Extracting "Difficult" Piles

At Chickamauga cofferdam, oil was poured down the interlocks to reduce the friction in the interlocks and to aid the loosening process. Due to the large number of damaged interlocks and warped piling, however, the effectiveness of this method was never definitely established.

At Fort Loudoun cofferdam (stage 2) grout was used to consolidate cell fill to stop leakage. This seeped into the piling interlocks, locking them in position so that pulling of the piles was almost impossible. Attempts were made to loosen the pile by drilling holes in the cell fill close to the face of the piles and using dynamite. This resulted only in such a deformation of the piling that it could not be pulled at all. As a result the piling in two cells had to burned off at water level and the bottom ends of the piling left in place.

The next method tried was to pull one pile in the cell. In some cases this took several hours of continuous extractor operation. A wood pole was then inserted in the opening to keep it open. If the adjacent pile would not respond to the extractor, the wood pole was removed from the opening and approximately three sticks of dynamite exploded along the edge of the piling. This usually broke the bond of the grout in the interlocks without much damage to the piling, although removal continued to be a rather slow process.

It should be noted here that piling difficult to extract may often be left in place by burning off at water level. This piling is often hard to pull because it is badly deformed and will have very little value for reuse. It may be more economical to avoid removal if no ill effect to streamflow, navigation, etc., will result. Usually the piling in the upstream arm of the cofferdam can be cut off at low water without any ill effects except in the case of interference with the flow of water to the powerhouse intakes. Shore tie-ins and cutoff walls which will not pull may often be burned off below grade and the bottom ends left in place. The piling in the downstream arm of the cofferdam, however, will usually have to be entirely removed regardless of the difficulties involved because any interference with streamflow at this point may be dangerous.

Piling in the upstream arm of Kentucky stage 2 cofferdam was burned off at low water level to gain time and avoid a delay in the dam closure schedule.

CHAPTER 9

TESTS

In attempting to design a steel sheet pile cofferdam the designer may find that a number of the essential factors are often unknown. These may include such factors as the characteristics of the fill material to be used, the results that may be expected during pile driving and pile pulling

effect of restrictions and streamflow on navigation to be expected during the several stages of construction.

Where difficulty is anticipated, the extent to which the difficulties may be expected during driving and pulling operations may be actually determined through practical field driving and pulling tests. In the case of large streams, the use of models may be justified in determining the expected stream behavior. Test to accurately determine the characteristics of the fill material are often omitted, however, using instead existing approximate data. Such practice may be dangerous as this type of data is often found to be either in error or not applicable at all. The added expense resulting from tests to determine these factors is usually justifiable, particularly where the size of the cofferdam and construction costs are large.

A brief outline of tests performed by TVA to determine similar data for the construction of Kentucky cofferdam follows.

Fill Material Characteristics for Kentucky Cofferdam

This cofferdam was of such magnitude that even a small improvement in procedure, resulting from a more accurate knowledge of the material to be used, could mean a substantial saving as well as possible improvement in the stability of the structure. This being true, then, additional expense in making tests to determine such coefficients of friction as fill on fill, fill on rock, fill on steel, and the weight of the material (both in the wet and dry stage) was warranted.

Tests were made at TVA Hydraulic Laboratory at Norris, Tennessee, on samples of sand and gravel sent from Kentucky Dam. A summary of the results of these tests is given in table 18.

The samples for these tests were small, however, and it was necessary to supplement with similar materials. The tests were conducted in three groups as follows: (a) fine angular sand, fairly uniform in size, (b) a mixture of quartz and angular fragments of gravel or chert, and (c) fine sand and fine gravel with a small percentage of clay. Later tests were made using the same methods but with larger samples and testing apparatus. The results are shown in table 19.

The weights of the fill material in a dry, submerged, and saturated state (tables 18 and 19), as determined by conventional laboratory methods, were rational and reasonably consistent throughout these and subsequent tests. For recommended values for design purposes see table 22, and for values used see figure 33. Tests (tables 18 and 19) conducted at Norris Laboratory for the purpose of determining the coefficients of friction of fill materials sliding on steel and fill materials sliding on rock were also sufficiently consistent and conclusive to warrant the adoption of 0.4 as the coefficient for the former and 0.5 as the coefficient for the latter. These figures were conservative, however, even in the face of the tests.

Coefficient of Internal Friction

For many years the angle of internal friction for cohesionless materials was determined by a simple test in which the material was placed in a cone-shaped pile, the angle of repose being consdered as equal to the angle of internal friction. As an alternate method for determining this angle, a box was filled with the material and then tilted until the material began to slide. Values obtained by the latter method were usually larger. Neither test, however, gave very accurate results. As a matter of fact, the shearing strength of the material depends on many factors. Among these are (1) grain size and shape, (2) material structure and density, (3) previous treatment of the material (loose, compact, placed

Table 18
Summary of Test Results - Coefficient of Internal Friction - Sand and Gravel Fill Material for Kentucky Cofferdam

		Group A	Group B	Group C
Unit weight of material, pounds per cubic foot	Poured dry — Dry density	86	90	90
	Saturated density	116	118	118
	Submerged density	54	56	56
	Poured into water — Dry density	93	92	93
	Saturated density	120	120	120
	Submerged density	58	58	58
Coefficient of internal friction	Dry	0.8	0.8	0.8
	Submerged	.7	.8	.8
Coefficient of friction on steel	Dry	.7	.6	.6
	Submerged	.6	.6	.6
Coefficient of friction on rock	Dry	.65	.65	.65
	Submerged	.65	.6	.6
Cohesion on rock, pounds per square foot	Submerged	80	150	0
Angle of repose, degrees	Poured dry	32	34	34
	Poured into water	26	30	29
Coefficient of permeability, feet per day		47	28	8

Editors Note

The equipment and test methods for determining the engineering properties of soils have improved considerably since TVA developed these "in house" methods. Their results may be surprisingly good however.

Table 19
Summary of Test Results - Coefficient of Internal Friction - Sand and Gravel Fill Material for Kentucky Cofferdam

Tests		Angle of Friction (Degrees)	Apparent Cohesion (Psf)	Coefficient of Friction
Dry, loose gravel	Core No. 1	29.6[a]	3	0.57[b]
	Core No. 1	24.3	6	.45[b]
	Core No. 2	23.9	2	.44[b]
	Steel	22.0	0	.40[c]
Dry, dense gravel	Core No. 1	25.7	10	.48[b]
	Core No. 2	25.3	4	.47[b]
	Steel	23.0	0	.42[c]
Wet sluiced gravel	Core No. 1	29.6	18	.57[d]
	Core No. 2	27.4	11	.52[d]
	Steel	23.3	3	.43[c]
Submerged test	Core No. 1	23.5[a]	5	.44
	Core No. 2	24.2[a]	0	.45
	Steel	18.8	0	.34
Dry, loose gravel larger than 3/4 inch	Core No. 1	25.1	9	.47
	Core No. 2	26.8	0	.50
	Steel	25.3	0	.47[c]
Dry, dense gravel larger than 3/4 inch	Core No. 1	35.9	0	.65
	Core No. 2	28.9	0	.55
	Steel	26.0	0	.49

a. Using first bag of gravel from test pit No. 8.
b. Sliding within gravel.
c. Sliding on steel.
d. Sliding on rock.

The angle of repose was found to be 31 degrees. Dry densities of material were as follows:

Dry loose: 109 lbs per cu ft
Dry tamped: 118 lbs per cu ft
Sluiced: 117 lbs per cu ft

Table 20
Summary of Test Results - Coefficient of Internal Friction - Sand and Gravel Fill Material for Kentucky Cofferdam

Test	Angle of Internal Friction (Degrees)	Cohesion (Pounds per Square Foot)	Coefficients of Friction
Dry loose	23.3	11	0.43
Dry dense	26.4	17	.50
Sluiced	26.1	34	.49
Submerged	25.4	6	.47

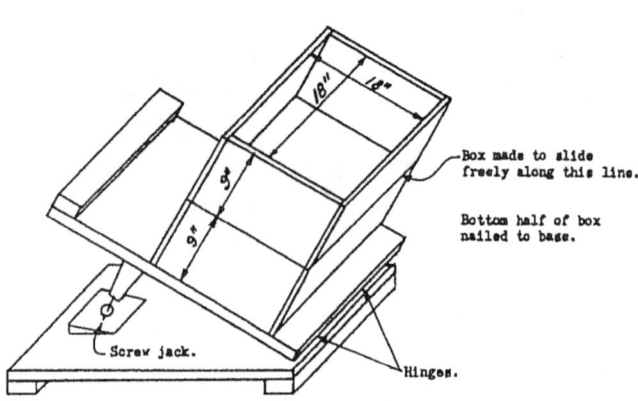

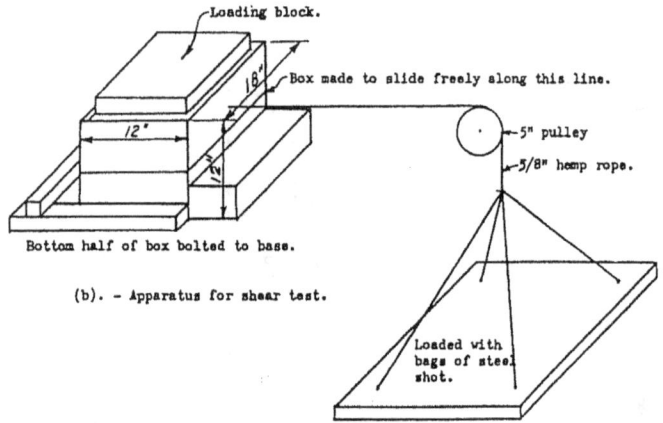

Figure 75
Test Apparatus for Coefficient of Friction Test on Cell Fill Material - Norris Laboratory

hydraulically, etc.), and (4) the water content. To properly interpret the results of shear tests, then, one must consider what happens during the test and how the results are affected by the facts noted above.

From Coulomb's law, when one body slides over another body the frictional force will equal the normal pressure multiplied by a constant (f), known as the coefficient of friction. Here (f) is assumed equal to tan $\emptyset$, where $\emptyset$ is the angle of slope with the horizontal of a plane such that the body is just on the point of sliding under its own weight. Experiments show, however, that for cohesionless materials the angle of internal friction is not a constant due to the interlocking of these irregularities on the surfaces of the interface and the variability of these irregularities. It has also been found that a vapor may act as a lubricant on the surfaces in contact, thereby reducing the values of $\emptyset$ and (f).

Considering sand, a relatively cohesionless material, the resistance to displacement is due first to sliding of grain on grain and second to the interlocking of the grains. It is important to note here that the interlocking action plays a more important part as the material becomes denser.[1]

Tests by TVA for Coefficient of Internal Friction -- The two sets of tests conducted at Norris (tables 18 and 19) were made without superimposed loading. Results were inconsistent and it was felt that additional tests were needed. A third set of tests was then conducted in which superimposed loads were used. Results are shown in table 20. Since these results were at considerable variance with previous results (tables 18 and 19), it was decided to make additional shear tests, using higher superimposed loading. A summary of these tests is given in table 21. The testing apparatus that was used is shown in figure 108, page 271. The five conditions under which the samples of material were tested are as follows:

1. Five percent moisture content and loose.
2. Five percent moisture content and rodded.
3. Dry and loose.
4. Dry and rodded.
5. Saturated.

Pressures on the fill material test sample amounted to approximately 1000 pounds per square foot. It was expected, however, that pressure on the fill in the bottom of the cells might go as high as 11,000 pounds per square foot due to the extreme height of the cells. Still further shear tests were therefore conducted in an attempt to determine the coefficient of internal friction of the fill material. These tests were conducted at Guntersville Dam, using testing equipment to simulate fill material under pressures approximating the pressures expected to occur in the cell. Tabular results are shown in table 22 and a series of curves plotted from the results are shown in figure 76.

The finally adopted coefficient of internal friction, 0.55, representing an angle of internal friction of 28° 50', is indicated by line "A" in figure 76. Professor Terzaghi considers this angle to be too small in the face of other tests with which he is familiar. The test results shown in figure 76 indicate a higher value for the coefficient of internal friction for the lower pressures (up to 2000 pounds). There was also a considerable

Table 21
Summary of Test Results - Angle of Internal Friction - Sand and Gravel Fill Material for Kentucky Cofferdam

Sample Number	Location Range	Location Station	Test Sample	Sliding Test Angle of Failure	Sliding Test Tangent	Shear Test Load, lb per sq ft Vertical	Shear Test Horizontal	Shear Test Coefficient of Friction
1	GH	30 + 15	58% sand, 42% 1/4" to 2" gravel	40° 20'	0.911	411	347	0.844
						718	504	.702
						1,052	789	.750
2	GH	27 + 60	48% sand, 52% 1/4" to 1-1/2" gravel	36° 40'	.744	386	328	.850
						707	582	.823
						1,053	728	.691
3	GH	26 + 00	40% sand, 60% 1/4" to 1-1/2" gravel	36° 10'	.731	388	288	.742
4	GH	23 + 65	44% sand, 56% 1/4" to 1-1/2" gravel	37° 10'	.758	709	539	.760
						1,093	761	.731
5	FG	17 + 39	24% sand, 76% 1/4" to 1" gravel	34° 40'	.692	380	282	.742
						701	541	.772
						1,047	719	.687
6	FG	20 + 79	29% sand, 71% 1/4" to 1" gravel	40° 10'	.844	400	290	.725
7	FG	23 + 79	29% sand, 71% 1/4" to 1" gravel	36° 20'	.735	707	492	.696
						1,041	665	.639
8	FG	26 + 19	42% sand, 58% 1/4" to 1" gravel	36° 20'	.735			
9	FG	29 + 24	42% sand, 58% 1/4" to 1" gravel	39° 30'	.824			
10	FG	30 + 54	43% sand, 57% 1/4" to 1-1/2" gravel	40° 00'	.839			
11	HJ	16 + 70	41% sand, 59% 1/4" to 1-1/2" gravel	36° 10'	.731			
12	HJ	20 + 10	42% sand, 58% 1/4" to 1-1/2" gravel	40° 10'	.844			
13	HJ	23 + 10	41% sand, 59% 1/4" to 1-1/2" gravel	38° 13'	.795	405	283	.699
						712	493	.692
14	HJ	25 + 50	46% sand, 54% 1/4" to 1-1/2" gravel	38° 50'	.805	1,046	671	.641
15	HJ	28 + 55	41% sand, 59% 1/4" to 1-1/2" gravel	41° 50'	.895			
16	HJ	29 + 60	62% sand, 38% 1/4" to 2" gravel	42° 00'	.900			

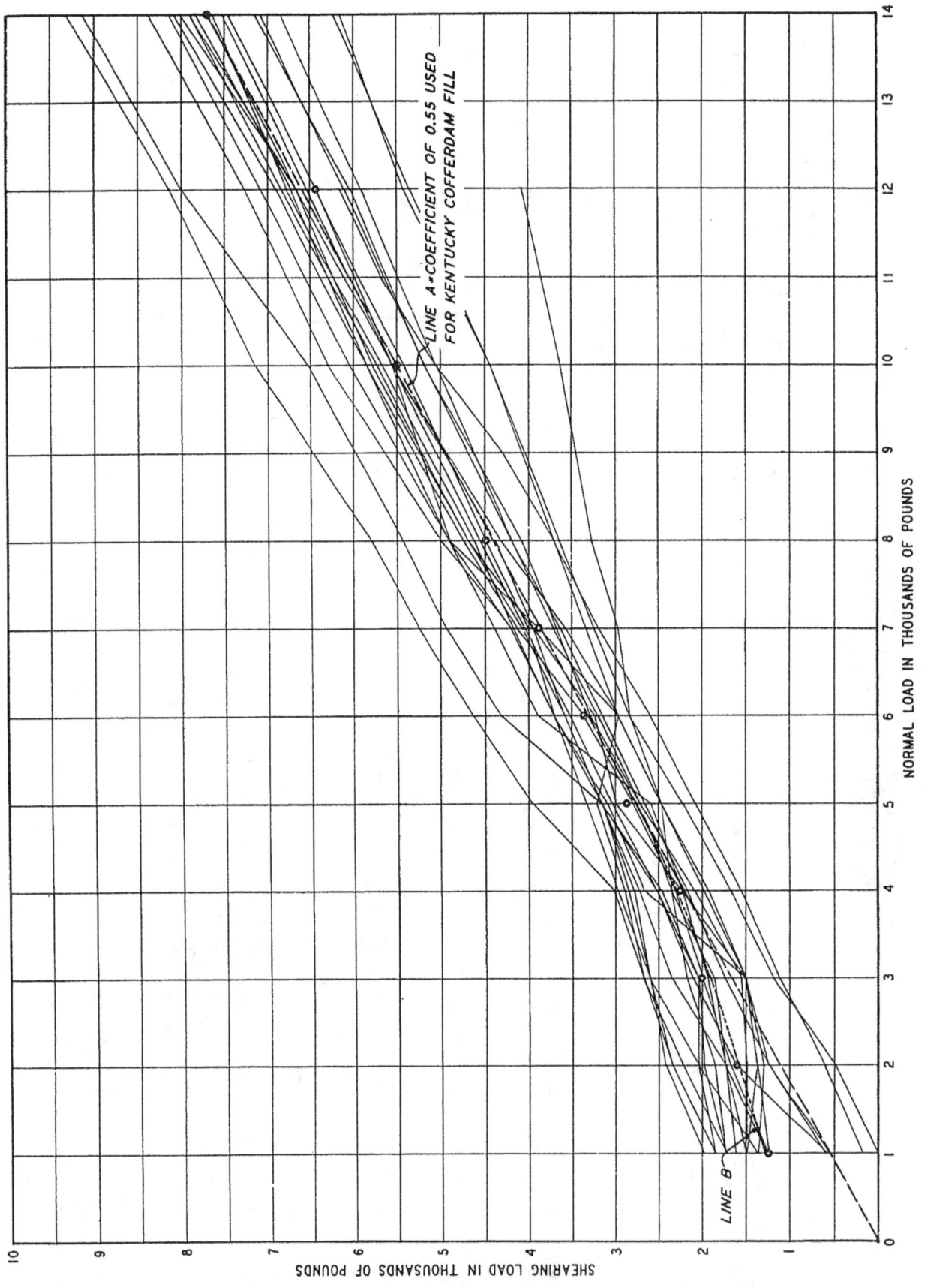

**Figure 76
Results from Series of Coefficient of Friction Tests Conducted at Guntersville Dam on River-Run Sand and Gravel Used as Cell Fill in the Kentucky Cofferdam**

Table 22
Summary of Test Results - Coefficient of Internal Friction - Sand and Gravel Fill Material for Kentucky Cofferdam

Sample Number	Weight of Material - Lb per Cu Ft				Saturated Sample		Specific Gravity
	Dry Loose	Dry Rodded	Saturated	Submerged	Approx Coefficient of Internal Friction	Approx Angle of Internal Friction	
1	107	114.8	132.1	69.6	0.54	28° 20'	
2	107.8	115	132.8	70.3	.51	27° 00'	
3	108.6	113.8	132.0	69.5	.47	25° 10'	
4	106.6	113.6	132.8	70.3	.52	27° 30'	
5	93.6	101.2	119.4	56.9	.45	24° 10'	2.40
6	98.8	110.4	128.4	65.9	.57	29° 40'	
7	101.2	113.6	132.4	69.9	.58	30° 10'	
8	103.6	116.2	131.4	68.9	.51	27° 00'	
9	105.2	117.2	132.8	70.3	.59	30° 30'	
10	104.4	115.2	133.8	71.3	.70	35° 00'	
11	102.8	114.4	134.6	72.1	.55	28° 50'	
12	102.8	114.4	134.6	72.1	.55	28° 50'	
13	104.6	117.4	134.2	71.7	.56	29° 10'	
14	107.2	117.8	131.2	68.7	.50	26° 30'	2.42
15	102.4	118.6	132.8	70.3	.55	28° 50'	2.43
16	103.0	113.2	131.0	68.5	.57	29° 40'	
Total	1659.6	1826.8	2106.3	1106.3	8.72		
Average	103.7	114.2	131.6	69.1	0.545		
Values Used	100	110	127.5	65	0.55[a]	28° 50'	

a. Also see figure 109 for determination of this coefficient.

Notes:

1. Compaction of material under columns headed "Dry Rodded" and "Saturated" was obtained by placing material in container in 3-inch layers, each layer being rodded 25 times with 5/8-inch round-nose rod. In the case of saturated material, the material was thoroughly wetted before placing in the container and, during placing, water was added to fill the voids completely. Compaction of material in this case appeared to be the maximum.

2. Submerged weight was determined by deducting 62.5 pounds from the saturated weight.

3. Tests indicated there was little difference between coefficients of internal friction between dry and saturated samples. Coefficients for saturated samples only are shown.

variation in the value of the coefficient of internal friction for the lower pressures (see line "B," figure 76). This may have been due to variations in test samples, cohesion, etc., as the relative effect of these factors would be greater in samples tested under lower pressures.

Fill Material for Hales Bar Upstream Cofferdam

In designing Hales Bar cofferdam it was felt that the same values used for the coefficient of internal friction, weight of material, etc., for Kentucky cofferdam would be close enough for all practical purposes and that special tests would not be necessary. However, as previously pointed out under the discussions on "Cell Fill" on pages 25, and 32, the need for a free draining fill material was considered as most important. A sieve analysis was made of the available supplies and a decision was reached to remove fines by screening. Such a process may not be economical in the case of extensive cofferdam operations but sieve analyses should be made so that a reasonable estimate of the saturation line in the cells may be made.

Test Cell for Kentucky Cofferdam

Steel sheet piling was tentatively designed for a penetration of as much as 110 feet. While driving to such depths is not without precedent, these depths have frequently been attended with such difficulty that the condition of the piling at the conclusion of driving operations and the cell stability to serve the purpose for which it was driven have been uncertain. A test cell, 20 feet in diameter, was therefore driven for the purpose of excavating to the full depth of penetration so as to expose the full length of the piling and examine the overburden and bedrock. The preliminary design of the bracing for this shaft for a depth of 95 feet indicated that a diameter larger than 20 feet would meet with structural difficulties. Piling of the same design as that to be used in the cofferdam, with web thickness of 3/8 inch, was mostly used. Some piling with 1/2-inch web thickness was used for comparison purposes.

Eight piles were first driven in a straight line to determine the difficulty which might be expected in driving the cell.

A 22-foot-high, 19-foot 7-inch-diameter timber template was used for setting the piling in the cell. The piles were driven in pairs, using a model 9B3 McKiernan-Terry double-acting steam pile hammer. The blows per foot that were required in this case increased rapidly after the penetration had passed 50 feet and reached as high as 600 blows per foot at a penetration of 80 feet. Jetting was tried but was of little value in improving driving conditions. Driving in pairs continued until rock was reached (after which each pile was driven separately until firmly seated on the rock), using a follower.

Excavation inside the cell met with difficulty and it became necessary to resort to grouting and finally to freezing before excavation was complete.

The 3/8-inch web thickness withstood driving conditions satisfactorily, resulting in only six interlock ruptures at a depth of 69 to 78 feet below ground level. Ruptures were caused by some of the piling leaving the shape of the true circle, thus producing vertical folds where the deflection angle exceeded the safe value of 10 degrees.

Other valuable information gained from sinking the test cell included condition of the overburden and bedrock, benefits resulting from grouting the overburden, shaft sinking by the freezing method, etc.

Model Studies

Model studies were made at the Norris Hydraulic Laboratory for a number of TVA cofferdams to gain information on river velocities, scour tendencies, required height of cofferdam cells to support river discharges through restricted channels, and other hydraulic conditions.

Location of Saturation Line in Cofferdam Cells

Tests were conducted in Pickwick, Guntersville, Chickamauga, and Kentucky cofferdams to determine the location of the saturation line in the cell fill after the cofferdams had been unwatered. Results from these tests are discussed under the heading "Saturation and Drainage of Cell Fill and Berms," page 26.

Cell Deflection

Tests were conducted to determine the amount of deflection, if any, in the upstream arm of Kentucky stage 2 cofferdam. Deflections were slight and amounted to a maximum inward deflection of 0.13 foot for the upstream arm and 0.09 foot for the downstream arm. Deflections were less near the shoreline where the overburden rose higher on the cell.

Tests for Strength of Piling Interlocks

Tests were made to determine the strength of the piling interlocks after the piling had been reused several times. Tests were conducted with the piling interlocks at 3- and 22-degree angles. The results obtained were inconclusive but did indicate that with a reasonable amount of inspection the piling could be reused several times.

At Chickamauga cofferdam, attempts were made to determine the stress in the piling, which in use in the cell, by using strain gages. The cells tested were constructed in the dry and were filled before water pressure was released against the cell. Gage points were placed at regular intervals on two of the cells and the adjoining connecting cells. Points were placed vertically on 12-foot centers. Readings were taken during filling of the cells and before and after the water pressure had been released against the cells. The readings, however, were erratic and inconclusive and a curve could not be plotted. One possible explanation for this is that the piling does not always interlock evenly over its full length, thereby making the tension across the face of the pile vary. In fact, in the case of a damaged interlock over a certain length of the pile, the tension across the face of the pile through this length might be very low with a correspondingly higher stress in the interlock immediately adjacent. The greatest value derived from this test, then, was that the stresses in this type of structure are of a very uncertain nature and that it is dangerous to rely implicitly on theoretical design.

Photoelastic Model Studies

An attempt was made to analyze the stresses in the cofferdam at Kentucky Dam by photoelastic model studies. The results obtained were not entirely conclusive, however. Fairly high concentration of stresses occurred at the bottom of the cofferdam, both at the upstream and downstream corners, which in general checked the results obtained by analytical methods.

ACKNOWLEDGMENTS

The contents of this report represent the efforts and experience of many individuals from engineering and construction personnel of TVA. The work related covers a period of cofferdam design and construction beginning with Pickwick Landing Dam in 1935 and extending to the addition of units 15 and 16 for Hales Bar Dam in 1950.

The report was produced by the Construction Plant Branch of TVA's Office of Engineering under the general direction of C. E. Blee and Geo. K. Leonard, successive chief engineers; Harry Wiersema, assistant to the chief engineer; and Ray E. Martin, J. F. Partridge, and Lee M. Ragsdale, successive construction plant engineers.

Material for the report was prepared by Hugh R. Kinzer and R. S. Blomfield, civil engineers. Editorial work was carried out by Jack W. Hind, civil engineer, and A. Lionel Edney and Dot H. Owens, editorial clerks, under the supervision of John C. Voorhees, civil engineer. Illustrations were reviewed and prepared for printing by Thomas Benson, civil engineer.

Among others who made significant contributions to the report are:

A. L. Paul   D. P. Tsagaris
A. J. Ackerman   J. W. Peerson
R. T. Colburn   O. R. Bengston
Archie J. Vincent   G. E. Cate
A. F. Hedman

BIBLIOGRAPHY

Ackerman, A. J., and Locher, Charles H. Construction Planning and Plant. 381 pp., McGraw-Hill Book Company, Inc., New York, 1940.

Baes, Louis. "Belval P" Flat Sheet Piles for Cellular Structures. (Papers presented to the Second International Conference on Soil Mechanics and Foundation Engineering - Rotterdam, June 21-30, 1948. English translation.)

Bengston, O. R. "Template Takes Trouble Out of Driving 70 Foot Cofferdam Cells," Engineering News-Record, January 17, 1952, pp. 34-35.

Chellis, Robert Dunning. Pile-Driving Handbook. 276 pp., Pitman Publishing Corp., New York, 1944.

Descans, L. Cellular Sheet Piling Structures. (Reprint from January and February 1954 issues of "L'Ossature Metallique," Monthly Review of Steel Applications Published by the Centre Belgo Luxembourgeois d'Information de l'Acier - 154, Avenue Louise, Brussels.)

Hedman, A. F. "Cofferdam Design for Kentucky Dam," Engineering News Record, January 1, 1942, p. 50.

Jacoby, Henry S., and Davis, lRowland P. Foundations of Bridges and Buildings. 535 pp., McGraw-Hill Book Company, Inc., New York, 1941.

Plummer, Fred Leroy, and Dore, Stanley M. Soil Mechanics and Foundations. 473 pp., Pitman Publishing Corp., New York, 1940.

Terzaghi, Karl. "Stability and Stiffness of Cellular Cofferdams" (with discussions), Transactions, American Society of Civil Engineers, vol. 110 (1945), p. 1083.

Terzaghi, Karl. Theoretical Soil Mechanics. 510 pp., John Wiley and Sons, Inc., New York, 1943.

U. S. Steel Corporation. Steel Sheet Piling (Handbook).

White, Lazarus, and Prentis, Edmund A. Cofferdams. 2d ed., rev. 311 pp. Columbia University Press, New York, 1950.

Footnotes

Chapter 1

1. Technical Reports
No. 21, Concrete Production and Control
No. 22, Geology and Foundation Treatment
No. 23, Surveying, Mapping, and Related Engineering

2. Karl Terzaghi, "Stability and Stiffness of Cellular Cofferdams," (with discussions), Transactions, American Society of Civil Engineers, vol. 110 (1945), p. 1083.

Chapter 2

1. Lazarus White and Edmund A. Prentis, Cofferdams, 2d ed. (New York: Columbia University Press), 1950.

2. Karl Terzaghi, "Stability and Stiffness of Cellular Cofferdams" (with discussions), Transactions, American Society of Civil Engineers, vol. 110 (1945), p. 1100.

Chapter 3

1. Henry S. Jacoby and Rowland P. Davis, Foundations of Bridges and Buildings, McGraw-Hill, 1941.

2. Louis Baes, "Belval P Flat Sheet Piles for Cellular Structures" (photoelastic tests, etc.). (Papers presented to the Second International Conference on Soil Mechanics and Foundation Engineering, Rotterdam, June 21-30, 1948. English translation.)

3. L. Descans, "Cellular Sheet Piling Structures." (Reprint from L'Ossature Metallique, Monthly Review of Steel Applications, Published by the Centre Belgo-Luxembourgeois d'Information de l'Acier - 154, Avenue Louis, Brussels, January-February, 1954.

4. A fourth manufacturer, U.S. Steel Corporation, inadvertently overlooked in the original printing, supplied section MP-101.

Chapter 4

1. Karl Terzaghi, "Stability and Stiffness of Cellular Cofferdams" (with discussions), Transactions, American Society of Civil Engineers, vol. 110 (1945), p. 1083.

2. Terzaghi, op. cit., p. 1088.

Chapter 5

1. Karl Terzaghi, "Stability and Stiffness of Cellular Cofferdams" (with discussions), Transactions, American Society of Civil Engineers, vol. 110 (1945), p. 1083.

2. Lazarus White and Edmund A. Prentis, Cofferdams, 2d ed. (New York: Columbia University Press), 1950.

3. Terzaghi, op. cit., p. 1083.
4. Terzaghi, op. cit., p. 1083.
5. Terzaghi, op. cit., pp. 1093-1095.
6. Terzaghi, op. cit., pp. 1093-1095.
7. Terzaghi, op. cit., pp. 1175-1178.
8. Terzaghi, op. cit., p. 1200.
9. Ibid., pp. 1200-1201.
10. Ibid., pp. 1176-1177.
11. Terzaghi, op. cit., p. 1201.
12. Terzaghi, op. cit., pp. 1161-1165.
13. Terzaghi, op. cit., p. 1177.
14. Terzaghi, op. cit., p. 1100.
15. Ibid., p. 1090.

16. Terzaghi, op. cit., p. 1099.
17. Ibid., p. 1175.
18. Terzaghi, op. cit., p. 1098.
19. Terzaghi, op. cit., pp. 1093-1096.
20. Terzaghi, op. cit., p. 1086.
21. Karl Terzaghi, Theoretical Soil Mechanics (New York: John Wiley and Sons, Inc.), 1943.
22. Terzaghi, "Stability and Stiffness of Cellular Cofferdams," p. 1083.
23. Terzaghi, "Stability and Stiffness of Cellular Cofferdams," p. 1088.

Chapter 6
NONE

Chapter 7

1. Lazarus White and Edmund A. Prentis, cofferdams, 2d ed. (New York: Columbia University Press), 1950.
2. Robert Dunning Chellis, Pile-Driving Handbook (New York: Pitman Publishing Corp.), 1944.
3. Robert Dunning Chellis, Pile-Driving Handbook (New York: Pitman Publishing Corp.), 1944.
4. Chellis, op. cit., pp. 3-4.
5. Ibid., p. 14.
6. Chellis, op. cit.
7. Ibid., p. 25.
8. Ibid.
9. Ibid., pp. 164-169, 190.

10. Chellis, op. cit., p. 164, table I.
11. Ibid., p. 164, table II.
12. Ibid., p. 164, table III.
13. Ibid., p. 21.
14. Chellis, op. cit.
15. Chellis, op. cit., p. 26.
16. Ibid., pp. 165-169, table IV.
17. Chellis, op. cit.
18. Ibid.
19. Ibid., pp. 45, 73.
20. Chellis, op. cit., p. 230.

Chapter 8
NONE

Chapter 9

1. Fred Leroy Plummer and Stanley M. Dore, Soil Mechanics and Foundations (New York: Pitman Publishing Corp.), 1940.

THEORETICAL MANUAL FOR DESIGN OF CELLULAR SHEET PILE STRUCTURES
(Cofferdams and Retaining Structures)

PREFACE

This report provides the derivations and describes the procedures for the design of cellular sheet pile cofferdams. The work was accomplished by the U.S. Army Engineer Waterways Experiment Station (WES) and sponsored under funds provided by the Civil Works Directorate, Office, Chief of Engineers (OCE), in an effort to update the Corps' Engineer Manuals.

The first draft of the manual was written by Dr. Mark Rossow, Department of Civil Engineering, Southern Illinois University at Edwardsville, under the direction of Mr. Reed Mosher, Engineering Applications Office (EAO), Scientific and Engineering Application Division (SEAD), Automation Technology Center (ATC), WES. Additional sections were written by Mr. Edward Demsky, Foundation Section, Geotechnical Branch, U.S. Army Engineer District, St. Louis, and Mr. Mosher. Example problems were developed by Mr. Demsky. The work accomplished at WES was under the general supervision of Dr. N. Radhakrishnan, A/Chief, Information Technology Laboratory (ITL), formerly chief, ATC, and under the direct supervision of Mr. Paul Senter, A/Chief, Information Research Division, ITL, formerly chief, SEAD. The technical monitor for OCE was Mr. Don Dressler. This manual was edited by Ms. Gilda Miller, Information Products Division, ITL, WES.

COL Allen F. Grum, USA, was the previous Director of WES. COL Dwayne G. Lee, CE, is the present Commander and Director. Mr. Robert W. Whalin is Technical Director.

CONVERSION FACTORS, NON-SI TO SI (METRIC) UNITS OF MEASUREMENTS

Non-SI units of measurement used in this report can be converted to SI (metric) units as follows:

Multiply	By	To Obtain
degrees	0.01745329	radians
feet	0.3048	metres
inches	2.54	centimetres
kip/inches	112.9848	newton-metres
pounds (force)	4.448222	newtons
pounds (force) per foot	14.5939	newtons per metre
pounds (force) per square foot	175.1268	pascals
pounds (force) per inch	47.88026	newtons per metre
pounds (force) per cubic foot	16.01846	kilograms per cubic metre
pounds (mass per cubic inch	27.6799	grams per cubic centimetre
seconds	4.848137	radians

THEORETICAL MANUAL FOR DESIGN OF CELLULAR SHEET PILE STRUCTURES (COFFERDAMS AND RETAINING STRUCTURES)

(May, 1987)

by Mark Rossow
Southern Illinois University
Edward Demsky
U.S. Army Engineer District, St. Louis
Reed Mosher
U.S. Army Engineer Waterways Experiment Station
Information Technology Laboratory
Department of the Army
Waterways Experiment Station, Corps of Engineers
PO Box 631, Vicksburg, Mississippi 39180-0631

"The information contained in the following report was reprinted with the express permission of the U.S. Army Corps of Engineers, Waterway Experiment Station at private expense.

The information is provided WITHOUT WARRANTY OF ANY KIND, EITHER EXPRESSED OR IMPLIED, including, without limitation, the IMPLIED WARRANTIES OF MERCHANTABILITY AND FITNESS FOR A PARTICULAR PURPOSE. The United States, the Waterways Experiment Station, and their agents and employees assume no risk as to the accuracy, quality, and performance of information. In no event shall the United States, the Waterways Experiment Station, and their agents assume any liability whatsoever to any party for any damages, direct, indirect, incidental, or consequential, including, without limitation, any lost profits or lost savings arising out of the use or inability to use information."

PART I: INTRODUCTION

Purpose
1. This manual is a companion volume to the planned Engineer Manual (EM), "Design of Cellular Sheet Pile Structures," and is intended to provide theoretical background for that EM to the reader. It is also designed to present the background for the computer program CCELL (X$\phi\phi$40) for the analysis/design of sheet pile cellular cofferdams.

Scope
2. The manual contains derivations and discussions of procedures used in CCELL. It includes several procedures mentioned in the technical literature but found inadequate, and therefore omitted from CCELL. Several numerical examples illustrating the use of the design methods and an extensive list of references on cellular cofferdams are included in the manual.

Design Methods Presented in Rational Form
3. Most of the design methods discussed in this report are expressed in terms of a factor of safety (FS)* as

$$FS = \frac{\text{Maximum available resisting force (or moment)}}{\text{Driving force (or moment)}} \quad (1)$$

That is, the design methods are based on a comparison of resisting effect to driving effect. For this comparison to be meaningful, the following two criteria must be satisfied:

a. Identification of a single free body must be possible.
b. Both the driving and resisting forces (or moments) must act on this free body.

Criteria and Design Procedures
4. Although these criteria may seem obvious, some statements of design procedures in the literature do not satisfy them, and these procedures had to be modified or reinterpreted for inclusion in the manual. Thus, several of the design procedures presented herein differ somewhat from the formulations in the references cited.

Limitations of the Manual
5. The question of what constitutes a minimum acceptable value of a safety factor for a given failure mode is as much a policy issue as a technical issue and thus is not treated herein; values of safety factors are available in the EM.

6. Consideration is limited primarily to failure modes involving soil-structure interactions. Other important potential failure modes, such as undermining or piping caused by excessive seepage, are not considered.

Basic Combinations
7. Not every possible combination of foundation conditions (e.g., bare rock, rock with overburden, deep-sand, clay, berm or no berm) is considered. Instead, one or sometimes two sets of conditions have been chosen for each failure mode, and the corresponding free body and acting forces are identified. The intent of this approach is to provide the reader with the basic analysis procedure to be used for a particular failure mode. Once the procedure is understood, modification for different foundation conditions should be straightforward.

Soil-Structure Interaction in a Cellular Cofferdam
8. A cofferdam cell consists of a flexible steel membrane enclosing a granular soil fill. The soil-structure interaction in a structure of this type is a complex process involving composite action of the fill and the membrane. For example, the gravity forces acting on the fill cause it to exert pressure on the membrane and as a result of the pressure, tensile forces are produced in the membrane. These forces stiffen the membrane against further extension, thus providing a confining effect on the fill. This effect, in turn, stiffens the fill and enables it to develop the large compressive and shearing stresses it needs to transmit the hydrostatic and gravity loads to the foundation. Hence, the fill serves as the principal load-bearing element in the structure, but could not perform its task without the aid of the steel membrane.

Two Possible Analogies
9. To clarify the behavior of a cellular cofferdam further, it is helpful to consider two analogies, one good and the other

---
* For convenience, symbols and abbreviations are listed in the Notation (Appendix B).

poor. The better analogy consists of a thin polyethylene bag, such as the type used to wrap sandwiches, filled with sand. When a distributed horizontal force of reasonable size is applied to the bag, its only resistance is through the mobilization of shear resistance within the sand. Thus, the sand will be seen to displace within the bag. This behavior is a valid comparison for a cofferdam cell. By contrast, a poor analogy would be a typical, kitchen-size metal can, filled with sand. When a horizontal force of reasonable magnitude is applied to the sand-filled can, the can tips over, or if sufficient friction is present between it and the surface upon which it rests, the can simply remains at rest with no change of shape. In either event, the external load is carried primarily by the shell (the can) rather than by the fill. Such behavior is not representative of a cofferdam cell. Of course, all analogies have limitations. The behavior of an actual cofferdam lies somewhere between that of a sand-filled bag and a can, although it is much closer to that of a bag.

State of the Art

10. Beginning with the construction of the first steel sheet-pile cellular cofferdam at Black Rock Harbor, near Buffalo, N.Y., in 1908 and lasting at least until the publication of Terzaghi's famous paper (Terzaghi 1945), most cellular cofferdams were designed as gravity walls. Terzaghi pointed out the error in this approach and introduced the concept of designing the fill on a vertical plane to prevent shear failure, an idea which had been used, but not published, by TVA engineers some time earlier (TVA 1957). In the same paper, Terzaghi discussed the possibility of slip between the fill and the sheet-pile walls, and of penetration of the inboard walls into the foundation. Several currently-used design rules concerned with these phenomena appear to have been derived, at least in part, from his discussions. Other types of internal-stability failure modes have also been hypothesized by Hansen (1953, 1957), Ovesen (1962), and Cummings (1957). Some notable cofferdam failures attributable to excessive underseepage or lack of bearing capacity of the foundation (ORD 1974) have given rise to yet more potential failure modes for the designer to consider.

Hypothesized Failure Modes

11. Thus, over the years a rather large number of hypothesized failure modes have accumulated. Several model studies (Cummings 1957, EM 1110-2-2906, Maitland 1977, Ovensen 1962, Rimstad 1940, TVA 1957, Kurata and Kitajima 1967) have been conducted to determine which failure modes are likely and which are improbable. With one recent exception (Maitland 1977, Maitland and Schroeder 1979, Schroeder and Maitland 1979), these studies have not been of great help. Indeed, some of the studies have actually hindered the understanding of cellular-cofferdam behavior. The use of relatively small models with overly stiff walls led the experimenters to postulate failure modes which are highly unlikely to appear in a full-sized cell. In addition to model studies, field measurements (Summary Report Lock and Dam No. 26 (Replacement) 1983; Khuayjarernpanishk 1975; Moore and Alizadeh 1983; Schroeder, Marker, and Khuyjarernpanishk 1977; Sorota, Kinner, and Haley 1981; Sorota and Kinner 1981; TVA 1957; Naval Research Laboratory 1979; White, Cheney, and Duke 1971) of full-sized cells have also been conducted. Although valuable data on operating conditions have been obtained, no instrumented, full-sized cell has failed, and thus no data are available on cell behavior during failure.

12. Given the plethora of hypothesized failure-modes and the lack of sound experimental data, it is not surprising that "most designers in this field still rely heavily on past practice and experience" (USS 1972). At the present time, theoretical calculations, alone or even in large part, do not suffice for cellular cofferdam design.

13. In the next 5 to 10 years, this situation may change as finite element programs, polished and specialized for everyday use by the cofferdam designer, are developed. Such programs are not available at present.

PART II: ANALYSIS OF FAILURE MODES

Conventional Simplifications and Equivalent Layout

14. The analysis of many failure modes is simplified if the original cofferdam geometry is replaced by an equivalent straight-walled cofferdam. The literature contains several different procedures for calculating the dimensions of this equivalent cofferdam. The procedure adopted herein is to choose the distance L between crosswalls in the equivalent layout that equals the average distance between crosswalls in the original cofferdam (Lacroix, Esrig, and Lusher 1970) as shown in Figure 1. The equivalent width b is then computed by equating plan areas of the original and equivalent layouts. For example, in the case of a circular cofferdam, this procedure leads to the equation

$$b = \frac{\text{Area of main cell and one arc cell}}{2L} \quad (2)$$

Average vertical slice

15. Another simplifying approximation made often is the calculation of resultant forces and moments. Included are those arising from the crosswall for a length L of the equivalent cofferdam and dividing these quantities by L to get the average force and moment per unit length of cofferdam. This procedure is equivalent to assuming that the behavior of the entire cofferdam can be represented by a single "average" vertical slice such as that shown in Figure 2.

Flat walls

16. An obvious consequence of analyzing the equivalent rather than the actual cofferdam is that the curvature of the walls is neglected. For certain choices of free body, this amounts to neglecting the effect of interlock tension.

17. For example, Figure 2c shows a free body consisting of unit widths of both the outboard and inboard walls. Typical forces which act on this free body are also shown. Note that the interlock tension is not included. In effect, the walls are assumed to be flat. This latter statement may be clarified by considering Figure 3, in which is shown a free body consisting of unit widths of opposing walls in an actual (not an equivalent) cofferdam. Because of the curvature of the cell walls, the interlock tensions T_o and T_i have components T_{ox} and T_{ix} acting the x-direction. This can be seen in Figures 3b and c. These components are neglected in the free body shown in Figure 2c.

Critique of Simplifications

18. Although statements are frequently found in technical literature that it is "correct" to analyze a cofferdam by replacing it with an equivalent rectangular layout, no studies have been published which estimate the error involved in making the approximation.

    a. The alternative to replacing the cofferdam with its rectangular equivalent is to perform a three-dimensional FE

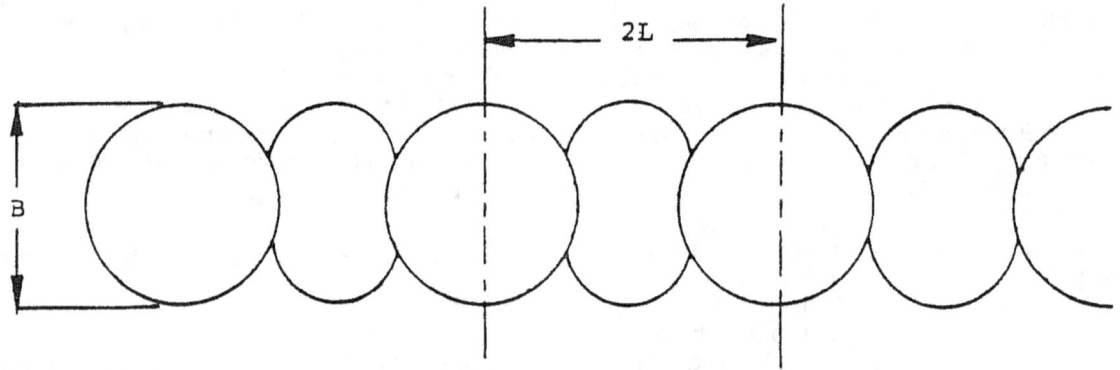

a. Circular cofferdam

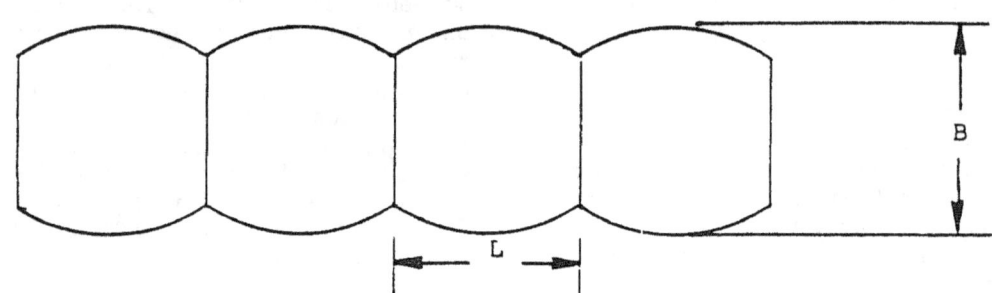

b. Diaphragm cofferdam

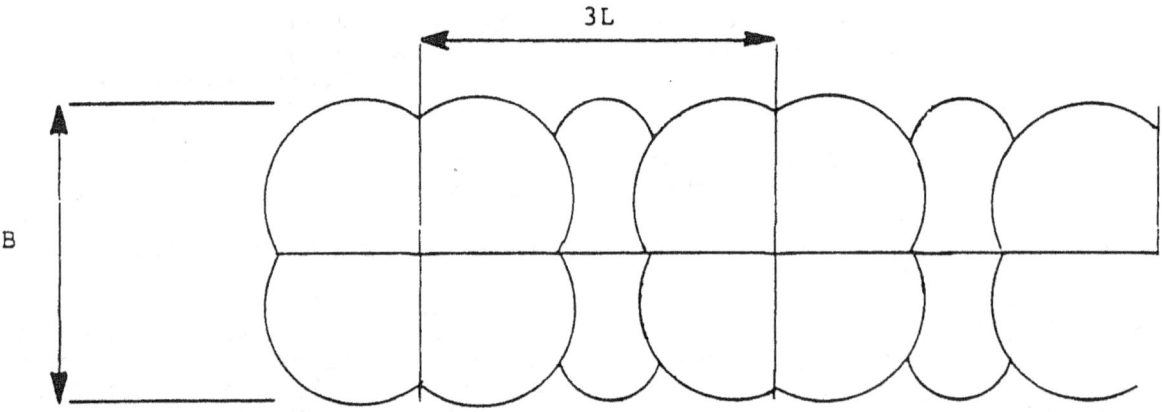

c. Cloverleaf cofferdam

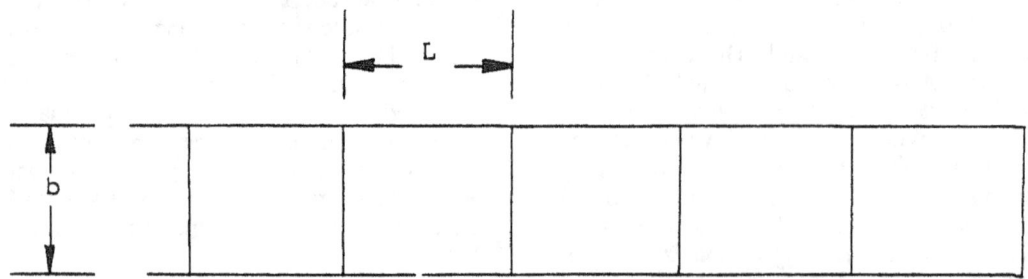

d. Equivalent rectangular cofferdam

**Figure 1
Actual cofferdams replaced by rectangular equivalent**

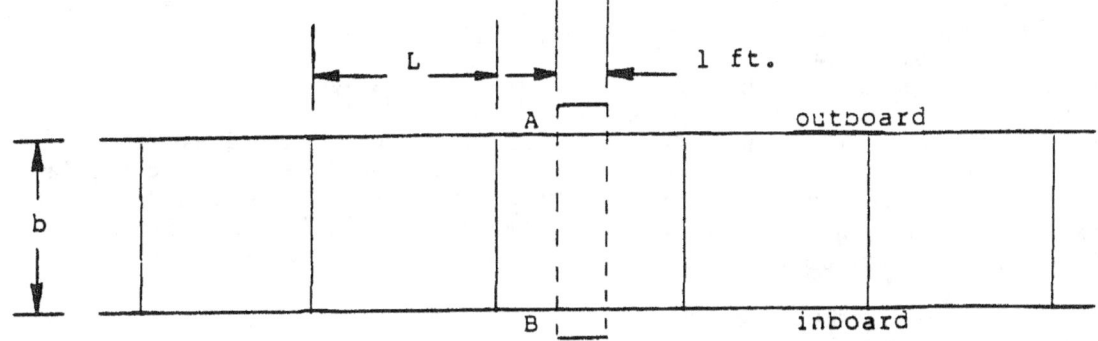

a. Portion of equivalent cofferdam selected for analysis

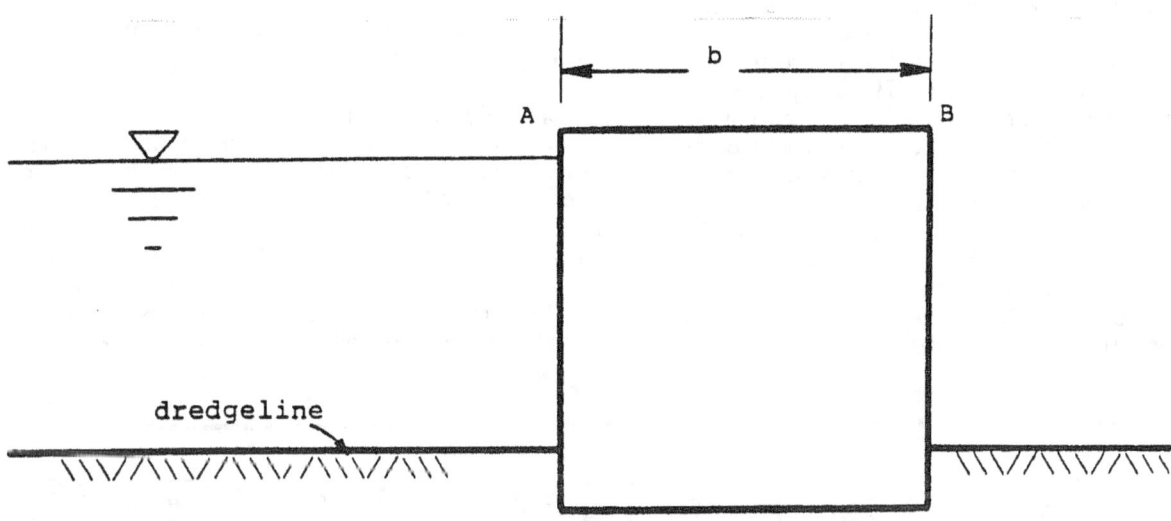

b. Elevation view of region in a

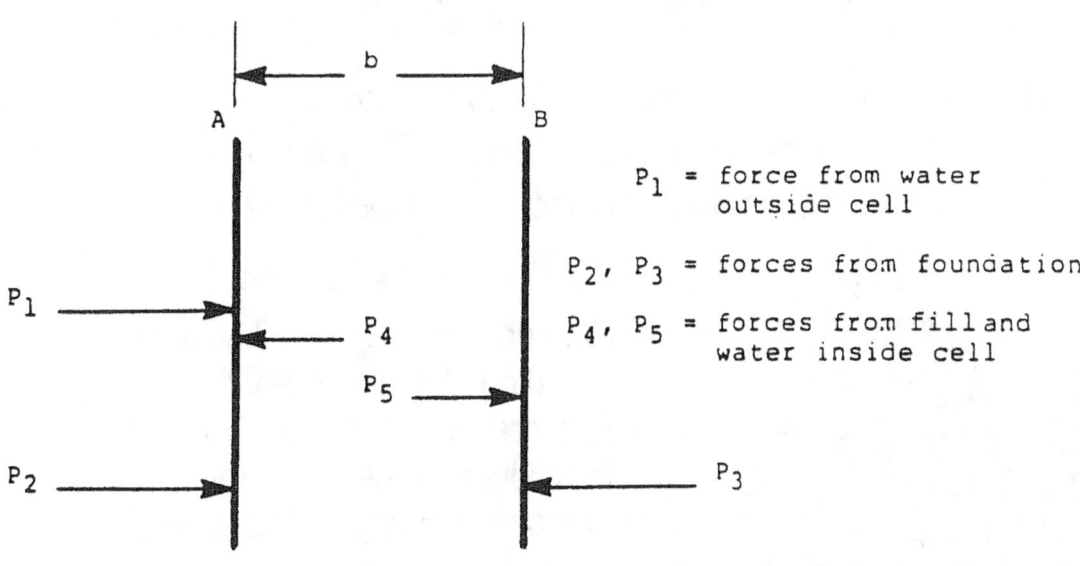

P_1 = force from water outside cell

P_2, P_3 = forces from foundation

P_4, P_5 = forces from fill and water inside cell

c. Free body comprised of unit widths of outboard and inboard walls

**Figure 2
Average vertical slice of cofferdam**

analysis. At the present time, this is not a feasible approach for a design office.

b. Similarly, analyses are based on an average vertical slice, not because the error in doing so is known to be small, but because of the lack of a feasible alternative.

c. For analyses based on a free body consisting of a single wall, the flat-wall assumption appears questionable. The component of interlock tension acting in the x-direction in Figures 3b and c is the primary means by which the wall resists the force from the fill and should not be neglected. This may be graphically demonstrated by considering the wall on the right in Figure 3c as a single free body. If the component of interlock tension 2Tix is neglected, moment equilibrium cannot be satisfied.

d. For analyses based on a free body consisting of both the inboard and outboard walls, the flat-wall assumption is somewhat more defensible, although the magnitude of the error implied by this assumption is not known. Consideration of the free body consisting of both walls in Figure 3c shows that the components of interlock tension will cancel each other provided that the magnitude of the interlock tension in the outboard wall equals that in the inboard wall, and the two tensions have the same line of action. This means that the resultant tensions in the inboard and outboard walls act at the same elevation. To the extent that these conditions are not satisfied, a net horizontal force and a moment arise from the interlock tensions acting on the free body.

e. Finally, point should be made of the development of FE models which use elastic springs to connect the outboard and inboard walls of a vertical slice of the cofferdam (Clough and Duncan 1977; Hansen and Clough 1982). In this way, the effect of wall curvature can be included.

Failure Modes and Example Problems

19. Detailed descriptions and discussions of ten failure modes follow in the next 10 parts. Further explanation through example problems, illustrated by step-by-step solutions, is presented in Appendix A.

PART III: BURSTING

Effects of Internal Lateral Stresses

20. The lateral stresses exerted by the fill and acting on the walls produce hoop forces which cause the interlocks to separate (Figure 4). The fill is lost and the cell may collapse.

21. The FS against bursting is shown by:

$$FS = \frac{t_{ult}}{t_{max}} \quad (3)$$

t_{ult} – maximum permissible interlock tension (per unit length of sheet) as specified by the sheet-pile manufacturer

t_{max} = maximum interlock tension (per unit length of sheet) existing in the cell wall

Critical Loading Cases

22. The following discussion on interlock tension is primarily concerned with the case of an isolated main cell, both after filling and at low water. This loading state usually represents the most important condition producing interlock tension in the main cell. The lateral forces associated with dewatering and the presence of a berm will also affect the interlock tension, although the effect does not appear to be great (Schroeder and Maitland 1979; White, Cheney, and Duke

---

AMERICAN UNDERWATER CONTRACTORS, INC.

When it comes to Underwater Work associated with the construction of Cellular Cofferdams, call the contractor who spent over 8 years at Lock 26!

Cell Inspections • Sheet Pile Cut Off
Welding • Burning • Blasting
Pile Restoration • Salvage • Recovery
Pipe Lines • Video Recording
Ultrasonic Testing • And More!

American Underwater Contrs. Inc.
219 Foerster Rd., St. Louis, MO 63042
PH) 314-739-5235   NITES) 314-739-3939

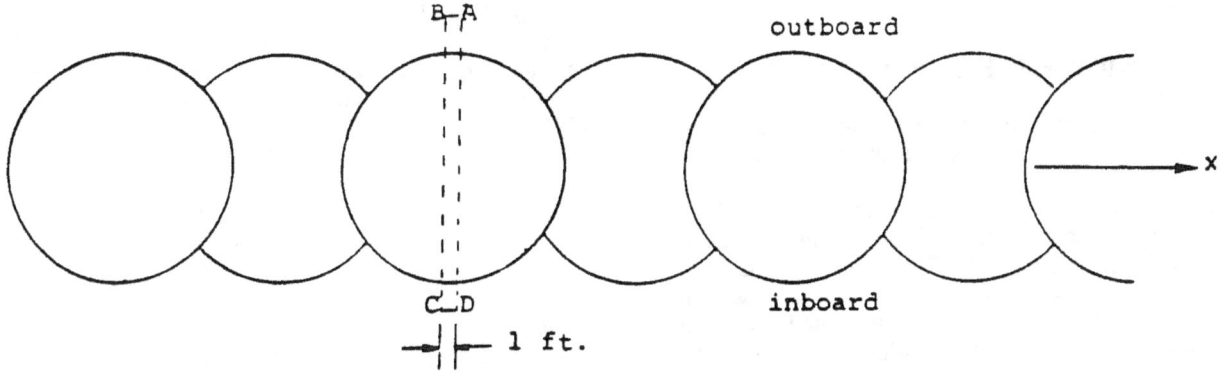

a. Portion of actual (not the equivalent) cofferdam selected for analysis

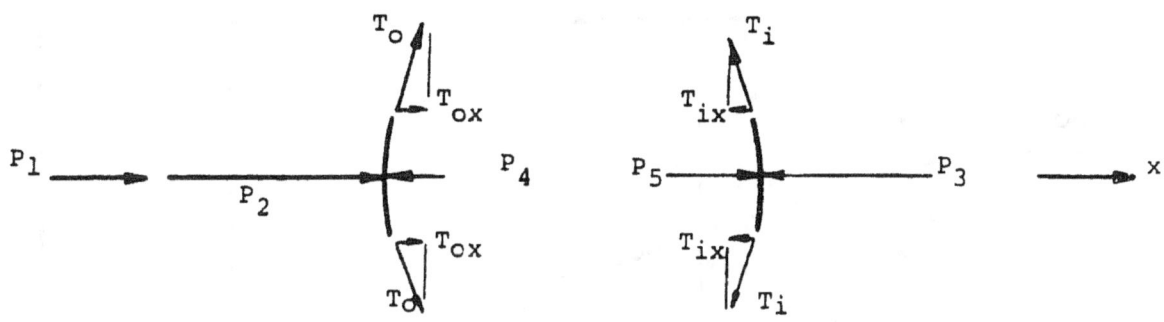

b. Plan view of free body ($P_1 - P_5$ defined in Figure 2)

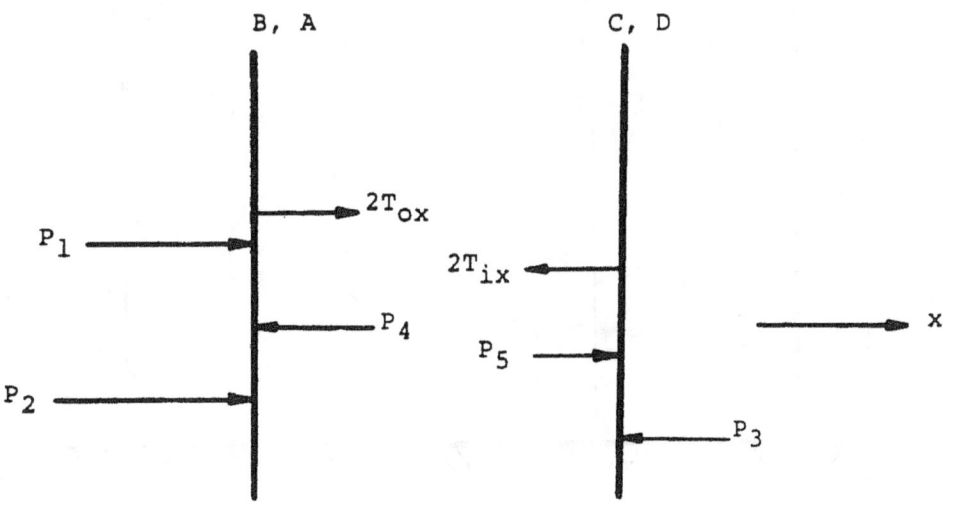

c. Elevation view of free body

**Figure 3
Free body showing effect of wall curvature**

1971; St. Louis District, CE 1983). In unusual circumstances such as anticipated removal of an interior berm during high water, a FE analysis may be necessary (Clough and Duncan 1977, 1978; Hansen and Clough 1982; St. Louis District, CE 1983) to resolve doubts over possible excessive interlock-tension.

Considerations in Interlock-Tension Calculations
Plane of fixity

23. The interlock tension and the lateral earth pressure acting on the cofferdam cell wall are each at a maximum at the same elevation. Both intuition and field measurements indicate that this elevation is close to the elevation at which maximum bulging of the cell occurs (Schroeder and Maitland 1979; St. Louis District, CE 1983). The location of the point of maximum bulging depends on the degree of restraint provided by the foundation acting on the embedded portion of the sheet-pile walls and may best be estimated by use of the concept of the plane of fixity.

24. The plane of fixity is defined as the plane below which the interlock tension in the sheet piling is small, or alternatively, as the plane of potential plastic hinges in the piling (Figure 5). Analytically, for a cell founded on a weak or a strong soil foundation, the plane of fixity may be located by using established results for the behavior of laterally-loaded piles. In deriving Equations 4 and 6, the assumption is made that the plane of fixity occurs at the point of zero rotation.

Sand Foundation

25. For cofferdam cells in a sand foundation, the depth-to-fixity d' (= distance from plane of fixity to dredgeline, Figure 5) is given by the equation (Schroeder and Maitland 1979)

$$d' = 3.1 \left(\frac{EI}{n_h} \right)^{1/5} \quad (4)$$

where
E = modulus of elasticity of the pile
I = moment of inertia of the pile section
n_h = constant of horizontal subgrade reaction

For Equation 4 to be valid, the embedment depth of the cofferdam cell, d, must satisfy the relation

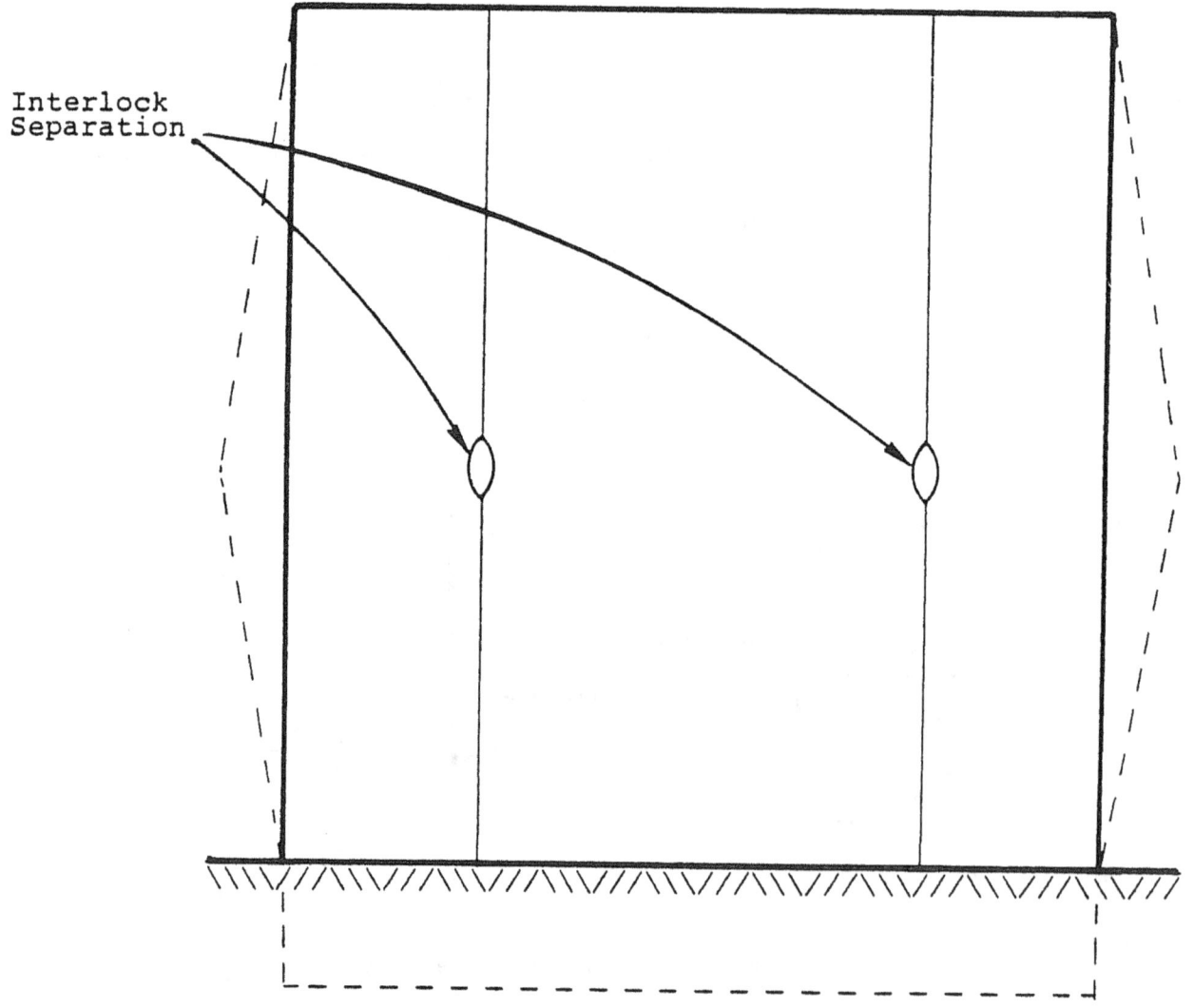

**Figure 4
Failure by interlock separation (Dismuke 1975)**

$$d \geq 5\left(\frac{EI}{n_h}\right)^{1/5} \quad (5)$$

Note that the effect of the foundation characteristics on the depth-to-fixity enters the calculations through the appearance of the parameter n_h.

Clay Foundation

26. For cofferdam cells in a clay foundation the depth-to-fixity is given by the equation

$$d' = 3.3\left(\frac{EI}{E_s}\right)^{1/4} \quad (6)$$

where E_s equals a horizontal spring modulus representing the behavior of the soil-pile system. For Equation 6 to be valid, the embedment depth of the cofferdam cell must satisfy the relation

$$d \geq 4\left(\frac{EI}{E_s}\right)^{1/4} \quad (7)$$

Equations 4 through 7 are derived from the theory of beams on elastic foundation (Hetenyi 1946). Thus, the above equations depend on the assumptions made in deriving that theory and also on the assumption that the bending response of the cofferdam cell can be represented by the theory of beams on elastic foundation.

27. The value of n_h for sheet-pile walls can be calculated by the following equation.

$$n_h = \frac{b_s}{d'} c_h \quad (8)$$

The value of E_s for sheet-pile walls can be calculated by the following equation

$$E_s = \frac{1}{3} b_s k_{sl} \frac{1}{d'} \quad (9)$$

Values of k_{sl} and c_h are given by Terzaghi (1955) in Tables 2 and 4. The terms in the above equations not previously defined are:

b_s = width of a single sheet pile
c_h = constant of horizontal subgrade reaction for anchored bulkhead with free earth support
k_{sl} = basic value of coefficient of vertical subgrade reaction

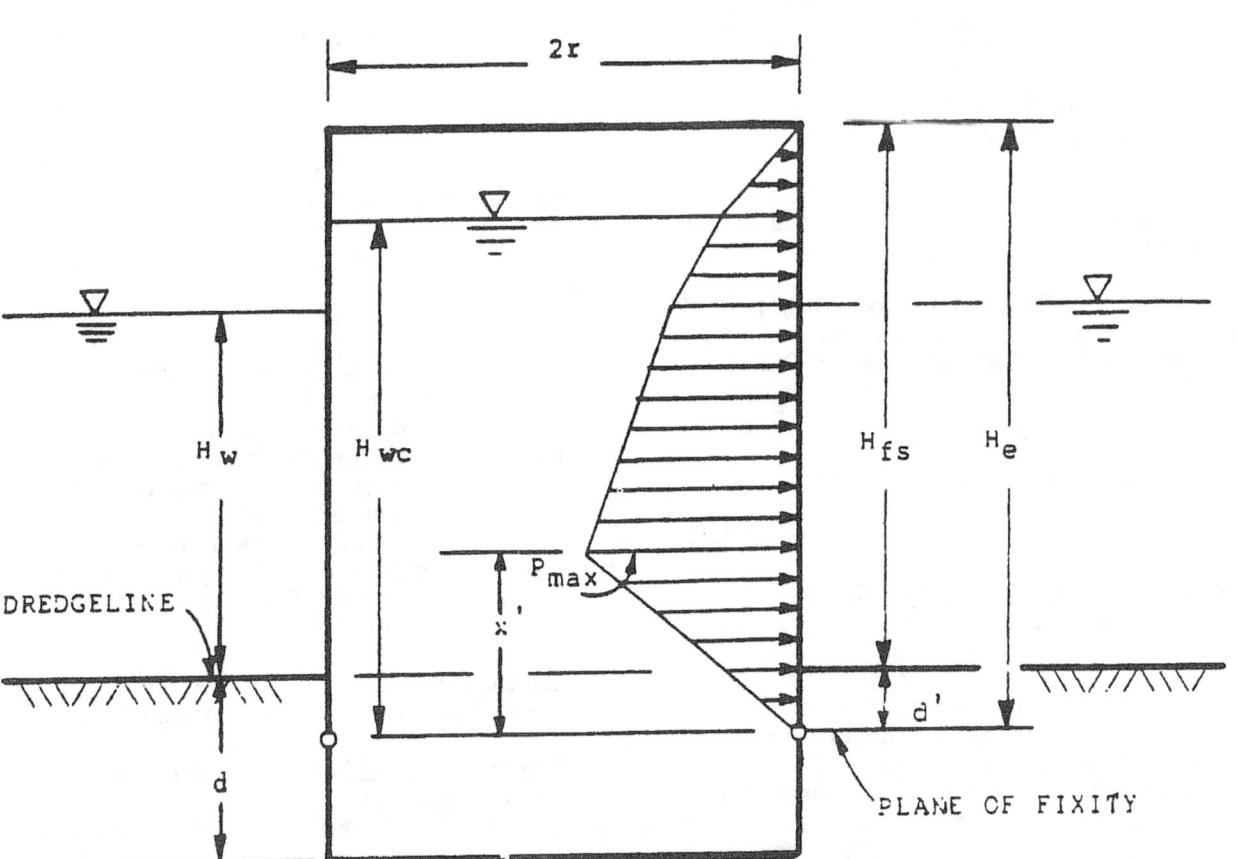

Figure 5
Definitions used in interlock force calculations

It should be noted that in Equation 9 the 1 term above the d' term has units of feet associated with it, thus d' should also have units of feet. The assumption is made in the above equations that the cofferdam walls are flat. The curvature of the cofferdam cells in neglected.

Alternate Method for Locating Plane of Fixity

28. Another method for finding the plane of fixity is derived by equating the internal and external pressure acting on the equivalent cofferdam. In this method the plane of fixity is assumed to occur at the point where the internal cell pressure is equal to the external pressure. This method assumes that active earth pressure is mobilized inside the cell and passive earth pressure is mobilized outside of the cell. The assumption of neglecting the interlock force as stated in paragraph 16 for the equivalent cofferdam is also made.

29. For a cofferdam in a sand foundation where the water level inside and outside of the cell is at different levels, the plane of fixity is given by the equation

$$d' = \frac{K_a[\gamma_m(H_{fs} + d - H_{w4}) + \gamma'(H_{w4} - d)] + \Delta H_w \gamma_w}{\gamma'(K_p - K_a)} \quad (10)$$

Terms not previously defined are:

K_a = active earth-pressure coefficient
K_p = passive earth-pressure coefficient
H_{fs} = vertical distance from dredgeline to top of cell (free-standing height)
H_{w4} = vertical distance from sheet-pile tips to water level inside of cell
ΔH_w = differential water head between the inside and outside of the cell, the water level inside of the cell minus the water level outside of the cell
γ_m = unit weight of moist fill
γ_w = unit weight of water
γ' = effective unit weight of soil

An equation similar to Equation 10 can be derived for cells in a clay foundation by equating internal and external pressures.

30. Once the plane of fixity has been located, the point of maximum bulging and interlock tension can be calculated from the empirical formula (Schroeder and Maitland 1979; St. Louis District, CE 1983)

$$x' = \frac{(H_{fs} + d')}{3} \quad (11)$$

where
x' = distance from the plane of fixity to the point of maximum interlock-tension

Rock Foundation

31. For the case of a cell founded on rock, where the embedment of the sheet-pile tips is sufficient to prevent radial displacement when the cell is filled, Equation 11 may still be applied by substituting $d' = o$. A plane of fixity cannot be said, strictly speaking, to exist since the slope of the sheet at the tips cannot be considered small. However, the rock foundation provides enough radial restraint to reduce interlock tension to near zero at the base of the cell. Note that substituting $d' = o$ in Equation 11 in this case gives $x' = H_{fs}/3$, a result similar to that given by the Tennessee Valley Authority (TVA) rule (TVA 1957), which specified that the maximum interlock tension be calculated at $H_{fs}/3$ or $H_{fs}/4$.

32. A final observation on the use of Equation 11 is that it is based on the assumption that at some point along the length of the sheetpiling, the radial displacement of the piling is restrained. It is inappropriate to use Equation 11 if this assumption is not valid. An example would be the case for a cell founded on very hard rock, into which piling penetration is very small, or in the case of a weak soil foundation for which no depth-to-fixity could be established (that is, Equations 5 or 7 are not satisfied). In these instances, the foundation provides little lateral restraint, and the point of maximum interlock stresses may be very close to the bottom of the piling.

Hoop-stress Equation

33. The interlock tension in the main or arc cell outside the crosswall is computed from the hoop-stress equation

$$t_{max} = P_{max} r \quad (12)$$

where
P_{max} = maximum lateral pressure acting against the wall
r = radius of the cell

34. The maximum pressure p_{max} is assumed to occur at the elevation of the point of maximum bulging and is calculated by summing the effective lateral-earth pressure and the difference in water pressure inside and outside the cell. Based on the water depths shown in Figure 5, the equation for P_{max} is

$$P_{max} = K\left\{\gamma_f(H_e - H_{wc}) + \gamma'_f\left[\frac{2H_e}{3} - (H_e - H_{wc})\right]\right\} + \gamma_w(H_{wc} - H_w - d') \quad (13)$$

This equation will work only when the differential water-level height inside the cell is above point at which P_{max} is being calculated. If the water level is the same, inside the outside of the cell, Equation 13 is suitable, regardless of the water level.

where
K = lateral earth-pressure coefficient
γ_f = unit weight of dry fill
γ'_f = submerged unit weight of fill
H_e = vertical distance from plane of fixity to top of cell (effective length of the sheet piles)
H_w = vertical distance from dredgeline to surface of water outside of cell
H_{wc} = vertical distance from plane of fixity to intersection of phreatic surface with center line of cell

The other terms retain their previous meanings. Selecting a value of K will be discussed in paragraph 39.

Interlock-Tension Calculations in Crosswall
Swatek's equation

35. When both the main cell and an adjacent arc cell are filled, the crosswall near the arc connection must provide sufficient tension to support the tension from the main cell and the arc cell. An equation for the interlock tension in the crosswall may be derived by considering the free body shown in Figure 6. Since the total force acting on a unit depth of wall

in this figure is $p_{max}L$, a balance of forces gives

$$t_{cw} = p_{max}L \quad (14)$$

in which t_{cw} is the interlock tension (per unit length of sheet) in the crosswall, and the pressure p_{max} is computed from Equation 13. Equation 14 is commonly referred to as Swatek's equation, since it was first used by Paul Swatek (Consulting Eng. Sewickley PA.).

TVA Secant Equation

36. An alternative equation, the TVA secant equation (TVA 1957), especially intended for use near the arc connection, is

$$t_{cw} = p_{max}L[\sec(\theta)] \quad (15)$$

where (Figure 7a) θ is the angle measured from the cofferdam axis to the connecting pile.

37. Equation 15 may be derived by referring to Figure 7b and noting that the resultant $p_{max}L$ is equilibrated by the inboard component of t_{cw}, giving

$$t_{cw}[\cos(\theta)] = P_{max}L \quad (16)$$

from which Equation 15 follows by dividing through by $\cos(\theta)$.

Rational Design Procedure to Avoid Bursting

38. Because the elevation of the point of maximum bulging (and hence, the point of maximum interlock tension) depends on the stiffness of the foundation, traditional design rules such as "the interlock tension should be calculated at one-third to one-fourth the free-standing height of the cell" are necessarily subject to interpretation and modification each time a cell is designed for different foundation conditions. However, if the plane-of-fixity concept is used, the effect of different foundation conditions is automatically taken into account: the maximum interlock tension is calculated at a lower level for weak soils compared to strong soils, and may even occur at or below the dredgeline (Schroeder and Maitland 1979). Thus, use of the concept provides a rational basis for considering the stiffness of the foundation when calculating the interlock tension.

Selecting a Value of K

39. Because of the take up of slack and stretching of the interlocks during cell filling, some movement of the fill occurs, and the earth-pressure coefficient is reduced from its at-rest value. The degree to which the pressure is reduced, however, is controversial, and the theoretical arguments, field data, and model studies reported in the literature give a wide range of values of K to choose from. It should be noted that no reliable direct measurements of soil pressure inside cells have been reported. (Several investigators have installed soil pressure

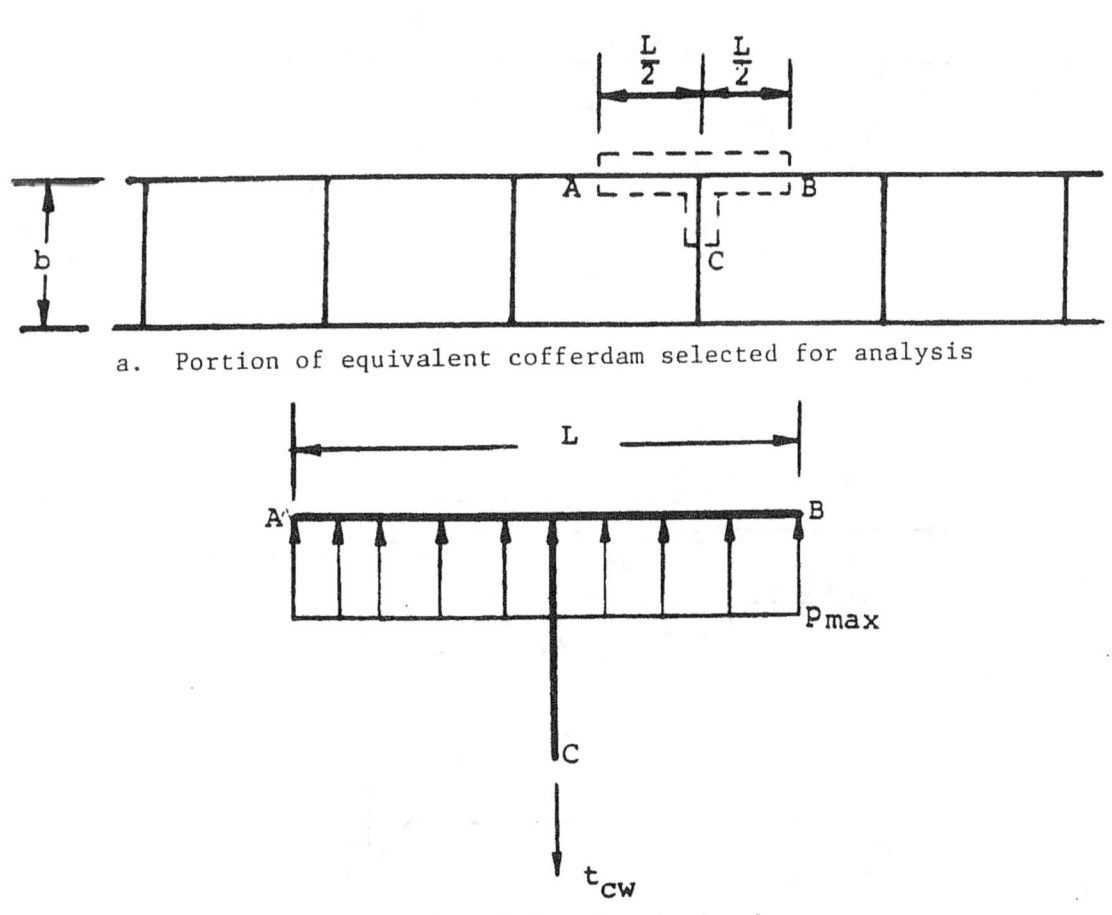

a. Portion of equivalent cofferdam selected for analysis

b. Free body of unit depth into plane of figure

Figure 6
Geometry and free body for derivation of Swatek's equation

cells, but little data were obtained which could be viewed with confidence.) Instead, soil pressure has generally been calculated from the hoop-stress equation, Equation 12, in which the interlock tension has in turn been calculated from the generalized Hooke's law for the steel sheet pile and from strains measured by strain gages. Thus, even soil-pressure values purportedly obtained experimentally are based on theoretical assumptions such as how the strain in the sheet pile is distributed across the cross section, or whether or not vertical strain and the associated Poisson effect are present.

40. Further complicating the question of what value of K to use is the fact that for a given fill material K will be influenced by a host of factors. Examples of these factors include the method and rate of filling, the presence of a surcharge, internal drainage conditions, and the method of compaction of the fill. In light of these uncertainties, K values are best chosen by relying on previous experience, rather than on theoretical arguments (Sorota, Kinner, and Haley 1981; Sorota and Kinner 1981). Values of 1.2 to 1.6 times the Rankine active coefficient have been proposed (Schroeder and Maitland 1979; St. Louis District, CE 1983); alternatively, Terzaghi (1945) proposed using a value of 0.4. Since some movement of the fill within the cell does occur, it would be overly conservative to use the at rest earth pressure coefficient.

41. Finally, it is important to see the uncertainty of the K value in proper perspective. For example, the interlock safety of the cell is much more strongly affected by the assumption made for the height of saturation within the cell than it is by

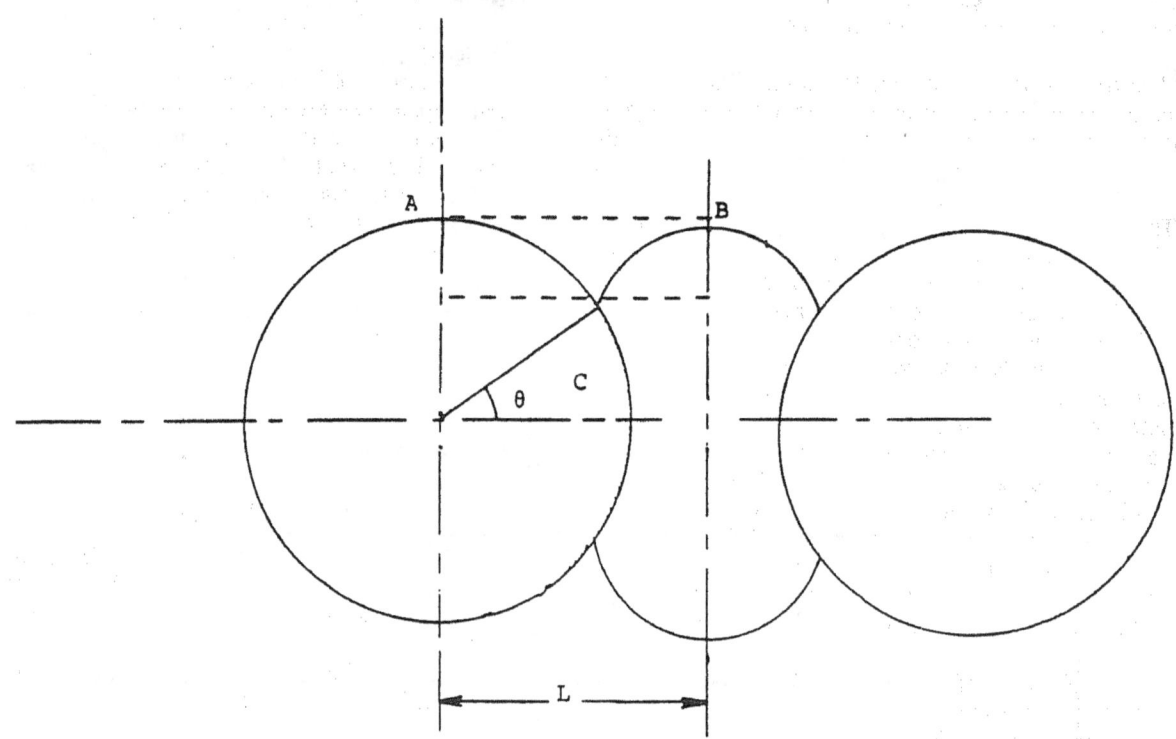

a. Region selected for analysis

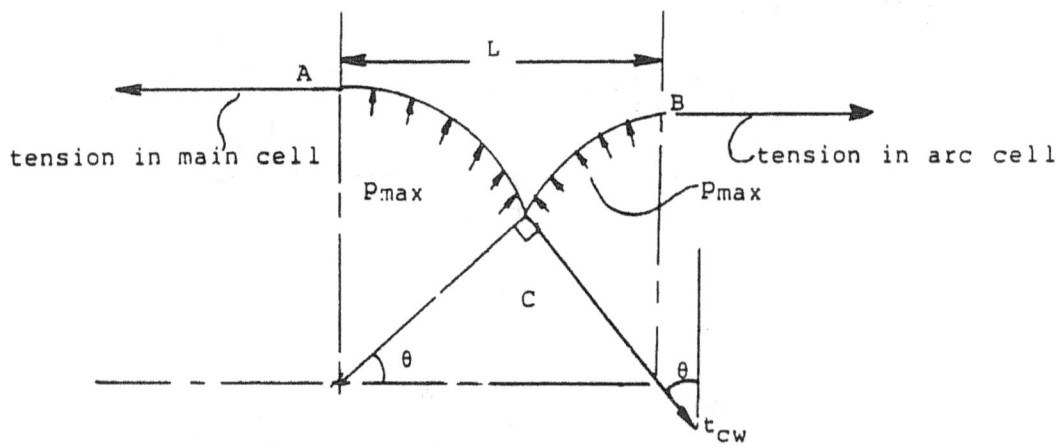

b. Free body of unit depth into plane of figure

Figure 7. Geometry and free body for derivation of TVA-secant formula

choosing K equal to, for instance, 0.4 or 0.5. Thus, for protection against bursting, much more attention should be paid to ensuring proper drainage of the cell than to lengthy deliberations about what value of K to use (Swatek 1970).

Secant Formula

42. Although the field and model-test data reported in the literature (St. Louis District, CE 1983; Schroeder and Maitland 1979; Sorota, Kinner, and Haley 1981; Sorota and Kinner 1981) are not completely consistent, at least some measurements indicate that the interlock tension in the crosswall near the arc connection may be as much as 20 percent higher than the main-cell tension. Thus, it appears reasonable to design for higher values of interlock tension in the crosswall than in the rest of the cell walls. The secant formula, Equation 15, was developed by TVA engineers (TVA 1957) to estimate the crosswall tension near the connecting pile, but for the following reasons its use is not recommended.

 a. The derivation of Equation 15 implies that forces are balanced in the inboard-to-outboard direction only; using the value of t_{cw} given in Equation 15 and summing forces in the direction of the axis of the cofferdam shows that equilibrium is violated in this latter direction.

 b. The angle θ appearing in Equation 15 corresponds to the angle which could be measured in the field before the main arc cells are filled. Since the sheet-pile walls can transmit only membrane forces (bending resistance is negligible), once the cells are filled the walls must deform and reorient themselves in order to accommodate the load from the fill. In particular, the connecting pile at the juncture of the main and arc cells must rotate and deform (some plastic yielding will be present (Grayman 1970) to equilibrate the three tensile forces meeting there, and thus the value of the angle θ of Figure 7b must change. Alternatively, inspection of free-body diagrams of the connecting pile (Dismuke 1975 and Swatek 1970) also show that θ must change under loading. Thus, in any derivation based on Figure 7b, both t_{cw} and θ should be considered as unknowns to be determined by equilibrium requirements.

43. The implication of these observations is that Equation 15 is based on premises which violate one of the fundamental principles upon which a cofferdam cell depends for its ability to carry its loads. Namely, large deformations of the sheet-pile skin are necessary to permit shearing resistance to develop in the fill. As a result, it is not surprising to find that both model and field data indicate that interlock tensions predicted by Equation 15 are overly conservative, and that its use is not recommended (Schroeder and Maitland 1979; Sorota and Kinner 1981; St. Louis District, CE 1983; Lacroix, Esrig, and Lusher 1970).

Swatek's Equation

44. In place of the TVA secant equation, Swatek's equation, Equation 14, is recommended for the following reasons:

 a. The approximation made in basing the derivation of Equation 14 on the equivalent rectangular cofferdam is consistent with the approximation made in analyzing other failure modes.

 b. Equation 14 predicts results in better agreement with measured field data (St. Louis District, CE 1983; Moore and Alizadeh 1983; Sorota, Kinner, and Haley 1981; and Naval Research Laboratory 1979).

 c. Equation 14 may be shown to yield good agreement with that obtained from an analysis which satisfies equilibrium and compatibility and is based on the actual positions of the loaded walls (Rossow 1984).

 d. Finally, most bursting failures which have occurred can be traced to sheets being driven out of interlock, to damage or fabrication errors (e.g., welding-related problems) associated with the connector pile, or to the extreme deformations required of a tee connector (Belz 1970; OCE 1974; Grayman 1970; ORD 1974). Most designers consider separation of the interlocks a prime candidate as a cause of cell failure. A striking fact, however, is that the literature contains no reports of failures for which underdesigning for interlock tension was identified as the principle cause, at least for Y rather than tee connections. This fact offers evidence for using a less conservative formula for the common-wall tension near a Y pile, such as Equation 14.

Equation Comparisons

45. A final observation may be made here to summarize the essential difference between the TVA secant equation and Swatek's equation. Swatek's equation is based on using a crude model of the cofferdam (that is, the equivalent rectangular cofferdam) to estimate an average interlock tension for the entire crosswall. The equations of statics are satisfied. In contrast, the TVA secant equation is based on a geometrical model which takes into account wall curvature, and an estimate is obtained for the crosswall interlock-tension at a specific point - adjacent to the Y. However, an equation of statics is violated, and, furthermore, the geometrical model is flawed, since it does not take into account the movement and rotation of the walls which occurs as the cell is filled.

PART IV: SLIP ON VERTICAL CENTER PLANE IN FILL

Effects of External Lateral Forces

46. The lateral force acting on the cell causes shear failure on vertical planes within the fill. Large distortions of cell shape occur and the cell may collapse towards the inboard side of the cofferdam. See Figure 8.

47. The FS against failure by slip on vertical center plane (Terzaghi 1945) is written as

$$FS = \frac{\text{Maximum available resisting force}}{\text{Driving Force}}$$

$$= \frac{S'_m + S''_m}{S' + S''} \quad (17)$$

where

S' = actual shearing force acting on vertical centerplane of cell

S'_m = maximum possible value of shearing force on vertical centerplane of cell

S'' = actual friction force from interlocks in crosswall

S''_m = maximum possible value of friction force from interlocks in crosswall

It should be noted that all quantities are calculated for a unit length of the cofferdam.

Cell Foundation

48. The discussion of this failure mode is based on the assumption that the cell is found on a rock, sand, or hard clay foundation. Thus, the foundation is able to resist the unequal pressure distribution resulting from the combined vertical and horizontal forces acting on the cell. As a consequence, shearing resistance on vertical planes within the fill can be mobilized. The magnitude of this resistance is an important

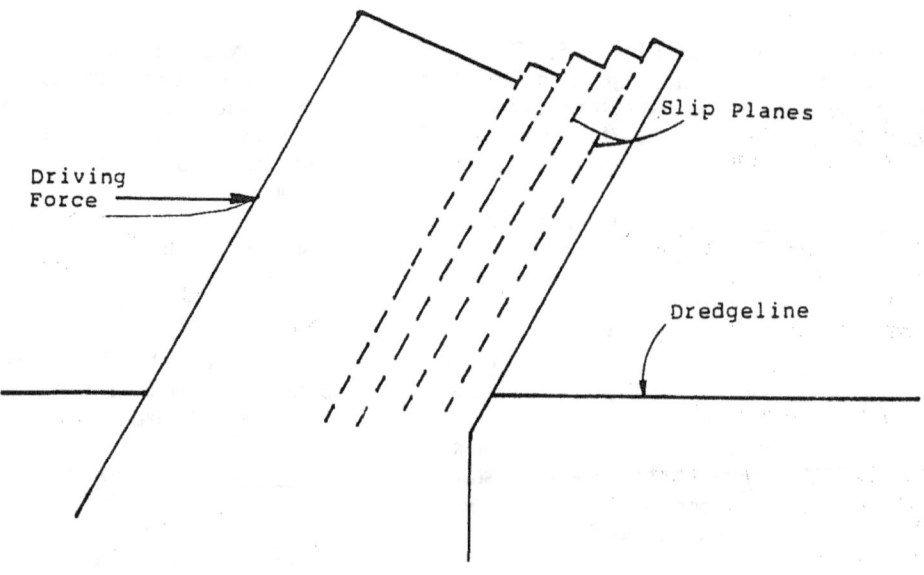

Figure 8
Failure by slip on vertical center plane

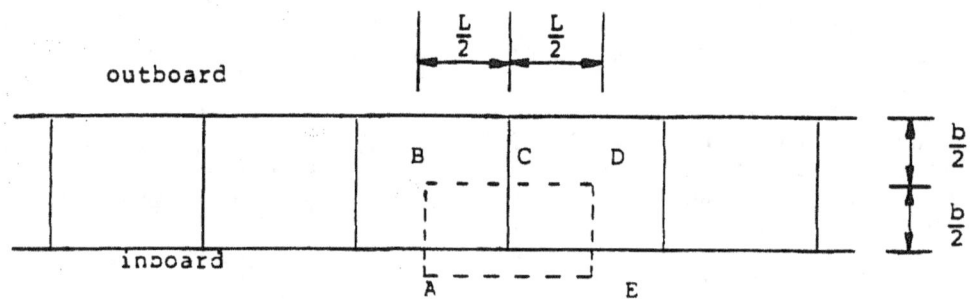

a. Portion of equivalent cofferdam selected for analysis

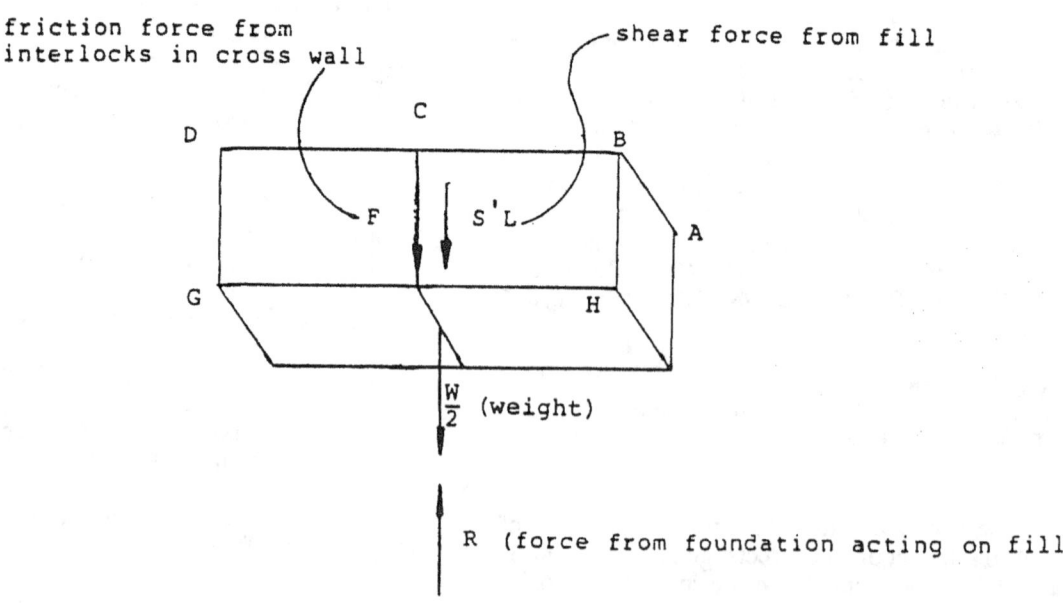

Figure 9
Free-body diagram showing driving forces

consideration in evaluating the stability of the cell. In contrast, the design of cofferdams founded on soft clay or other compressible soils tends to be governed by the bearing capacity of the foundation, rather than by considerations of internal stability. The stability calculations can still be made, however, based on the resisting moment provided by interlock friction (Jumikis 1971; Terzaghi 1945; USS 1972).

Considerations in Analysis of Failure by Vertical Shear on Center Plane

49. The individual driving forces S' and S" cannot be easily calculated; their sum, however, can be expressed in terms of the overturning moment. The relevant free body is shown in plan view in Figure 9a and in an isometric view in Figure 9b. Only vertical forces are shown. These forces are:

   a. F, the friction force in the interlock of the crosswall.
   b. S'L, the shear force acting on the center plane DGHB (S' is produced by the cell fill on one side of the center plane acting on the fill on the other side.).
   c. W/2, the weight of half the contents of the cell.
   d. R, the upward reaction from the foundation.

Summing vertical forces and equation to zero gives

$$S'L + F + \frac{W}{2} - R = 0 \qquad (18)$$

Dividing through by L and using the definition of S" as friction force per unit length, namely,

$$S" = \frac{F}{L} \qquad (19)$$

gives, after some re-arrangement,

$$S' + S" = \frac{\left(R - \frac{W}{2}\right)}{L} \qquad (20)$$

Thus, the driving force (S' + S") has been expressed in terms of the weight and the upward reaction R on the inboard half of the cell.

50. In turn, R may be expressed in terms of the overturning moment. Figure 10a shows the portion of the cofferdam selected for analysis, and an elevation view is shown in part b of that figure. Also shown in Figure 10b is the distributed force from the foundation which acts upward on the cell. Since only the weight W of the cell (no lateral forces) is considered in this sketch, the distributed force is uniform. In Figure 10c, the overturning effect of lateral forces has been included through the presence of the overturning moment. If the symbol M denotes moment per length, then the magnitude of the overturning moment is ML. The force distribution has now been altered and is assumed to vary linearly across the base of the cell (USS 1979). As shown in Figure 10d, this latter force distribution may be replaced by the sum of a uniform distribution and a linearly varying symmetric distribution. But these two distributions may be replaced by a pair of concentrated loads of magnitude W/2 and a couple defined by two forced of unspecified magnitude Q. Thus R, the vertical reaction from the foundation acting upward on the inboard half of the cell, can be expressed in terms of Q as

$$R = Q + \frac{W}{2} \qquad (21)$$

51. It remains to express Q in terms of the overturning moment. Since in Figure 10d each force Q represents the resultant of a triangular distribution, each force must act through the centroid of the triangle; thus the distance between the Q forces is 2b/3, and the magnitude of the couple produced by the foundation pressures is 2bQ/3. Since the cell is in equilibrium, this couple must balance the overturning moment ML; that is,

$$ML = \frac{2bQ}{3} \qquad (22)$$

52. Equation 22 may be solved for Q in terms of the overturning moment and the result substituted in Equation 21 to yield an expression for R in terms of M. Substituting this latter expression into Equation 20 then yields

$$S' + S" = \frac{3M}{2b} \qquad (23)$$

Overturning Moment

53. The lengths used in the calculation of the overturning moment M are defined in Figure 11a. Here, the quantities H_{w1}, H_{w2}, and H_{w3} are the vertical distances from the sheet-pile tips to the intersection of the phreatic surface with the inboard sheeting and the center line, respectively. H_{wo} is the vertical distance from the tips of the sheet-pile to the water level outside of the cofferdam. The forces are defined in Figure 11b. Here, P_w and P_{w1} represent the resultants (per unit length of cofferdam) of the water pressure on the exterior faces of the outboard and inboard walls of the cell, and are given by the relations

$$P_w = \frac{\gamma_w (H_{wo})^2}{2} \qquad (24)$$

and

$$P_{w1} = \frac{\gamma_w (H_{w1})^2}{2} \qquad (25)$$

P'_a, the horizontal effective force of the foundation soil acting on the outboard sheeting, is calculated using the Rankine active earth pressure coefficient K_a:

$$P'_a = \frac{K_a \gamma'_s d^2}{2} \qquad (26)$$

in which γ'_s is the submerged unit weight of the foundation soil.

54. The calculation of P'_p, the horizontal effective force from the berm and foundation soil acting on the inboard sheeting, is problematical. One approach is to calculate it using the sliding-wedge theory, the Coulomb theory modified

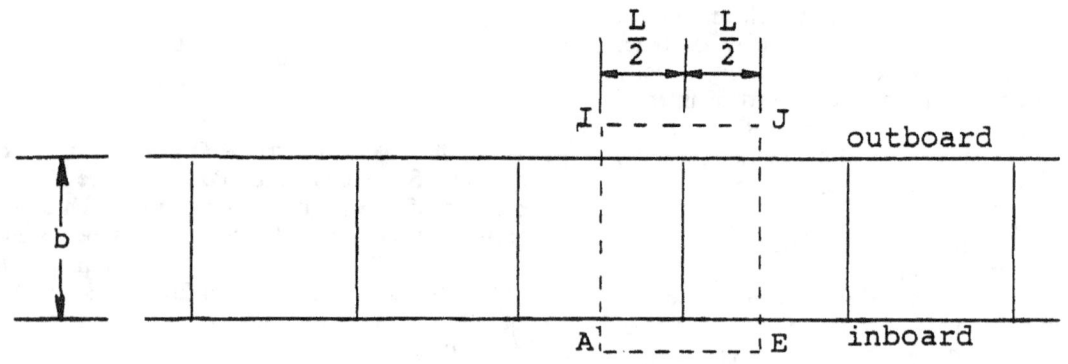

a. Portion of equivalent cofferdam selected for analysis

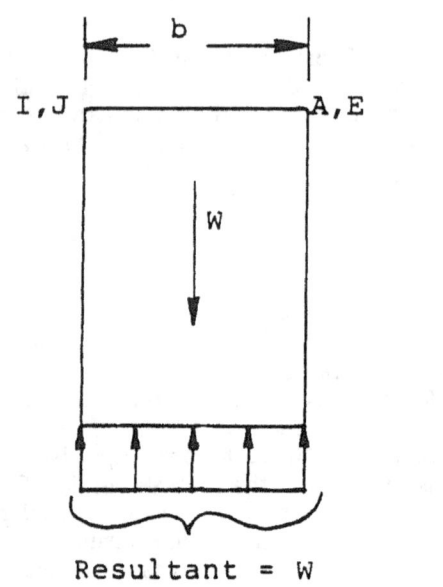

b. Free body with no overturning moment

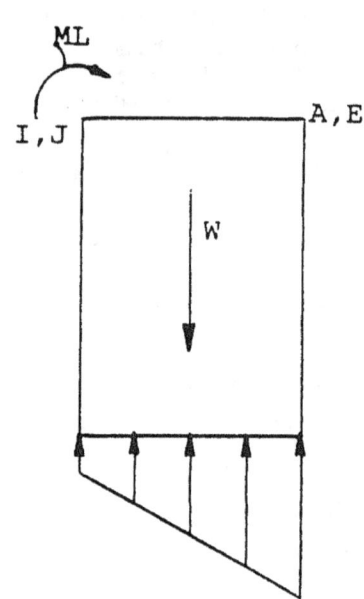

c. Free body with overturning moment

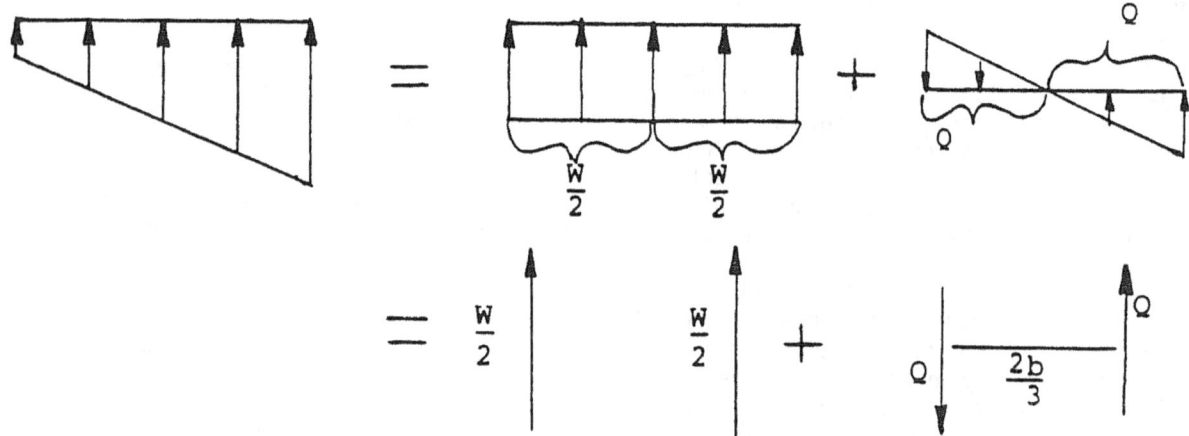

d. Statically equivalent reactions

**Figure 10
Pressure distribution from foundation acting on cell**

by the presence of the back slope of the berm or the friction circle method. However, under certain conditions this procedure may lead to such a large value of P'$_p$ that the overturning moment and, hence, the factor of safety become negative. Clearly, this is physically unrealistic, since the passive resistance of the berm and foundation acting on the inboard wall is mobilized only in response to the driving forces P_w and P'_a.

55. Since horizontal equilibrium must be maintained, it can be seen from Figure 11b that the value of P'$_p$ cannot exceed the following equation:

$$P'_p = P'_a + P_w - P_{w1} - T^* \qquad (27)$$

where T^* is the horizontal shear force on the base of the cofferdam per unit length of cofferdam. The value of P'$_p$ for use in Equation 28 can be calculated as follows: calculate the maximum passive earth force P*$_p$ acting on the cofferdam. This may be calculated using the trial-wedge method, Coulomb theory modified by the presence of the back slope of the berm, or the friction-circle method. Compare P*$_p$ to the results of Equation 27 with T^* taken as zero. Let P'$_p$ be the smaller of these two terms. Assume that H'$_p$, the moment arm at which P'$_p$ acts about the base of the cell, for P'$_p$ is the same as for P*$_p$. When P'$_p$ is taken as the value of P*$_p$, equilibrium is maintained by increasing T^* so that Equation 27 is satisfied. Note that T^* must be less than $N' \tan \phi$ to prevent sliding along

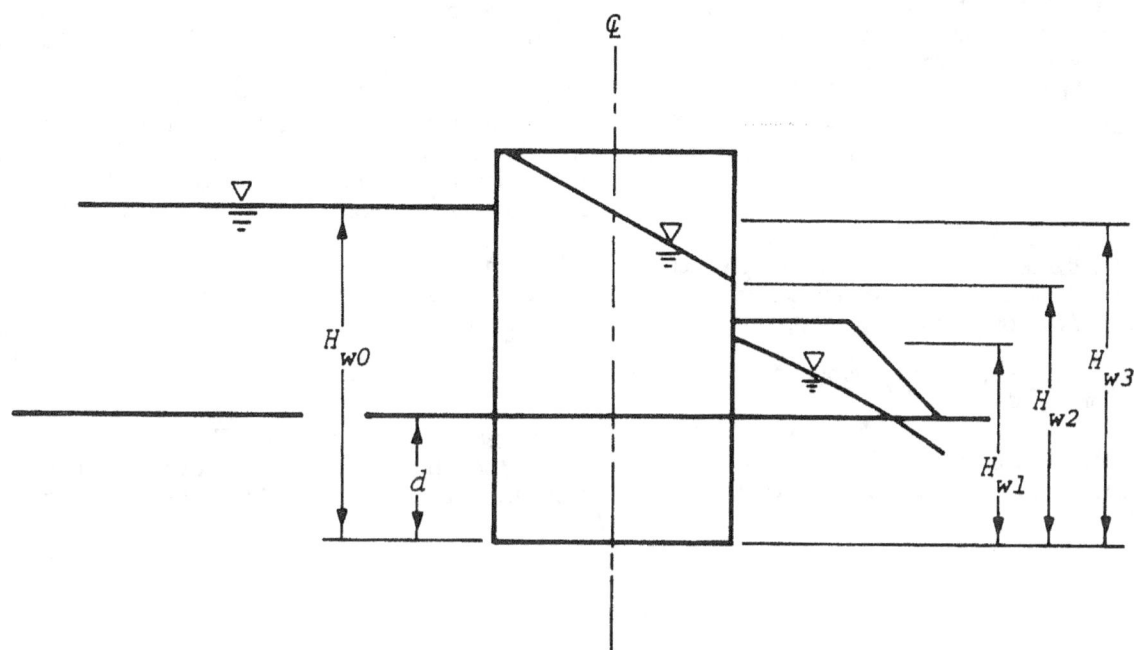

a. Definition of lengths used in M calculation

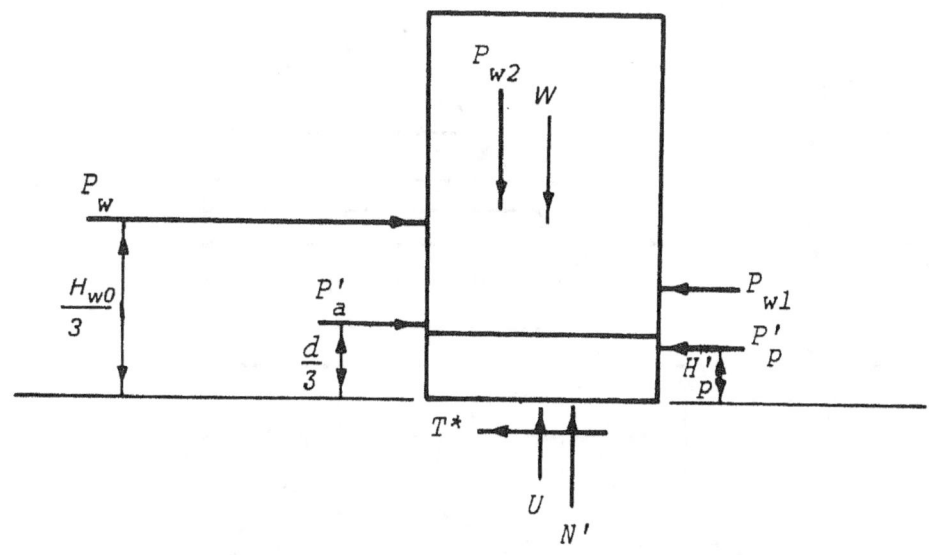

b. External forces considered in calculation of overturning moment

Figure 11
Calculation of overturning moment

the base of the cell where N' is the resultant effective soil force acting on the base of the cofferdam per unit length of cofferdam. The T* term does not enter into the calculation of the overturning moment since it passes through the point about which the moment is summed.

56. In terms of the quantities which have now been defined, the over-turning moment per unit length of cofferdam can be calculated by summing moments about the point where the center line of the cell intersects the base.

$$M = \frac{P_w H_{wo}}{3} + \frac{P'_a d}{3} - P'_p H'_p - P_{w1} \frac{H_{w1}}{3} \quad (28)$$

57. In deriving Equation 28, the assumption is made that water forces P_{w2} and U (Figure 11) act through the same point, thus cancelling each other in determining the overturning moment. If the cofferdam designer does not believe this to be the case, the P_{w2} and U terms should be included in Equation 28. The weight of the contents of the cell along the center line of the cell thus has no contribution to the overturning moment.

Calculation of S'_m

58. The maximum possible value of the shear on the vertical centerplane of the cell may be estimated as the product of the effective normal force P'_c acting on the center plane times the coefficient of friction of the fill. That is,

$$S'_m = P'_c \tan(\phi) \quad (29)$$

where ϕ equals the angle of internal friction of fill. The normal force P'_c may be computed from the pressure diagram shown in Figure 12. The question of what value to use for the lateral earth pressure coefficient K will be deferred to paragraph 60.

Calculation of S''_m

59. Figure 13 shows the relevant free body. T_{cw} is the resultant tensile force in the crosswall. If f denotes the coefficient of friction of the interlock (steel-on-steel), then the maximum possible friction force in the crosswall is fT_{cw}. Thus, the maximum friction force S''_m per unit length of wall is given by the equation

$$S''_m = \frac{fT_{cw}}{L} \quad (30)$$

The tension T_{cw} may be found by summing forces on the free body shown in Figure 13b. In Figure 13b the pressure diagram goes to zero at the plane of fixity or the tip of the sheet pile. This is a choice left up to the cofferdam designer. There is still some tension in the crosswall below the plane of fixity because, as mentioned earlier, the plane of fixity is calculated at the point of zero slope, not zero deflection. Sometimes there is tension in the crosswall all the way to the tip of the sheetpiling. But this is not always the case, especially for cofferdam cells with deep embedment into the foundation. It should be noted that in some cases for cofferdams on rock, the pressure diagram may not go to zero at the tip of the sheet piling. The pressure diagram will still decrease from the P_{max} value, but it may not be zero at the tip.

Discussion of Vertical-Shear Failure Mode

60. The staircase-type surface shown at the top of the cell in Figure 8 has been observed in mooring cells which have

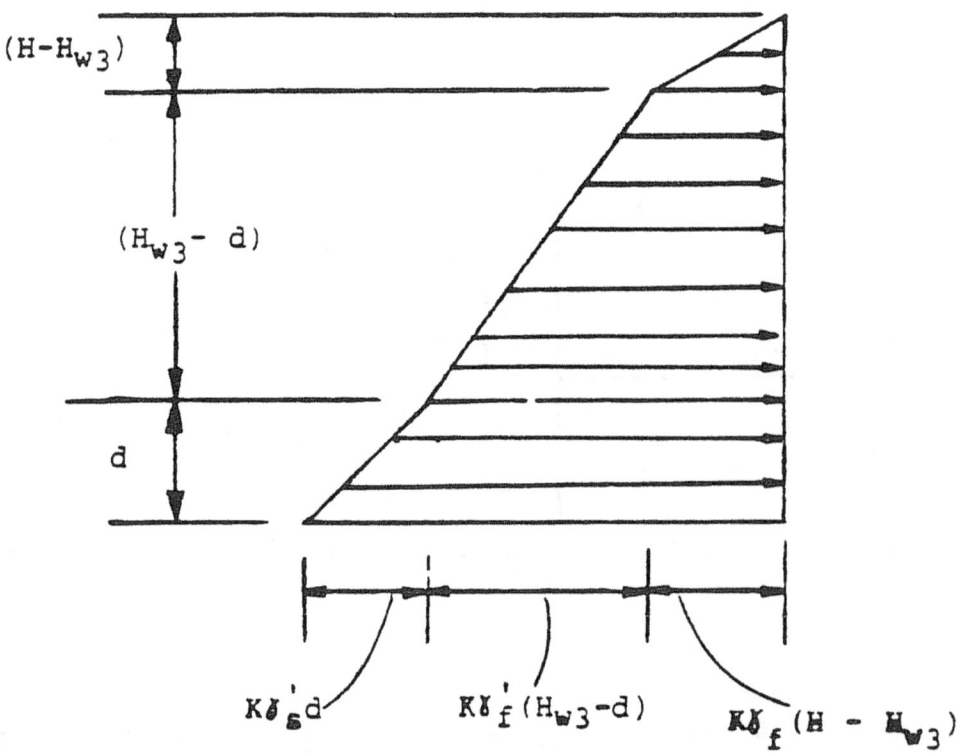

Figure 12
Pressure diagram for calculation of P'_c effective force on center plane

been struck by barges. This suggests that the vertical-shear failure mode does occur in a full-sized cell.

61. In the most realistic model tests which have been conducted to date (Maitland and Schroeder 1979), the vertical-shear mode was found to be the actual mode of failure under lateral load. Controversy remains, however, as to the appropriate value to use for the lateral earth pressure coefficient K. In his original work, Terzaghi (1945) suggested use of the Rankine coefficient, K_a. In a discussion of Terzaghi's paper, however, Krynine (1945) pointed out that it is incorrect to use the Rankine coefficient. The reason for this is that the assumed failure plane (the vertical plane) cannot be a principal plane because shear acts on it. Using a Mohr's circle analysis, Krynine derived the equation

$$K = \frac{\cos^2 \phi}{2 - \cos^2 \phi} \quad (31)$$

62. It was subsequently pointed out by Cummings (1957) and Esrig (1970) that Krynine's expression for K had the physically unrealistic property that K decreased with increasing values of the angle of friction ϕ, and this result led Esrig to doubt the value of the entire vertical-shear failure mode. Based on their model studies, however, Schroeder and

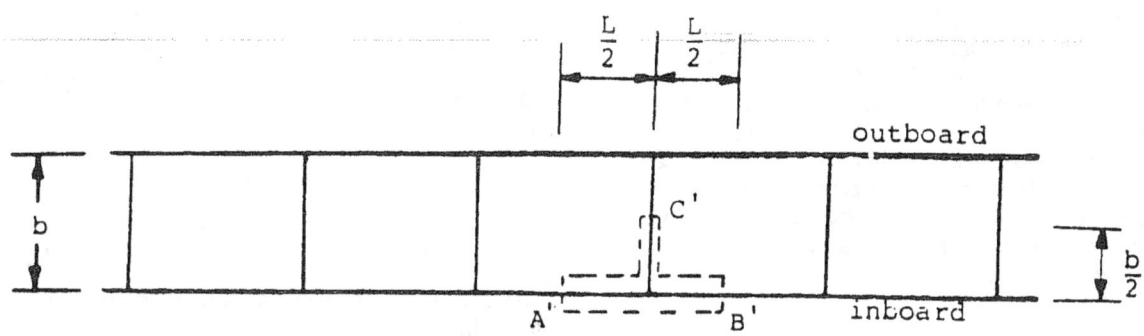

a. Portion of equivalent cofferdam selected for analysis

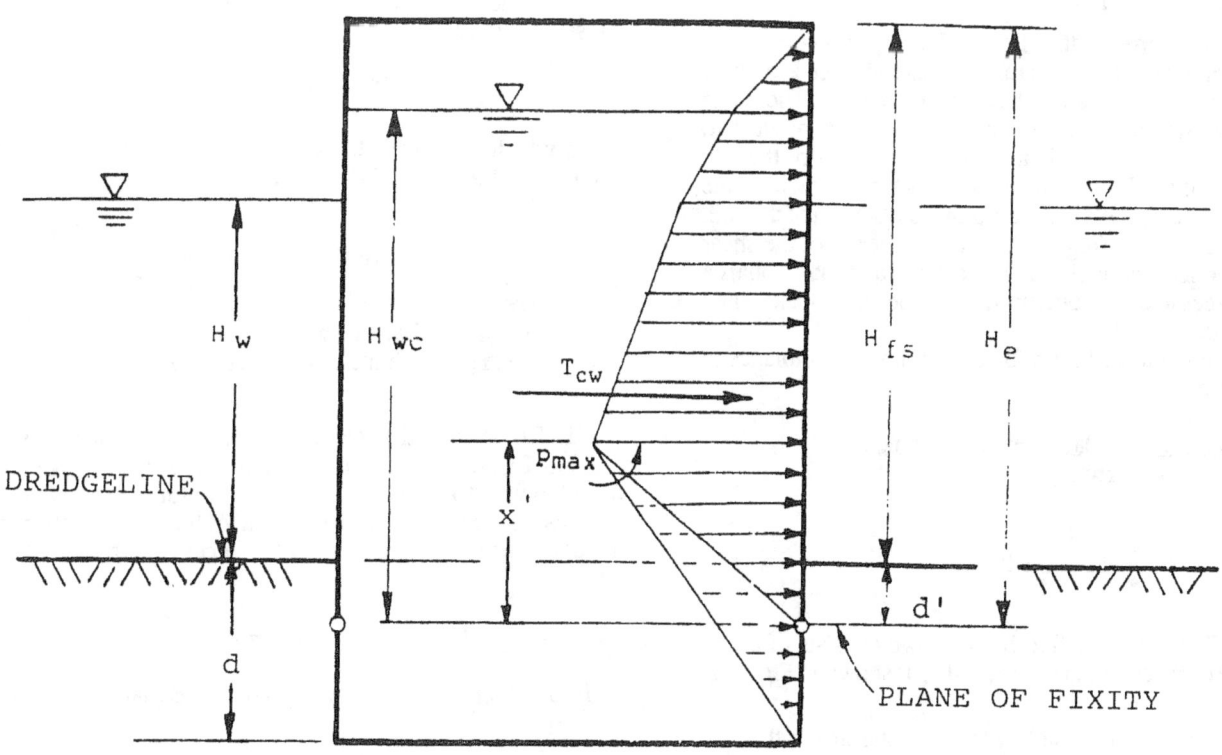

b. Free body for calculation of force in cross wall

**Figure 13
Free body for calculation of T_{cw}**

Maitland (Maitland and Schroeder 1979; Schroeder and Maitland 1979) argued that the overturning moment applied to the cell tends to compress the fill significantly on the inboard side and, as a result, the lateral-earth pressure coefficient is appreciably increased. They suggest using the empirical value $K = 1$. This approach yielded calculated values of ultimate overturning moment in good agreement with values determined experimentally in their model study. Using $K = 1$, however, does not appear sufficiently justified by experience at this time, especially since currently used safety factors are based on much lower values of K. Thus, it is recommended to use the Krynine earth pressure coefficient when calculating P'_c. Schroeder and Maitland also suggest that pressure calculations be based on the effective height H_e of the cell.

63. For some fill materials, interlock friction $S" = fT_{cw}/L$ may account for 30 to 40 percent of the total resisting force (Terzaghi 1945).

64. The approximation of the base pressure distribution by a straight line in Figure 10c is valid only if the base remains in compression over its entire length. This condition will be satisfied only if the resultant force acting on the cell from the foundation acts within the middle third of the base. If the resultant does not act within the middle third, the entire analysis for shear on the vertical center plane is questionable.

65. The free body used to calculate the overturning moment, Figure 11b, neglects the interlock tensions occurring in the actual curved-wall cell. The error involved in using free bodies which neglect the curvature of the walls has been discussed in paragraph 18.

PART V: SLIP ON HORIZONTAL PLANES IN FILL (CUMMINGS' (1957) METHOD)

Horizontal Plane Sliding Due to Lateral Forces

66. Under the action of the resultant lateral force P, a plane of rupture forms. This is illustrated in Figure 14. The plane extends from the toe of the cell at B upward at an angle ϕ to the outboard wall at A. ϕ is the angle of friction of the fill. Shear failure of the fill occurs on horizontal planes within the triangle bounded by the plane of rupture, the bottom of the cell, and the outboard wall (region AOB). As a result of the shear failure of the fill, the cell tilts excessively and may collapse through excessive deformation of the interlocks and consequent loss of fill.

67. The FS against failure by slip on horizontal planes in fill is shown as:

$$FS = \frac{\text{Maximum available resisting moment}}{\text{Driving moment}}$$

$$= \frac{M_f + M_{shear}}{M} \quad (32)$$

where (Figure 15, the free body shown consists of a unit width of the outboard and inboard walls, as shown in Figure 2)

- M_f = moment caused by the friction force in the interlocks of the crosswall
- M_{shear} = moment caused by the pressure of that portion of the fill which fails in shear on horizontal planes

Consideration in Horizontal Shear Calculations
Discussion of the theory

68. Figure 16a shows a vertical slice of the fill within a cell. The slice is assumed to be of unit thickness into the plane of the figure. Line AGJN defines the assumed phreatic surface, and line BQLO, which makes an angle ϕ with the base of the cell, defines the plane of rupture. According to Cummings (1957), as the cell begins to tilt under lateral pressure, the fill above the rupture plane begins to slide down the plane. This motion is, of course, inhibited to a large extent by the confining effect of the sheet-pile envelope. The fill below the rupture plane (within the prism BEO) is transformed into a passive state by the combined effect of the lateral forces acting on the cofferdam exterior and the weight of the fill above the rupture plane, is assumed to remain in an active state.

Resisting Moment Acting on Outboard Wall

69. Figure 16a shows the pressure distribution P_1, of the outboard sheet-pile wall acting on the fill. By Newton's Third Law, this pressure is equal and opposite to the pressure of the fill acting on the wall. Above point B, where the plane of rupture intersects the wall, P_1 is the sum of the pressure produced by water inside the cell and by fill in the active state. That is,

$$o < y < y_B$$

$$P_1(y) = (\gamma_w + \gamma'_f K_a) y \quad (33)$$

in which y is the distance measured downward from the top of the cell.

70. Below point B, the pressure distribution is more complex since the resisting shear on the horizontal sliding planes within the fill must be taken into account. In Figure 16b, region ABCQGF has been drawn isolated from the rest of the cell, and the shear F* is shown acting to resist the lateral pressure. F* can be calculated from the relation

$$F^* = W_e \tan(\phi) \quad (34)$$

in which W_e equals the effective weight of the fill above the base of the free body, that is,

$$W_e = \gamma_f V_{AGF} + \gamma'_f V_{ACQG} \quad (35)$$

where
- V_{AGF} = volume of prism AGF
- V_{ACQG} = volume of prism ACQG

71. To develop the relation between the shear F* and the pressure P_1, consider the infinitesimal region CDLK extending from the wall to the plane of rupture (Figure 16c). Note that the pressure P_2 acting on the right end of the free body has been included. Summing the horizontal forces acting in the figure gives

$$P_1 \, dy - P_2 \, dy - dF^* = 0 \quad (36)$$

from which it follows, through use of Equation 34, that

$$P_1 = P_2 + \frac{dF^*}{dy}$$

$$= P_2 + \frac{dW_e}{dy} \tan(\phi) \quad (37)$$

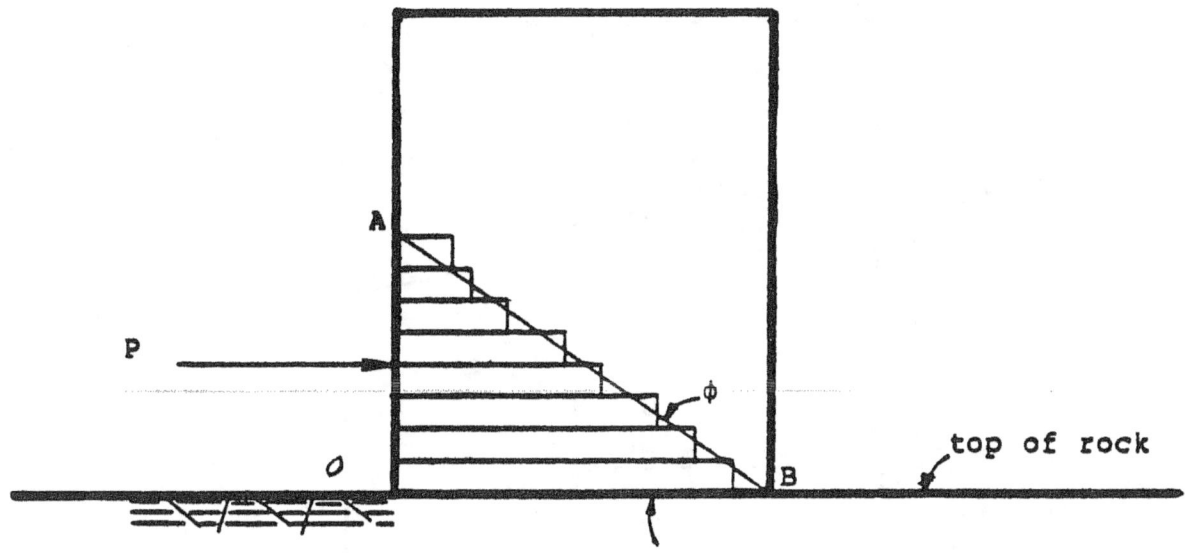

**Figure 14
Failure by sliding on horizontal planes within fill**

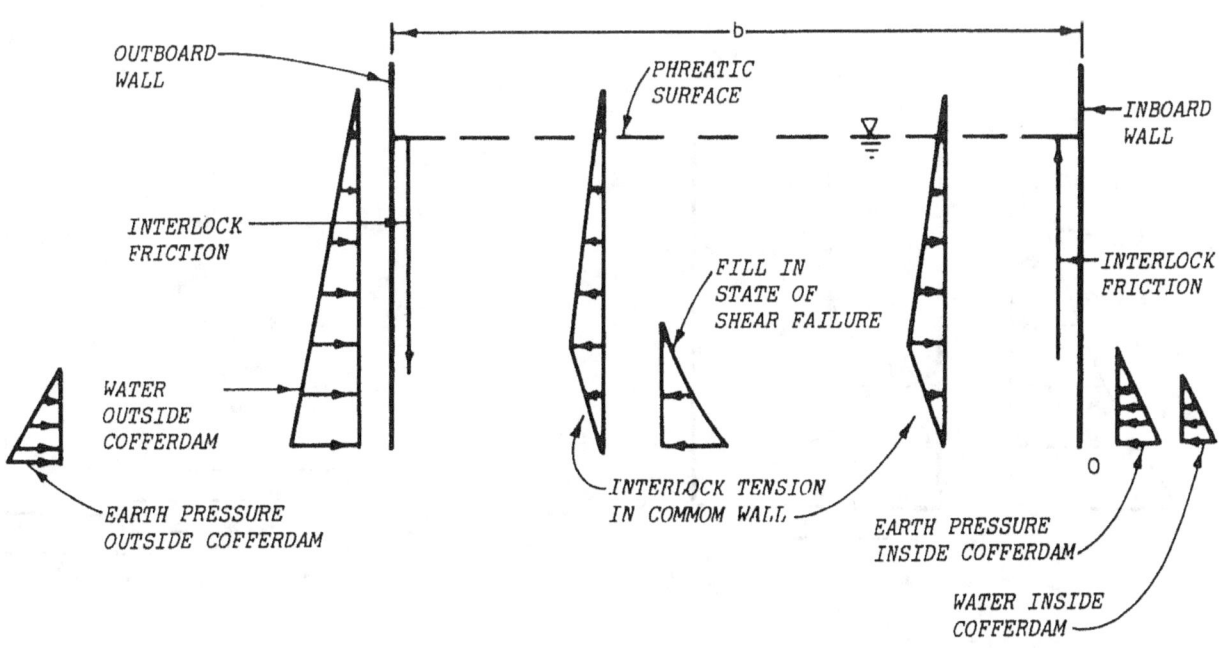

**Figure 15
Free body for calculation of moments in Cummings' Method. Moments are summed about point 0**

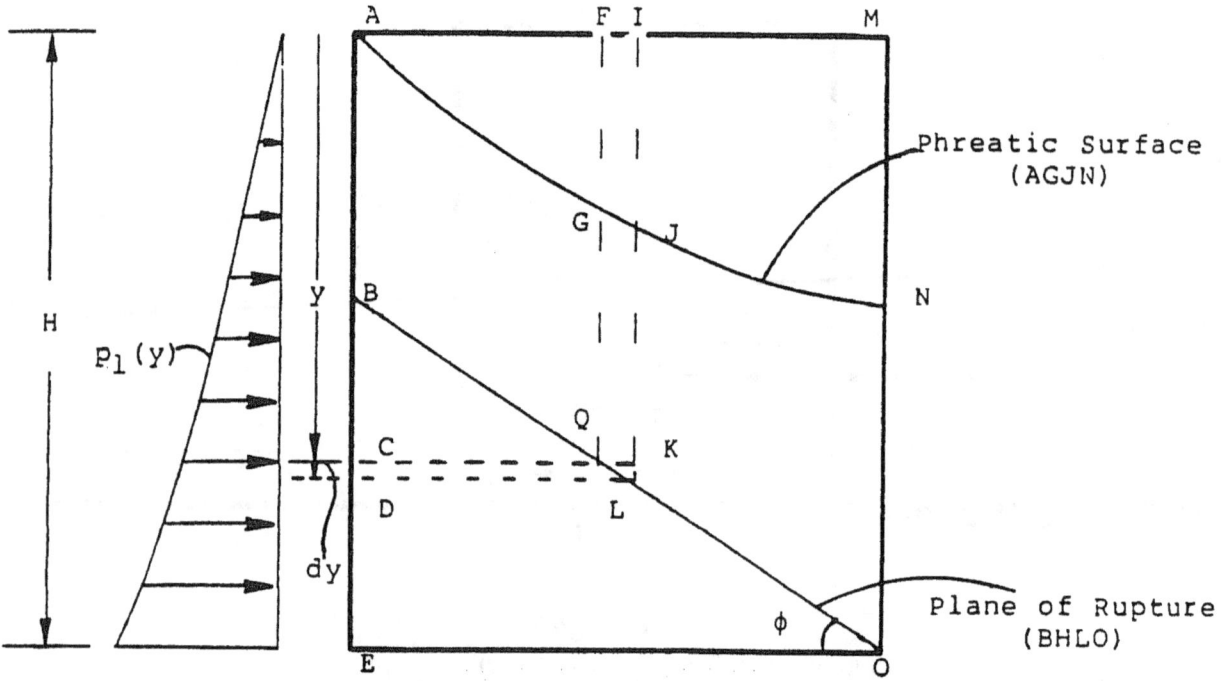

a. Cross section of cell fill, with pressure P_1 from sheetpiling

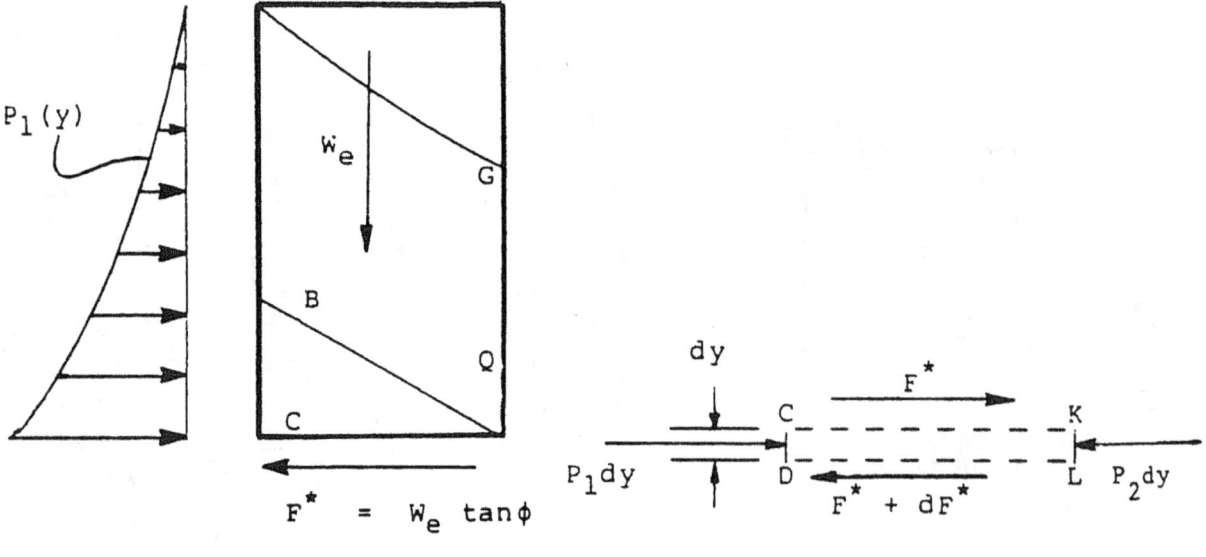

b. Shear force F^* on horizontal plane CH

c. Free body for relating P_1, F^*, and P_2

**Figure 16
Lateral forces on fill**

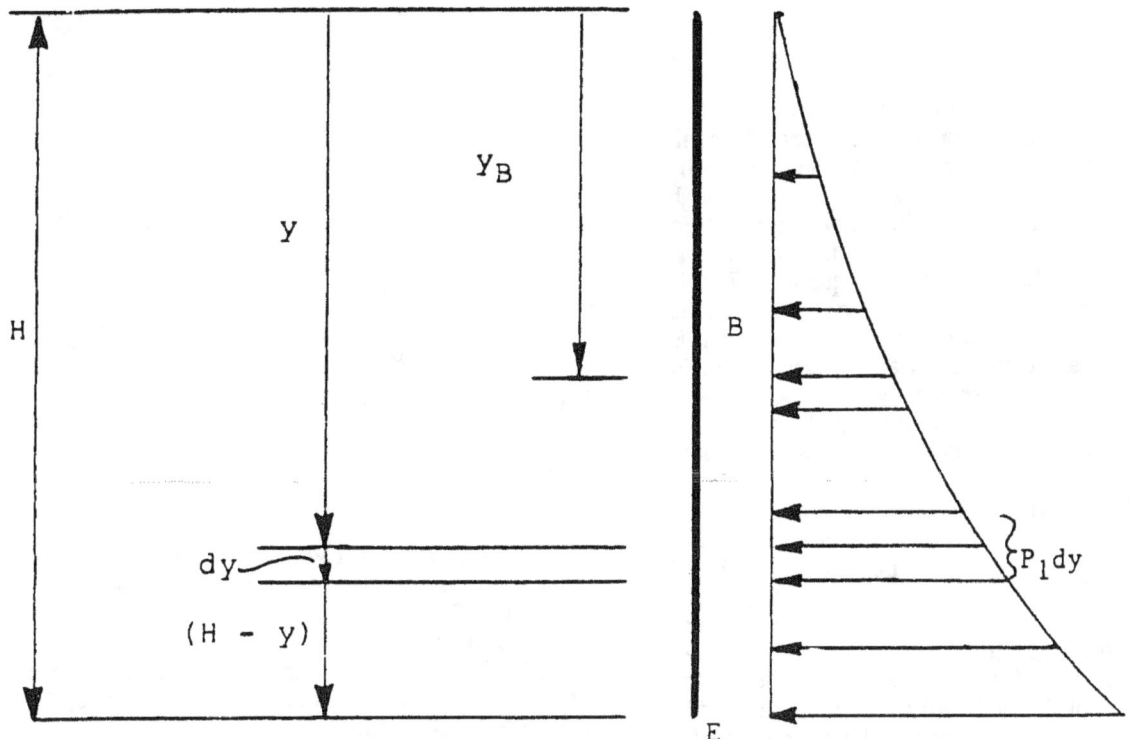

Figure 17
Resisting pressure acting on outboard wall

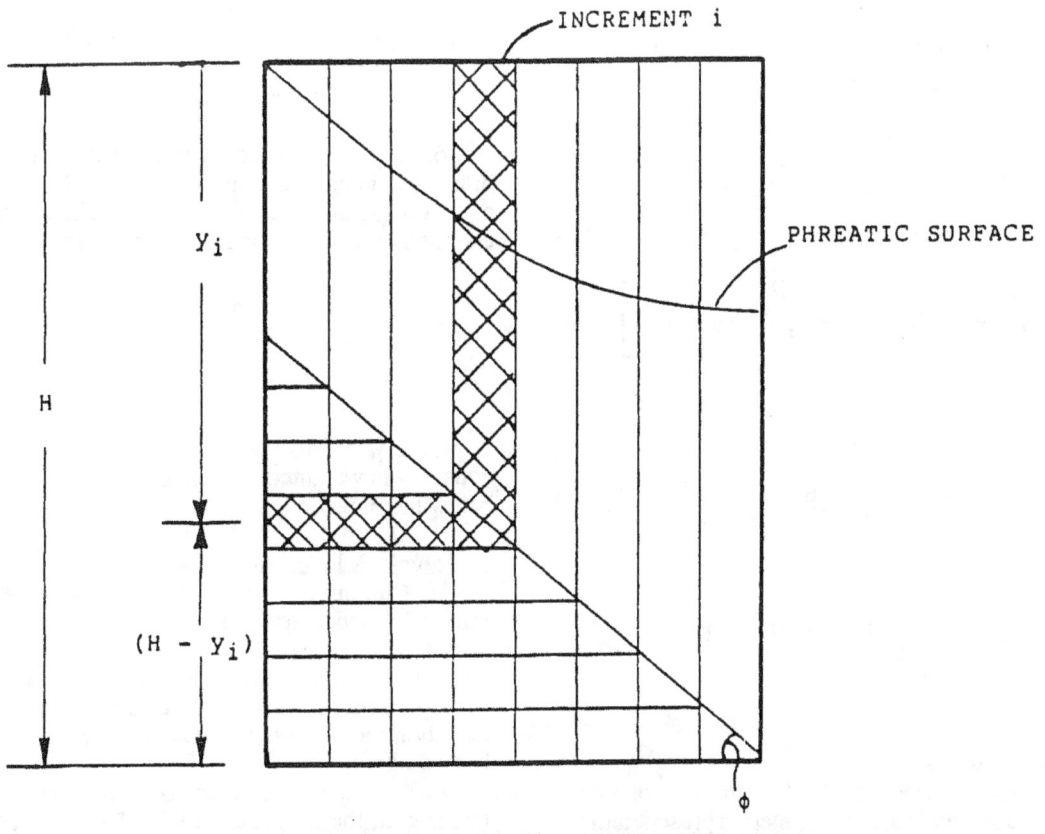

Figure 18
Increments used in computing M_shear

72. At this point in the derivation, to arrive at Cummings' expressions for the resisting moment, the following equation for P_2 must be assumed:

$$P_2 = (\gamma_w + \gamma_f' K_a) y \quad (38)$$

That is, the lateral pressure acting at points on the rupture plane is the sum of the hydrostatic pressure and the active earth pressure corresponding to a completely saturated cell - the downward slope of the phreatic surface is ignored, even though it is included in the calculation of the effective weight W_e given by Equation 35. Since the pressure P_2 contributes to the resisting pressure P_1 of the cell through Equation 37, it is conservative to assume that the unit weight of the fill has the submerged value γ_f' over the entire depth y. However, it is nonconservative to assume that the water pressure term in Equation 38 is based on the entire depth y.

73. With P_2 now defined, the pressure P_1 of the wall on the fill below point B can be written using Equations 37 and 38:

for $y_B < y < H$

$$P_1 = (\gamma_w + \gamma_f' K_a) y + \frac{dW_e}{dy} \tan(\phi) \quad (39)$$

Pressure Distribution Defined

74. Equations 33 and 39 together completely define the pressure distribution of the outboard wall acting on the fill. The equal and opposite pressure acting on the outboard wall is shown in Figure 17.

The total moment about the base produced by this pressure is

$$M^* = \int_0^H (H - y) P_1 \, dy$$

which can be written, with the use of Equations 33 and 39, as

$$M^* = \int_0^{y_B} (H - y)(\gamma_w + \gamma_f' K_a) y \, dy +$$

$$\int_{y_B}^H (H - y) \left[(\gamma_w + \gamma_f' K_a) y + \frac{dW_e}{dy} \tan(\phi) \right] dy$$

$$= \int_0^H (H - y) \gamma_w y \, dy + \int_0^H (H - y) \gamma_f' K_a y \, dy +$$

$$\int_{y_B}^H (H - y) \frac{dW_e}{dy} \tan(\phi) \, dy \quad (40)$$

Representative Integrals

75. Examination of the right-hand side of the last equation shows that the three integrals appearing there represent terms in the expression for the resisting moment in Equation 32:

$$M_{wo} = \int_0^H (H - y) \gamma_w y \, dy \quad (41)$$

$$M_{ao} = \int_0^H (H - y) \gamma_f' K_a y \, dy \quad (42)$$

$$M_{shear} = \int_{y_B}^H (H - y) \frac{dW_e}{dy} \tan(\phi) \, dy \quad (43)$$

The effective weight W_e, which appears in the last integral, depends on the location of the phreatic surface within the cell (Equation 35 and Figure 16).

Since this surface may vary in a nonlinear manner within the cell, Cummings suggests that an "incremental method" (in effect, a low-order numerical integration rule) be used to evaluate Equation 43:

$$M_{shear} = \tan(\phi) \sum_i \left[(\Delta W_e)_i (H - y_i) \right] \quad (44)$$

where
$(\Delta W_e)_i$ = effective weight of the cross-hatched region shown in Figure 18
y_i = distance measured downward from the top of the cell to the midpoint of increment i in Figure 18.

76. Equation 44 could be replaced by a more accurate numerical integration method such as Simpson's rule. The equation given for M_{shear} by Cummings, rearranged to be consistent with notation in this report, is

$$M_{shear} = \frac{Hb^2 \gamma_e}{6} \left[3 \tan^2 \phi - \frac{b}{H} \tan^3 \phi \right] \quad (45)$$

where γ_e equals the effective unit weight of soil which is equal to the weighted average of γ_m above phreatic line and γ, below the phreatic line.

Interlock Friction

77. Cummings' method also considers the contribution of interlock friction to resisting tilting of the cell. Figure 19a shows two portions of the cofferdam walls which will be analyzed for the effect of interlock friction. In Figure 19b, these portions of the walls have been isolated and shown in an elevation view. Three forces have been included in the free body diagram:

(a) T_{cw}, the interlock tension in the crosswall; (b) LP^*, the resultant of the fill pressure P^* (P^* is force per unit length of cofferdam acting on the main-cell walls); and (c) F, the friction force from the interlock of the cross-wall sheet-pile adjacent to the connector pile. Following steps similar to those leading to Equation 30 leads to an expression for the friction force per unit length of cofferdam:

$$\frac{F}{L} = fP^* = f\frac{T_{cw}}{L} \quad (46)$$

78. The moment per unit length of cofferdam, M_f, can be calculated by multiplying the friction force F/L of Equation 46 by the moment arm b (Figure 15) to obtain

$$M_f = fP^*b = \frac{fbT_{cw}}{L} \quad (47)$$

Comments on Horizontal Shear Method

79. The technical literature contains no reports of the failure of a full-sized cell by sliding on horizontal planes in the fill. The only positive experimental evidence for such a failure mode comes from Cummings' own model studies (1957), reported in the paper describing his theory, and from TVA experiment using a cigar box with the top and bottom replaced by glass (TVA 1957). The walls of these models were all relatively stiff, and their small size raises serious concern that surface effects in the fill were far more important than they would be in a full-sized cell. Indeed, it is interesting to observe Erzen (1957) comments in a discussion published simultaneously with Cummings' paper. Erzen expressed concern that the stress distribution in the fill in Cumming's models was strongly influenced by the proximity of the walls, and as a result, he doubted Cummings' demonstration that distinct zones (active and passive) exist in the fill.

80. Heyman (1957), another discussor of Cummings' paper, state that additional tests on cellular cofferdams were

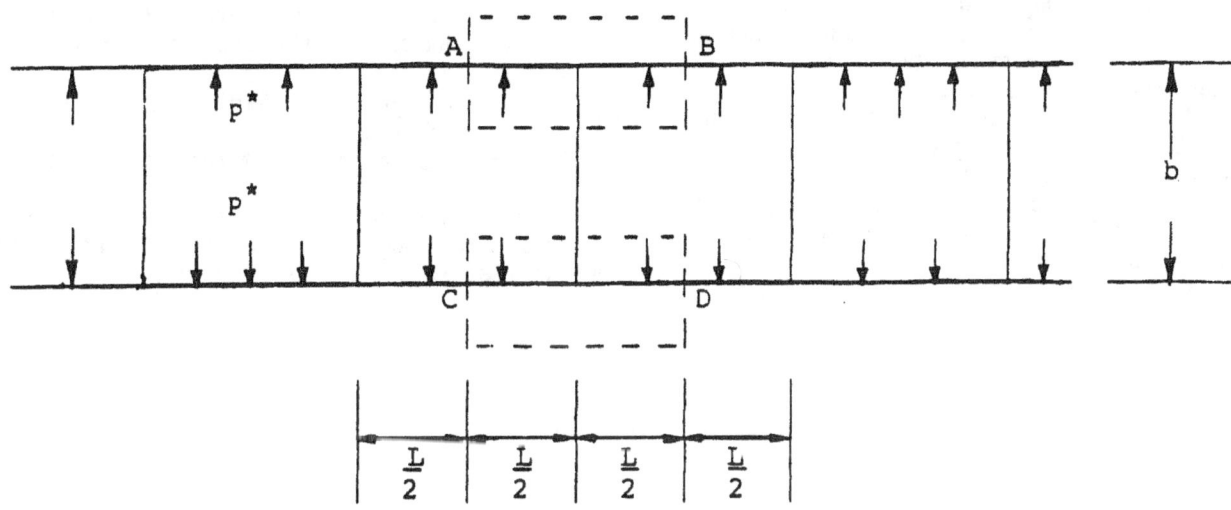

a. Portion of equivalent cofferdam selected for analysis

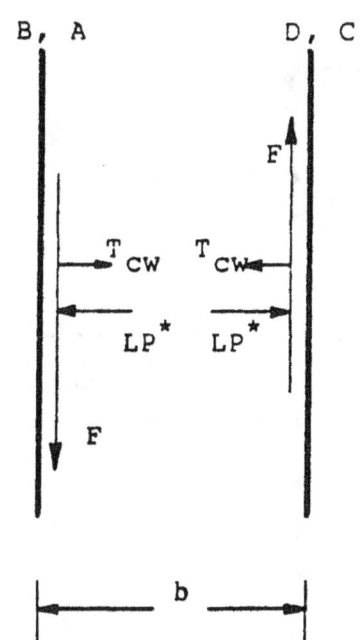

b. Elevation of portions AB and CD

Figure 19. Free-body diagram for calculation of resisting moment produced by friction in interlocks

needed to determine where the actual plane of failure occurs. Aside from Cummings' own studies, no such studies have subsequently been reported in the literature.

81. In summary, then, the basic assumptions upon which Cummings' method rests - that is, a plane of rupture exists which makes an angle ϕ with the base and divides the fill into an active region and a region in which sliding occurs on horizontal planes - are not well supported by experimental evidence.

82. Even while Cummings' models provide the only experimental support for the horizontal-shear failure mode, the subsequent model studies of Maitland and Schroeder (Maitland 1977; Maitland and Schroeder 1979) provided evidence against it. These studies showed that Cummings' method predicted maximum resisting moments significantly larger than those actually observed at failure.

83. Figure 15, which was used in the derivation of the FS, shows all forces which are considered in Cummings' method. Note that these forces are considered to act on a single free "body" consisting of a unit width of the outboard wall and the inboard wall. Thus, although Cummings' method is often characterized as being based on a criterion of internal shear-failure of the fill, this characterization is misleading: the internal shear-failure mechanism is used only as a device to permit the calculation of the pressure of the fill on the wall. The actual FS is based on the assumption that the outboard and inboard walls behave as a rigid unit and a sum of all moments which act on this unit. The uncertainties involved in neglecting the interlock tensions for this flat-walled free body have been discussed previously paragraph 18d.

PART VI: SLIP BETWEEN SHEETING AND FILL

Vertical Sheeting Slip from Overturning Moment

84. Under the action of the overturning moment, the sheeting slips vertically upward relative to the fill, and fill runs out at the heel. This is illustrated in Figure 20.

85. The FS against slip between sheeting and fill (Jumikis 1971); Lacroix, Esrig, and Lusher 1970; NAVDOCKS DM7 1971: USS 1972) is expressed as:

$$FS = \frac{\text{Maximum available resisting moment}}{\text{Driving moment}}$$

$$= \frac{b\{P'_a \tan(\delta) + [P_s + P_s(b/L)] \tan(\delta)\}}{M} \quad (48)$$

where
P_s = horizontal effective-force (per unit length) of the cell fill and foundation material within an equivalent cofferdam cell acting on a cell wall
δ = angle of friction between sheeting and soil

and the other quantities retain their previous meanings. All moments are computed with respect to the toe of the cell and are expressed per length of cofferdam.

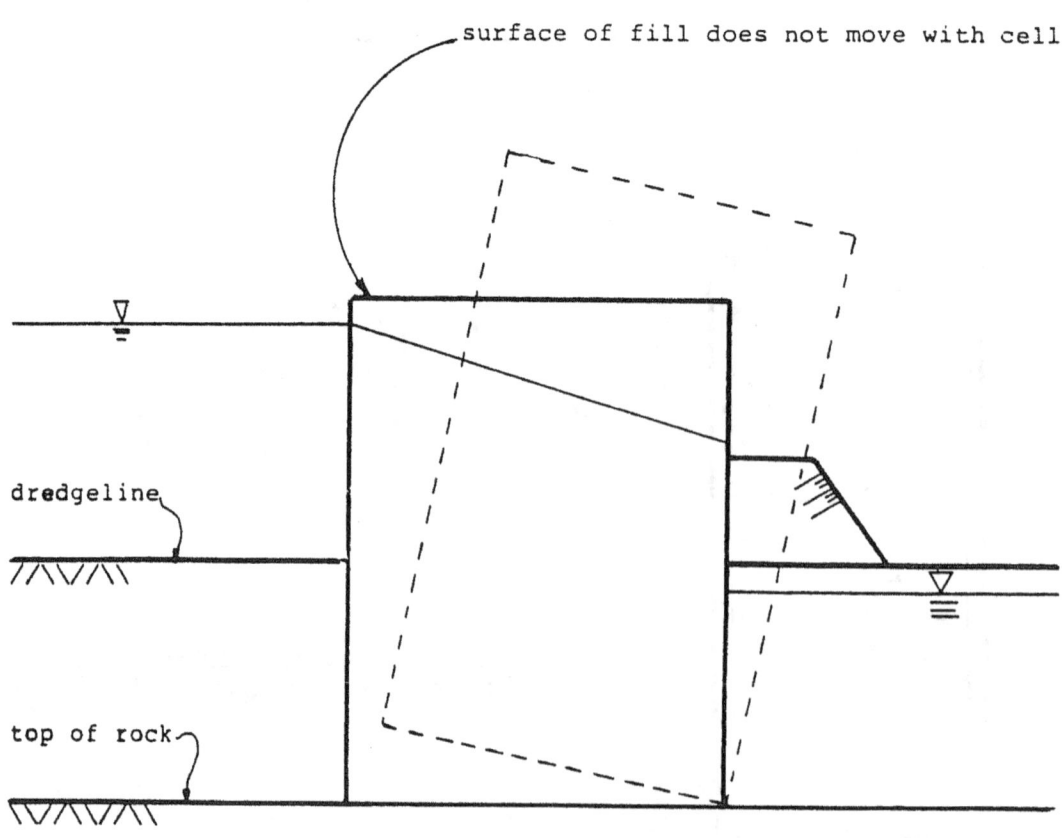

Figure 20. Illustration of slip between fill and sheeting

Considerations in Calculations for Slip Between Fill and Wall
Moment calculations

86. Figure 21a shows the portion of the cofferdam selected as the free body, and Figure 21b shows the forces (per unit length of cofferdam) which are considered in the analysis. The driving forces acting on the body are given separately in an elevation view, Figure 22a. The factor of L has been introduced to convert from force-per-unit length to force. The resisting moment is considered to arise solely from friction forces. The effective forces which produce these friction forces are shown in Figure 22b. The friction forces themselves are illustrated in Figure 22c. Summing moments about point 0 in Figure 22c yields the resisting moment. Dividing the result through by L to obtain moment-per-length then yields the expression appearing in the numerator of Equation 48. It should be noted that the moment arm for the friction force on the crosswall is b/2. However, the 2 in the denominator cancels with the factor of 2 introduced to account for the fact that friction acts on both sides of the crosswall.

87. The P_s is calculated using an earth-pressure coefficient between $1.2K_a$ and $1.6K_a$. The assumption is made in this failure mode derivation that the inboard sheeting does not slip with respect to the fill and the foundation material.

Alternative Failure Mode

88. The failure mode discussed thus far in PART VI is based on the assumption that the fill slips with respect to the entire sheet-pile shell of the cell - that is, all the walls act together as a unit. An alternative slip failure mode is conceivable. The crosswall may remain stationary relative to the fill, while the outboard wall alone slips upward (Figure 23). For this failure mode, the relevant free body is the portion of the outboard wall shown in Figure 24; the driving forces acting on this free body are shown in Figure 24b, and the resisting forces in Figure 24c. The arrow labeled fT_{cw} in the latter figure represents the interlock-friction force from the crosswall.

89. In attempting to calculate a FS based on Figure 24, a difficulty arises. The ratio of resisting and driving moments

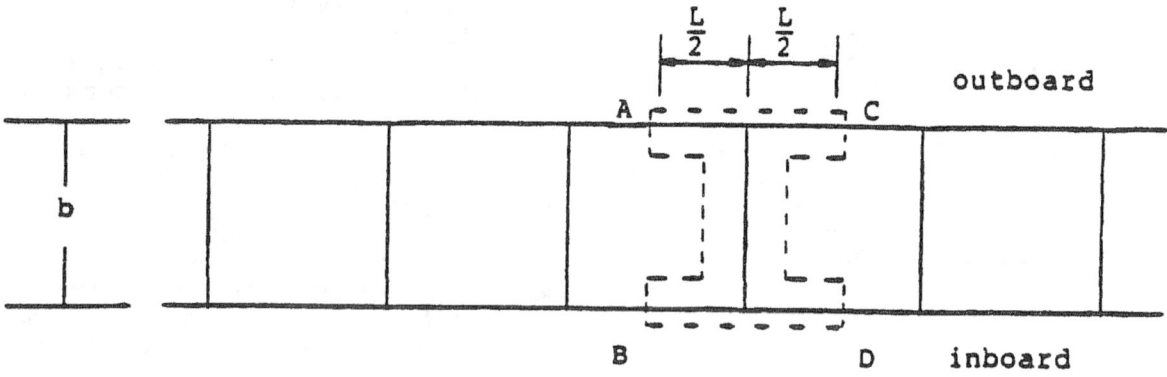

a. Portion of equivalent cofferdam selected for analysis

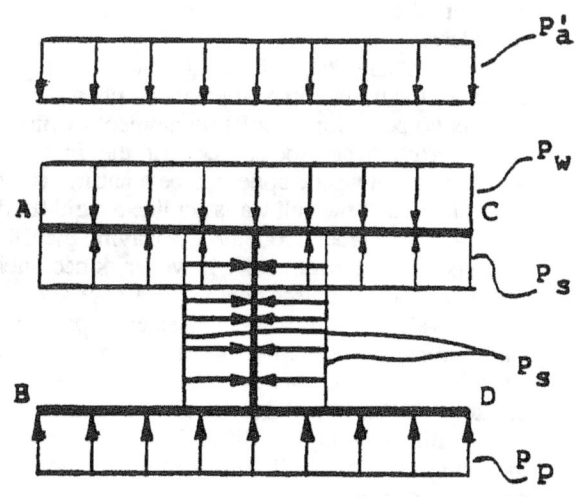

b. Forces considered in analysis

Figure 21. Free body for analysis of slip between fill and sheeting

depends strongly on the point about which the moments are computed, but no point is the obvious choice. Although it is less apparent, this same comment also applies to the slip failure mode previously discussed in which the entire sheet-pile shell acts as a unit. The basic problem is that the design procedure does not identify the driving forces which actually cause the failure. This matter is discussed further in the next paragraphs.

Comments on Failure Mode

90. The error involved in using free bodies which neglect the curvature of the walls has been discussed in paragraph 18. No reports exist in the literature of a full-sized cell failing by this mode (ORD 1974; Grayman 1970). If overburden is present on the rock foundation, the fill cannot escape the cell until the bottom of the piling is above the dredgeline. Of course, scour might have removed the overburden.

91. Apparently, the possibility of this failure mode originated from model studies performed by TVA engineers (TVA 1957). The cell walls in their models were very stiff. The effect of this excessive stiffness can be seen in Figure 25, which has been reproduced from the TVA manual (1957). The front portion of the walls is seen to act like a rigid body. Furthermore, if bulging were present, the illustrator apparently did not consider it pronounced enough to include in the sketch. These observations raise doubts as to whether the TVA models were flexible enough to permit the fill to develop its maximum shearing resistance. That is, the failure mode observed in the TVA experiments may occur only if the cell walls are very rigid. In a full-sized cell, the large deformations occurring in the fill may lead to a different failure mode such as a shear failure in the fill before slip between the fill and the walls can happen.

92. The TVA model studies quite possibly exhibited slip between the walls and the fill because of the loading device used (a string wrapped around the cell). The loading device kept the load horizontal and did not allow it to remain normal to the wall as tilting occurred, as would be the case with water pressure on the outboard wall. The string may thus have exerted an upward friction force on the walls which contributed significantly to their upward motion.

Contribution of Cell Bulging

93. The TVA report and other discussions of slip between the fill (Jumikis 1971; Lacroix, Esrig, and Lusher 1970) and the wall neglect the contribution of cell bulging toward preventing this failure mode. This effect may be discussed qualitatively by considering Figure 26, in which is shown a portion of a sheet-pile wall near the point of maximum bulging. According to the figure, as the wall tends to move upward, the normal component of the earth pressure acting on the lower side of the bulge tends to increase, and thus, the resistance to slip includes both friction and normal contributions. The derivations of safety factors based on straight rigid walls ignore this effect. Furthermore, the original TVA experiments were based on models with walls so rigid that significant bulging may not have occurred.

94. Since the free body is the entire sheet-pile shell, it appears arbitrary to base the resisting-moment calculation on the friction forces alone; the resisting moment produced by the differential head of water within the cell is ignored. Or, in the case of the alternative free body, Figure 24, the resisting moment of the horizontal force coming from the fill is not considered.

95. The magnitude of the FS is strongly dependent on the choice of the point about which moments are computed. Yet, there is no point for calculating moments which is clearly to be preferred over others. Use of the inboard toe as the reference point would appear to be a natural choice - if it can be assumed that the cell walls act like a rigid shell, rotating in a rigid-body manner about the toe, leaving the fill at rest. This assumption is questionable, however, since such a rotation could only occur if large bending stresses were transmitted by the cell walls. But, as was emphasized in paragraph 8, the cell walls transmit primarily membrane rather than bending stresses. Thus, a rigid-body rotation of the whole shell appears unlikely. On the other hand, if a rigid-body motion of only part of the piling occurs, it is difficult to know beforehand where the center of rotation will be and, thus, at which point moments should be summed.

96. The most important effect likely to cause pullout has been ignored. As the wall tilts toward the interior of the cofferdam, the normal component of the force from the fill inclines slightly upward, and tends to push the wall up, causing it to "ride up" on the fill (Figure 27). The implication of this observation is that a rational design procedure to prevent slip between the sheeting and the fill must be based on an analysis which accounts for the movement of the outboard wall under load. Such an analysis will, unfortunately, be nonlinear since

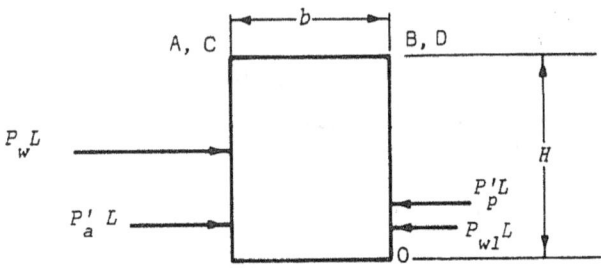

a. Driving force producing overturning moment about O

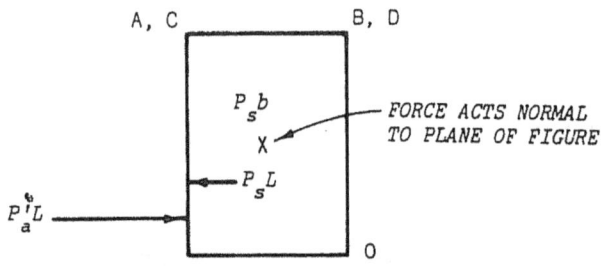

b. Effective forces producing friction forces shown in c

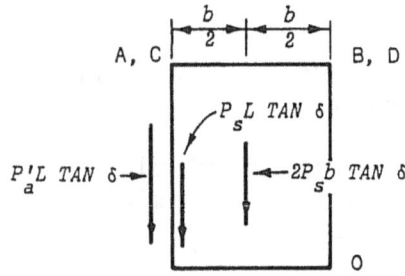

c. Friction forces producing resisting moment about O

Figure 22
Elevation views of free body of Figure 21 resultant forces shown

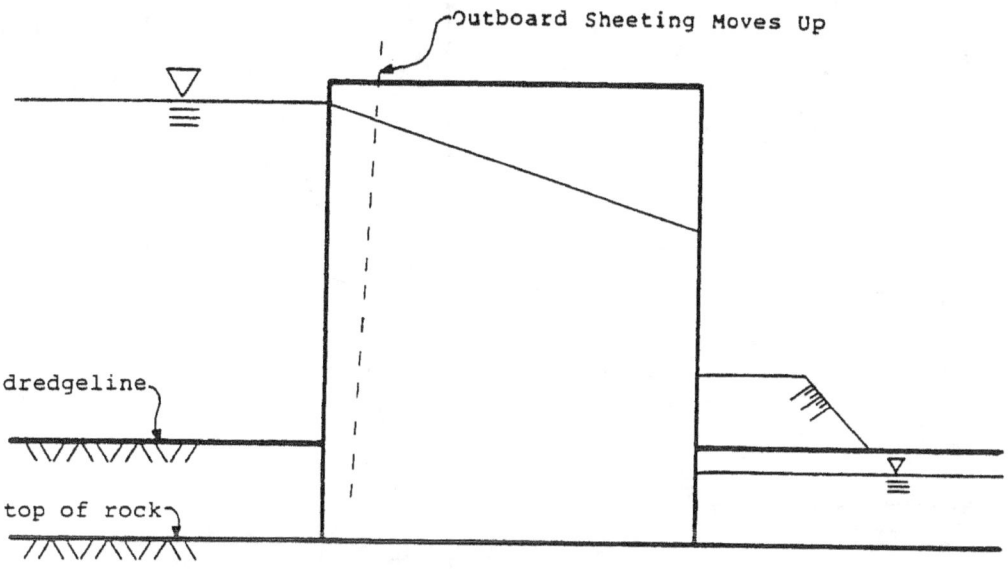

**Figure 23
Alternative failure mode in which outboard
sheeting slips relative to fill and rest of piling**

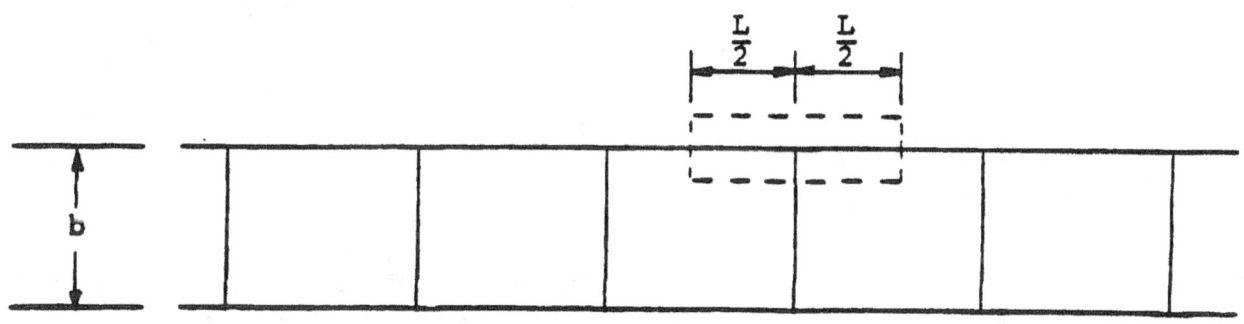

a. Portion of equivalent cofferdam selected for analysis

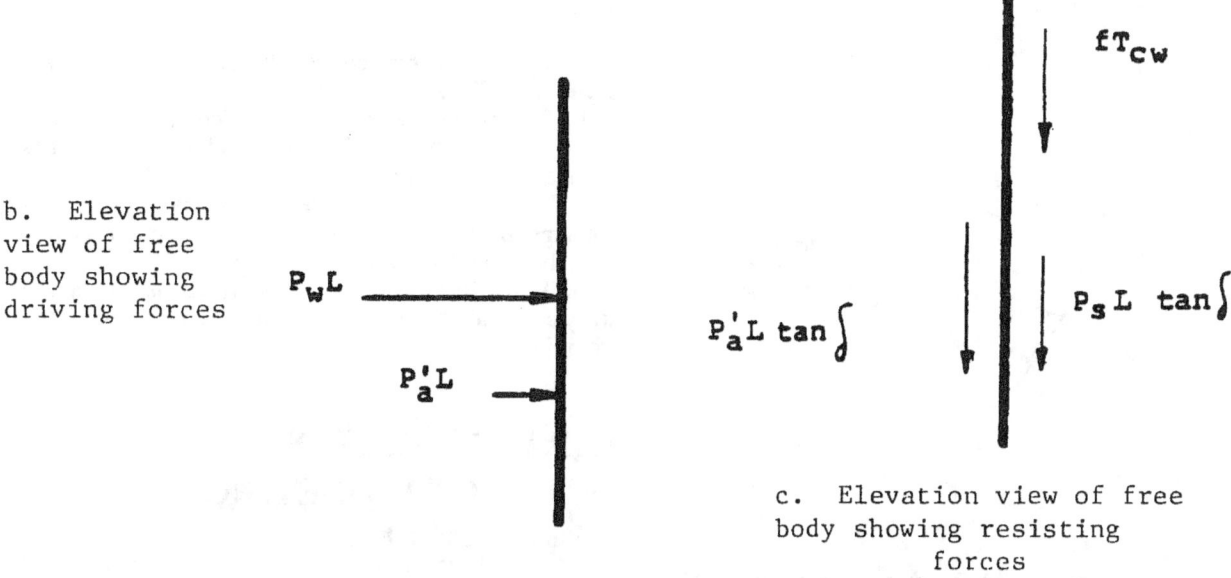

b. Elevation view of free body showing driving forces

c. Elevation view of free body showing resisting forces

**Figure 24
Driving and resisting forces acting on outboard sheeting**

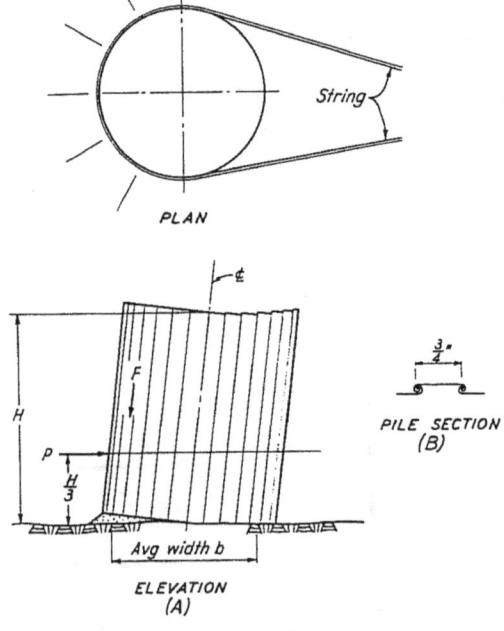

Figure 25
TVA model test showing slip between outboard sheets and fill (TVA Technical Monograph No. 75)

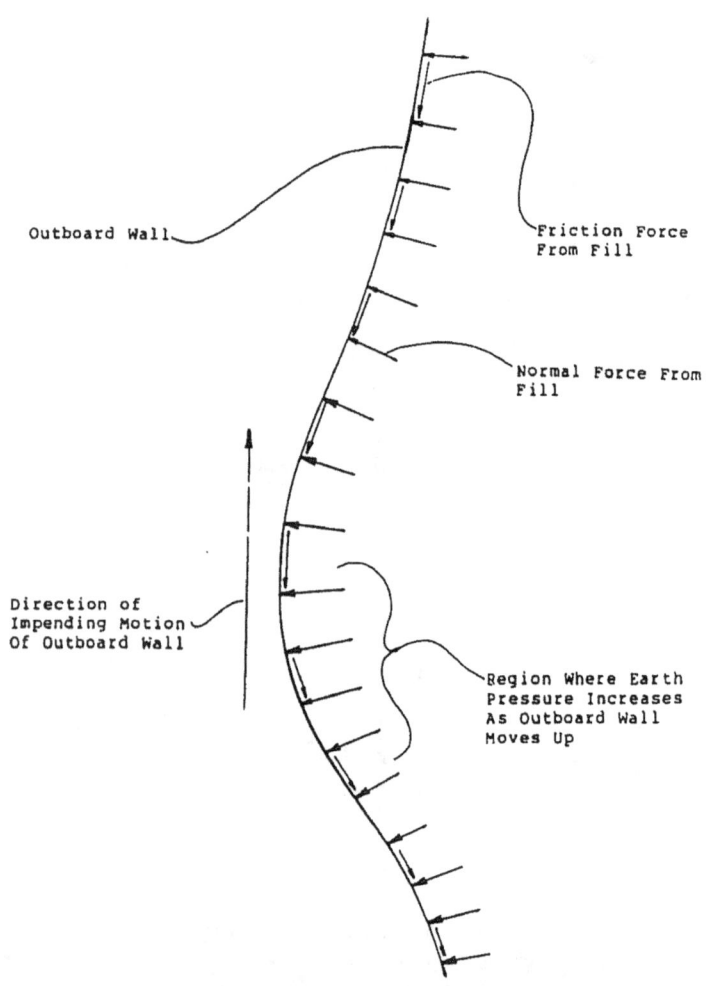

Figure 26
Effect of bulging in preventing slip between wall and fill

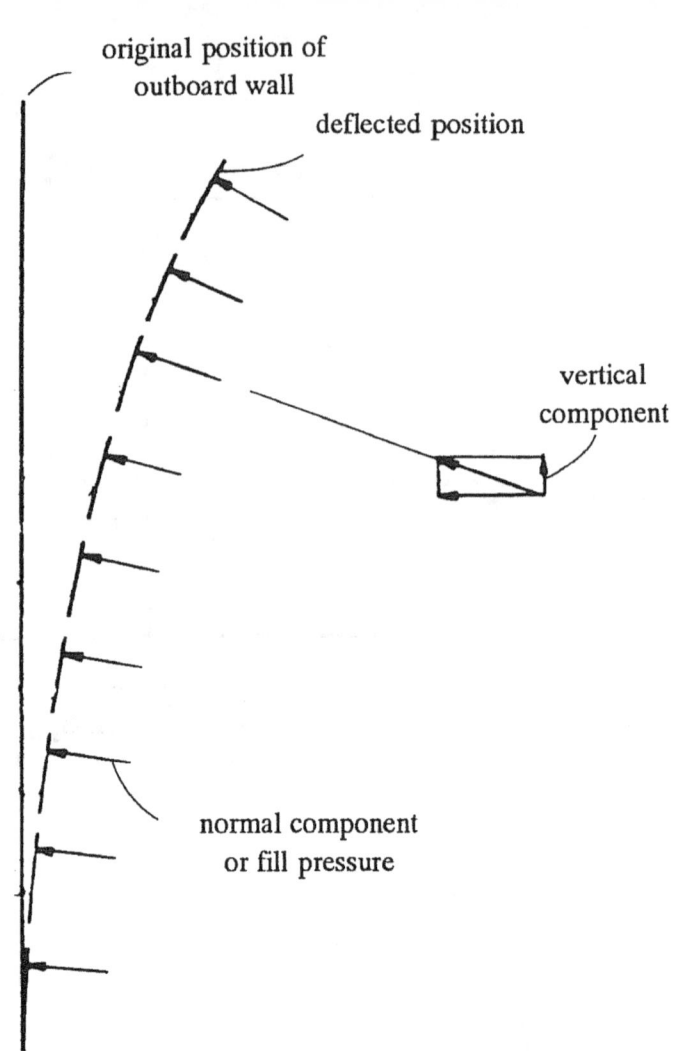

Figure 27
Development of force component which tends to produce upward slip of outboard wall

the deflected position of the sheeting is not known beforehand.

97. The above observations may be summarized as follows:

a. It is not established that slip between fill and wall can occur in a full-sized cell.

b. The only model studies in which slip between fill and wall occurred were the TVA studies and serious questions exist about the validity of the TVA models.

c. The existing design rule is inadequate since it ignores a fundamental mechanism (the change in direction of the force from the fill) by which slip might occur.

d. To clarify the mechanism by which slip might occur, an analysis should be performed which accounts for rotation of the piling.

PART VII: PULLOUT OF OUTBOARD SHEETING

Rotation About the Toe

98. The lateral forces acting on the cell cause it to rotate about the toe. These forces cause the outboard sheeting and the crosswall to pullout from the foundation. The failure mode resembles that shown in Figure 23, except that it is assumed that the common wall sheeting moves up with the outboard sheeting, and they both slip with respect to the cell fill and

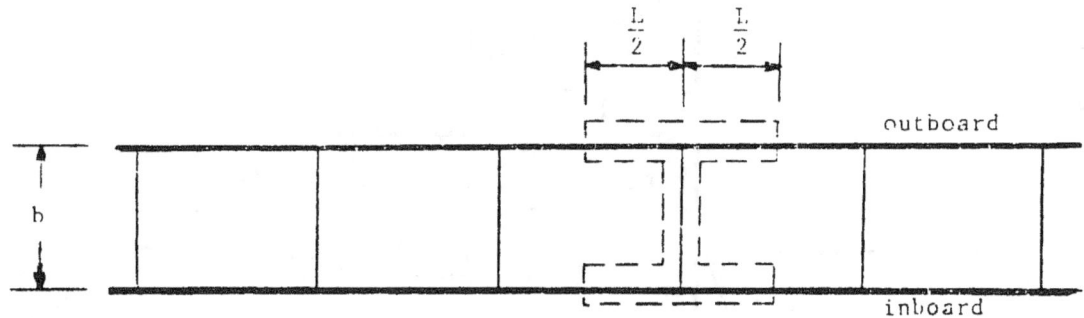

a. Portion of equivalent cofferdam selected for analysis

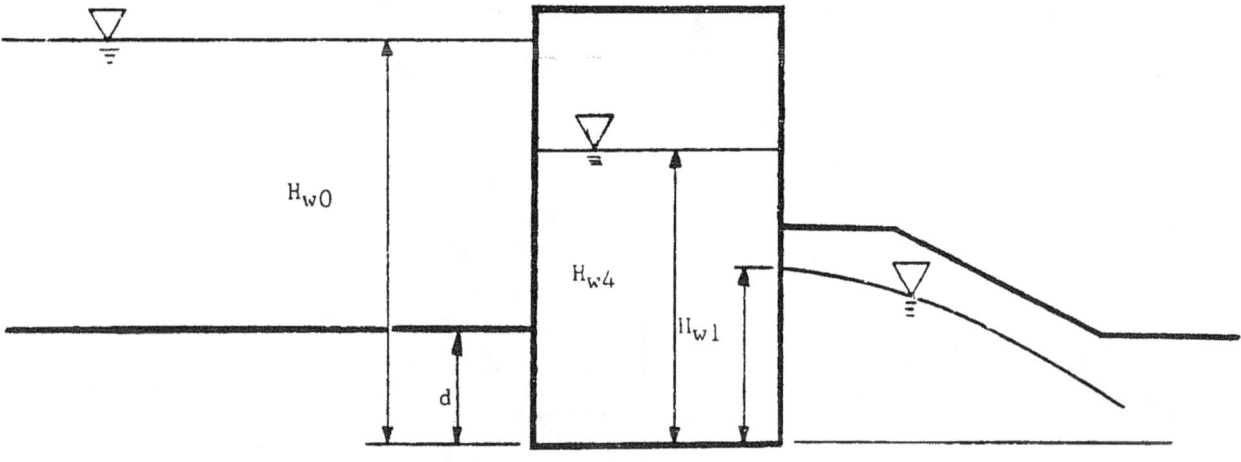

b. Definition of lengths used in pull out of outboard sheeting analysis

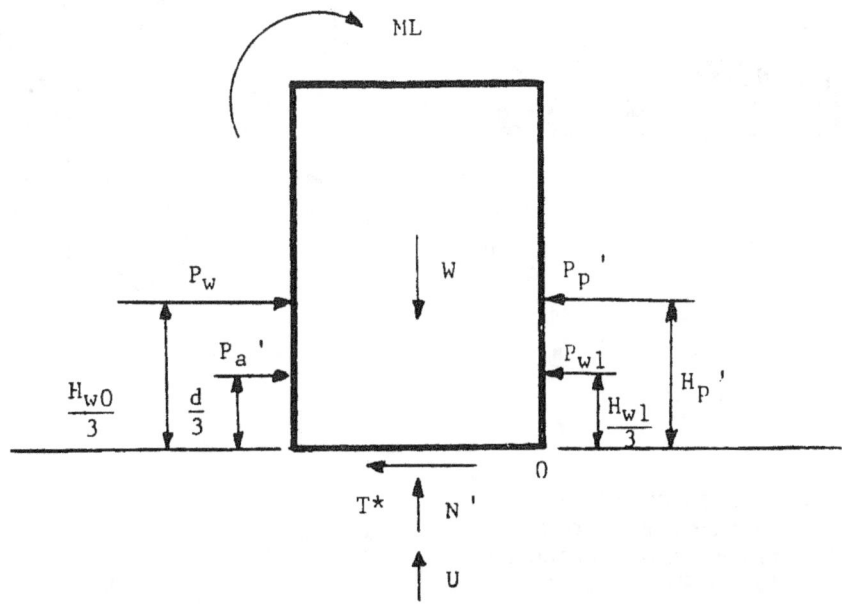

c. External forces considered in calculations overturning moment

**Figure 28
Free-body diagrams for pullout calculations**

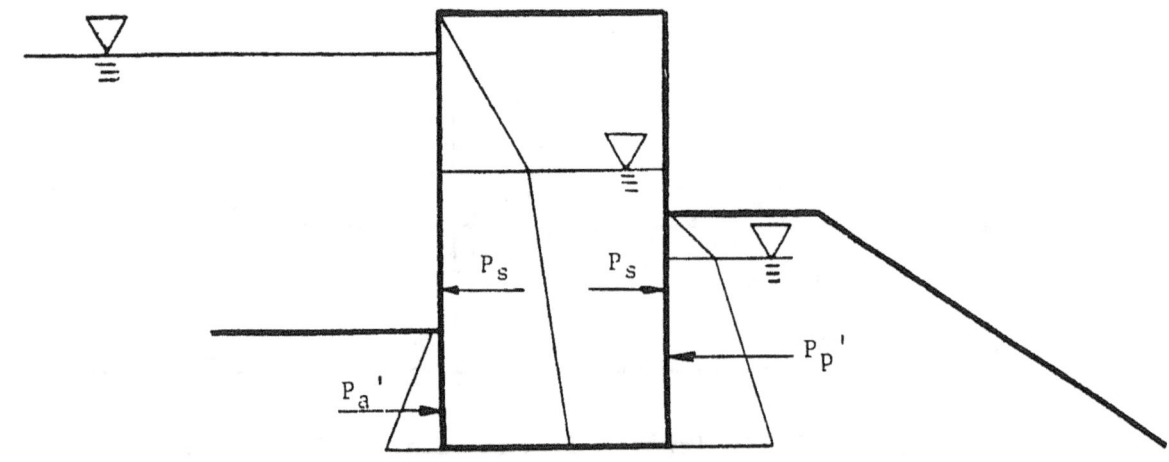

d. Effective soil pressure acting on sheet-pile walls

e. Forces resisting pull out of equivalent cofferdam

**Figure 28 (Continued)
Free-body diagrams for pull out calculations**

foundation material. The inboard sheeting is assumed not to slip with respect to the fill or foundation material and to rotate into the berm as the outboard and common wall sheets pullout.

99. The FS against pullout (DM7 1971; USS 1972) can be written as:

$$FS = \frac{\text{Maximum available resisting moment}}{\text{Driving moment}} = \frac{M_r b}{M L}$$

$$= \frac{b(Q_{uo}L + 0.5 Q_{uc} b)}{L\left(\dfrac{P_w H_{wo}}{3} + \dfrac{P'_a d}{3} - P'_p H'_p - \dfrac{P_{wl} H_{wl}}{3}\right)}$$

where

Q_{uo} = ultimate sheet-pile pullout capacity (per unit length of cofferdam) of outboard

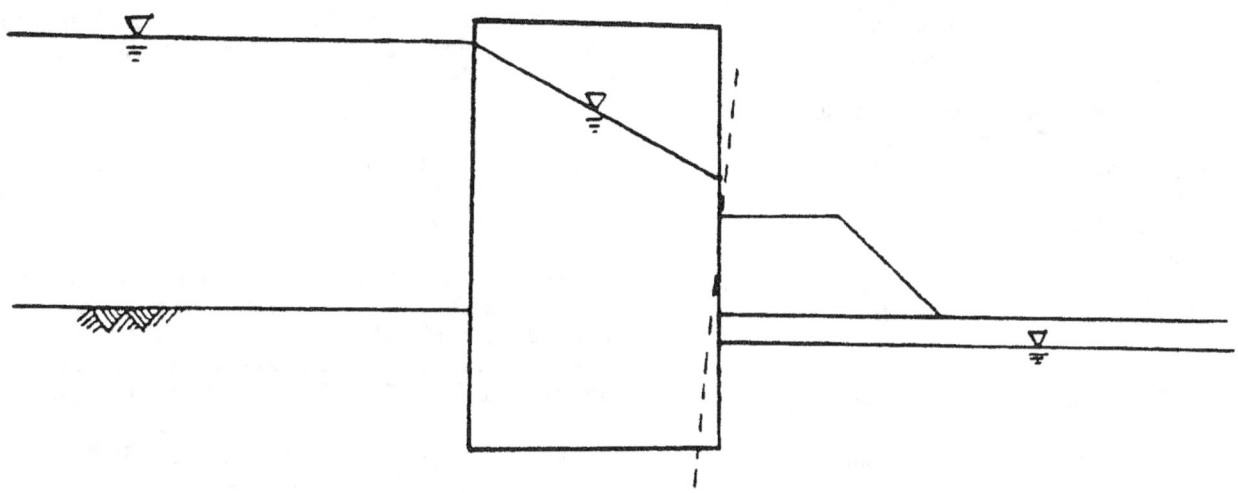

**Figure 29
Failure by excessive penetration of inboard sheeting (plunging)**

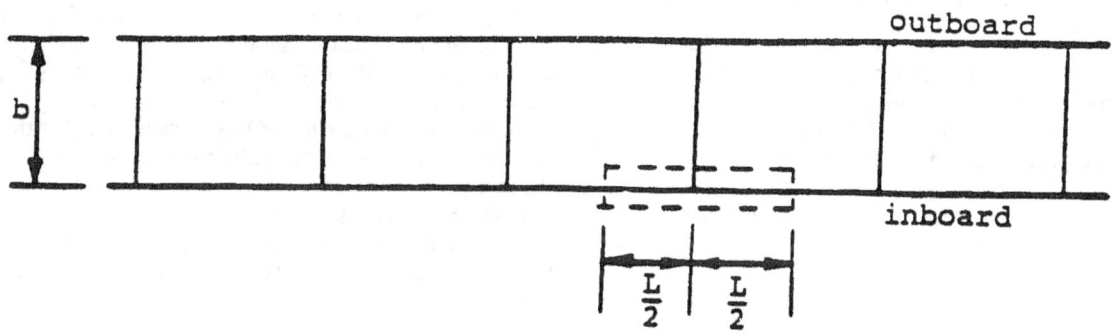

a. Portion of equivalent cofferdam selected for analysis

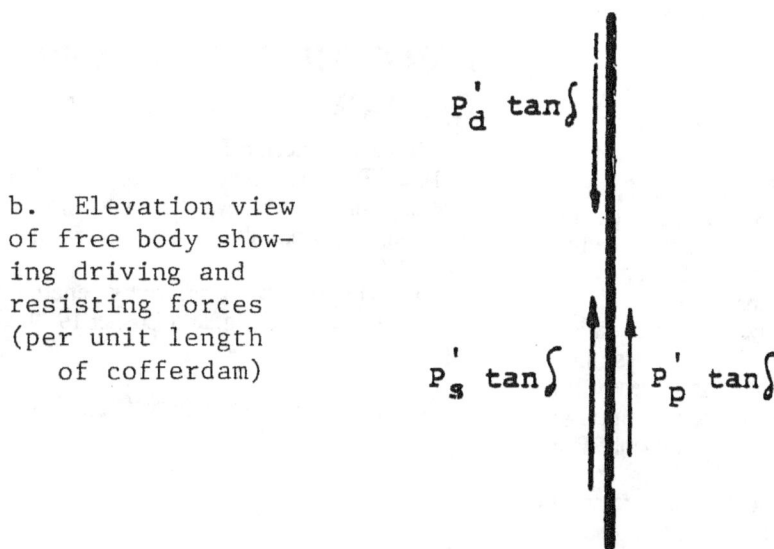

b. Elevation view of free body showing driving and resisting forces (per unit length of cofferdam)

**Figure 30
Free-body diagrams for penetration calculations**

Q_{uc} = ultimate sheet-pile pullout capacity (per unit length) of common wall

H_{wo} = vertical distance from sheet-pile tips to the intersection of water level outside of cofferdam with outboard sheeting (Figure 28b)

H'_p = vertical distance from sheet-pile tips to line of action of P'p (Figure 28c)

M_r = the resisting moment (per unit length of cofferdam) due to pull out of the equivalent cofferdam outboard and commonwall sheeting

M = the overturning or driving moment (per unit length of cofferdam) due to external forces on the cofferdam

The other quantities retain their previous meanings.

Considerations for Calculation for Pullout

Calculation of the Driving Moment

100. The portion of the cofferdam to be isolated as a free body is shown in Figure 28a. The vertical distances used in calculating cofferdam pull out are shown in Figure 28b. The overturning moment ML referred to as the driving moment (calculated with respect to point 0) is shown acting in Figure 28c along with the external forces which cause overturning. The forces shown in Figure 28c and not previously defined are:

U = resultant uplift force due to water pressure acting on the base of the cofferdam (per unit length of cofferdam)

N' = resultant effective soil force acting on the base of the cofferdam (per unit length of cofferdam)

T^* = horizontal shear force on base of cofferdam (per unit length of cofferdam)

101. The assumption is made that the weight of the contents of the cell (W), the resultant of the uplift force (U) and effective soil force (N') act through the same point. This assumption allows the W, N', and U terms to be neglected when summing moments about point 0. From horizontal equilibrium the following equation is obtained:

$$P_w + P'_a = P'_p + P_{wl} + T^*$$

Rearranging terms yields the equation:

$$P'_p = P_w + P'_a - P_{wl} - T^* \quad (50)$$

Equation 50 must always be satisfied. The terms P_w, P'_a, and P_{wl} are easily calculated (see Part IV). Let P^*_p be the horizontal effective-force (per unit length of cofferdam) acting on the inner cofferdam wall when the berm and foundation exert full passive earth pressure against the cofferdam. It is assumed that P'_p is the smaller of P^*_p and the result of Equation 50 with T^* taken as zero. When P'_p equals P^*_p and is less than Equation 50 with T^* taken as zero, then T^* is increased to maintain horizontal equilibrium. Since T^* passes through point 0, it does not contribute to the overturning moment. The justification for the above assumptions is that for the outer and common wall to pullout, the inner wall will have to rotate a large amount, mobilizing something approaching full passive pressure in the berm and foundation. At the same time the passive pressure mobilized must satisfy horizontal equilibrium as given by Equation 50. Note that T^* must be less than N' tan ϕ to prevent sliding along the base of the cell.

Calculation of Resisting Moment

102. The pull-out capacity Q_u of a pile depends on the skin friction arising from the material into which the pile is embedded and the skin friction between the cell fill and the sheet-piling. Figure 28d shows the earth pressure forces acting on the equivalent cofferdam walls. The force not previously defined is:

P_s = horizontal effective-force (per unit length) of the cell fill and foundation material within an equivalent cofferdam cell acting on a cell wall

The force P_s is calculated using an earth pressure coefficient K between $1.2K_a$ and $1.6K_a$. It should be noted that the force P_s acts on both sides of the common wall. The forces resisting pullout of the equivalent cofferdam section and being analyzed are shown in Figure 28e. The force not previously defined is

Q_{ui} = ultimate sheet-pile capacity (per unit length of cofferdam) of the inboard sheeting

The ultimate sheet-pile pullout capacities are computed as follows:

$$Q_{uo} = (P'_a + P_s) \tan \delta \quad (51)$$

$$Q_{uc} = 2 P_s \tan \delta \quad (52)$$

103. It should be noted that the value of Q_{ui} is not needed to calculate the resisting moment about point 0 in Figure 28e because it passes through point 0. In developing Equation 49, the assumption is made in calculating the resisting moment that there is no interlock slip. It is also assumed that the outboard and common wall sheeting slip with respect to the cell fill in the analysis. If the assumption is made that the fill moves up with the cell during pull out then the method of calculating the resisting moment should be revised to take into account the weight of the fill. However, it seems unlikely that the fill would move up with the cell during pullout.

Comments on the Design Procedure for Preventing Pull Out

104. No full-sized cell nor model-test failures by this mode have been reported in the literature.

PART VIII: PENETRATION OF THE INBOARD SHEETING (PLUNGING)

Effects of Friction Downdrag

105. The friction force from the fill drives the inboard sheeting further downward into the foundation, leading to tilting and also possible loss of fill from the top of the cell (Figure 29).

106. The FS against failure by penetration of the inboard sheeting (Lacroix, Esrig, and Lusher 1970; USS 1972) can be expressed as:

$$FS = \frac{\text{Maximum available resisting force}}{\text{Driving force}}$$

$$= \frac{(P'_p + P'_s) \tan(\delta)}{P'_d \tan(\delta)} \quad (53)$$

in which (Figure 30) P'_s is the horizontal effective force of the foundation soil acting on the interior of the inboard sheeting below the dredgeline. The other quantities retain their previous meanings. The forces are calculated for a unit length of cofferdam. Penetration is of concern for cells founded on deep soil foundations.

Considerations in Calculations for Penetration

107. The free body used in the factor of safety calculation is a portion of the inboard wall, as shown in Figure 30. The FS is based on a simple comparison of the downward friction force from the fill and the upward friction forces from the foundation soil.

108. The calculation of P'_d was described in paragraph 58. Lacroix, Esrig, and Lusher (1970) recommend using a value of $K = 0.4$ in this calculation. The calculation of P'_p was described in paragraphs 54 and 55 with exception that P'_p should not be less than the at-rest earth pressure. The calculation of P'_s requires the construction of a flow net to account for the effects of the hydraulic gradient (DM7 1971).

Comments on the Design Procedure for Preventing Penetration

109. Concern for preventing penetration apparently originated from the Terzaghi (1945) paper on cofferdams, in which he expressed the view that the friction force from the inboard wall acting on the fill contributed significantly to the overall resisting moment of the cell (Terzaghi 1945). Thus, assessing the wall's resistance to penetration was considered important in establishing the stability of the entire cell itself. Terzaghi was led to this view by his assumptions about the pressure distribution from the foundation acting upward on the base of the cell and by what he admitted was a "very crude approximation": the assumption that the lateral earth-pressure coefficient has the same value at all points in the fill. From their model studies, Schroeder and Maitland (1979) concluded that K does in fact vary significantly within the fill. This observation plus the additional observation that no failures by the penetration mode have been reported in model studies or in the field indicate that the need to design against sheet-pile penetration cannot be considered well-established.

110. The force $P'_d \tan(\delta)$ which causes the inboard wall to be driven downward is caused by settlement of the cell fill. As the inboard wall is driven downward into the foundation, the outboard wall and crosswall remaining stationary, friction forces must act in the interlock connecting the inboard wall and the crosswall. Since these forces aid the inboard wall in resisting downward movement, neglecting them, as is done in the expression for the FS, is conservative.

111. As was discussed in paragraph 93, and shown in Figure 26, cell bulging of the outboard wall tends to decrease the possibility of failure by slip occurring between the fill and the sheeting. This is true since the pressure on the sheeting from the fill in the bulge acts primarily in the direction of the resisting forces, or downward. However, bulging of the inboard wall tends to increase the possibility of failure by penetration because the driving direction is downward in this case. A nonlinear analysis would be required to evaluate this phenomenon, due to the unpredictable magnitude and location of the bulging.

112. Based on model studies and field observations, Schroeder and Maitland (1979) concluded that the ability of the sheet-pile walls to mobilize passive resistance is limited to the region above the plane of fixity. If this recommendation is accepted, the at-rest value of the lateral earth-pressure coefficient should be used to calculate effective forces acting below the plane of fixity.

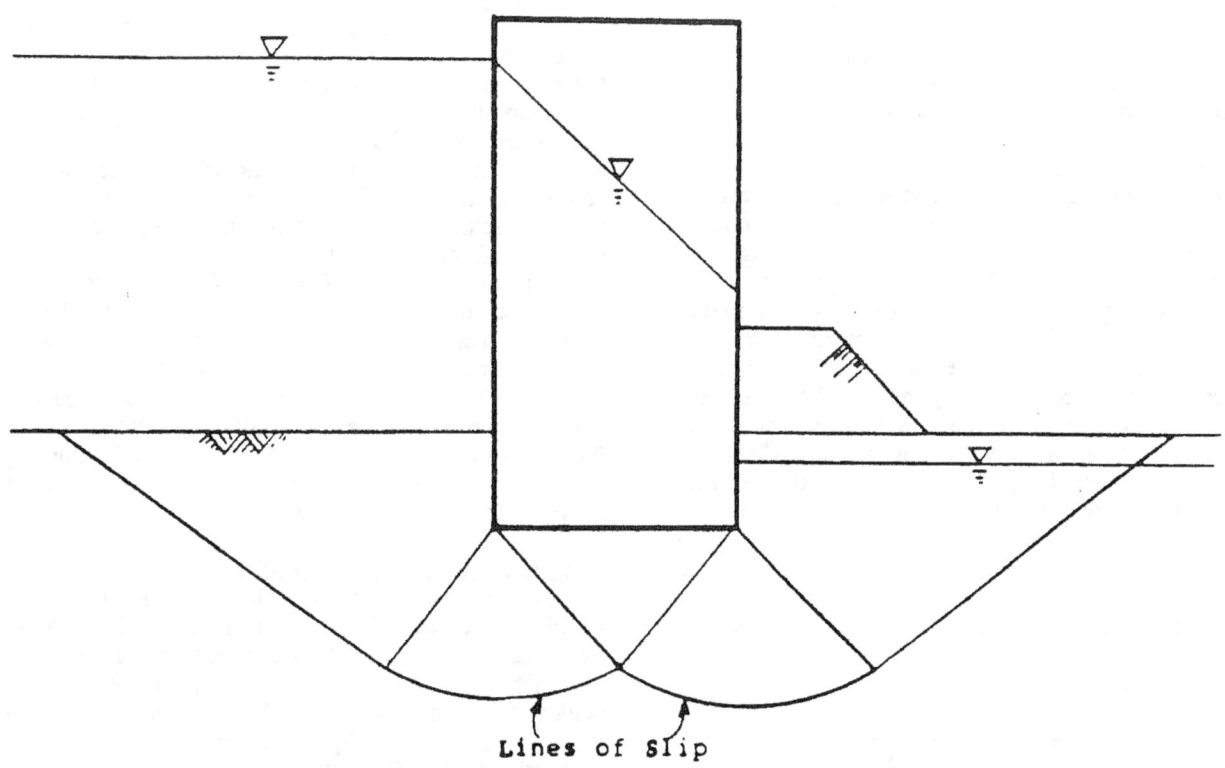

**Figure 31
Bearing failure of foundation**

PART IX: BEARING FAILURE OF FOUNDATION

Effects of Lateral Forces on Bearing Capacity

113. Lateral forces acting on the cell combine with the weight of the cell to produce an eccentric bearing force which exceeds the bearing capacity of the foundation. Foundation material is pushed downward and out from underneath the cell (Figure 31).

114. The FS against bearing failure of foundation is shown below as:

$$FS = \frac{\text{Ultimate bearing capacity}}{\text{Effective bearing pressure}} = \frac{q_{ult}}{q_{eff}} \quad (54)$$

where

q_{ult} = ultimate bearing capacity, calculated by dividing the effective width of the cofferdam into the total vertical load for which the the foundation has the capacity

q_{eff} = effective bearing pressure, calculated by dividing the effective width into the resultant vertical force acting on the foundation

The cofferdam is analyzed as a strip footing of width b.

Considerations in Calculations for Avoiding Bearing Failure of Foundation
Use of CBEAR

115. A comprehensive discussion of the theory and the calculations required for avoiding bearing failure is given in the user's guide to the CBEAR computer program (Mosher and Pace 1982), and thus will not be given here. Instead, the following remarks will be confined to indicating how the bearing-capacity design problem is formulated for the special case of a cellular cofferdam.

Cell Foundation Action

116. Because both vertical and horizontal forces act on the cell, the resultant force from the cell acting on the foundation is eccentric and inclined, that is, it does not act through the center of the base. Figure 32a shows a typical cell and foundation, while Figure 32b shows the isolated cell and the external forces acting on it. The resultant of these forces is shown acting on the foundation in Figure 32c at a distance e from the center of the cell. The magnitude of the vertical component of the resultant and the value of the eccentricity e are required if a bearing capacity analysis is to be performed.

117. The eccentricity is used to reduce the width b of the cofferdam to its effective value:

$$B' = b - 2e \quad (55)$$

118. The calculations should be based on all applicable water forces and the weights of fill and foundation soil.

PART X: SLIDING INSTABILITY

Effects of Lateral Force on Sliding

119. The horizontal forces acting on the cell cause it to slide on its base, or together with a portion of the foundation underneath the cell (Figure 33).

120. The FS against sliding instability for a cofferdam founded on rock, is computed from the equation:

$$FS = \frac{\text{Maximum available resisting force}}{\text{Driving force}}$$

$$= \frac{W_e f^* + P_{min}}{P_w + P'_a} \quad (56)$$

where

f^* = coefficient of friction of fill on rock

P_{min} = the smaller of (1) $P^*_p + P_{wl}$ or (2) the friction force acting from the rock on the bottom of the berm the other quantities retain their previous meanings. The sliding-stability of a cofferdam founded on soil is analyzed by the methods of wedge used for slope stability. FS is computed from the equation

$$FS = \min = \frac{\text{Maximum available shear resistance}}{\text{Shear force required to maintain equilibrium}}$$

in which the minimum is taken over all possible failure surfaces, and "shear" refers to shear forces acting on the failure surface. To obtain the FS for a given failure surface, a limit-equilibrium analysis is performed similar to that performed for slope stability.

Consideration in Calculations for Sliding Instability
Rock foundation

121. A common recommendation in the literature (Lacroix, Esrig, and Lusher 1970; USS 1972; Beiz 1970; TVA 1957) is to take the coefficient of friction f^* equal to $\tan(\phi)$ unless the rock surface is smooth, in which case a value of 0.5 is advised.

Soil Foundation

122. The wedge method is recommended for analyzing the slope-stability problem corresponding to a cellular cofferdam founded on soil. Since a comprehensive discussion of the theory and the calculations required for applying the wedge method is given in ETL 1110-2-256 and EM 1110-2-1902, it will not be given here. Instead, the following remarks will be confined to indicating how the wedge method is applied to the special case of a cellular cofferdam. To begin the analysis, a set of surfaces in the foundation are chosen which are candidates for the actual failure surface. Figure 34 shows a reasonable choice for a typical cell with berm. The wedge method is then used to compute a FS for each trial surface in turn. For example, in the figure a FS in computed for surface 9-4-5-8, then for 9-4-5-7, for 9-4-5-6, and so on until all combinations of surfaces have been considered. (Each trial failure surface must include the surface 4-5 bounding the structural wedge.) The minimum FS found by this procedure is the FS against sliding for the cell.

Comments on Failure by Sliding

123. For a cofferdam founded on a rock foundation, an additional degree of conservatism is implicit in the FS expression, Equation 56, since the resisting force produced by the slight penetration of the sheet piles into the rock is neglected. Even if the rock surface is hard and penetration is very small, natural irregularities are usually present which contribute towards the resistance to sliding.

124. A conservative assumption which may be made is that only normal forces (no friction forces) are transmitted across the vertical boundaries of the wedges. In situations where sliding instability appears to be a significant possibility, a

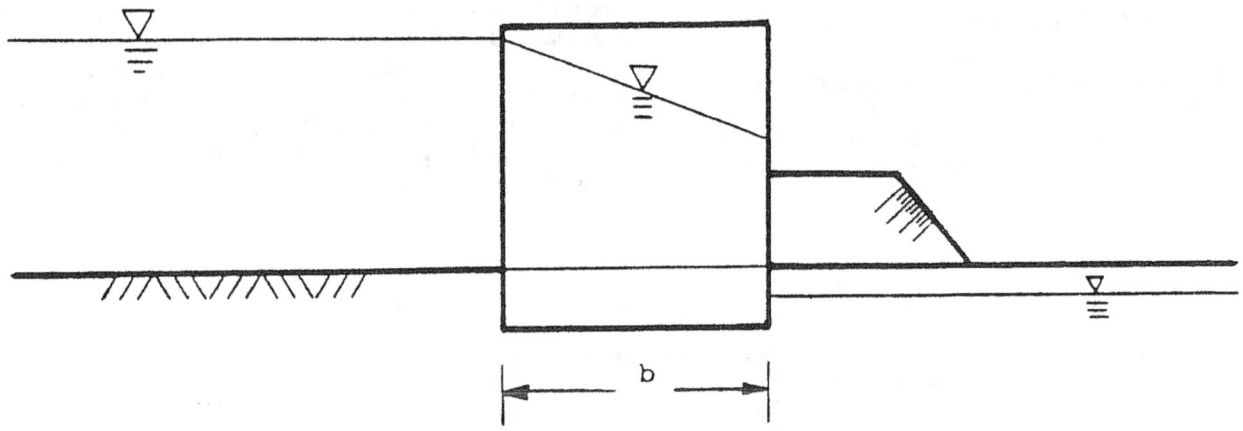

a. Cell founded on soil

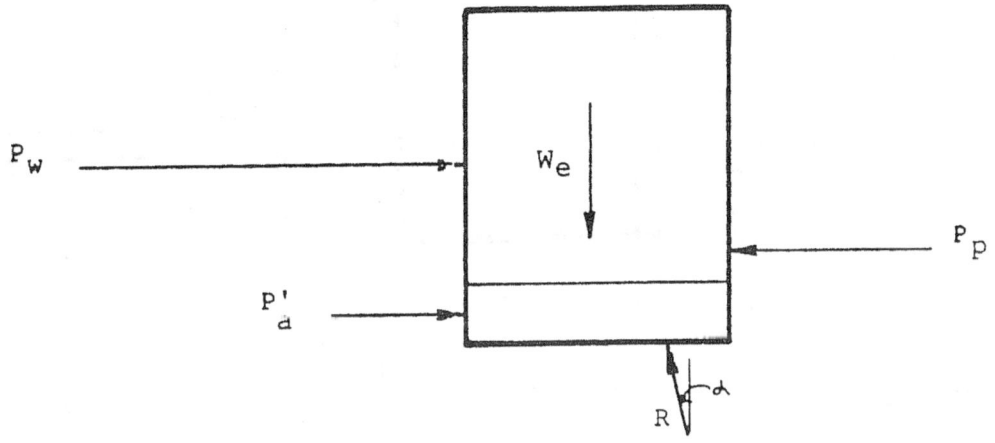

b. External forces acting on cell

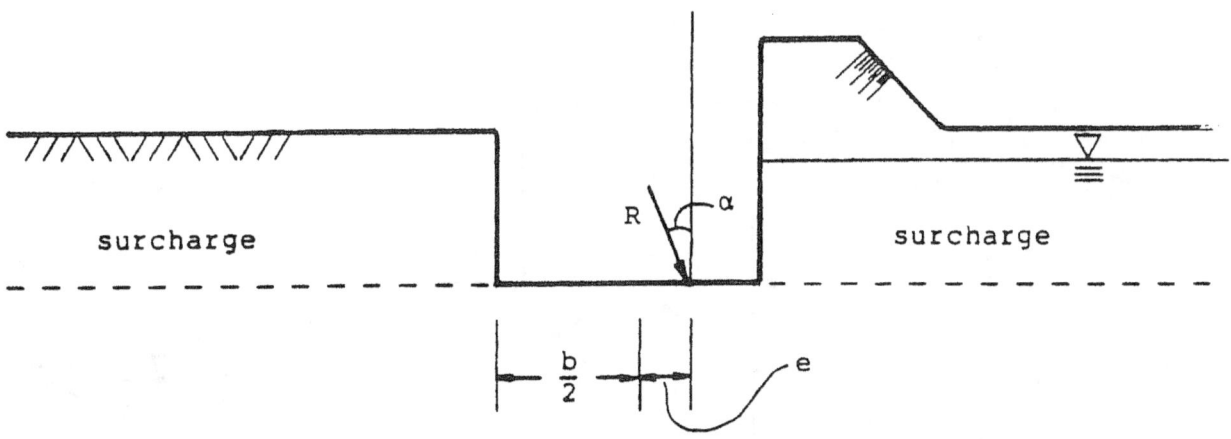

c. Force bearing on foundation

**Figure 32
Calculation of bearing force R**

refined analysis may be considered which includes shear on all vertical boundaries.

125. The earth pressures on the cell walls computed by the wedge method are not necessarily good approximations to the actual earth pressures, since the former pressures may correspond to a relatively high FS against sliding.

126. If a weak stratum is present in the foundation soil, it should be included in the trial failure surfaces.

127. The wedge method gives an upper bound to the FS. Thus, there is no guarantee that the lowest FS has been found by the procedure described above. The reliability of the procedure depends on the analyst's experience and ability in predicting failure surfaces.

PART XI: SLIP ON CIRCULAR FAILURE SURFACE (HANSEN'S METHOD)

Alternative Mode of Failure

128. A circular failure surface forms between the tips of the outboard and inboard walls as shown in Figure 35. That portion of the cell above the surface rotates as a rigid body about the center of the circle. Sliding occurs between the fill and the walls and between some of the sheets in the crosswall (Lacroix, Esrig, and Lusher 1970; Ovesen 1962).

129. The FS against slip on circular failure surface can be computed from the equation:

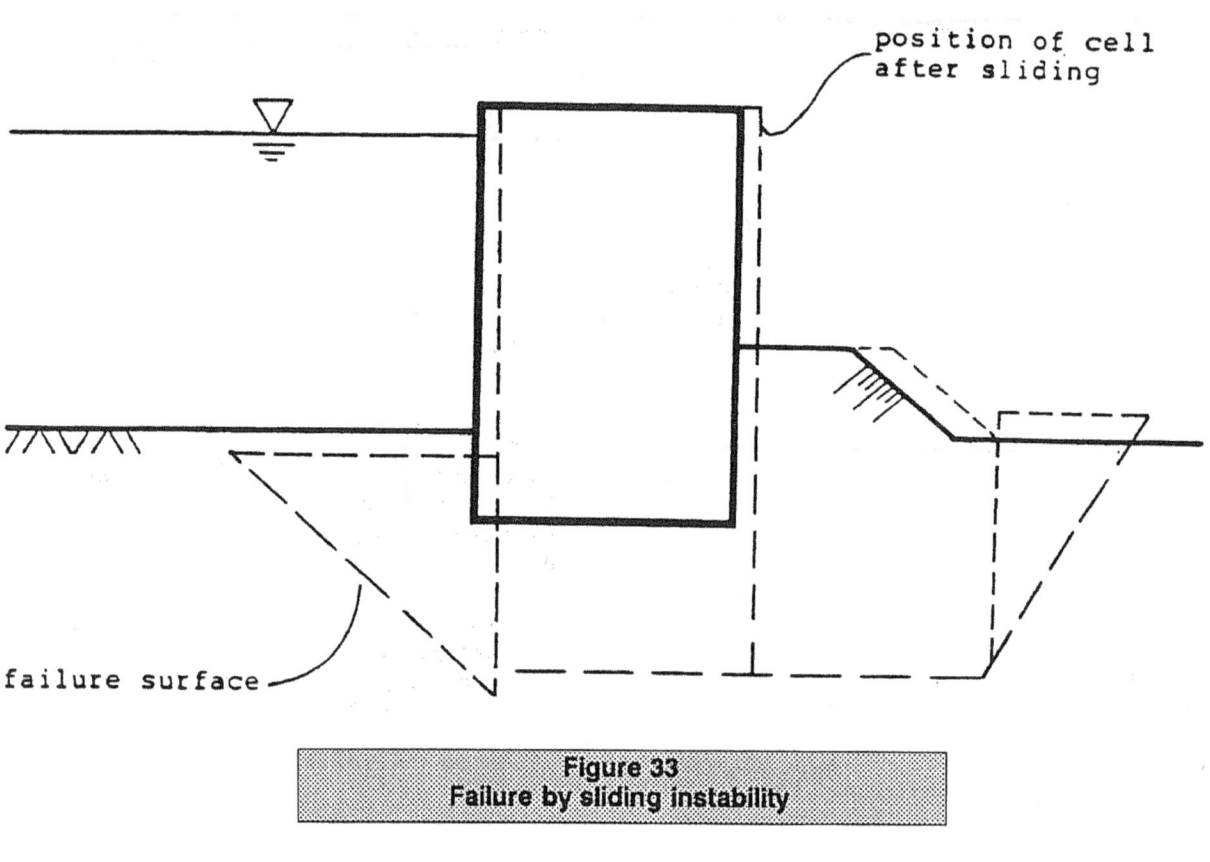

Figure 33
Failure by sliding instability

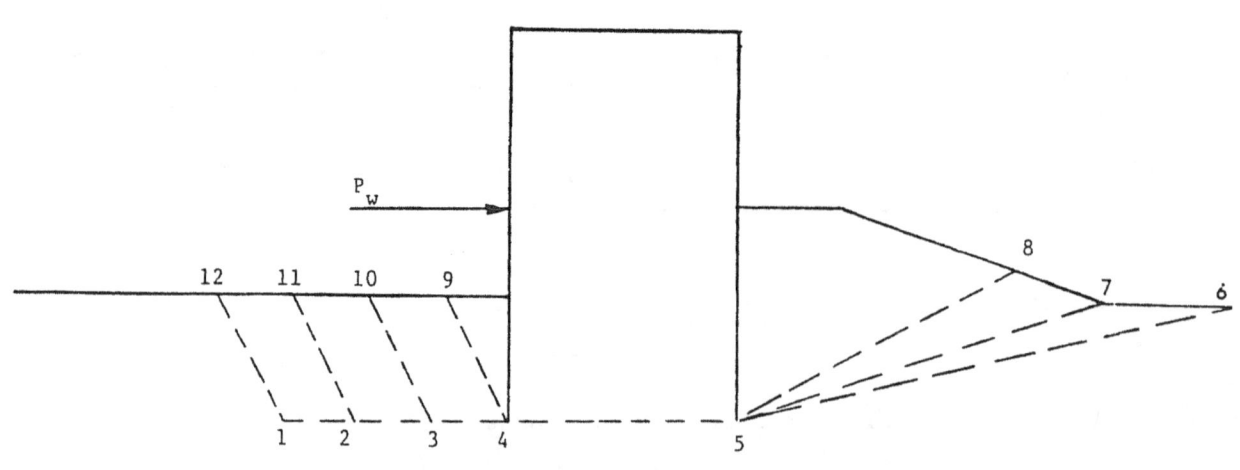

Figure 34
Trial failure surfaces for application of wedge method

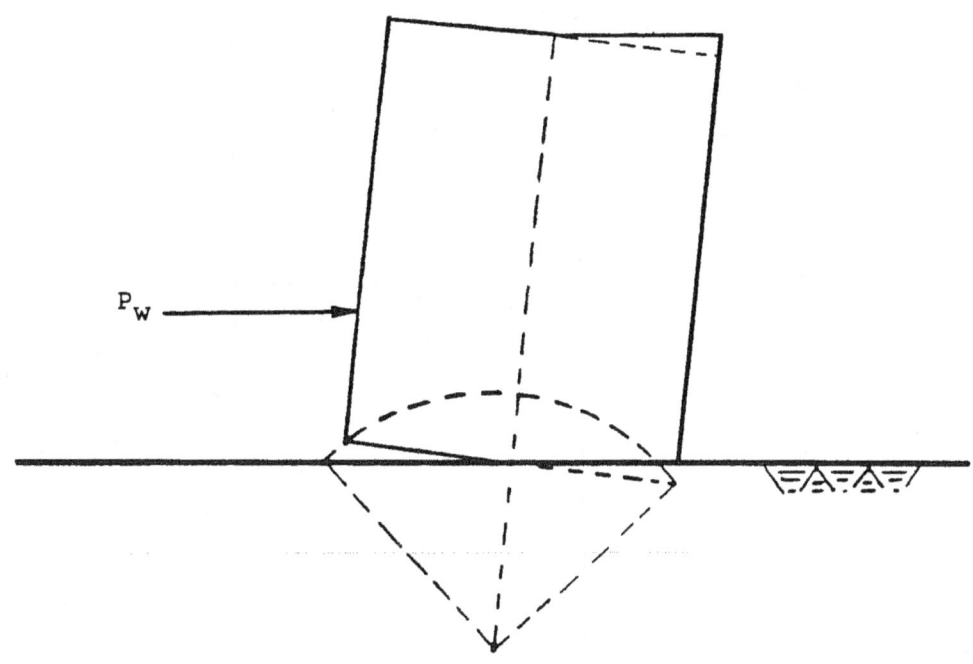

**Figure 35
Failure by sliding on a circular rupture surface
(rock foundation)**

$$FS = \frac{\text{Maximum available resisting moment}}{\text{Overturning moment}}$$

$$= \frac{M_W}{M'} \qquad (57)$$

where
M_W = moment of effective weight of fill above the failure surface
M' = moment caused by the driving forces (Figure 11)

The moments are computed with respect to the center of the failure circle.

Considerations in Calculations for Hansen's Method

130. Hansen proposed two different methods for evaluating the stability of a cofferdam cell: the equilibrium method and the extreme method. The calculations required when using the equilibrium method are complicated. Fortunately, essentially the same results may be obtained by use of the extreme method, which is computational simpler.

131. The extreme method is based on approximating the failure circle by a logarithmic spiral which obeys the equation

$$r = r_A e^{\theta \tan(\phi)} \qquad (58)$$

where (Figure 36)
r = radial distance from point 0
θ = angle measured counterclockwise from line OA
r_A = radius corresponding to $\theta = 0$
e = base of natural logarithms
ϕ = angle of internal friction of fill

132. The logarithmic spiral defined by Equation 58 has the property that the resultant of the friction and normal force at each point on the rupture surface defined by the spiral passes through the pole of the spiral. Thus, if the moments acting on the free body are computed with respect to the pole, the resultant force acting on the rupture surface will not appear in the moment expression.

133. The analysis procedure is not straightforward. A sketch such as Figure 36 is made to scale, and a logarithmic spiral is plotted which passes through the inboard and outboard tips of the walls, but is otherwise arbitrary. The moments M_W and M' are computed (with respect to the pole of the spiral) and the FS evaluated from Equation 57. In evaluating M_W, which depends on the effective weight W' of the fill above the failure surface, work can be saved by applying the following equation for the cross-hatched area A' shown in Figure 36:

$$A' = \frac{(r_B)^2 - (r_A)^2}{4 \tan(\phi)} - \frac{ab}{2} \qquad (59)$$

where
r_B = distance from pole to tip of outboard sheet
a = vertical distance from pole to base of cell

134. The pole for the most critical failure spiral may be found by repeating the above procedure with many different assumed spirals, and searching for the pole that yields the minimum value for the FS. A more direct approach is to make use of the fact that the pole of the failure spiral is on the locus of poles of those Logarithmic spirals which pass through the tips of the sheet piles. The pole of the failure spiral can be found by drawing the tangent to this locus from the intersection of the force W' and the resultant of the driving forces (Figure 36).

Failure Modes for Cofferdams on Sand

135. Hansen hypothesized that two distinct failure modes are present for cofferdams on sand and must be investigated.

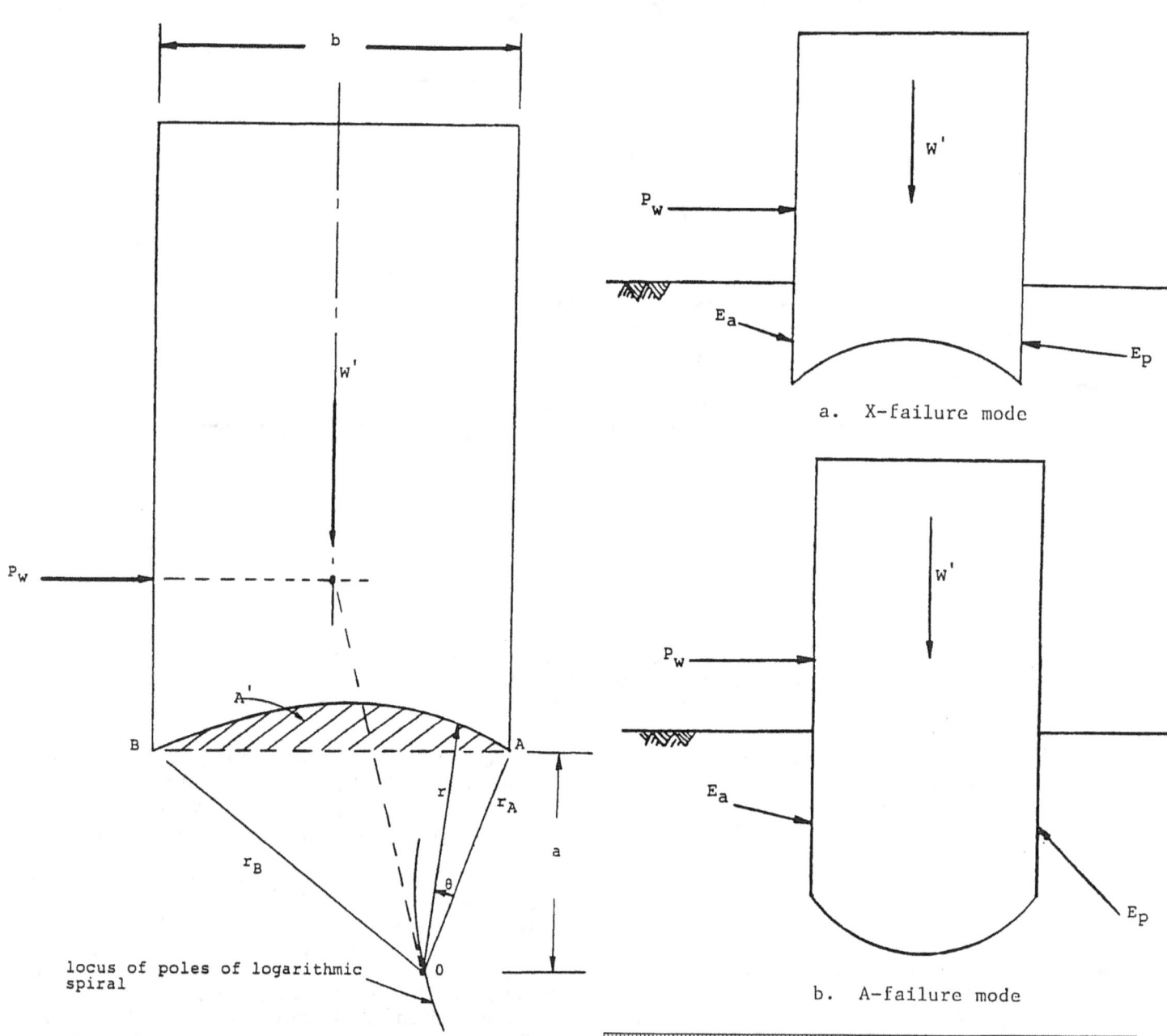

Figure 36
Free body for Hansen's method, rock foundation

Figure 37
Possible failure modes for cofferdams on sand (Hansen's method)

an "X-mode" and an "A-mode" (Figure 37). Furthermore, the lateral forces E_a and E_p from the foundation soil must be taken into account in evaluating the moments appearing in the FS equation. Whether these additional moment contributions are to be added to the driving or to the resisting moment must be determined for each particular case. Ovesen (1962) and Hansen (1957) present tables and charts for calculating the lateral forces. If the sheetpiles are embedded to great depth in the foundation, then the possibility must be considered that plastic hinges form in the walls (Hansen 1957).

Comments on Hansen's Method

136. The method is based on a highly theoretical approach to soil mechanics. The following assumptions are made:

a. The fill is homogeneous, isotropic, cohesionless, obeys Coulomb's failure law, and follows the constitutive law for a rigid-plastic material.

b. In the rupture state, the dilatation is constant (same value of the ratio of volumetric-strain to maximum shear everywhere).

c. The axes of principal stress and principal deformation coincide.

137. If these assumptions are made, an analytical solution can be found. However, these are not the assumptions which are made by contemporary finite element analysts, for whom analytical simplicity is not a concern (Clough and Duncan 1977; Clough and Duncan 1978; Duncan et al. 1980; Duncan and Chang 1970). In particular, Hansen's assumptions do not allow the stress-strain law to vary from point to point in the fill according to the stress and deformation history which has been experienced locally. Thus, for example, the fill near the tip of the inboard wall is assumed to behave the same as the fill near the tip of the outboard wall, even though the compressive stress near the base of the inboard wall is much higher (Figure 10). Since the behavior of cohesionless material like sand is known

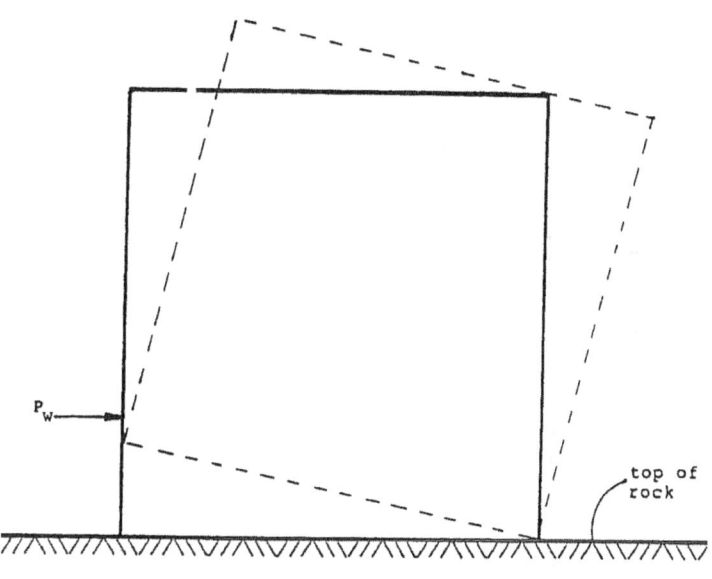

**Figure 38
Failure by overturning**

to be strongly history dependent, Hansen's neglect of this feature is questionable.

138. Hansen's entire analysis is predicated on the existence of circular failure surfaces. Unfortunately, the model cells used by Ovesen (1962) to demonstrate the circular shape of the failure surface were only 20 cm high and 15 cm wide. Furthermore, the walls were made of two glass panes and two brass sheets which were "sufficiently rigid for the elastic deformations to be ignored" (Ovesen 1962). It appears that the walls in these models would carry a much higher proportion of the external load than would be carried by the walls of a full-sized cell, and thus the approximately circular rupture figures observed in these model tests may not represent what would happen in the field. It should be noted that Ovesen also carried out tests with larger models (72 cm in diameter), but did not observe the failure surface because of the nature of the test set up.

139. Hansen's method, as described above, gives a conservative estimate of the FS, since it does not include the stabilizing effects of the friction between the wall and fill and also, in the case of a cell founded on rock, neglects the reaction from the rock. Ovesen shows how these effects may be included.

140. The complexity of applying Hansen's method, especially for cells founded on soil, argues against its everyday use by the practicing engineer. It is based on highly theoretical concepts from solid mechanics with which most soils engineers have little familiarity. It requires computations sufficiently complicated that a computer program would be helpful, if not strictly necessary. Given these aspects of the method, most designers would probably prefer to use other methods.

PART XII: OVERTURNING

Cause of Overturning

141. The lateral force acting on the cell causes it to rotate about the toe and tip over into the interior of the cofferdam (Figure 38). The resultant of the weight of the fill and the lateral forces acting on the cell shall lie within the middle third of the cell base.

Considerations in Overturning Calculations
Assumed force distribution on cell base

142. If no lateral force were to act on the cell, the force distribution from the foundation acting upward on the cell would be the uniform pressure ACEB shown in Figure 39. When lateral force P_w is present, the pressure from the foundation is assumed to change to the linearly varying distribution CEBD. To investigate the possibility of overturning, the resultant R of the lateral force P_w and the weight W of the cell contents is computed, and the intersection of R with the cell base noted. For the conditions assumed in the sketch, the resultant intersects in the base at a distance B/6 from the center, and the foundation and cell are beginning to lose contact at the heel, since the distributed force there is zero.

143. From 1908 until the publication of Terzaghi's paper (1945) on cofferdams, most cellular-cofferdam designs were based on the criterion of resistance to overturning. This criterion is now recognized as fundamentally incorrect (Belz 1970; TVA 1957; Cummings 1957; Schroeder and Maitland 1979). The fact that many cofferdams designed according to the criterion did not fail may be attributed to (a) cell heights were usually relatively low, compared to many modern cofferdams, and (b) cells filled with granular material are inherently quite stable because of the high shear resistance of the fill. It is important to realize that notable failures have occurred in cells filled by material deficient in shear resistance, such as clay. The requirement of designing against overturning did not prevent these failures (Cummings 1957).

Erroneous assumption

144. The basic difficulty with the approach of designing against overturning is that it is founded upon the assumption that a cellular cofferdam acts as a rigid, gravity-block structure which remains intact as it tips over. But, in fact, cofferdams are far from rigid, and failure by overturning is very unlikely to occur, since other failure modes, such as failure of the fill in shear, would occur first. These statements may be illustrated by consideration of Figure 39, in which overturning is shown to have proceeded to such an extent that the resultant now passes through the toe of the cell. Almost the entire base of the cell has lost contact with the foundation. But this situation is extremely unlikely, since once the base of the cell has lost contact with the foundation, the only means of supporting the weight of the fill is through arching. Arching is defined as vertical shear forces transmitted from the cell walls into the interior of the fill, and arching across the entire width of the cell is physically unrealistic. In place of arching, were the cell to reach the condition shown in Figure 39, a shearing failure of the fill would probably occur first. Indeed, the weakness of the overturning design rule is that it does not consider the shear resistance of the fill (Cummings 1957).

145. For the reasons just discussed, use of the criterion of designing against overturning is not recommended (Belz 1970; TVA 1957; Cummings 1957; Schroeder and Maitland 1979). However, the resultant base force must intersect the base at a distance less than b/6 from the center of the cell or the slip on vertical center plane in fill analysis presented in Part III will not be correct. If the resultant force falls outside of the distance b/6 from the center of the cell, the cell base pressure will go into tension. Since soil cannot take tension, the base pressure must be revised to reflect no tension pressure and the slip on vertical center plane in fill analysis must be revised. Also, if the resultant force falls outside of the distance b/6 from the center of the cell, the bearing area in the bearing capacity analysis is greatly reduced. For these reasons, even though the overturning analysis may not be fundamentally correct, the criteria should still be satisfied to make other analyses valid.

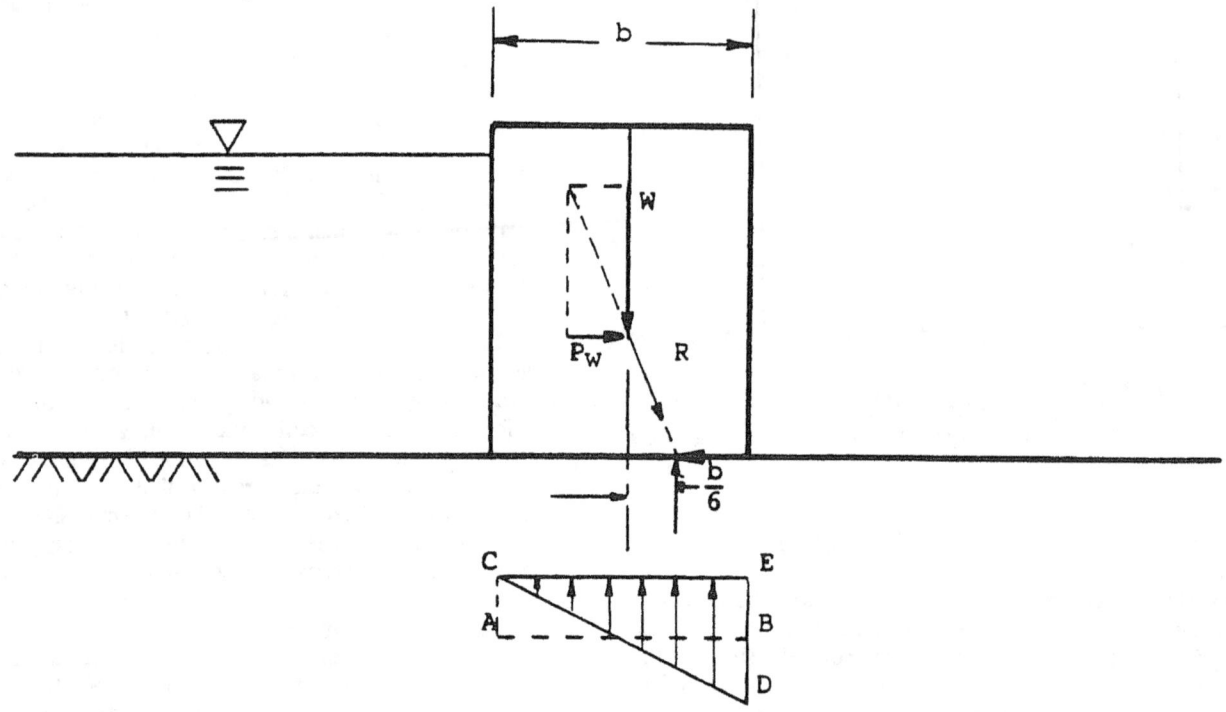

a. Stress distribution on base of cell

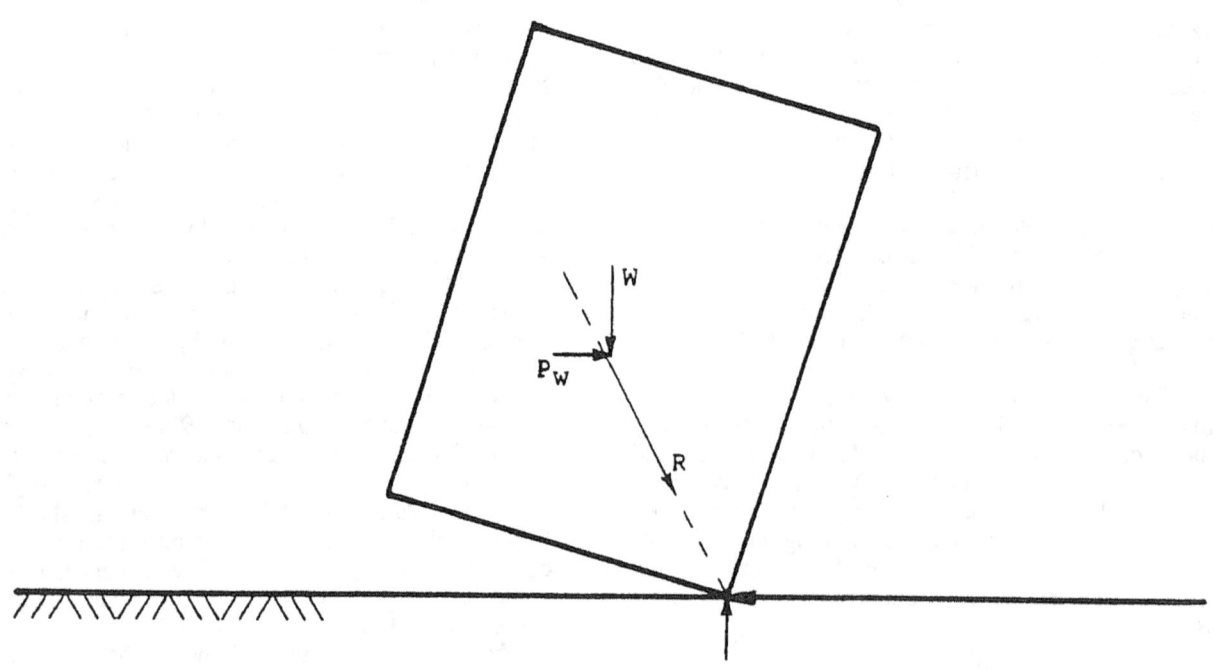

b. Impending tipping

Figure 39
Figure from foundation acting on base of cell

APPENDIX A: EXAMPLE PROBLEMS

Example Problem 1

Assumptions

1. cofferdam on bare rock
2. cell completely saturated
3. $H = 55$ ft*
4. $r = b/[2(0.875)]$ (initial estimate)
5. $L = 1.5b/2$ (initial estimate)
6. $b = 30$ ft (initial estimate)
7. $\theta = 45°$ (initial estimate)

Data for fill

$$\gamma'_f = 65 \text{ lbs/ft}^3$$
$$\phi = 28.83° \text{ (implies } K_a = 0.3493\text{)}$$
$$\tan \delta = 0.4$$

Data for sheet piles

$$t_{ult} = 16.0 \text{ k/in.} = 192{,}000 \text{ lb/ft}$$
$$f = 0.3$$

------------------------------------Bursting------------------------------------

$K = 1.2 K_a = 1.2(0.3493) = 0.419$

Water level outside cell at $H/2$

Equation 13:

$$P_{max} = 0.419\{0 + 65[2(55/3) - 0]\} + 62.4[55 - (55/2)]$$

$$= 2{,}714.6 \text{ lb/ft}^2$$

Equation 12:

$$t_{max} = P_{max} r$$

$$= 2{,}714.6(b)/[2(0.875)]$$

---

* A table of factors for converting non-SI units of measurement to SI (metric) units is presented on page 212.

$$= 1{,}551\, b\, (\text{lb/ft})$$

Equation 3:

$$FS = (t_{ult})/t_{max}$$

$$= 192{,}000/(155b)$$

For $b = 30.0$, $FS = 4.1$
For $b = 60.0$, $FS = 2.1$

Crosswall
Equation 14:

$$t_{cw} = p_{max} L$$

$$= 2{,}714.6(1.5b/2)$$

$$= 2{,}035.9b\ \text{lb/ft}$$

$$FS = 192{,}000/(2{,}035.9b)$$

For $b = 30$, $FS = 3.1$
For $b = 60$, $FS = 1.6$

Alternative - "TVA Secant Equation"
Equation 15:

$$t_{cw} = p_{max} L \sec(\theta)$$

$$= 2{,}714.6(1.5b/2)\sec(45)$$

$$= 2{,}879.3(b)\ \text{lb/ft}$$

$$FS = 192{,}000/(2{,}879.3b)$$

For $b = 30.0$, $FS = 2.2$
For $b = 60.0$, $FS = 1.1$

------------------------Slip on vertical centerplane------------------------

Equation 28:

$$M = (55)^3(62.4)/6 = 1{,}730{,}300 \text{ ft-lb/ft}$$

Equation 31:

$$K = \frac{\cos^2(28.83)}{2 - \cos^2(28.83)} = 0.623$$

Figures 12 & 30:

$$P'_c = P'_d = \gamma'_f K H^2/2$$

$$= 65(.623)(55)^2/2$$

$$= 61{,}217.2 \text{ lb/ft}$$

Equation 29:

$$S'_m = 61{,}217.2 \tan(28.83)$$

$$= 33{,}696 \text{ lb/ft}$$

Figure 13:

$$T_{cw} = 22.5\left[\tfrac{1}{2} \times 2{,}465 \times 27.5 + \tfrac{1}{2} 9.17(2{,}465 + 2{,}714.7) + \tfrac{1}{2} 2{,}714.7 \times 18.33\right]$$

$$= 1856765 \text{ lbs}$$

Equation 30:

$$S''_m = \frac{0.3 \times 1{,}856{,}765}{22.5}$$

$$= 24{,}755.1 \text{ lb/ft}$$

Equation 17:

$$FS = \frac{(33{,}696 + 24{,}755.1)(2b)}{3(1{,}730{,}300)}$$

$$= 0.0225b$$

For $b = 30.0$, $FS = 0.68$
For $b = 60.0$, $FS = 1.35$

----------------------Slip on horizontal planes in fill----------------------

Equation 28:

$$M = (55)^3(62.4)/6 = 1{,}730{,}300 \text{ ft-lb/ft}$$

Equation 46:

$$p^* = \frac{T_{cw}}{L} = \frac{1{,}856{,}765}{22.5} = 82{,}523$$

Equation 47:

$$M_f = 0.3(82{,}523)30 = 742{,}706 \text{ ft-lb/ft}$$

For completely saturated fill, the integral in Equation 43 reduces to (Cummings 1957) ($c = b \tan \phi = 30.0 \tan(28.83) = 16.51$; $a = H - c = 55 - 16.51 = 38.49$):

Equation 45:

$$M_{shear} = 65 \left[\frac{38.49 \times 16.51^2}{2} + \frac{16.51^3}{3} \right]$$

$$= 438{,}484 \text{ ft-lb/ft}$$

Equation 32:

$$FS = \frac{742{,}706 + 438{,}484}{1{,}730{,}300} = 0.69$$

FS increases with b

---------------------------Slip between fill and sheets---------------------------

$$P'_a = 0$$

$$P_s = (0.419)65(55)^2/2 = 41{,}193 \text{ lb/ft}$$

Equation 48:

$$FS = \frac{b\{[41{,}193 + 41{,}193(b/(0.75b))]0.4\}}{1{,}730{,}300}$$

$$= 0.0222b$$

For b = 30, FS = 0.67
For b = 60, FS = 1.33

---------------------------------Sliding on base---------------------------------

Equation 56:

$$W_e = 65(55)b = 3{,}575 \text{ lb/ft}$$

$$P_w = 62.4(55)^2/2 = 94{,}380 \text{ lb/ft}$$

$$FS = \frac{\tan(28.83)(3{,}575b)}{94{,}380}$$

For b = 30.0 , FS = 0.63
For b = 60.0 , FS = 1.3

Example Problem 2

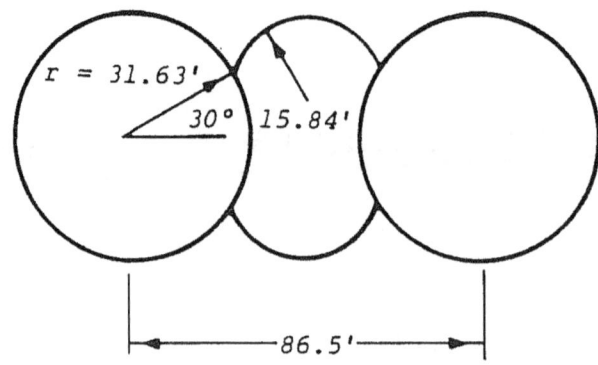

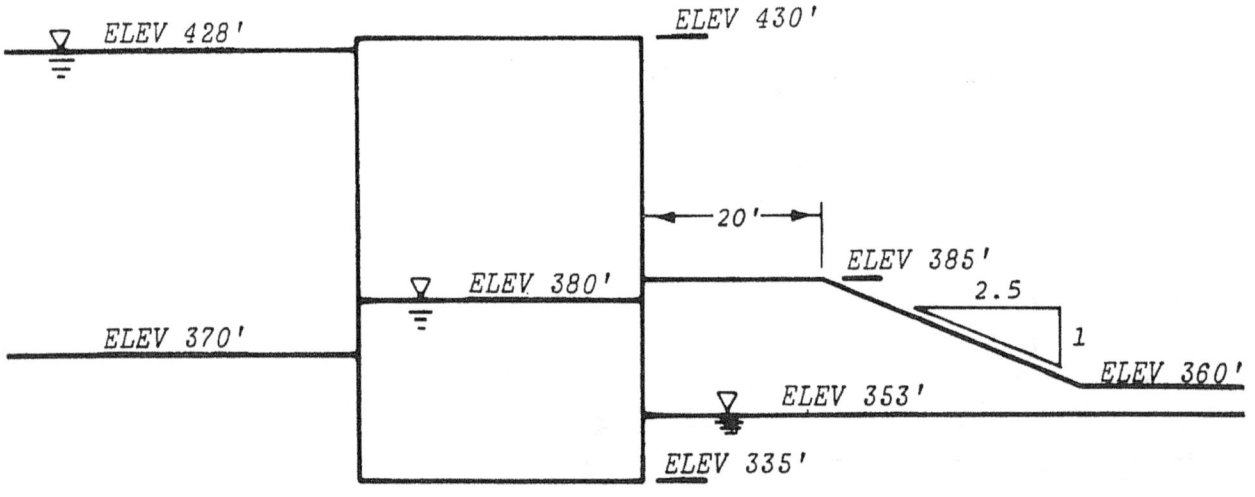

Foundation and fill material: sand

$\phi = 35°$
$\gamma_{sat} = 131 \text{ lb/ft}^3$
$\gamma_m = 120 \text{ lb/ft}^3$
$\gamma' = 68.6 \text{ lb/ft}^3$
$c = 0$
$\delta = 2/3\ \phi = 23.3°$
$L = 43.3 \text{ ft}$
$b = 54.9 \text{ ft}$

Sheet pile properties

$$PS - 32$$
$$t_{ult} = 16{,}000 \text{ lb/in.} = 16 \text{ k/in.}$$
$$I = 3.6 \text{ in.}^4$$
$$b_s = 15 \text{ in.}$$
$$f = 0.3$$

Bursting (Part III)

$$K_a = \tan^2\left(45 - \frac{\phi}{2}\right) = 0.27$$

$$K = 1.6 K_a = 0.43$$

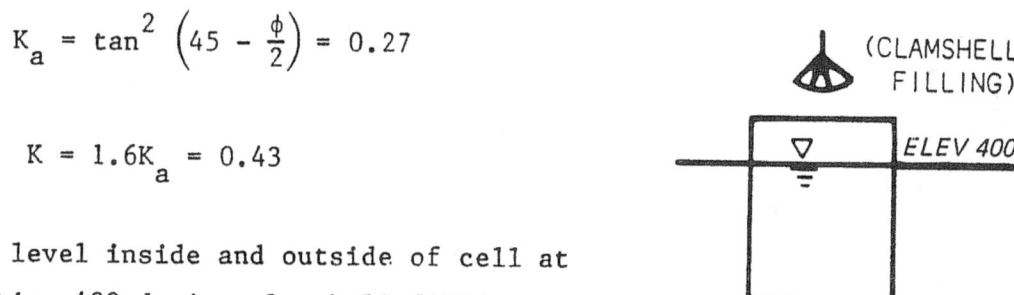

Water level inside and outside of cell at elevation 400 during clamshell filling.

Note: If cell were filled hydraulically, water level inside cell would be at elevation 430, top of cell. With clamshell filling, water level inside and outside of the cell are near the same level for sand-filled cells on a sand foundation with good drainage.

Calculate plane of fixity

$$\ell_h = 5 \text{ tons/ft}^3 = 5.79 \text{ lb/in.}^3 \qquad \text{From Terzaghi 1955}$$

$$n_h = \frac{b_s}{d'} \ell_h \qquad \text{(Eq 8)}$$

$$d' = 3.1 \sqrt[5]{\frac{EI}{n_h}} \qquad \text{(Eq 4)}$$

Equations 4 and 8 can be rearranged to give

$$d' = 4.1 \sqrt[4]{\frac{EI}{b_s \ell_h}}$$

$$d' = 4.1 \sqrt[4]{\frac{30,000,000(3.6)}{(15)(5.79)}} = 136.9 \text{ in.} = 11.4 \text{ ft}$$

Alternate method for plane of fixity

$$d' = \frac{K_a\left[\gamma_m(H_{fs} + d - H_{w4}) + \gamma'(H_{w4} - d)\right] + \Delta H_w \gamma_w}{\gamma'(K_p - K_a)} \quad \text{(Eq 10)}$$

$$K_p = \tan^2\left(45 + \frac{\phi}{2}\right) = 3.69$$

$$H_{fs} = 60 \text{ ft}$$

$$d = 35 \text{ ft}$$

$$H_{w4} = 65 \text{ ft}$$

$$\Delta H_w = 0$$

$$d' = \frac{0.27[120(60 + 35 - 65) + 68.6(65 - 35)] + 0}{68.6(3.69 - 0.27)}$$

$$d' = 6.5 \text{ ft}$$

$$6.5 \text{ ft} < d' < 11.4 \text{ ft}$$

Terzaghi's values of ℓ_h are taken at an ultimate loading. Since the cell is not loaded at ultimate loading, we can use higher values of ℓ_h. Scott recommends to double, at least, Terzaghi's values of ℓ_h and K_{sl}.

$$d' = 4.1 \sqrt{\frac{30,000,000(3.6)}{(15)(2)(5.79)}} = 9.6 \text{ ft}$$

$$6.5 \text{ ft} < d' < 9.6 \text{ ft}$$

Take $d' = 8 \text{ ft}$

$$x' = (H_{fs} + d')/3 = (60 + 8)/3 = 22.7 \text{ ft} \quad \text{(Eq 11)}$$

Point of maximum interlock tension is

22.7 - 8 = 14.7 ft above dredge line.

Alternately, point of maximum interlock tension is

$$H_{fs}/4 = 60/4 = 15 \text{ ft above dredge line}$$

Use 15 ft above dredge line as point of maximum interlock tension. Check cell embedment

$$d \geq 5[EI/n_h]^{1/5} \qquad \text{(Eq 5)}$$

$$n_h = \frac{b_s}{d'} \ell_h = \frac{1.25 \text{ ft}}{8 \text{ ft}} (5.79) \frac{\text{lb}}{\text{in.}^3} = 0.9 \frac{\text{lb}}{\text{in.}^3} \qquad \text{(Eq 8)}$$

$$d \geq 5 \sqrt[5]{\frac{30,000,000(3.6)}{2(0.9)}} = 179.7 \text{ in.} = 15 \text{ ft}$$

See note about doubling Terzaghi's values

$$d = 35 \text{ ft} > 15 \text{ ft}$$

$$P_{max} = K[120(30) + 68.6(15)]$$

$$P_{max} = 0.43[120(30) + 68.6(15)] = 1,990 \text{ lb/ft}^2$$

$$t_{max} = P_{max} r \qquad \text{(Eq 12)}$$

$$t_{max} = 1,990(31.63) = 62,943.7 \text{ lb/ft} = 5.2 \text{ k/in.}$$

$$FS = \frac{t_{ult}}{t_{max}} = \frac{16 \text{ k/in.}}{5.2 \text{ k/in.}} = 3.1 \qquad \text{(Eq 3)}$$

Check common wall

$$t_{cw} = P_{max} L \qquad \text{(Eq 14)}$$

$$t_{cw} = 1,990(43.3) = 86,167 \text{ lb/ft} = 7.2 \text{ k/in.}$$

$$FS = \frac{t_{ult}}{t_{cw}} = \frac{16 \text{ k/in.}}{7.2 \text{ k/in.}} = 2.2$$

Slip on Vertical Center Plane (Part IV)

Water level outside of cofferdam at Elev 428 ft
Water level inside cofferdam cell at Elev 380*
Water level inside of cofferdam at Elev 353 ft

$$P_w = \frac{1}{2}(62.4)(93)^2 = 269,849 \text{ lb/ft}$$

$$P'_a = \frac{1}{2}(0.27)(35)^2(68.6) = 11,345 \text{ lb/ft}$$

$$P_{w1} = \frac{1}{2}(62.4)(18)^2 = 10,109 \text{ lb/ft}$$

$$K_p = \tan^2\left(45 + \frac{\phi}{2}\right) = 3.69 \text{ (for level backfill)}$$

$$K_{p'} = \frac{\cos^2 \phi}{\left[1 - \sqrt{\frac{\sin\phi \sin(\phi - \beta)}{\cos\beta}}\right]} = 1.72$$

(For a scoping backfill with a scope of 1 to 2.5, $\beta = 21.8°$)

---

* After I worked this example, I remembered a rule of thumb for selecting the water level in a sand-filled cell on a sand foundation with good drainage. The water level inside the cell is horizontal and is the average of the water level inside and outside of the cofferdam.

$$\text{Water level inside of the cell} = \frac{428 + 353}{2} = 390.5 \text{ ft}$$

so I could have used Elev 390.5 ft instead of Elev 380.

Examining the geometry of the berm and the failure surfaces for the above assumptions indicates K_p is closer to 3.69 than 1.72, so use $K_p = 3.0$. Exact values of P'_p can be calculated using the trial wedge method.

$$P^*_p = \frac{1}{2}(3.0)(120)(32)^2 + 18(3.0)(120)(32) + \frac{1}{2}(18)^2(3.0)(68.6)$$

$$P^*_p = 184,320 + 207,360 + 33,340 = 425,020 \text{ lb/ft}$$

$$P'_p = P_w + P'_a - P_{wl} - T^* \qquad \text{(Eq 27)}$$

Let $T^* = 0$

$$P'_p = 269{,}849 + 11{,}345 - 10{,}109 - 0 = 271{,}085 \text{ lb/ft}$$

$$P'_p = 271{,}085 < 425{,}020 = P^*_p$$

Use $P'_p = 271{,}085$ lb/ft

$H_{wo} = 93$ ft

$d' = 35$ ft

$H_{wl} = 18$ ft

$$H'_p = \frac{184{,}320(32/3 + 18) + 207{,}360(9) + 33{,}340(18/3)}{425{,}020}$$

$H'_p = 17.3$ ft

$$M = 269{,}849 \left(\frac{93}{3}\right) + 11{,}345 \left(\frac{35}{3}\right) - 271{,}085(17.3) - 10{,}109 \left(\frac{18}{3}\right) \qquad \text{(Eq 28)}$$

$M = 3{,}747{,}253$ ft lb/ft

$$S' + S'' = \frac{3M}{2b} = \frac{3(3{,}747{,}253)}{2(54.9)} \qquad \text{(Eq 23)}$$

$S' + S'' = 102{,}384$ lb/ft

$$K = \frac{\cos^2 \phi}{2 - \cos^2 \phi} = \frac{\cos^2 35}{2 - \cos^2 35} = 0.50 \qquad \text{(Eq 31)}$$

$$P'_c = \tfrac{1}{2}(0.5)(50)^2(120) + \tfrac{1}{2}(0.5)(45)[2(50)(120) + 45(68.6)]$$

$P'_c = 244,729$ lb/ft

$S'_m = P'_c \tan \phi = 244,729 \tan 35°$ (Eq 29)

$S'_m = 171,361$ lb/ft

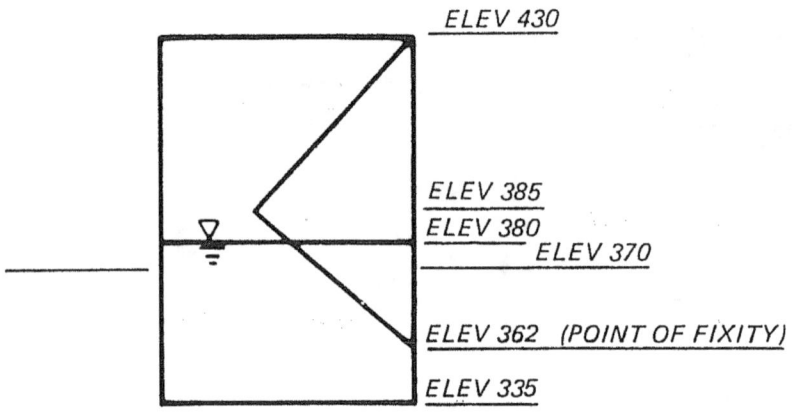

$K = 0.43$

$T_{cw} = 43.3 \left[\frac{1}{2} (0.43)(45)^2 (120) + \frac{1}{2} (0.43)(45)(23)(120) \right]$

$T_{cw} = 3,418,448$ lb

$S''_m = \frac{fT_{cw}}{L} = \frac{0.3(3,418,448)}{43.3}$ (Eq 30)

$S''_m = 23,684$ lb/ft

$FS = \dfrac{S'_m + S''_m}{\frac{3M}{2b}} = \dfrac{171,361 + 23,684}{102,384}$ (Eq 17)

$FS = 1.91$

Slip on Horizontal Planes in Fill (Part V)

$M = 3,747,253$ ft-lb/ft (Eq 28)

$$\gamma_e = \frac{120(50) + 68.6(45)}{95} = 95.65 \text{ lb/ft}^3$$

$$M_{shear} = \frac{Hb^2 \gamma_e}{6}\left(3 \tan^2 \phi - \frac{b}{H} \tan^3 \phi\right) \qquad \text{(Eq 45)}$$

$$M_{shear} = \frac{95(54.9)^2(95.65)}{6}\left(3 \tan^2 35° - \frac{54.9}{95} \tan^3 35°\right)$$

$M_{shear} = 5,808,342$ ft-lb/ft (Cummings Method)

Alternate method for calculating M_{shear} (Incremental Method)   (Eq 44)

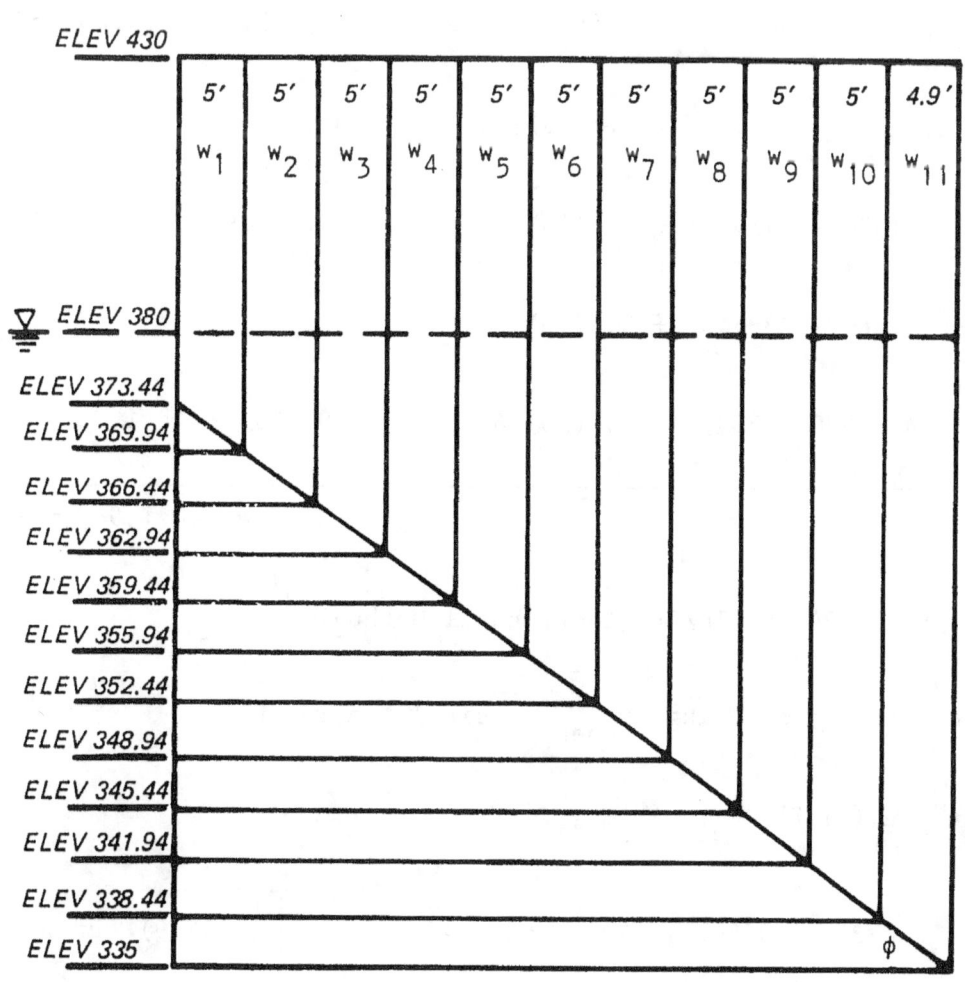

Item	Factor (Weight)	(Wt)(tan φ)	Arm	Moment
W_1	$\overbrace{[5(430 - 380)(120)]}^{30,000}$ $+ [5(380 - 369.94)(68.6)]$	23,422.35	36.69	859,366
W_2	$30,000 + [5(380 - 366.44) + 3.5(5)]68.6$	25,103.55	33.19	833,187
W_3	$30,000 + [5(380 - 362.94) + 3.5(10)]68.6$	26,784.74	29.69	795,239
W_4	$30,000 + [5(380 - 359.44) + 3.5(15)]68.6$	28,465.94	26.19	745,523
W_5	$30,000 + [5(380 - 355.94) + 3.5(20)]68.6$	30,147.14	22.69	684,039
W_6	$30,000 + [5(380 - 352.44) + 3.5(25)]68.6$	31,828.34	19.19	610,786
W_7	$30,000 + [5(380 - 348.94) + 3.5(30)]68.6$	33,509.54	15.69	525,765
W_8	$30,000 + [5(380 - 345.44) + 3.5(35)]68.6$	35,190.74	12.19	428,975
W_9	$30,000 + [5(380 - 341.94) + 3.5(40)]68.6$	36,871.93	8.69	320,417
W_{10}	$30,000 + [5(380 - 338.44) + 3.5(45)]68.6$	38,553.13	5.19	200,091
W_{11}	$30,000 + [4.9(380 - 335) + 3.44(50)]68.6$	39,859.66	1.72	68,559

$\Sigma = 6,071,947$ ft-lb/ft

$M_{shear} = 6,071,947$ ft lb/ft (Incremental Method)

Alternate method for calculating M_{shear} (single increment)

$$M_{shear} = \tan(\phi) \, \Sigma[(\Delta W_e)_i \, (H - y_i)] \qquad \text{(Eq 44)}$$

$$M_{shear} = \tan 35° \, [50(54.9)(120) + 68.6(45)(54.9)] \left[\frac{1}{2}(54.9)\tan 35°\right]$$

$M_{shear} = 6,713,872$ ft lb/ft (single increment)

Use Eq 45 for M_{shear}

$T_{cw} = 3,418,448$ lb

$$M_f = \frac{fbT_{cw}}{L} = \frac{0.3(54.9)(3,418,448)}{43.3} \quad \text{(Eq 47)}$$

$M_f = 1,300,273$ ft-lb/ft

$$FS = \frac{M_f + M_{shear}}{M} \quad \text{(Eq 32)}$$

$$FS = \frac{1,300,273 + 5,808,342}{3,747,253}$$

$FS = 1.90$

Slip Between Sheeting and Fill (Part VI)

$M = 3,747,253$ ft-lb/ft

$P'_a = 11,345$ lb/ft

$K = 1.6K_a = 1.6(0.27) = 0.43$

$P_s = \frac{1}{2}(0.43)(50)^2(120) + \frac{1}{2}(45)[2(0.43)(50)(120) + 0.43(68.6)(45)]$

$P_s = 645,000 + 145,967 = 210,467$ lb/ft

$$FS = \frac{b\left[P'_a \tan\delta + P_s + P_s \frac{b}{L} \tan\delta\right]}{M} \quad \text{(Eq 48)}$$

$$FS = \frac{54.9\left\{11,345 \tan 23.3° + 210,467 + 210,467\left[\left(\frac{54.9}{43.3}\right)\right]\tan 23.3°\right\}}{3,747,253}$$

$FS = 2.08$

Pullout of Outboard Sheeting (Part VII)

From Part IV

P_w = 269,849 lb/ft

P'_a = 11,345 lb/ft

P_{wl} = 10,109 lb/ft

P'_p = 271,085 lb/ft

H_{wo} = 93 ft

d = 35 ft

H_{wl} = 18 ft

H'_p = 17.3 ft

$ML = L\left(P_w H_{wo}/3 + P'_a d/3 - P'_p H'_p - P_{wl} H_{wl}/3\right)$

ML = 43.3[269,849(93)/3 + 11,345(35)/3 − 271,085(17.3) − 10,109(18)/3]

ML = 162,256,048 ft-lb

From Part VI

P_s = 210,467 lb/ft

$Q_{uo} = (P'_a + P_s) \tan \delta$ \hfill (Eq 51)

Q_{uo} = (11,345 + 210,467) tan 23.3° = 951,527 lb/ft

$Q_{uc} = 2P_s \tan \delta$ \hfill (Eq 52)

$$Q_{uc} = 2(210,467) \tan 23.3 = 181,283 \text{ lb/ft}$$

$$M_r b = b\left(Q_{uo}L + \frac{1}{2} Q_{uc}b\right)$$

$$M_r b = 54.9\left[95,527(43.3) + \frac{1}{2}(181,283)(54.9)\right]$$

$$M_r b = 500,278,306 \text{ ft-lb}$$

$$FS = \frac{M_{rb}}{ML} \qquad \text{(Eq 49)}$$

$$FS = \frac{500,278,306}{162,256,048} = 3.08$$

Same value of FS as in Part VI because these two equations are the same.

Penetration of Inboard Sheeting (Part VIII)

From Part IV

$$P'_p = 271,085 \text{ lb/ft}$$

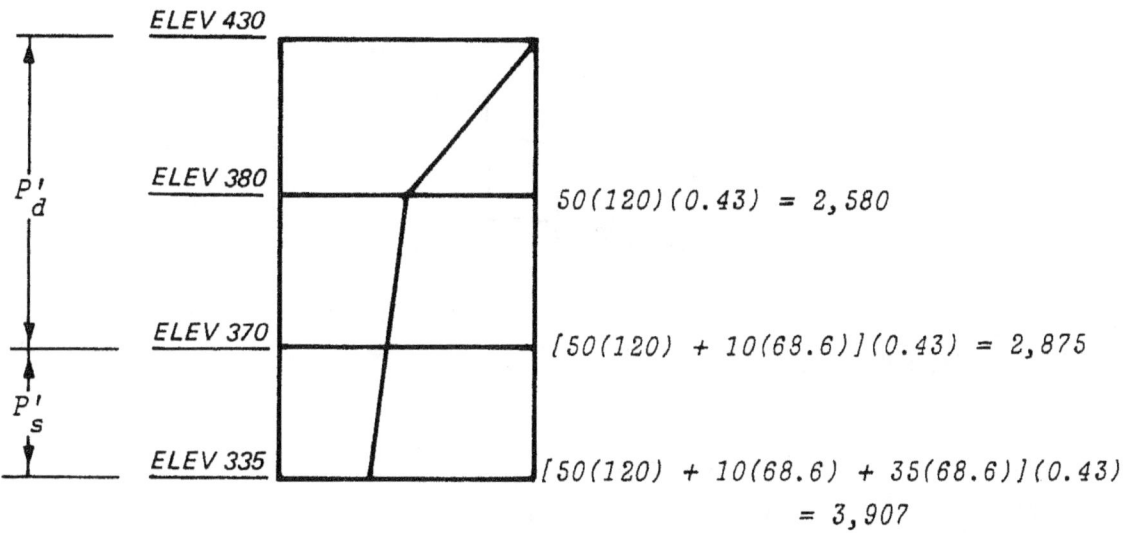

$$P'_d = \frac{1}{2}(0.43)(50)^2(120) + \frac{1}{2}(0.43)(10)[2(50)(120) + 10(68.6)]$$

$$P'_d = \frac{1}{2}(50)(2,580) + \frac{1}{2}(10)(2,580 + 2,875)$$

$$P'_d = 91,775 \text{ lb/ft}$$

$$P'_s = \frac{1}{2}(35)[2,875 + 3,907] = 118,685 \text{ lb/ft}$$

$$FS = \frac{(P'_p + P'_s) \tan \delta}{P'_d \tan \delta} \tag{Eq 53}$$

$$FS = \frac{(271,085 + 111,685) \tan 23.3°}{91,775 \tan 23.3°}$$

$$FS = 4.17$$

Bearing Failure of Foundation (Part IX)

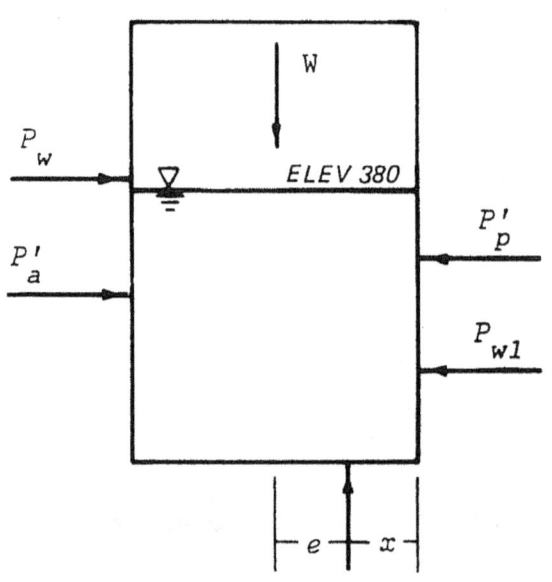

From Part IV

$$P_w = 269,849 \text{ lb/ft}$$

$$P'_a = 11,345 \text{ lb/ft}$$

$$P_{w1} = 10,109 \text{ lb/ft}$$

$P'_p = 271{,}085$ lb/ft

$H_{wo} = 93$ ft

$d = 35$ ft

$H_{wl} = 18$ ft

$H'_p = 17.3$ ft

$W = 54.9[50(120) + 45(68.6)] = 498{,}876$ lb/ft

$R_v = W = 498{,}876$ lb/ft

$R_H = 269{,}849 + 11{,}345 - 271{,}085 - 10{,}109 = 0$

Find the eccentricity

$$R_v x = 498{,}876 \left(\frac{54.9}{2}\right) + 271{,}085(17.3) + 10{,}109 \left(\frac{18}{3}\right)$$

$$- 269{,}849 \left(\frac{93}{3}\right) - 11{,}345 \left(\frac{35}{3}\right)$$

$R_v x = 9{,}946{,}893$ ft-lb/ft

$x = \dfrac{9{,}946{,}893}{498{,}876} = 19.9$ ft

$e = \dfrac{54.9}{2} - 19.9 = 7.51$ ft

Effective base width due to eccentricity

$B' = b - 2e = 54.9 - 2(7.51)$ (Eq 55)

$B' = 39.88$ ft

$$q_{act} = \frac{498,876}{39.88} = 12,509 \text{ lb/ft}^2$$

For $\phi = 35°$

$$N_\phi = \tan^2\left(45 + \frac{35}{2}\right) = 3.69$$

$$N_q = e^{\pi \tan 35°}(3.69) = 33.3$$

$$N_\gamma = (33.3 - 1)\tan[(1.4)(35)] = 37.2$$

$$\zeta_q = \zeta_\gamma = 1 + 0.1(3.69)\left(\frac{39.88}{\text{large number}}\right) = 1$$

$$\zeta_{qd} = \zeta_{\gamma d} = 1 + 0.1\left(\frac{35}{39.88}\right)\tan\left(45 + \frac{35}{2}\right) = 1.17$$

$$\zeta_{qi} = \zeta_{\gamma i} = 1$$

$$\zeta_{qt} = \zeta_{\gamma t} = 1$$

$$\zeta_{qg} = \zeta_{\gamma g} = (1 - \tan \beta)^2 = ? \text{ Say} = 1$$

$$q_{ult} = 1(1.7)(1)(1)(1)(35)(68.6)(33.3)$$

$$+ \frac{1(1.17)(1)(1)(1)(39.88)(68.6)(37.2)}{2}$$

$$q_{ult} = 93,545 + 59,536 = 153,081 \text{ lb/ft}^2$$

$$FS = \frac{q_{ult}}{q_{eff}} = \frac{153,081}{12,509} \qquad \text{(Eq 54)}$$

$$FS = 12.2$$

Sliding Stability (Part X)

See Figure 34

A sliding stability analysis was made of an equivalent cofferdam as shown in Figure 34 using a computer program. The results of the analysis are as follows:

Sliding Surface	FS
9 - 4 - 5 - 7	2.28
10 - 3 - 5 - 8	2.46
11 - 2 - 5 - 8	2.60

$$FS = 2.28$$

Overturning (Part XII)

From Part IX

$$e = 7.51 \text{ ft}$$

$$b = 54.9 \text{ ft}$$

$$\frac{b}{6} = \frac{54.9}{6} = 9.15 \text{ ft}$$

$$7.51 \text{ ft} < 9.15 \text{ ft}$$

APPENDIX B: NOTATION

- a   Vertical distance from pole to base of cell (Figure 36)
- A'   Area between circular failure surface and base of cell (Figure 36)
- b   Equivalent width of cofferdam (width of fictitious straight-walled cofferdam of same plan area) (Figure 1)
- b_s Width of a single sheet-pile
- B   Total width of cellular cofferdam (Figure 1)
- c_h Constant of horizontal subgrade reaction for anchored bulkhead with free earth support
- d   Depth of embedment (Figure 5)
- e   Eccentricity of bearing force. Also, base of natural logarithms
- E   Modulus of elasticity of the sheet-pile
- E_s A horizontal spring modulus representing the behavior of the soil sheet-pile system
- f   Coefficient of friction of interlocks (steel-on-steel)
- f*   Coefficient of friction of fill on rock
- F   Resultant friction force in interlocks of crosswall
- F*   Shear force on horizontal planes within the fill (Figure 16)
- FS   Factor of safety
- H   Vertical distance from sheet-pile tips to

H_e	Vertical distance from plane of fixity to top of cell (effective length of the sheet piles) (Figure 5)	P'_p	Horizontal effective force (per unit length of cofferdam) of the foundation soil acting on the outside of the inboard sheeting
H'_p	Vertical distance from sheet-pile tips to line of action of P'_p (Figure 28c)	P^*_p	Horizontal effective force (per unit length of cofferdam) of berm and foundation soil on inboard sheeting, calculated using passive earth-pressure coefficient
H_w	Vertical distance from dredgeline to surface of water outside of cell (Figure 5)		
H_{fs}	Vertical distance from dredgeline to top of cell (free-standing height) (Figure 5)	P_s	Horizontal effective force (per unit length) of cell fill and foundation material within an equivalent cofferdam) cell acting on a cell wall (Figure 28)
H_{wc}	Vertical distance from plane of fixity to intersection of phreatic surface with center line of cell (Figure 5)		
H_{wo}	Vertical distance from sheet-pile tips to water level outside of cofferdam (Figure 28)	P'_s	Horizontal effective force (per unit length of cofferdam of the foundation soil acting on the interior of the inboard wall (Figure 30)
H_{w1}, H_{w2}	Vertical distances from sheet-pile tips to intersection of phreatic surface with inboard sheeting (Figure 11)	P_w	Resultant force (per unit length of cofferdam) from water pressure acting on exterior of outboard wall (Figure 11)
H_{w3}	Vertical distances from sheet-pile tips to intersection of phreatic surface with cell center line (Figure 11)	P_{wl}	Resultant force (per unit length of cofferdam (from water pressure acting on exterior of inboard wall (Figure 11 and Equation 25)
H_{w4}	Vertical distance from sheet-pile tips to water level inside of cell	P_{max}	Maximum lateral pressure acting against the wall
I	Moment of inertia of the sheet-pile section	P_1, P_2	Pressures used in derivation of Cumming's method (Figure 16)
k_{sl}	Basic value of coefficient of vertical subgrade reaction	Q	Force from foundation acting on half of fill (Figure 10)
K	Lateral earth-pressure coefficient		
K_a	Active earth-pressure coefficient	Q_{uc}	Ultimate sheet-pile pullout capacity (per unit length) of the common wall
K_p	Passive earth-pressure coefficient		
L	Average distance between crosswalls (Figure 1)	Q_{uo}	Ultimate sheet-pile pullout capacity (per unit length of cofferdam) of outboard sheeting
M	Overturning moment (per unit length of cofferdam) (Equation 22)	r	Radius of cell. Also, radial distance (polar coordinate)
M'	Moment caused by the driving forces and effective weight about the center of the circle of rupture (Figure 36)	R	The resultant force of the horizontal and vertical forces acting on the cell
M^*	Total moment acting on outboard sheet pile in Cummings' method	S'	Vertical shearing force (per unit length of cofferdam)
M_{ao}, M_{ai}	Moments caused by active pressure of the fill acting on the inside of the outboard and inboard walls, respectively	S''	Friction force (per unit length of cofferdam) from interlocks
		S'_m	Maximum possible value of shearing force on vertical center plane of cell
M_f	Moment caused by the friction force in the interlocks of the crosswall	S''_m	Maximum possible value of friction force from interlocks in crosswall
M_r	The resisting moment (per unit length of cofferdam) due to pullout of the equivalent cofferdam outboard and commonwall sheeting	t_{cw}	Interlock tension (per unit length of sheet) in crosswall
		t_{max}	Maximum interlock tension (per unit length of sheet) existing in the cell walls
M_{wo}	Moment caused by water pressure acting on the inside of the outboard wall	t_{ult}	Maximum permissible interlock tension (per unit length of sheet) as specified by the sheet-pile manufacturer
M_{shear}	Moment caused by the pressure of that portion of the fill which fails in shear on horizontal planes		
		T	Resultant tensile force in interlock of a single sheet pile
M_w	Moment of effective weight of fill above circular failure surface	T^*	Horizontal shear force on base of cofferdam (per unit length of cofferdam)
M_L	Overturning moment		
n_h	Constant of horizontal subgrade reaction	T_{cw}	Resultant tensile force in crosswall
N'	Resultant effective soil force acting on the base of the cofferdam (per unit length of cofferdam)	U	Resultant uplift force due to water pressure acting on the base of the cofferdam (per unit length of cofferdam)
P^*	Horizontal total force (per unit length of cofferdam) acting on inside of cell, in Cumming's method (Figure 19)	W	Weight of contents of cell
		W'	Effective weight of fill above circular failure surface (Figure 37)
P'_a	Horizontal effective force (per unit length of cofferdam) of foundation soil on outboard sheeting, calculated using active earth-pressure coefficient (Figure 11 and Equation 26)	W_e	Effective weight of fill
		x'	Distance from the plane of fixity to the point of maximum interlock tension
P'_c	Horizontal effective force (per unit length of cofferdam) acting on center plane of cell (Figure 12)	y	Distance measured downward from the top of the cell
		δ	Angle of friction between soil and sheetpiling
P'_d	Horizontal effective force (per unit length of cofferdam) of the fill acting on the inboard wall (Figure 30)	Δ	Small increment

ΔH_w Differential water head between inside and outside of the cell the water level inside of the cell minus the water level outside of the cell
γ' Effective unit weight of soil
γ_e Effective unit weight of soil = weighted average of γ_m above the phreatic line and γ' below the phreatic line
γ_f Unit weight of dry fill
γ'_f Submerged unit weight of fill
γ_m Unit weight of moist fill
γ'_s Submerged unit weight of foundation soil
γ_w Unit weight of water
θ Angle measured from the cofferdam axis to the connecting pile; also, angle in polar coordinate system
ϕ Angle of internal friction of fill

REFERENCES

Belz, C. A. 1970. "Cellular Structure Design Methods," Proceedings, Conference on Design and Installation of Pile Foundations and Cellular Structures, H. Y. Fang and T. D. Dismuke, eds., Envo Publishing Co., pp 319-338.

Clough, G. W., and Duncan, J. M. 1977. "A Finite Element Study of the Behavior of the Willow Island Cofferdam," Technical Report No. CE-218, Department of Civil Engineering, Stanford University, Stanford, Calif.

_____. 1978 (Dec). "Finite Element Analyses of Retaining Wall Behavior," Journal, Soil Mechanics and Foundations Division, American Society of Civil Engineers, Vol 97, No. SM2, pp 1657-1672.

Cummings, E. M. 1957 (Sep). "Cellular Cofferdams and Docks," Journal, Waterways and Harbors Division, American Society of Civil Engineers, Vol 83, No. WW3, Paper No. 1366, pp 13-45.

Dismuke, T. D. 1975. "Cellular Structures and Braced Excavations," Foundation Engineering Handbook, J. F. Winterkorn and H. Fang, eds., Reinhold, New York, pp 451-452.

Duncan, J. M., and Chang, C. Y. 1970 (Sep). "Nonlinear Analysis of Stress and Strain in Soils," Journal, Soil Mechanics and Foundations Division, American Society of Civil Engineers, Vol 96, No. SM5, Paper No. 7513, pp 1625-1653.

Duncan, J. M., et al. 1980 (Aug). "Strength, Stress-Strain, and Bulk Modulus Parameters for Finite Element Analyses of Stresses and Movements in Soil Masses," Report No. UCB/GT/80-01, University of California, Berkeley, Calif.

Erzen, C. Z. 1957 (Sep). Discussion of "Cellular Cofferdams and Docks," by E. M. Cummings, Journal, Waterways and Harbors Division, American Society of Civil Engineers, Vol 83, No. WW3, pp 37-43.

Esrig, M. 1970 (Nov). "Stability of Cellular Cofferdams Against Vertical Shear," Journal, Soil Mechanics and Foundations Division, American Society of Civil Engineers, Vol 96, No. SM6, Paper No. 7654, pp 1853-1862.

Grayman, R. 1970. "Cellular Structure Failures," Proceedings, Conference on Design and Installation of Pile Foundations and Cellular Structures, H.Y. Fan and T.D. Dismuke, eds., Envo Publishing Co., Lehigh Valley, pp

Hansen, J. B. 1953. Earth Pressure Calculations, Danish Technical Press, Institution of Danish Civil Engineers, Copenhagen.

_____. 1957. "The Internal Forces in a Circle of Rupture," Bulletin No. 2, Danish Geotechnical Institute, Copenhagen.

Hansen, L.A., and Clough, G. W. 1982. "Finite Element Analyses of Cofferdam Behavior," Proceedings, Fourth International Conference on Numerical Methods in Geomechanics, Edmonton, Canada, Vol 2, pp 899-906.

Headquarters, Department of the Army. 1958 (Jul). "Design of Pile Structures and Foundations Manual," EM 1110-2-2906, Washington, DC.

_____. 1070 (Apr). "Stability of Earth and Rock-Fill Dams," EM 1110-2-1902, Washington, DC

_____. 1974 (Apr). "Cellular Sheetpile Cofferdam Failures," Washington, DC.

_____. 1981 (Jun). "Sliding Stability for Concrete Structures," ETL 1110-2-256, Washington, DC.

Headquarters, Department of the Navy. 1971 (Mar). Soil Mechanics, Foundation and Earth Structures, NAVDOCKS Design Manual DM7, Bureau of Yards and Docks, Washington, DC.

Hetenyi, M. 1946. "Beams on Elastic Foundation," University of Michigan Press, Ann Arbor, Mich.

Heyman, S. 1957 (Sep). Discussion of "Cellular Cofferdams and Docks," by E. M. Cummings, Journal, Waterways and Harbors Division, American Society of Civil Engineers, Vol 83, No. WW3, pp 34-37.

Jumikis, A. R. 1971. Foundation Engineering, Intext Educational Publishers, Scranton, Pa.

Khuayjarernpanishk, T. 1975. Behavior of a Circular Cell Bulkhead During Construction, Ph.D. Dissertation, Oregon State University, Corvallis, Oreg.

Krynine, D. P. 1945. Discussion of "Stability and Stiffness of Cellular Cofferdams," by K. Terzaghi, Transactions, American Society of Civil Engineers, Vol 110, Paper No. 2253, pp 1175-1178.

Kurata, S., and Kitajima, S. 1967 (Sep). "Design Method for Cellular Bulkhead Made of Thin Steel Plate," Proceedings, Third Asian Regional Conference on Soil Mechanics and Foundation Engineering, Haifa, Vol I, Divisions 1-7.

Lacroix, Y., Esrig, M. I., and Lusher, U. 1970 (Jun). "Design, Construction and Performance of Cellular Cofferdams," Proceedings, ASCE Specialty Conference on Lateral Stresses in the Ground and Earth Retaining Structures, American Society of Civil Engineers, pp 271-328.

Maitland, J. K. 1977. Behavior of Cellular Bulkheads in Sands, Ph.D. Dissertation, Oregon State University, Corvallis, Oreg.

Maitland, J. K., and Schroeder, W. L. 1979 (Jul). "Model Study of Circular Sheetpile Cells," Journal, Geotechnical

Engineering Division, American Society of Civil Engineers, Vol 105, No. GT7, pp 805-821.

Matlock, H., and Reese, L. 1969. "Moment and Deflection Coefficients for Long Piles," Handbook of Ocean and Underwater Engineering, J. J. Myers, C. H. Holm, and R. F. McAllister, eds., McGraw-Hill, New York.

Moore, B. H., and Alizadeh, M. M. 1983 (Sep). "Design of Cellular Cofferdam Instrumentation," Proceedings, International Symposium of Field Measurements in Geomechanics, pp 503-512.

Mosher, R. L., and Pace, M. E. 1982 (Jun). "User's Guide: Computer Program for Bearing Capacity Analyses of Shallow Foundations (CBEAR)," US Army Engineer Waterways Experiment Station, Vicksburg, Miss.

Naval Research Laboratory. 1979 (Jan). "Trident Cofferdam Analysis," NRL Memorandum Report 3869, Naval Research Laboratory, Washington, DC.

Ovesen, N. K. 1962. "Cellular Cofferdams, Calculation Methods and Model Tests," Bulletin No. 14, Danish Geotechnical Institute, Copenhagen.

Rimstad, I. A. 1940. Zur Bemessung des Doppelten Spunwandbauwerkes, Akademiet for de Tekniske Videnskaber, Copenhagen.

Rossow, M. P. 1984 (Oct). "Sheetpile Interlock Tension in Cellular Cofferdams," Journal, Geotechnical Engineering Division, American Society of Civil Engineers, Vol 110, No. GT10, pp 1446-1458.

Schroeder, W. L., and Maitland, J. K. 1979 (Jul). "Cellular Bulkheads and Cofferdams," Journal, Geotechnical Engineering Division, American Society of Civil Engineers, Vol 105, No. GT7, pp 823-838.

Schroeder, W. L., Marker, D. K., and Khuayjarernpanishk, T. 1977 (Mar). "Performance of a Cellular Wharf," Journal, Geotechnical Engineering Division, American Society of Civil Engineers, Vol 103, No. GT3, Paper No. 12790, pp 153-168.

Scott, R. F. 1981. "Foundation Analysis," Prentice-Hall, Inc., Englewood Cliffs, New Jersey.

Sorota, M. D., and Kinner, E. B. 1981 (Dec). "Cellular Cofferdam for Trident Drydock: Design," Journal, Geotechnical Engineering Division, American Society of Civil Engineers, Vol 197, No. GT12, Paper No. 16758, pp 1643-1655.

Sorota, M. D., Kinner, E. B., and Haley, M. X. 1981 (Dec). "Cellular Cofferdam for Trident Drydock: Performance," Journal, Geotechnical Engineering Division, American Society of Civil Engineers, Vol 107, No. GT12, Paper No. 16733, pp 1657-1676.

Swatek, E. P. 1970. "Summary-Cellular Structure Design and Installation," Proceedings, Conference on Design and Installation of Pile Foundations and Cellular Structures, H. Y. Fang and T. D. Dismuke, eds., Envo Publishing Co., Lehigh Valley, pp 413-423.

Tennessee Valley Authority. 1957 (Dec). "Steel Sheet Piling Cellular Cofferdams on Rock," TVA Technical Monograph No. 75, Vol 1, pp 61-62, 66-67, and 69.

Terzaghi, K. 1945. "Stability and Stiffness of Cellular Cofferdams," Transactions, American Society of Civil Engineers, Vol 110, Paper No. 2253, pp 1083-1119.

Terzaghi, K. 1955 (Dec). "Evaluation of Coefficients of Subgrade Reaction," Geotechnique, Vol 5.

United States Steel. 1972. USS Sheet Piling Handbook, Pittsburgh, Pa., p 70. (NEW PILE BUCK STEEL SHEET PILING DESIGN MANUAL)

US Army Engineer Division, Ohio River, 1974 (Apr). "An Analysis of Cellular Sheet Pile Cofferdam Failures," Cincinnati, Ohio, under sponsorship of the Civil Works Engineering Studies Program.

US Army Engineer District, St. Louis. 1983 (Nov). "Instrumentation Data Analyses and Finite Element Studies for First Stage Cofferdam," summary report for study of Lock and Dam No. 26 (Replacement).

White, A., Cheney, J. A., and Duke, M. 1971 (Aug). "Field Study of a Circular Bulkhead," Journal, Soil Mechanics and Foundations Division, American Society of Civil Engineers, Vol 87, No. SM4, Paper No. 2902, pp 89-124.

VIBRATORY PILE DRIVING EQUIPMENT

(1989)

Don C. Warrington, Vulcan Iron Works Inc.
©Copyright 1989, Don C. Warrington

EDITORS NOTE

Vibratory installation equipment came on the scene over forty years ago with the development of some very cumbersome vibrators for installing sheet piling in Russia. Development of more useful models did not get going until the 1960's when their advantage for rapid installation of sheet piling became apparant and demand for rental equipment or ownership soared. Today there are over fifty vibratory drivers/extractor models offered by eight manufacturers. This method of installation has dominated the sheet pile field, however it's influence in foundation pile driving is still awaiting a more positive method for determination of dynamic capacity. Pile Buck has had the good fortune to obtain this state of the art paper on this important subject from one of the countries leading authorities. We offer if so that our readers will have complete information on all currently available driving systems.

INTRODUCTION

Vibratory hammers have been used in the installation of deep foundations for about forty years now. They are most commonly used in the installation of sheet piling walls and caissons, although they are also being used in the installation of concrete and wood piling and also of offshore platform piling. They provide an efficient and relatively quiet method of installing piles for deep foundations. The scope of their application increases on a continuous basis.

This article discusses the various aspects of vibratory hammers and their application in foundation installation, including both practical examples and a discussion of the drivability and capacity of piles driven by vibration.

HISTORICAL DEVELOPMENT

Development in the U.S.S.R.

The first vibratory pile driver used was in the Soviet Union, a model BT-5 developed and first used under the direction of D.D. Barkan. This hammer had a dynamic force of 214 kN and the eccentrics rotated at 41.67 Hz, powered with 28 kW of power. Used in the construction of the Gorki hydroelectric development, the hammer drove 3700 sheet piles 9-12 m long in 2-3 minutes each.

As is the case in most of the world, Soviet made vibratory pile drivers can be divided into two groups:

1) Low frequency machines, with a vibrator frequency of 5-10 Hz, used primarily with piles with high mass and toe resistance, such as concrete and large steel pipe piles. An example of this type of machine is the VPM-170 low frequency driver, which is the largest made in the U.S.S.R.. This machine produces a maximum dynamic force of 1,700 kN at its maximum frequency of 9.17 Hz and eccentric moment of 510 kg-m. This machine is designed primarily to drive caissons up to two (2) meters in diameter and is bolted to the pile rather than clamped.

2) High frequency machines, with a vibrator frequency of 10-25 Hz, used for piling such as sheet piles, small pipe piles, etc.. An example of this type is shown in Figure 1, a B-402 vibratory hammer actually driving Larssen type sheet pile on a tunnel construction job in Leningrad. This unit has a maximum dynamic force of 270 kN while operating at its rated oscillation frequency of 23.8 Hz and its maximum eccentric moment of 12 kg-m.

The Soviets first licensed their technology to the Japanese, who with several concerns have developed an extensive array of vibratory hammers. This technology has since spread worldwide, with such concerns as PTC in France, Mueller and Tünkers in Germany, and Hera in the Netherlands being well established in the field.

Vibratory Hammers in the U.S.

The first American made hydraulic vibratory was

Figure 1
Soviet B-402 Vibratory Pile Driver

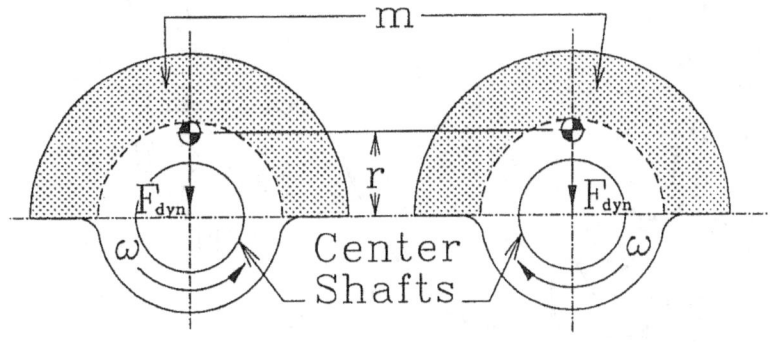

**Figure 2
Eccentric Configuration**

introduced by MKT in 1969, although both Vulcan and Foster had introduced Japanese and French vibratory pile drivers respectively in the early 1960's. The unique practices in the U.S. have lead to the evolution of a distinctive style of vibratory hammer in the U.S. and Europe, and today this is embodied to a greater or lesser degree in the Vulcan, ICE, MKT, Foster, H&M and PSI/Casteel units that are on the market. These characteristics include slim throat hammers for sheet pile driving, hydraulic drive, high power motors, pumps, and engines, and (in most cases) soft, rubber spring suspensions.

Of special interest in the U.S. is the Guild/Bodine resonant driver, first introduced in the early 1960's. As seen above, the original vibratory hammer rotated its eccentrics at 41.67 Hz; however, later machines did not operate at this frequency, but branched into two directions. On the one hand, most vibratory hammers operate in one of the two ranges that the Soviet drivers do, although the upper limit for the "high frequency" drivers has been extended in the U.S. to at least 28 Hz. On the other hand, the resonant driver operated at frequencies starting at 120 Hz and ranging upward, thus inducing mechanical resonance into the pile and facilitating driving. Although in principle this concept has held great potential, the mechanical complexity of this machine has withheld it from extensive use.

Concerning higher frequencies, though, one recent trend in both the U.S. and elsewhere is the development of vibratory pile drivers that operate in the 30-40 Hz range. These have been developed simultaneously both in Europe (ICE, Tünkers, PTC) and in the U.S. (Vulcan). The primary advantage of these machines is their lowered transmission of ground excitation to neighboring structures. Their installation characteristics relative to more conventional machines at lower frequencies have not been established.

OVERVIEW

Basic Concepts

A vibratory pile driver is a machine that installs piling into the ground by applying a rapidly alternating force to the pile. This is generally accomplished by rotating eccentric weights about shafts. Each rotating eccentric produces a force acting in a single plane and directed toward the centerline of the shaft. Figure 2 shows the basic setup for the rotating eccentric weights used in most current vibratory pile driving/extracting equipment. The weights are set off center of the axis of rotation by the eccentric arm. For a rotating body, the force exerted on the center shaft is given by the equation

$$F_{dyn} = (2\pi\theta)^2 mr/1000 \quad (1)$$

where
F_{dyn} = dynamic force of eccentrics, kN
θ = frequency of vibrations, Hz
m = eccentric mass, kg
r = eccentric moment arm, m

If we define

$$K = mr \quad (2)$$

where
K = eccentric moment, kg-m

we can substitute to

$$F_{dyn} = (2\pi\theta)^2 K/1000 \quad (3)$$

If only one eccentric is used, in one revolution a force will be exerted in all directions, giving the system a good deal of lateral whip. To avoid this problem, the eccentrics are paired so the lateral forces cancel each other, leaving us with only axial force for the pile. Machines can also have several pairs of smaller, identical eccentrics synchronized and obtain the same effect as with one larger pair. Thus, the "m" term means the sum of all the eccentric weights, the eccentric arm length for all being equal.

Equipment Details

Although there are many variations in design and construction, the vast majority of vibratory hammers are of the configuration shown in Figure 3. Briefly, there are two main components of the system: the exciter, which produces the actual vibrating force, and the power pack, which provides the usable energy for the motor(s) on the hammer to spin the eccentrics.

The Exciter

The exciter itself is divided into three parts:

1)Vibrator case: This contains the eccentric weights and does the actual vibration. Generally the eccentrics are mounted to and synchronized by a gear system; however, in some vibratory hammers, the gear is made to be an eccentric and thus gear and eccentric are one piece. Other schemes of synchronization are a)there are no gears, and most of the time the amplitude of the system synchronizes their rotation, each eccentric driven by its own motor (Tünkers), or b)the gears are synchronized by a chain and each eccentric is driven individually (H&M). In any case, the dynamic force generated by the eccentrics is transmitted to the case by the use of antifriction bearings, which also facilitate rotation.

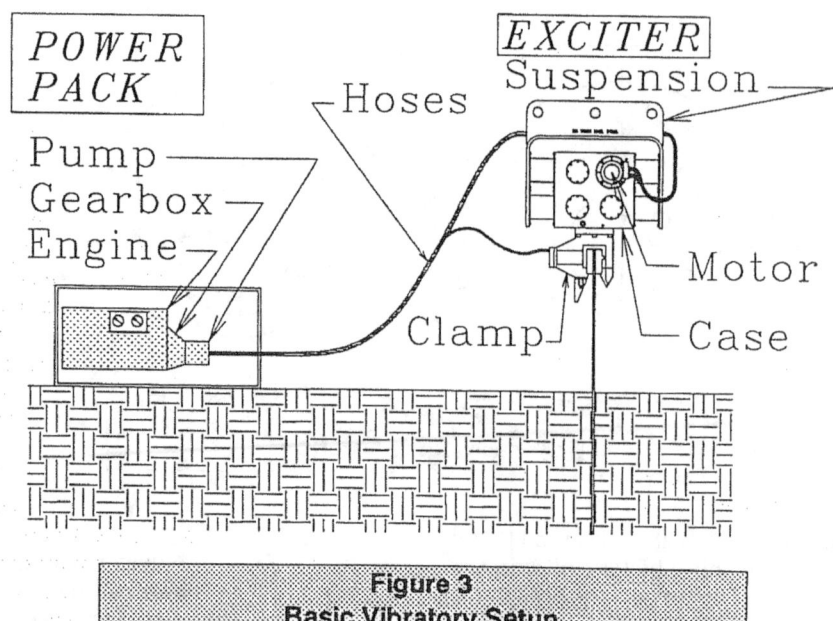

**Figure 3
Basic Vibratory Setup**

**Figure 4
Hydraulic Power Pack**

a pinion, or through belts or chain drives. For the latter two the motor is mounted on the static weight; a pinion drive requires that the motor be mounted directly to the vibrator case.

2) *Clamp:* This connects the vibrator case to the pile and thus transmits the vibrator's power from the vibrator case to the pile. Generally speaking, most clamps pinch the pile using a hydraulic cylinder and jaws, thus making a frictional connection. A few vibrators actually bolt or pin the pile to the vibrator case, as was done with the old Vulcan or MKT impact extractors.

3) *Suspension:* This is connected to the vibrator case by rubber or metal springs. In driving this provides additional weight to the system to force the pile into the ground without degrading the vibration of the system, although with most units additional bias weight can be attached to the suspension. In extraction the suspension system transmits static pull while dampening out vibration and thus protects the crane boom. For this to be effective the springs must be sufficiently soft and the bias weight sufficiently heavy to insure a suspension natural frequency that is much lower than the vibrator's operating frequency.

The Power Pack

A few vibrators, such as the Guild/Bodine resonant drivers, some of the early Soviet vibrodrilling machines, and some Japanese units, drive rotating eccentrics straight from diesel or gasoline engines by mechanical couplings. However, most vibratory hammers transmit energy from the prime mover to the eccentrics through either electric or hydraulic systems. Since construction sites are usually remote, transportable power sources have been developed for vibratory hammers. These are referred to as power packs or generator sets (for electric units).

Electric systems: These usually employ three-phase induction motors driven at a single frequency, which has encouraged the development of many systems to vary the eccentric moment and thus the driving force. In some cases

three-phase mains, obviating the need for a generator set. The hammer thus only requires a switchbox to control it. A separate, small power pack, driven with an electric motor, is required to operate the hydraulic clamp, if there is one. This can either be on the ground or mounted on the static overweight.

Hydraulic systems: For a variety of reasons, in the U.S. hydraulic systems have become dominant, and the major manufacturers, employ hydraulic drive almost exclusively. These systems use a diesel engine to drive a hydraulic pump, which in turn drives the motor on the exciter. Some power packs have the engine connected to a gearbox and others make a direct connection. Also, some units have separate pumps for the hydraulic clamps and some integrate these into the main power source. Both the frequency and the force can be varied either by using variable displacement pumps in the power pack or by simply varying the engine speed. These units can employ air, electric, or manual controls A typical hydraulic power pack is shown in Figure 4.

Applications

Vibratory pile drivers have been used to drive and extract virtually any type of piling used, although the specialty of vibratory and impact hammers is somewhat different. In the United States, the main application of vibratory hammers is the installation of steel, non-displacement piles where a definite bearing capacity is not required. This last condition is because of the state of the art of vibratory capacity prediction; this will be addressed later in the article. Vibratory hammers are also sometimes not as effective as impact ones in stiff, cohesive soils, or for displacement piles that develop a good deal of toe resistance.

Steel Piles

Sheet Piling: Figure 5 shows a hydraulic vibratory hammer installing sheet piling, in this case a Vulcan 2800, which generates 890 kN of dynamic force at 40 Hz with 14.1 kg-m of eccentric moment. The sheeting is set up according to normal American practice, namely to set the wall in place using a template and then to drive the pile to the desired depth. This practice requires that the vibratory hammer be no wider at the throat than about 355mm, as the hammer must clear the adjacent piles. In driving sheeting in this way, it is also normal to drive the sheets two at a time, using a jaw with two sets of teeth and a recess between them large enough to accommodate the interlock.

Figure 1 shows an alternate method of driving sheeting with a vibratory hammer. Here the sheets are set as they are driven. As a rule, in this case the sheets are driven one at a time. No matter how sheet piling are driven, though, they should not be driven at penetration velocities of less than 5 mm/sec.

H-Beams: Figure 6 shows a vibratory hammer driving an H-Beam. The conditions are similar to driving sheeting; however, when the pile's angle is critical, the vibratory hammer can be mounted in a set of leaders much as is done with an impact hammer. In addition to a bearing application (where the beam might be impacted to refusal), vibrated H-Beams are used for soldier beams and in slurry wall construction.

Caissons and Pipe Pile: Figure 7 depicts a hammer driving a caisson. Caissons are a versatile item, extensively used with drilled shafts. To drive these, a special device called a caisson beam is employed. This is a horizontal slide with a set of two

Figure 5

Figure 6

clamps attached to it. The clamps affix the pile to the hammer on opposite sides of the caisson. The clamps are locked to the slide during use but can be moved along the slide to enable a caisson beam setup to drive a variety of pile.

The equipment setup for caissons is duplicated with pipe pile. Generally it is best to vibrate pipes open ended, although some closed ended installation is done. An example application of this is the installation of pipe piles for offshore structures, such as petroleum production platforms. For some of these two caisson beams with two sets of clamps are used, the beams being configured in an "x" arrangement.

Concrete Piles

Concrete pile installation with vibratory hammers is rare in the U.S. but more common abroad. It is done with both prismatic (square and octagonal) and cylinder pile. As concrete pile is always a displacement pile, the vibratory hammer must develop some toe impact by raising and lowering the pile during the vibration cycle, thus allowing penetration. This is generally accomplished using low frequency vibrators with high amplitudes.

Wood Piles

As it is almost exclusively a bearing pile, wood is rarely vibrated in the U.S. Extraction of wood piles, however, is common, and the vibratory hammer is an effective tool for this purpose. The wood pile can be extracted intact in this manner. Special wood clamps are used for this purpose.

EQUIPMENT OPERATION

As we have seen, vibratory hammers vary from manufacturer to manufacturer and model to model in their construction and operation; however, there are things which are common to all vibratory hammers. This section deals with some of these items. The guidelines below presuppose equipment with a remote power unit equipped with an engine. Equipment powered from an external mains will be similar except for the lack of an engine.

Safety

Safety is basically common sense. There are standard safety rules, but each situation is different. Your common sense and experience will be your best guide to safety. Always be alert to problems and correct deficiencies promptly.

Since Soviet machines have been described in detail, the following from the operation manual of the B-402 may prove of interest:

"Persons are allowed to operate the Vibratory Pile Driver who have reached eighteen (18) years, passed the training and the knowledge check in the safety of building, handling, and piling works, and received the certificate for the right of operating the Vibratory Pile Driver, studied the present Technical Description and Operating Instructions, learned by practice the operation of the Vibratory Pile Driver, and have an experience in driving and removing piles.

Work supervisors must carry out a detailed briefing of the persons who are to operate the Vibratory Pile Driver on rules and safe techniques prior to the works. The persons who have not been present at the safety briefing are not allowed to work."

The one thing you must do FIRST

First, take time to take your operating manual, go to the exciter and power pack, and review all of the operating and safety features in the manual, and to likewise familiarize anyone else working on or with the equipment with these features.

Things you should NEVER do

Never allow unauthorized or unqualified people to either operate, maintain, or come within thirty (30) meters of the equipment.

Never allow anyone to stand directly under or within three (3) meters of the hammer during operation. Failure to do so could result in *injury or death* by being struck by falling parts, rocks, or dirt that the hammer has picked up by being laid on the ground.

Never operate the power pack's engine in a closed area. The breathing of the fumes can be fatal.

Never smoke or use open flame when servicing batteries. Proper ventilation is necessary when charging batteries. On units with a power pack enclosure, all of the doors of the unit must be open during battery charging.

Never smoke when filling fuel tank or hydraulic reservoir, or for that matter while anywhere near exciter, hoses, or power pack. Diesel fuel, gasoline, and hydraulic fluid are very flammable.

Never adjust or repair the unit while it is in operation, except with the main motor and clamp controls provided for that purpose. If you need to make any other adjustments, shut the entire system down first.

Never attempt to operate the engine with the governor linkage disconnected.

Never store flammable liquids near the engine.

Never unclamp the exciter from the pile when there is any line pull on the suspension or when the hammer is still

**Figure 7
Vibratory Hammer Driving Caissons**

vibrating.

Things you should ALWAYS do

Always store oily rags in containers. If these get into an hydraulic system, you will have a mess.

Always remove all tools from unit before starting.

Always be sure that, with hydraulic systems, all pressure is out of the system and that all pressure gauges read zero before you start working on the hydraulics of the system. The high pressure fluid in hydraulic lines can be very dangerous if it escapes, such as in a hose break or loosening of a fitting or component. Even when you think the system pressure is zero, you should proceed with extreme caution, assuming all the while that all of the lines are still fully pressurized. This means primarily that you should open all fittings and connections slowly until you see that there is no pressure on that particular connection. Keep your face and body away from the potential line of fire of any fluid while you are working with hydraulic connections.

Always make sure that you make any hose fittings or connections very tight when you reassemble them. Failure to do so can result in the hoses coming loose, resulting in hoses flying around, hydraulic fluid spraying everywhere, and people being injured or killed by all of this. This also applies to any bolted or otherwise fastened connection on either the exciter or power pack.

Always make sure that electrical systems are properly grounded during operation. Also, make sure that they are not connected to any power source in any way and that there is no voltage of any kind in the system before servicing same.

Always make sure that electrical connections and wiring are tight and completely insulated to prevent shock if accidentally touched. This is especially important in waterfront or marine situations; uninsulated wire can result in widespread electrocution of wild and human life. Any electrical fuse, breaker, or control boxes must be *closed* before any kind of operation.

Always be sure to wear gloves and other protective clothing while working on any part of the system, or even better to wait until the system has cooled down. Hydraulic components, electrical wiring and switchgear, and the engine get very hot during operation.

Always make sure that the pile is firmly gripped by the jaws when clamping.

Basic Operation

Rigging

To permit lifting of the hammer, a wire rope must be secured from the crane line to the lifting hole or pin on the suspension. In choosing the wire rope for any unit, a generous safety factor should be used. Several turns of a smaller diameter cable will usually last longer than one turn of a large diameter cable. Make sure that the wire rope assembly you use has at least double the capacity of the suspension.

Engine Start-Up

Read the engine start-up procedure in the manufacturers operation manual. Follow any applicable instructions.

Do not start the power pack if the temperature of the hydraulic oil is below -18° C.. The case oil and/or fluid temperature should be at least 0° C. before starting the exciter at a slow speed. After starting engine, let it run slowly for at least five minutes.

In Use

Complete all applicable preparation for operation as required periodic maintenance before operation. The user should also be thoroughly familiar with the inner workings and control operation of the power pack/generator and exciter before operation. An understanding of how the unit works will give the operator a feel for what is happening and will also be invaluable in troubleshooting the unit should a problem arise.

Do not operate the exciter at full speed if the temperature of the hydraulic fluid and/or case oil is below 15° C.. If the temperature of the hydraulic fluid or case oil is below that level, set the engine at a moderate speed and start the hammer. Allow the exciter to run until the fluid and/or oil reaches the required temperature. Full speed operation is now permissible.

Make sure the exciter is positioned parallel to the pile and that the full length of the jaw will make contact with the pile when clamped. Engage the "clamp close" control. The clamp will close in a couple of seconds. The operator should make sure that the pile is firmly gripped by the jaws.

Once adequate clamping pressure has been reached, the hammer is ready to vibrate. Engage the "start" control. The exciter case and pile will begin vibrating.

If driving, the combined weight of the hammer and pile will force the pile into the ground. As the driving resistance increases, the drive pressure or amperage will increase until it reaches the maximum power output of the power pack. Any increased resistance may cause the hammer to slow down somewhat. The hammer may be operated in this condition for short periods of time; however, extended periods will cause the unit to overheat, causing possible damage to the vibrator components. Also, if the conditions warrant, with hydraulic units, the frequency can be varied by adjusting the engine speed.

If extracting, the crane must exert a net pull on the hammer-pile system. This will cause the pile to move upward. When extracting, in general the best procedure is to start the exciter without crane pull, allowing it to come up to speed, loosen the soil and to drive a little. Once this loosening has taken place, extraction is easier. It is *very important* that the crane not exert an upward pull greater than the rated capacity of the exciter's suspension.

Shutdown

Once driving or extracting is complete, engage the "stop" control. The exciter will stop in a couple of seconds. Make sure there is no net crane pull on the hammer, and the exciter has stopped vibrating. Engage the "clamp open" control, and remove the hammer from the pile. When you're ready to stop for a longer time, also allow the engine to idle for five (5) minutes to cool, reduce engine speed to idle, and turn engine start switch to off.

THEORY OF OPERATION

Vibrator Mechanics

Before we consider the actual method by which vibratory hammers move piles through soil, we must first consider the mechanics of the hammer/pile system itself. Generally speaking, many of the traditionally measured quantities for vibratory hammers such as amplitude, acceleration ratio, etc., are computed for the "free-hanging" case, i.e., with only the mass of the system taken into account and no soil resistance. For most conventional vibratory hammers, one can consider the entire system a rigid mass. This is because the relatively low frequency vibrations of most vibratory hammers do not bring the distributed mass and elasticity of the system into play. By definition, with the sonic pile drivers the resonant properties of the system become significant, and the analysis becomes more complicated.

effects of gravity, the equation of motion is

$$x'' = 1000 F_{dyn} \sin(2\pi\theta t)/M \quad (4)$$

where
x'' = instantaneous acceleration of system, m/sec^2

The solution of this equation is

$$x = K \sin(2\pi\theta\tau)/M \quad (5)$$

where
x = system displacement, m/sec

In the process of integrating Equation (4), we can derive three very important quantities. The first is the ratio of accelerations, or the peak acceleration during a vibratory cycle; it is

$$n = F_{dyn}/W_{dyn} \quad (6)$$

where
n, n_1, n_2 = ratio of maximum acceleration of system to acceleration due to gravity, g's
W_{dyn} = vibrating weight of system, kN = $gM/1000$
g = acceleration of gravity, m/sec^2

The second is the peak velocity, which is

$$v_{dyn} = gn/(2\pi\theta) \quad (7)$$

where
v_{dyn} = peak dynamic velocity during the cycle, m/sec

These quantities are important because the power transmitted to the soil must be done in an efficient manner from a high energy source through the pile-soil interface to a low one in the soil. As the vibratory excitation is dynamic, it must be done through these quantities. Minimum values for n have been established from 1.5 to 9, but there is no consensus on this.

Finally, the maximum displacement is

$$x_{max} = K/M \quad (8)$$

where
x_{max} = maximum displacement of system (zero to peak), m or mm

Since the acceleration, velocity, and displacement of the system solved from Equation (4) are all sinusoidal with respect to time, these quantities are measured from the zero line of the sine wave. Customarily, the maximum cycle displacement of the vibrator, called the amplitude, is measured from peak to peak and is expressed as

$$A = 2x_{max} \quad (9)$$

where
A = amplitude of system (peak to peak), m or mm

The instantaneous torque driving the eccentrics is

$$T_{inst} = (1000 F_{dyn}/(2\pi\theta))^2 \sin(4\pi\theta t)/(2M) \quad (10)$$

where
$T, T_{inst}, T_{max}, T_{rms}$ = motor torque, N-m

The maximum instantaneous torque is

$$T_{max} = (1000 F_{dyn}/(2\pi\theta))^2/(2M) \quad (11)$$

Looking forward to the power requirements, normally one would use an root mean square (rms) value to match an application to a motor, so

$$T_{rms} = T_{max}/\sqrt{(2)} \quad (12)$$

From the torque the power is simple to compute, given by

$$N = \pi\theta T_{rms}/500 \quad (13)$$

where
N = motor power, kW

Adequate power is *essential* for successful vibratory driving because, among other reasons, maintenance of the vibratory frequency is impossible without it. An underpowered machine will slow down and thus reduce its own driving capability

Soil Reaction

To drive piles by impact, it is necessary for the hammer to generate high forces to move the pile blow by blow; on the other hand, vibratory hammers impart energy to the pile-soil system continuously rather than incrementally. Impact drivers also impart their motive force in one direction, namely down, which is where one wants the pile to go. Vibrators, however, are inherently bidirectional in their force generation, so either during driving or extraction half of the force is going the wrong way. Yet vibrators, under the right conditions, are effective for both operations. How can this be?

It is generally agreed that, when the soil is excited by vibration, the resistance is reduced so that the system can either drop by its own weight or come out of the ground from the pull of the crane. The pile is not actually forced into the ground by the vibrating force but by the net applied static force on the system. For driving, gravity acts to push the pile downward into the soil; however, if the crane were to exert a net upward force on the system, the pile moves upward. This enables a vibratory hammer to act as an extractor, which is important in many applications.

The key to solving this problem lies in the soil response to vibration. There are several explanations that have been tendered to explain the mechanism of vibratory driving. The explanation of the resistance reduction varies, though, and three of the mechanisms proposed are as follows:

Thixotropy: This method states that vibratory hammers overcome soil resistance by reducing it through a process called thixotropy. This is observed mostly in clays. Gumenskii and Komarov (1959) describe this process:

> By thixotropy (from the Greek "thixis" -- shaking -- and "trope" -- change) of dispersed systems in general, or of soils in particular, we mean their liquefaction during jarring by some mechanical action (shaking, stirring, etc.). Under such circumstances, at a constant temperature, there occurs a transition from a gel to a sol, which,

Thixotropy should thus be considered to involve two aspects: liquefaction and solidification. These constitute a reversible process, since it is repeated many times.

From this the vibration of the pile, through the liquefaction process, reduces the soil resistance and allows the pile to drop by its own weight.

Amplitude: This concept is most succinctly stated by Erofeev et.al. (1985):

Centrifugal force created by the vibration exciter with the turning of the shafts with unbalanced loads, or eccentrics, attached to them cause the pile to vibrate. The characteristics of these vibrations depend on the static moment of the eccentrics, the frequency of the vibrations as determined by the angular velocity (ω), the weight of the vibratory pile driver-pile system, and the properties of the soil. The amplitude of the system's vibrations is decisive for the insertion of the pile. At a low vibrational amplitude, displacement of the soil with respect to the side surface of the element being inserted does not exceed the limit of its elastic deformation and the pile is not sunk into the ground. As the amplitude of the vibrations increases, residual deformation of the soil occurs and the pile begins to slip relative to the soil, i.e., it is sunk into the ground.

This principle, of course, is also applied to wave equation analysis of impact driving, as no set of the pile is possible without exceeding the elastic limit, or quake, of the soil.

Dynamic Force: This concept is based on the hypothesis that the force generated by the rotating eccentrics breaks the bond between the pile and the soil, thus making vibratory driving (or extraction) possible. A general adjunct to this hypothesis is the modeling of the soil resistance as being dry frictional or Coulombic in nature. Starting with a version of Equation (4) modified to include Coulombic soil resistance, Tseitlin et. al. (1987) describe a mathematical model to compute vibratory penetration. In this discussion, having made the preceding assumptions, they go on as follows:

Experimental research has established that the amplitude of the vibrations A_g of the ground surrounding the pile is comparable to the amplitude of the pile being sunk only during the initial stage in which the pile vibrates together with the ground. As breakthrough progresses, the ground vibrations diminish, while those of the pile increase; in the final stage of breakthrough, the ratio of the amplitude of the pile's vibrations to that of the ground vibrations reaches two to three orders of magnitude, which also permits us to consider the ground surrounding the pile immobile during the sinking process...Experimental data attesting that the elastic component of lateral ground resistance is two orders of magnitude less than the plastic component exemplify the fact that the elastic component of lateral ground resistance is negligible during vibrational sinking at high speeds. As far as the viscous component of ground resistance during vibrational sinking is concerned, it is nonlinear in nature (i.e., it has a soft characteristic)...and it changes little with increases in speed even at low vibration speeds (5-10 cm/sec).

Although each of these models is presented as universally applicable, it is more likely that each of them or a combination thereof is descriptive for various soil conditions. It is also likely that our understanding of the mechanism of vibratory continues.

DRIVABILITY AND CAPACITY PREDICTION

The ability of vibratory machines to drive piling is well demonstrated; however, one major obstacle to the expanded use of these machines is the lack of an accepted method to relate the driving performance of a hammer/pile/soil system to either the resistance of the soil to driving or the static capacity of the pile. This section will discuss some of the methods developed in the past to determine the drivability of vibratory hammers and ultimately the bearing capacity of the piles driven.

Overview of Methods

The methods presented below date from the very earliest application of vibratory technology to pile driving to the present; they take a wide variety of approaches and will give a wide variety of results. For convenience, we will break down these methods into four groups:

1) Parametric methods: Certain characteristics are tested against some kind of standard to determine drivability.

2) Energy methods: Drivability is determined based on the energy flow through the system, along with other considerations.

3) Methods from Laboratory and Model Tests: These methods are derived from correlations gathered from tests in a laboratory setting, usually driving a pile through a soil tank.

4) Time Dependent Nonlinear Methods: These seek to apply numerical integration techniques to the direct solution of the equation(s) of motion of the vibrating system. These include the wave equation techniques popular with impact hammers.

One important observation at this point is that, in general, vibrated piles have lower bearing capacity than impact driven ones. This is because impact driving produces soil compacting at the toe that impact driven piles do not.

Most of the formulae below are reproduced from their sources; however, when necessary they have been altered into a uniform units (SI) and notation system.

Parametric Methods

Tünkers Dynamic Force Method: Basically, this method employs the formula

$$F_{dyn} \geq \sigma A_s \quad (14)$$

where

σ = Unit Soil Shaft Resistance, kPa
A_s = Shaft Area of soil, m^2

Values for the factor σ are given in Table 1. The formula is only applicable when $x_{max} \geq 2.38$mm. To compute the shaft area for sheet piling, this method employs the following computation:

$$A_s = 2.8 l d_{inter} \quad (15)$$

where
l = Pile Length, m
d_{inter} = Width of Sheet piling, m (from interlock to interlock).

Beta Method: A given vibratory hammer is suitable for driving a given pile when

$$F_{dyn} + W_{dyn} + W_{st} \geq \beta_o R_{so} + \beta_i R_{si} + \beta_t R_t \quad (16)$$

where
W_{st} = Non-Vibrating Weight of System, kN
β = Beta Factor for Soil Resistance (general)

β_o = Beta Factor for Soil Resistance (inside shaft)
β_t = Beta Factor for Soil Resistance (toe)
R_{si} = Inside Pile Shaft Soil Resistance, kN
R_{so} = Outside Pile Shaft Soil Resistance, kN
R_t = Pile Toe Soil Resistance, kN.

Suggested values for β are given in Table 2. For extraction, this formula is altered to read

$$F_{dyn} + F_{ext} - W_{dyn} - W_{st} \geq \beta_o R_{so} + \beta_i R_{si} + \beta_t R_t \quad (17)$$

where
F_{ext} = Extraction Force of Crane, kN.

Savinov and Luskin Method: This method was developed in the USSR by two of the pioneers in the development of vibratory pile driving equipment. Presented below is a reformulated version, done for simplicity and clarity. The steps are as follows:

1) Computation of the required minimum dynamic weight: To insure sufficient weight for pile sinking, the minimum dynamic weight of the system is computed by the formula

$$W_{dyn} \geq p_o A_t \quad (18)$$

where A_t = Toe Area of Pile, m^2

p_o = Toe Pressure of System, kPa
(as given in Table 3)

Although the method calls for the weight computed above to be dynamic, there is also the possibility of having part of this weight to be static.

2) Determination of the soil resistance: For vibration purposes, the soil resistance is determined by the following formula:

$$F_{cr} = Z \sum_{i=1}^{i=k} \sigma_i l_i \quad \text{........(Piling in general -- 19a)}$$

$$F_{cr} = \sum_{i=1}^{i=k} \sigma_i l_i \quad \text{....................(Sheet Piling -- 19b)}$$

Table 1
Values of σ for Tünkers Method

SPT Value, Blows/30cm		Soil Resistance σ
Cohesionless Soil	Cohesive Soil	kPa
0-5	0-2	9.86
5-10	2-5	11.87
10-20	5-10	12.83
20-30	10-20	14.84
30-40	20-30	15.80
40+	30+	16.76

Table 2
Values of β_n for Beta Method

Type of Soil	Value for β
Round Coarse Sand	0.10
Soft Loam/Marl, Soft Loess, Stiff Cliff	0.12
Round Medium Sand, Round Gravel	0.15
Fine Angular Gravel, Angular Loam, Angular Loess	0.18
Round Fine Sand	0.20
Angular Sand, Coarse Gravel	0.25
Angular/Dry Fine Sand	0.35
Marl, Stiff/Very Stiff Clay	0.40

where   n = i, o, or t depending upon relative position of pile and soil in question.

Table 3
Values of Pile Toe Weight Pressure for Savinov and Luskin Method

Type of Pile	Pressure p_o, kPa
For Saturated Sandy and Loose-Clayey Soils	
Small Diameter Steel Pipe and other Piles w/$A_t \leq 150$ cm²	150-300
Closed End Pipe Piles, $A_t \leq 800$ cm²	400-500
Square and Rectangular Reinforced Concrete Piles, $A_t \leq 2000$ cm²	600-800

where
- F_{cr} = Critical Force for Driving, kN
- Z = Pile Perimeter, m
- σ_i = Soil Element Shaft Resistance, kPa or kN/m
- l_i = Pile Element Length, m

The length of the pile is first divided into segments of length l_i, then the soil resistance σ for each segment is taken from Table 4, depending upon the type of soil in the given segment.

3) Computation of the dynamic force of the eccentrics: The dynamic force is first computed to meet the following two criterion:

a) Soil resistance factor: The dynamic force should be greater should be greater than the soil resistance, as expressed by the formula

$$F_{dyn} \geq \aleph \, F_{cr}/\psi \qquad (20)$$

where
- ψ = Pile Factor (0.8 for concrete piling and 1 for all other piling.)
- $\aleph$ = Soil Resilience Coefficient (should be between 0.6 and 0.8 for vibration frequencies between 5 and 10 Hz and 1 for all other frequencies.)

b) System acceleration factor: The peak cycle acceleration should fall within values such as

$$n_1 \leq F_{dyn}/W_{dyn} \leq n_2 \qquad (21)$$

Values for n_1 and n_2 are given in Table 5. The method seems to favor Equation (21) over (20) in case of conflict.

4) Compute the necessary frequency to insure a minimum peak vibration velocity by the equation

$$\theta = 1000 F_{dyn}/(2\pi v_{dyn} M) \qquad (22)$$

Table 4
Soil Resistance for Savinov and Luskin Method

Type of Soil and Pile	Separation Resistance σ For Piles, kPa	For Sheets, kN/m
1) Saturated Sandy and Visco-Plastic Clay Soils		
Steel Tubes	6	
Reinforced Concrete Piles	7	
Open-Ended Pipe Piles	5	
Sheet Piles, Light (Heavy) Sections		12 (14)
2) The same as (1), but with Interlayers of Compact Clay or Gravelly Soils		
Steel Tubes	8	
Reinforced Concrete Piles	10	
Open-Ended Pipe Piles	7	
Sheet Piles, Light (Heavy) Sections		17 (20)
3) Stiff Plastic Clay Soils		
Steel Tubes	15	
Reinforced Concrete Piles	18	
Open-Ended Pipe Piles	10	
Sheet Piles, Light (Heavy) Sections		20 (25)
4) Semi-Hard and Hard Clay Soils		
Steel Tubes	25	
Reinforced Concrete Piles	30	
Open-Ended Pipe Piles	20	
Sheet Piles, Light (Heavy) Sections		40 (50)

Table 5
Values for Ratio of Accelerations "n" for Savinov and Luskin Method

Type of Pile	n_1	n_2
Steel Sheet Piling	2	6.67
Light Piles	1.67	3.33
Heavy and Pipe Piles	1	2.5

Velocity v_{dyn} should fall between 0.5 and 0.8 m/sec.

5) Compute the eccentric moment using rigid body vibratory mechanics by the equation

$$K = 1000F_{dyn}/(2\pi\theta)^2 \qquad (23)$$

6) Check for adequate amplitude against the recommended values for x_{max} as shown in Table 6. Amplitude is computed using the equation

$$x_{max} = 1000K\psi/M \qquad (24)$$

7) Compute the power of the driving motors: This is done using the formula

$$N = K\theta^3(3.2\times10^{-7}D + .079K/M) \qquad (25)$$

where
 D = Diameter of Bearing Race, mm

Equation (25) was developed with the following assumptions:

a) Efficiency of the power transfer from motor to vibration exciter is 90%.

b) Coefficient of rolling friction in the bearings is 0.001.

c) Of the power actually sent into the soil, 15% of it is lost in the soil mass.

The Savinov and Luskin method is unique in that it uses parameters of empirical and theoretical derivation (ratio of accelerations, dynamic velocity, soil resistance, soil and pile material factors, and pile toe pressure) and combines their use using standard, free-hanging, rigid-body vibratory mechanics. The result is a hammer that is optimized for the pile to be driven. This method is iterative; it may require several cycles to get the resulting hammer to fit the parameters to the best extent possible.

Evaluation: The main advantage of the parametric methods is their relative simplicity of formulation and computation. The parametric methods make integrating experience-developed factors into the calculation very simple as well.

There are two main disadvantages of parametric methods. The first one is that none of the methods either adequately take into account all of the variables present in the vibratory installation of piles or account for the interaction between these variables. For instance, except for the Savinov and Luskin method, none of the methods take into account either the effect of power availability and input into the system or of frequency. Also none can be considered really valid at frequencies higher than 25-30 Hz. The second shortcoming of parametric methods is the lack of any consideration of installation velocity either as an input variable or as a result. This is important for two reasons; first, computing velocities for a number of system combinations is the only comprehensive way to compare different systems; second, any scheme to use vibratory drivers with bearing piles will probably use installation velocity as an acceptance criterion, much as the blows per meter (foot) are used now with impact hammers

Energy Methods

Energy methods to compute drivability are based on the assumption that, during most vibratory driving, the power put into the system by the vibrator equals the power taken out by the soil resistance. Thus energy methods are steady state methods, and do not take into account transient effects. There are two energy and power sources of a system: 1) the driving motor of the vibrator, 2) the potential energy of the system falling through the gravity field. The sink or destination for this energy is the resistance of the pile acting against the sinking pile. Mathematically, this energy and power flow is expressed by the formula

$$R_uV_{sys} = N + (W_{dyn}+W_{st})V_{sys} \qquad (28)$$

This equation can be reformulated in two ways; first, to compute penetration velocity,

Table 6
Amplitude Requirements for Savinov and Luskin Method

Type of Pile and Soil	x_{max} @ Frequency, mm		
	5-12 Hz	13-17 Hz	18-25 Hz
1) Steel Sheet Piling, Open Ended Pipe Piles, and Other Piles with $A_t \leq 150$ cm²			
Sandy Soils		8-10	4-6
Clayey Soil		10-12	6-8
2) Closed End Steel Pipe Piles, $A_t \leq 800$ cm²			
Sandy Soil		10-12	6-8
Clayey Soil		12-15	8-10
3) Reinforced Concrete Piles, Square or Rectangular Section, $A_t \leq 2000$ cm²			
Sandy Soil		12-15	
Clayey Soil		15-20	
4) Reinforced Concrete Cylinder Piles of Large Diameter, Driven with Soil Plug Removed			
Sandy Soil		6-10	4-6
Clayey Soil		8-12	6-10

$$V_{sys} = N/(R_u - W_{dyn} - W_{st}) \tag{29}$$

and for bearing capacity of the pile,

$$R_u = W_{dyn} + W_{st} + N/V_{sys} \tag{30}$$

where

R_u = Soil Resistance, kN
V_{sys} = Pile Penetration Velocity, m/sec

These are ideal equations; in practice, these methods add factors to account for actual conditions.

Davisson Method: This formula was proposed to predict the bearing capacity of piles driven with the Bodine BRD-1000 resonant vibratory pile driver. Making an analogy with the dynamic formulae, in the place of impact a full cycle of the eccentrics is considered as the time of energy transfer from hammer system to soil; thus, the loss factor is based on a per cycle basis. Generalizing, the formula is

$$R_u = (N + (W_{dyn} + W_{st})V_{sys})/(V_{sys} + \theta S_l/1000) \tag{31}$$

where
S_l = Soil Loss Factor, mm/cycle
and for penetration velocity, we can rearrange it to read

$$V_{sys} = (N - R_u \theta S_l/1000)/(R_u - W_{dyn} - W_{st}) \tag{32}$$

Values for S_l are given in Table 7. The formula is mainly intended for field use, and so all of the variables are taken from actual data.

Snip/Soviet Methods: These are placed with the energy methods because of their format and their involvement of horsepower, deadweight, and (indirectly) penetration velocity. They were developed for precast concrete cylinder pile. They are

$$R_u = (\lambda - 30000 V_{sys}/(A_o\theta))(245N/(A_o\theta) + W_{dyn} + W_{st})/F \tag{33a}$$

and

$$R_u = (245\lambda N/(A_o\theta) + W_{dyn} + W_{st})/F \tag{33b}$$

where
λ = Soil Coefficient

Equation (33a) is for penetration velocities of 0.5-1.67 mm/sec, and (33b) for 0.05-0.5 mm/sec. Values for λ vary with soil conditions and are given in Table 8. Factor of safety F is generally 2.

Evaluation: Energy methods have three main advantages. They are inherently simple. They are able to incorporate many factors into the analysis. They give as a result a penetration velocity (or conversely a bearing capacity), which allows meaningful comparison of different systems with each other.

The main disadvantage of energy methods lies in one of their advantages, namely their simplicity. They may not take into account all of the factors necessary in a meaningful way. This is in part due to their lack of broad field correlation. The Davisson method has extensive field documentation, but it has only been done for the Bodine hammer, which operates at frequencies well above most any other vibratory pile driving machine. Conversely, the Snip/Soviet method is well tested for penetration velocities that would be considered low in the U.S. These methods need more field development under a wide variety of conditions if they are to reach their full potential.

Methods from Laboratory and Model Tests

The use of laboratory and model testing to establish the drivability of vibrated piling represents an attempt to establish a reliable correlation based on an actual physical situation but in a controlled environment. Generally it involves setting up a tank full of soil and driving the piling through the tank, either horizontally or vertically. The various parameters of the system can then be varied to produce data for correlation purposes.

Bernhard's Method: To compute the static bearing capacity, the method employs the following formula:

$$R_u = \phi Nl/(V_{sys} l_{soil}) \tag{34}$$

where
l_{soil} = Length of Soil Penetration by Pile, m
ϕ = Soil Loss factor (suggested value is 0.1)

The tests that established the formula were run in a frequency range of 50-5250 Hz.

Schmid and Hill's Formula: They propose from statistical data reduction that the penetration velocity of the pile can be estimated by the formula

$$V_{sys} = 0.417 g n^{0.75}((W_{st} + W_{dyn})/R_u + 0.0036n - 0.018)/\theta \tag{35}$$

The test pile was driven through sand exclusively. The test setup limited the frequency to a maximum of 30 Hz. Schmid went on to develop another method, his Toe Impulse Formula, which is

$$R_u = \alpha(W_{dyn} + W_{st})/(\theta\sqrt{(2V_{sys}/(\theta n' g))} \tag{36}$$

Table 7
Loss Factors for Davisson Method

Soil @ Pile Tip	Loss Factor S_l, mm/cycle*	
	Closed-End Pipe Pile	H-Beam
Loose Silt, Sand, or Gravel	0.24	−0.21
Medium Dense Sand or Sand and Gravel	0.76	0.76
Dense Sand or Sand and Gravel	2.44	2.13

*Values developed for Bodine BRD-1000 Resonant Driver. Formula not applicable in rock.

where
- α = a coefficient, taken to be 0.67
- n' = acceleration in excess of the minimum acceleration to effect driving, g's

O'Neill's Formula: This formula came from tests run on a vibrator driving a miniature pile in a sand tank. It is

$$R_u = 0.050N'/(V_{sys}(\sigma'_h/92.8-0.486)(1.96D_r-1.11)(1.228-0.19d_{10})) \quad (37)$$

where
- σ'_h = horizontal effective stress, kPa
- D_r = relative density
- d_{10} = grain size, mm
- N' = power actually delivered to the pile top, kW

The theoretical power the vibrator generates can be computed by the equation

$$N = \theta(4000W_{st}+2K(2\pi\theta)^2(1+\theta^2/(\theta^2+\theta_n^2)))(K\theta^2/(1000M(\theta^2+\theta_n^2))) \quad (38)$$

where
- θ_n = natural frequency of suspension with respect to the vibrator mass, Hz

and this is related to the power actually delivered to the pile top by the formula

$$N' = (0.25+0.063n)N \quad (39)$$

The peak acceleration can be computed by the formula

$$n = (3.54-2.186D_r)(8.99+2.76d_{10})((39.37V_{sys})^{(1.71-\sigma'_h/85.1)}) \quad (40)$$

Evaluation: The advantage of laboratory and model test formulas are that the results they give have their root data taken from a controlled physical environment. The various parameters of the system can thus be taken into account in a physically realistic and truly interactive fashion.

The main weakness of these laboratory derived formulae is that the conditions produced in the laboratory may not include all that is actually experienced in actual vibratory pile driving. They should be used only when the original conditions under which they were derived are present in the field. Also, none of the formulae above is comprehensive in its parameter inclusion. Nevertheless, because of their virtues, laboratory and model tests remain an important constituent of vibratory pile driving research.

Time Dependent Nonlinear Methods

The newest method to be applied to vibratory installation of piling is that of time dependent nonlinear methods. These methods seek to actually solve the equations of motion of the vibratory system through numerical integration. These methods divide themselves into two categories: 1) methods which consider the distributed mass and elasticity of the system (wave equation techniques), and 2) those which don't (rigid body techniques).

VIBEWAVE Method: This involves using a modified version of the TTI program. This model uses a finite difference mass-spring-dashpot model which is solved using a modified version of Euler's Method.

TNOWAVE Method: This again involves the modification of a wave equation analyzer for use with vibratory hammers. TNOWAVE uses the method of characteristics to solve the wave equation.

Piecewise Integration Techniques: These are rigid body techniques which are applied to the system. The pieces are determined by changes in the variables, especially the reversal of the soil frictional force. Since the equations for these techniques can be formulated dimensionlessly, parametric studies can be performed using these techniques. In addition to solving longitudinal vibratory motion, these have been applied for longitudinal-rotational and impact-vibrational type drivers.

VIBDRIVE Method: This was developed by the author for the VIBDRIVE analysis program. It is a rigid body technique that uses a variation of Euler's technique (different from TTI) to solve the equations of motion. A Coulombic soil model is used for the shaft resistance and a constant resistance plug model is used for the toe.

Evaluation: Assuming that they are properly set up, time dependent non-linear methods are the most complete method available to analyze the vibratory installation and extraction of piles. They can take into account all system variables through their thorough modeling of the system. This is especially important with the wave equation methods at higher frequencies, as both distributed mass and elasticity in the system become more important.

The main weakness of these methods is the accuracy of the constituent components of the model. These must be both thoroughly understood and accurately simulated for meaningful results. These conditions have not been met yet; the popular Smith model for soil response cannot be applied to vibratory soil excitation.

CONCLUSION

Vibratory hammers have proven themselves a versatile tool to install and extract many kinds of piling. They are limited principally by very stiff soil conditions where impact is required and a lack of an accepted method of determining the drivability and bearing capacity of the piles they drive. These will be addressed as the technology progresses and the versatility of this equipment will continue to be enhanced.

ACKNOWLEDGEMENTS

No work such as this can ever come to reality without the help of others; these include Messrs. E.A. Narozhnitskii and M.L. Pevzner of the Leningrad Experimental Works of Construction Machinery; Ms. Sherrill Gardner of the Naval Civil Engineering Laboratory; Mr. Richard Nelson of Pile Equipment, Inc.; Dr. David M. Rempe, consulting geotechnical engineer; Mr. L.V. Erofeev of VNIIstroidormash, the USSR Ministry for Constructional Road-Building and Municipal Machinery; Messrs. Kurt Winters and Bill Harrison of Vulcan Iron Works Inc.; Dr. Michael O'Neill of the University to Houston; and Mr. Christopher Smoot of *Pile Buck* for enabling the publication of this and related works.

Table 8
Soil Coefficient for Snip/Soviet Formula

Soil Type	λ
Saturated Sand	4.0-7.5
Moist Sand	3.0-4.5
Dry Sand	2.5-4.0
Sand Clay	2.5-5.0
Silty Clay	2.2-4.5
Clay	2.0-4.5

Finally this author would like to make an important "vibratory" acknowledgement by citing Haggai 2:6-7: "This is what the Lord Almighty says: 'In a little while I will once more shake the heavens and the earth, the sea and the dry land. I will shake all nations, and the desired of all nations will come, and I will fill this house with glory,' says the Lord Almighty."

REFERENCES AND FURTHER READING

BILLET, P., and SIFFERT, J.G. (1985) "Détermination Expérimentale du Frottement Latéral en Vibrofonçage." (Experimental Determination of Shaft Friction during Vibrodriving). Proceedings of the International Symposium on Penetrability and Drivability of Piles, International Society for Soil Mechanics and Foundation Engineering, pp. 89-92.

CHUA, K.M., GARDNER, S., and LOWERY, L.L. (1987) "Wave Equation Analysis of a Vibratory Hammer-Driven Pile." *Proceedings of the Nineteenth Annual Offshore Technology Conference*, Dallas, TX. OTC 5396, pp. 339-345.

DANCE, D.R. and RIDER, D.J. (1987) "Resonant Pile Driving." Presented at the Twelfth Annual Meeting of the Deep Foundations Institute, Hamilton, Ontario, Canada, 14-16 October 1987.

DAVISSON, M.T. (1970) "BRD Vibratory Driving Formula." *Foundation Facts*, Vol. VI No. 1.

DONDELINGER, M., and SOMMERFIELD, W.J. (1989) "The World in Steel Sheet Piling." *Pile Buck*, First August Issue 1989, pp. 6A-23A.

EROFEEV, L.V., SMORODINOV, M.I., FEDOROV, B.S., VYAZOVIKII, V.N., and VILLUMSEN, V.V. (1985) *Mashiny I Oborudanie Dlya Ustroistva Osnovanii I Fundamentov* (Machines and Equipment for the Installation of Shallow and Deep Foundation). Second Edition, Mashinostrenie, Moscow, pp. 95-111.

GARDNER, S. (1987) "Analysis of Vibratory Driven Pile." Naval Civil Engineering Laboratory Technical Note N-1779, Port Hueneme, CA.

GARRETT, R.E. (1988) "New Vibratory Hammer Proves Ideal for Aluminium Sheeting." *Pile Buck*, Jupiter, FL, Second March Issue 1988, pp. 10-14.

GUMENSKII, B.M., and KOMAROV, N.S. (1959) *Vibroburenie Gruntov* (Soil Drilling By Vibration). Ministry of Municipal Services of the RSFSR, Moscow.

HIRSCH, T.J., CARR, L. and LOWERY, L.L., Jr. (1976) "Pile Driving Analysis - Wave Equation User's Manual, TTI Program." Federal Highway Administration Project FWHA-IP-76-13 (4 vols.)

JONKER, G. (1987) "Vibratory Pile Driving Hammers for Pile Installations and Soil Improvement Projects." Proceedings of the Nineteenth Annual Offshore Technology Conference, Dallas, TX. OTC 5422, pp. 549-560.

MIDDENDORP, P., and JONKER, G. (1988) "Prediction of Vibratory Hammer Performance by Stress Wave Analysis." Presented at the Third International Conference on the Application of Stress-Wave Theory to Piles, Ottawa, Ontario, 25-27 May 1988.

O'NEILL, M.W., and VIPULANANDAN, C. (1989) "Appropriate Field Measurements for Testing Capacity - Prediction Methods for Piles Installed by Vibration." Presented at the Fourteenth Annual Members' Conference of the Deep Foundations Institute, Baltimore, MD, 9-11 October 1989.

O'NEILL, M.W., and VIPULANANDAN, C. (1989) "Laboratory Evaluation of Piles Installed with Vibratory Hammers." NCHRP Report 316. Washington: Transportation Research Board, National Research Council.

SCHMID, W.E., and HILL, H.T. (1966) "The Driving of Piles by Longitudinal Vibrations." Princeton Soil Engineering, Research Series No. 4, June 1966.

SMART, J.D. (1970) "Vibratory Pile Driving." Doctoral Thesis, University of Illinois at Urbana-Champaign.

TSEITLIN, M.G., VERSTOV, V.V., and AZBEL, G.G. (1987) *Vibratsionnaya Tekhnika I Tekhnologiya V Svainykh I Burovikh Rabotakh* (Vibratory Methods and the Technology of Piling and Boring Work). Stroiizdat, Leningradskoe Otdelenie, Leningrad.

WARRINGTON, D.C. (1989) "Theory and Development of Vibratory Pile Driving Equipment." Proceedings of the Twenty-First Annual Offshore Technology Conference, Dallas, TX. OTC 6030, pp. 541-550.

ADDITIONAL BOOKS, PUBLICATIONS AND SPECIFICATION CHARTS AVAILABLE FROM PILE BUCK®

In addition to this book, our Pile Buck® "Series" of (6) Marine Construction Handbooks consist of:

COASTAL CONSTRUCTION
■
HARBORS, PIERS & WHARVES
■
BULKHEADS, MARINAS & SMALL BOAT FACILITIES
■
PROTECTION, INSPECTION & MAINTENANCE OF MARINE STRUCTURES
■
MATERIALS & EQUIPMENT FOR MARINE CONSTRUCTION

Also available:

PILE BUCK® NEWSPAPER *(Published Twice Monthly)*
PILE BUCK® STEEL SHEET PILING DESIGN MANUAL
PILE BUCK® PILE HAMMER SPECIFICATIONS CHART
PILE BUCK® STEEL SHEET PILING SPECIFICATIONS CHART
PILE BUCK® PRODUCT DIRECTORY

For an up-to-date descriptive brochure regarding the available Pile Buck® Publications please call, FAX or write:

Pile Buck®, Inc.
Attn: Publications
P.O. Box 1056
Jupiter, Florida 33568-1056

PH: (407) 744-8780   FAX: (407) 575-9748

PILE HAMMER SPECIFICATIONS

DIESEL HAMMERS

ENERGY RANGE (FT.LB.)	MODEL	MANUFACTURER	SINGLE/DOUBLE ACTING	BLOWS PER MIN.	PISTON WEIGHT LBS.	TOTAL WEIGHT LBS.	TOTAL LENGTH FT.-IN.	MAXIMUM STROKE FT.-IN.	WIDTH BETWEEN JAWS (IN.)	FUEL USED (GPH)
750,000-384,000	D100-30	DELMAG	SINGLE	36-50	77,161	165,345	34'10"	9'8"	CAGE	22.46
300,000-157,740	D100-13	DELMAG	SINGLE	34-45	23,612	45,357	20'4"	12'8"	36"	7.93
280,000	K150	KOBE	SINGLE	45-60	33,100	80,500	29'8"	8'6"	CAGE	16-20
225,000-126,190	D80-23	DELMAG	SINGLE	36-45	19,500	37,739	20'4"	11'6"	36"	6.60
165,000-78,960	D62-22	DELMAG	SINGLE	36-50	14,600	27,055	19'6"	11'4"	32"	5.28
149,600-88,000	MH80B	MITSUBISHI	SINGLE	42-60	17,600	43,600	19'6"	14'10"	42"	8-12
141,000-63,360	MB70	MITSUBISHI	SINGLE	38-60	15,840	46,000	19'6"	8'11"	42"	7-10
135,100-79,500	MH72B	MITSUBISHI	SINGLE	38-60	15,900	44,000	19'6"	14'10"	42"	7-10
127,500-79,000	DE150/110	MKT	SINGLE	40-50	15,000	29,500	19'10"	10'9"	32"	7
117,000-62,500	D55	DELMAG	SINGLE	36-47	12,100	26,300	17'9"	9'8"	32"	5.50
107,170-52,260	D46-32	DELMAG	SINGLE	37-53	10,143	19,602	17'4"	10'7"	30"	4.23
105,600	K60	KOBE	SINGLE	42-60	13,200	37,500	24'3"	8'0"	42"	6.5-8.0
100,000-40,000	200S	ICE	SINGLE	53-70	20,000	33,600	17'0"	5'0"	32"	4.00
93,500-66,000	DE-150/110	MKT	SINGLE	40-50	11,000	24,550	17'10"	10'9"	32"	5.7
92,752	KC45	KOBE	SINGLE	39-60	9,920	24,700	17'11"	9'4"	—	4.5-5.5
91,100	K45	KOBE	SINGLE	39-60	9,900	25,600	18'6"	9'2"	36"	4.5-5.5
87,000-43,500	D44	DELMAG	SINGLE	37-56	9,500	22,300	15'10"	9'2"	32"	5.50
85,400-50,200	MH45	MITSUBISHI	SINGLE	42-63	10,500	24,600	17'11"	11'5"	36"	4-6
84,000-37,840	M43	MITSUBISHI	SINGLE	40-60	9,460	22,660	16'3"	8'10"	37"	4-6
83,880-40,900	D36-32	DELMAG	SINGLE	36-53	7,938	17,397	17'4"	10'7"	30"	3.04
79,500	J44	IHI	SINGLE	42-70	9,720	21,500	14'10"	8'2"	37"	6.86
79,000	B-500	BERMINGHAMMER	SINGLE	33-55	6,900	16,250	17'10"	13'0"	—	2.0
79,000	K42	KOBE	SINGLE	40-60	9,260	24,000	17'8"	8'6"	36"	4.5-5.5
73,000-40,150	3,400	F.E.C.	SINGLE	40-60	7,500	14,600	16'0"	—	26"	3.3-5.0
72,182	KC35	KOBE	SINGLE	39-60	7,720	17,400	16'10"	9'4"	—	3.2-4.3
70,800	K35	KOBE	SINGLE	39-60	7,700	18,700	17'8"	9'2"	30"	3.0-4.0
70,000-40,000	70S	ICE	SINGLE	38-55	7,000	13,500	16'8"	10'0"	26"	2.10
70,000-36,100	1070	ICE	DOUBLE	64-68	10,000	21,500	17'10"	7'	30"	3.5
39,900-35,380	D30-32	DELMAG	SINGLE	36-52	6,615	13,252	17'3"	10'7"	26"	2.64
35,600-38,600	MH35	MITSUBISHI	SINGLE	42-60	7,720	18,500	17'3"	11'	32"	3.4-5
64,600-29,040	M35	MITSUBISHI	SINGLE	40-60	7,260	16,940	13'2"	8'	32"	3.4-5
53,500	J35	IHI	SINGLE	72-70	7,730	16,900	14'6"	8'3"	32"	4.76
53,000-34,650	3000	F.E.C.	SINGLE	40-60	6,600	13,200	15'6"	10'6"	26"	2.8-4.2
50,100	K32	KOBE	SINGLE	40-60	7,050	17,750	17'8"	8'6"	30"	2.75-3.5
59,500-42,000	DE70/50B	MKT	SINGLE	40-50	7,000	14,700	16'11"	10'9"	26"	3.3
58,250-29,480	D25-32	DELMAG	SINGLE	37-52	5,513	12,149	17'3"	10'7"	26"	2.11
54,250-23,800	D30	DELMAG	SINGLE	39-60	6,600	12,300	14'3"	8'2"	26"	2.90
53,750	B-400	BERMINGHAMMER	SINGLE	37-60	5,000	15,000	14'10"	10'9"	—	1.5
51,518	KC25	KOBE	SINGLE	39-60	5,510	12,130	16'10"	9'4"	—	2.4-3.2
50,700	K25	KOBE	SINGLE	39-60	5,510	13,100	17'6"	9'3"	26"	2.5-3.0
50,000-27,500	2,500	F.E.C.	SINGLE	40-60	5,500	12,100	15'6"	10'6"	20"	2.8-4.2
50,000-25,100	660	ICE	DOUBLE	84-88	7,564	24,480	17'4"	6'7"	30"	3.25

DIESEL HAMMERS - CONTINUED

ENERGY RANGE (FT.LB.)	MODEL	MANUFACTURER	SINGLE/DOUBLE ACTING	BLOWS PER MIN.	PISTON WEIGHT LBS.	TOTAL WEIGHT LBS.	TOTAL LENGTH FT.-IN.	MAXIMUM STROKE FT.-IN.	WIDTH BETWEEN JAWS (IN.)	FUEL USED (GPH)
48,500-24,500	D22-23	DELMAG	SINGLE	38-52	4,850	11,400	17'2"	10'	26"	1.60
46,900-27,550	MH25	MITSUBISHI	SINGLE	42-60	5,510	13,200	16'8"	10'9"	26"	2.4-4
45,000-20,240	M23	MITSUBISHI	SINGLE	42-60	5,060	11,220	14'1"	8'10"	26"	2.4-3.7
44,800	DE50C	BSP	SINGLE	42-54	4,980	10,300	14'4"	9'	26"	2.68
42,800-20,540	D19-32	DELMAG	SINGLE	37-53	4,190	7,800	15'6"	10'3"	20"	1.45
42,500-30,000	DA-55C	MKT	SINGLE	40-50	5,000	17,000	17'4"	10'6"	26"	2.7
42,500-30,000	DE70/50B	MKT	SINGLE	40-50	5,000	12,700	16'11"	10'9"	26"	3.3
41,300	K22	KOBE	SINGLE	40-60	4,850	12,350	17'6"	9'2"	26"	2.0-2.75
40,300	B-300	BERMINGHAMMER	SINGLE	37-60	3,750	9,520	14'0"	10'10"	—	1.0
40,200-18,870	D16-32	DELMAG	SINGLE	36-52	3,528	7,386	15'6"	11'5"	20"	1.45
40,000-25,400	640	ICE	DOUBLE	74-77	6,000	14,460	15'7"	6'8"	26"	3.0
40,000-16,000	40S	ICE	SINGLE	38-55	4,000	7,500	15'¼"	10'2"	20"	1.20
39,700	D22	DELMAG	SINGLE	42-60	4,850	11,200	14'2"	8'2"	26"	2.90
39,100	J22	IHI	SINGLE	42-70	4,850	10,800	14'0"	10'0"	26"	3.2
38,200-31,200	DA-55C	MKT	DOUBLE	78-82	5,000	17,000	17'4"	—	26"	3.0
36,000-24,000	DE-40	MKT	SINGLE	40-50	4,000	11,275	15'	10'6"	26"	3.0
34,000-24,000	DA-45C	MKT	SINGLE	40-50	4,000	14,200	15'1"	10'6"	26"	2.5
31,320-15,660	D12-32	DELMAG	SINGLE	36-52	2,820	6,260	15'6"	11'2"	20"	0.95
31,000-17,700	520-30	ICE	DOUBLE	80-84	5,070	13,400	13'9"	5'11"	26"	1.35
30,700-18,500	DA-45C	MKT	DOUBLE	78-82	4,000	14,200	15'1"	—	26"	2.8
29,250-12,000	B-225	BERMINGHAMMER	SINGLE	39-60	3,000	8,730	14'0"	9'10"	—	1.0
28,100-16,550	MH15	MITSUBISHI	SINGLE	42-60	3,310	8,400	16'1"	10'3"	26"	1.3-2
28,050-19,800	DE-33/30/20B	MKT	SINGLE	40-50	3,300	7,750	15'11"	10'6"	20"	2.0
27,100-14,900	1500	F.E.C.	SINGLE	40-60	3,300	7,225	14'2"	10'11¾"	20"	1.5-2.3
27,100	D15	DELMAG	SINGLE	42-60	3,300	6,615	13'11"	8'2"	20"	1.75
27,000	DE30C	BSP	SINGLE	42-54	3,000	7,600	14'2"	9'	26"	1.7
26,300-17,700	520-26	ICE	DOUBLE	80-84	5,070	12,545	13'6"	5'3"	26"	1.35
26,000-11,800	M14S	MITSUBISHI	SINGLE	42-60	2,970	7,260	13'6"	8'9"	26"	1.3-2.2
25,200-16,800	DE33/30/20B	MKT	SINGLE	40-50	2,800	7,250	8'	10'0"	20"	2.0
24,400	K13	KOBE	SINGLE	40-60	2,860	7,300	16'8"	8'6"	26"	.75-2.0
23,800-16,800	DA-35C	MKT	SINGLE	40-50	2,800	10,800	17'	10'6"	20"	1.7
23,000	B-23	BERMINGHAMMER	DOUBLE	80	2,800	9,940	16'0"	4'6"	20"	1.9
22,500-12,375	1200	F.E.C.	DOUBLE	40-60	2,750	6,540	14'0"	10'9¼"	20"	1.5-2.3
22,500	D12	DELMAG	SINGLE	42-60	2,750	6,050	13'6"	8'2"	20"	1.75
22,500-9,000	422	ICE	DOUBLE	76-82	4,000	10,375	13'11"	5'8"	22"	1.0
22,500-9,000	30S	ICE	SINGLE	44-67	3,000	6,250	12'4"	7'5"	20"	0.80
21,000-15,600	DA-35C	MKT	DOUBLE	78-82	2,800	10,800	17'	10'6"	20"	2.7
18,100-7,700	440	ICE	DOUBLE	88-92	4,000	9,840	13'6"	4'8"	20"	1.16
18,000-9,435	D8-22	DELMAG	SINGLE	38-52	1,762	4,220	15'5"	10'2"	20"	1.00
18,000-7,500	312	ICE	DOUBLE	100-105	3,857	10,375	10'9"	4'8"	26"	1.1
18,000-8,600	B-200	BERMINGHAMMER	SINGLE	39-58	2,000	6,940	13'9"	10'2"	20"	1.9
17,000-12,000	DE33/30/20B	MKT	SINGLE	40-50	2,000	6,450	15'11"	10'6"	20"	2.0
10,500-6,300	D6-32	DELMAG	SINGLE	39-52	1,322	3,810	12'6"	7'11"	20"	0.70
9,350-6,600	DA-15C	MKT	SINGLE	40-50	1,100	4,825	13'11"	10'6"	20"	1.0
9,100	D5	DELMAG	SINGLE	42-60	1,100	2,730	12'6"	8'4"	20"	1.32
8,800	DE10	MKT	SINGLE	40-50	1,100	3,100	12'2"	9'	20"	0.90
8,200-6,600	DA-15C	MKT	DOUBLE	78-92	1,100	4,825	13'11"	—	20"	1.8
8,100-4,060	180	ICE	DOUBLE	90-95	1,725	4,546	11'3"	4'9"	20"	0.65
3,630-1,625	D4	DELMAG	SINGLE	50-60	836	1,360	7'9"	4'4"	—	.21
1,815-868	D2	DELMAG	SINGLE	60-70	484	792	6'9"	3'8"	—	.13

HYDRAULIC VIBRATORY DRIVERS / EXTRACTORS

DRIVING FORCE (TONS)	MODEL	MANU-FACTURER	FRE-QUENCY VPM	ECCENTRIC MOMENT	AMPLI-TUDE IN.	MAX. HYDRAULIC H.P.	ENGINE H.P.	MAX. EXTRACTION TONS	PULL FORCE TONS	SUSPENDED WEIGHT LB.	SHIPPING WEIGHT LB.	HEIGHT (W/CLAMP) FT. & IN.	WIDTH (Thickness) FT. & IN.	LENGTH FT. & IN.	THROAT WIDTH IN.
190	V-140	MKT	1,400	14,000	1.0	1,400	1,800	150	300	53,000	96,000	14'6"	4'0"	11'10"	48"
144	110H2	PTC	1,350	9,550	13/16	584	760	88	2 x 165	32,900	57,250	12'9"	2'11"	7'7"	—
144	110H1	PTC	1,350	9,550	11/8	565	760	132	4 x 120	39,050	63,400	10'1"	2'6"	12'4"	—
100	5600	VULCAN	2,400	—	.50	408	475	51	177	16,000	27,000	12'3"	1'2"	12'	14"
99	60H1	PTC	1,650	5,210	7/8	523	660	88	2 x 165	29,200	53,550	13'0"	2'3"	7'7"	14"
82	V-36	MKT	1,600	5,000	3/4	550	650	80	100	18,800	36,300	13'1"	1'2"	12'0"	14"
76	4,800	CASTEEL/PSI	1,600	4,800	1.22	499	600	66	196	16,200	32,700	8'	—	7'10"	14"
67	4,600	VULCAN	1,600	4,600	1.25	474	600	66	176	16,000	32,500	8'0"	1'6"	6'3"	14"
66	50H3	PTC	1,650	4,340	7/8	394	510	44	220	23,300	38,750	14'	1'1"	7'10"	—
63.6	1,412	ICE	400-1,200	8,000	1-1½	—	650	100	200	31,700	52,300	12'10"	3'5"	8'0"	32"
62	23HF1	PTC	2,400	2,000	7/8	302	465	44	220	12,850	30,500	9'10"	2'6"	7'2"	13"
60	V-30	MKT	1,600	4,400	1.00	510	550	80	100	15,000	32,500	11'1"	1'1"	9'4"	13"
48	23HF1	PTC	2,300	2,000	7/8	248	330	44	220	12,850	25,600	9'10"	2'6"	7'2"	13"
45.4	812	ICE	400-1,600	4,000	½-1	—	503	50	100	15,600	30,500	9'9"	2'0"	8'0"	14"
43/121	HVB130.05 Dual	DELMAG/TUNKERS	1,600/1,200	3,991/6,074	.79/1.18	547/362	625	55.1	154	12,600	32,300	5'9"	—	6'10"	17.32"
34	4,150	FOSTER	900-1,500	4,166	5/16-11/4	570	570	55	145/200	16,495	31,995	6'6"	2'3"	8'0"	12"
16	V-20	MKT	1,650	3,000	.66	310	350	60	75	12,500	31,500	12'2"	1'2"	8'0"	14"
10/83	HVB100.05 Dual	DELMAG/TUNKERS	1,600/1,200	2,950/3,731	.70/1.14	359/283	502.8	55.1	121	10,580	24,580	5'5"	—	6'10"	15.74"
100	2,800	VULCAN	2,400	—	.50	204	250	25	87	8,300	18,800	6'6"	1'	9'8"	12"
38	2,400	CASTEEL/PSI	1,600	2,400	1.02	255	365	33	96	8,300	17,800	7'6"	—	3'9"	14"
34	2,300	VULCAN	1,600	2,300	1	237	335	33	87	8,200	19,700	7'9"	1'6"	3'9"	14"
34	13HF1	PTC	2,300	1,130	7/8	177	255	33	120	7,150	18,600	6'10"	2'6"	5'8"	13"
33	25H2	PTC	1,650	2,170	15/16	253	325	44	120	9,350	16,900	8'11"	1'1"	7'6"	14"
31.8	612	ICE	400-1,200	4,000	.50-1.0	—	250	40	100	13,700	22,500	9'5"	1'9"	7'11"	12"
30	V-17	MKT	1,600	2,200	.75	254	280	60	70	12,000	31,500	10'0"	1'1"	9'0"	13"
28	V-16	MKT	1,750	1,800	.47	161	210	40	75	9,250	24,000	11'2"	1'2"	6'6"	14"
28/67	HVB70.05 Dual	DELMAG/TUNKERS	1,630/1,100	2,065/3,879	.74/1.33	290/200	324.5	27.5	91	7,050	18,850	5'3"	—	5'7"	11.81"
27	HVB70.04	DELMAG/TUNKERS	2,000	1,362	.47	258	324.5	27.5	91	7,050	18,850	5'3"	—	5'7"	11.81"
19	25H1	PTC	1,500	2,170	15/16	185	255	44	120	9,350	16,300	8'11"	1'1"	7'6"	—
16.5	416-L	ICE	400-1,530	2,000	.25-.75	—	250	40	100	9,900	18,700	8'5"	1'5"	7'11"	14"
26/55	HVB60.05 Dual	DELMAG/TUNKERS	1,500/1,000	2,065/3,879	.67/1.18	251/191	324.5	27.5	91	7,050	18,850	5'3"	—	5'7"	11.81"
25	1,800	FOSTER	1,000-1,600	1,800	5/16-3/4	—	220	30	90/120	11,000	24,375	6'8"	1'9"	8'0"	12"
27.5	416	ICE	400-1,500	1,800	¼-1	—	250	40	100	13,100	22,000	8'9"	1'10"	8'0"	12"
19.1	1,400	VULCAN	—	2,400	.38	123	175	25	50	4,350	10,350	8'8"	1'11"	3'8"	12"
16	V-14	MKT	1,500	1,500	.32	140	210	40	75	10,000	29,500	11'1"	1'2"	5'3"	14"
14	1,200	CASTEEL/PSI	1,600	1,200	.82	134	175	33	50	6,500	14,000	6'6"	—	3'9"	14"
14/39	HVB40.05 Dual	DELMAG/TUNKERS	1,500/1,000	1,346/2,616	.67/1.22	178/136	253.4	27.5	91	5,700	17,100	4'11"	1'8"	5'3"	11.81"
14	HVB40.04	DELMAG/TUNKERS	2,000	781	.35	188	253.4	27.5	91	5,700	17,100	4'11"	1'	5'3"	11.81"
14	13H1	PTC	1,700	1,085	11/16	112	155	22	60	5,400	12,250	6'5"	1'1"	4'7"	13"
12	1,150	VULCAN	1,600	1,150	.75	131	175	33	50	6,500	14,000	6'6"	1'6"	3'9"	14"
12	7HF1	PTC	2,300	565	5/8	102	150	17	60	3,800	10,650	5'9"	1'7"	3'7"	13"
10	V-5B	MKT	1,600	1,100	.75	95	175	40	62	7,200	11,200	7'8"	1'1"	6'4"	13"
7	1,000	FOSTER	1,000-1,600	1,000	.25-.75	—	185	20	90	6,094	10,500	5'11"	1'8"	5'7"	12"
6.4	216	ICE	400-1,600	1,000	¼-¾	—	155	30	50	4,825	12,450	6'6"	1'	3'11"	12"
3	HVB30	DELMAG/TUNKERS	1,800	712	.79	112	158.2	13.2	33	2,100	9,900	34"	34"	3'6"	14.50"
10.0	V-5	MKT	1,450	1,000	.50	59	175	30	62	5,200	9,200	94"	1'2"	6'10"	14"
—	H-75A	H & M	1,900	—	.25-.50	75	143	15	62	4,000	9,000	5'7"	—	3'1"	14"
9	600	CASTEEL/PSI	1,750	673	.625	65	102	16	28	2,900	6,200	4'6"	—	3'	14"

HYDRAULIC VIBRATORY DRIVERS/EXTRACTORS - CONTINUED

DRIVING FORCE TONS	MODEL	MANU- FACTURER	FRE- QUENCY VPM	ECCENTRIC MOMENT	AMPLI- TUDE IN.	MAX. HYDRAULIC H.P.	ENGINE H.P.	MAX. PULL EXTRACTION TONS	PILE CLAMP FORCE TONS	SUSPENDED WEIGHT LB.	SHIPPING WEIGHT LB.	HEIGHT (W/CLAMP) FT. & IN.	WIDTH (Thickness) FT. & IN.	LENGTH FT. & IN.	THROAT WIDTH IN.
27	HVB24	DELMAG/TUNKERS	2,100	434	.55	86	131.4	13.2	27	2,100	9,400	3'3"	—	3'6"	14.50"
23	6H1	PTC	1,800	—	.75	47	—	22	—	3,180	6,580	4'1"	1'1"	3'8"	14"
20	300	CASTEEL/PSI	2,000	350	.725	30	44	11	25	1,189	3,989	3'9"	—	2'	14"
17	400	VULCAN	2,400	400	.50	42	75	10	6	1,100	4,100	3'2"	10"	2'2"	10"
17	HVB16	DELMAG/TUNKERS	2,100	278	.39	64	131.4	13.2	27	2,000	9,300	3'3"	—	3'4"	14.50"
15	V-2	MKT	1,600	400	.75	44	50	8	25	2,400	4,400	5'0"	1'1"	4'0"	13"
12	HVB10	DELMAG/TUNKERS	2,100	191	.31	43	131.4	13.2	27	2,000	9,300	3'3"	—	3'4"	14.50"

AIR / STEAM HAMMERS

RATED ENERGY (FT.-LBS.)	MODEL	MANUFACTURER	TYPE	STYLE	BLOWS PER (MIN.)	WT. OF STRIKING PARTS (lbs)	TOTAL WEIGHT (LBS.)	HAMMER LENGTH (FT.-IN.)	JAW DIMENSIONS	INLET PRESSURE (PSI)	INLET SIZE (NPT)
1,800,000	6300	VULCAN	SGL.-ACT	OPEN	42	300,000	575,000	30'0"	22"x144"	235	2@6"
1,582,220	MRBS 12500	MENCK	SGL.-ACT	OPEN	36	275,580	540,130	35'9"	CAGE	171	2@6"
1,200,000	2000E6	CONMACO	SGL.-ACT	OPEN	40	200,000	490,000	35'6"	CAGE	155	2@6"
1,200,000	1750E6	CONMACO	SGL.-ACT	OPEN	40	175,000	465,000	35'6"	CAGE	135	2@6"
1,050,000	MRBS 8000	MENCK	SGL.-ACT	OPEN	38	176,370	330,690	30'10"	CAGE	171	8"
867,960	1500E5	CONMACO	SGL.-ACT	OPEN	42	150,000	283,000	30'6"	14½"x120"	135	2@6"
750,000	5150	VULCAN	SGL.-ACT	OPEN	46	150,000	275,000	26'3½"	22"x120"	175	2@6"
750,000	850E6	CONMACO	SGL.-ACT	OPEN	40	85,000	173,600	25'2"	18¾"x100"	160	2@4"
510,000	5100	VULCAN	SGL.-ACT	OPEN	48	100,000	197,000	27'4"	22"x120"	150	2@5"
500,000	MRBS 4600	MENCK	SGL.-ACT	OPEN	42	101,410	176,370	27'5"	CAGE	142	6"
499,070	700E5	CONMACO	SGL.-ACT	OPEN	43	70,000	152,000	23'2"	18¾"x100	130	2@4"
350,000	MRBS 3000	MENCK	SGL.-ACT	OPEN	42	66,135	108,025	25'0"	CAGE	142	5"
325,480	3100	VULCAN	SGL.-ACT	OPEN	60	100,000	195,500	23'3"	18¾"x88"(M)	130	3@4"
300,000	560	VULCAN	SGL.-ACT	OPEN	47	62,500	134,060	23'0"	18¾"x88"(M)	150	2@5"
300,000	450E5	CONMACO	SGL.-ACT	OPEN	45	45,000	103,000	23'3"	14"x80"	130	2@4"
225,000	540	VULCAN	SGL.-ACT	OPEN	48	40,900	102,980	22'7"	14"x80"(M)	130	2@5"
200,000	MRBS 1800	MENCK	SGL.-ACT	OPEN	44	38,580	64,590	22'5"	CAGE	142	4"
189,850	360	VULCAN	SGL.-ACT	OPEN	62	60,000	124,830	19'0"	18¾"x88"(M)	130	2@4"
180,000	300E5	CONMACO	SGL.-ACT	OPEN	40	30,000	58,400	20'10"	11¼"x56"	135	4"
150,000	530	VULCAN	SGL.-ACT	OPEN	42	30,000	57,680	20'5"	10½"x54"	125	3"
150,000	60X	RAYMOND	SGL.-ACT	OPEN	60	60,000	85,000	22'7"	14"x80"	165	3"
150,000	340	VULCAN	SGL.-ACT	OPEN	60	40,000	98,180	18'7"	14"x80"	120	2@3"
120,000	520	VULCAN	SGL.-ACT	OPEN	42	20,000	47,680	20'5"	11¼"x37"	102	3"
100,000	200E5	CONMACO	SGL.-ACT	OPEN	46	20,000	48,000	19'1"	11¼"x56"	110	4"
100,000	40X	RAYMOND	SGL.-ACT	OPEN	64	40,000	62,000	19'1"	11¼"x56"	135	3"
93,340	MRBS 850	MENCK	SGL.-ACT	OPEN	45	18,960	27,890	19'8"	CAGE	142	3"
90,000	030	VULCAN	SGL.-ACT	OPEN	54	30,000	53,470	16'4"	11¼"x56"	150	3"
90,000	300	CONMACO	SGL.-ACT	OPEN	52	30,000	55,390	16'10"	11¼"x56"	150	3"
81,250	8/0	RAYMOND	SGL.-ACT	OPEN	40	25,000	34,000	19'4"	10¼"x25"	135	3"
75,000	30X	RAYMOND	SGL.-ACT	OPEN	70	30,000	52,000	19'1"	—	150	3"
62,500	125E5	CONMACO	SGL.-ACT	OPEN	41	12,500	22,000	18'0"	9¼"x26"	100	2½"

AIR/STEAM HAMMERS - CONTINUED

RATED ENERGY (FT.-LBS.)	MODEL	MANUFACTURER	TYPE	STYLE	BLOWS PER (MIN.)	WT. OF STRIKING PARTS (lbs)	TOTAL WEIGHT (LBS.)	HAMMER LENGTH (FT.-IN.)	JAW DIMENSIONS	INLET PRESSURE (PSI)	INLET SIZE (NPT)
60,000	200	CONMACO	SGL.-ACT	OPEN	55	20000	44,560	15'0"	11¼"x56"	110	3"
60,000	512	VULCAN	SGL.-ACT	OPEN	41	12,000	23,480	18'5"	9¼"x26"	100	2½"
60,000	S-20	MKT	SGL.-ACT	CLOSED	60	20,000	38,650	15'5"	—x36"	150	3"
60,000	020	VULCAN	SGL.-ACT	OPEN	59	20,000	41,670	14'8"	11¼"x37"	120	3"
57,500	115E5	CONMACO	SGL.-ACT	OPEN	42	11,500	21,000	17'9"	9¼"x26"	100	2½"
56,875	5/0	RAYMOND	SGL.-ACT	OPEN	44	17,500	26,450	16'9"	10¼"x25"	150	3"
50,000	510	VULCAN	SGL.-ACT	OPEN	41	10,000	21,480	18'5"	9¼"x26"	83	2½"
50,000	100E5	CONMACO	SGL.-ACT	OPEN	47	10,000	19,500	17'9"	9¼"x26"	100	2½"
48,750	160	CONMACO	SGL.-ACT	OPEN	50	16,250	33,200	13'10"	11¼"x42"	100	3"
50,000	200-C	VULCAN	DIFFER.	OPEN	95	20,000	39,000	13'11"	11¼"x37"	142	4"
48,750	016	VULCAN	SGL.-ACT	OPEN	58	16,250	30,250	13'11"	11¼"x32"	120	3"
48,750	4/0	RAYMOND	SGL.-ACT	OPEN	46	15,000	23,800	16'1"	——	120	2½"
48,750	150-C	RAYMOND	DIFFER.	OPEN	95-105	15,000	32,500	15'9"	——	120	3"
45,200	MRBS 500	MENCK	SGL.-ACT	OPEN	48	11,020	15,210	16'8"	—x26"	142	2½"
44,000	MS-500	MKT	SGL.-ACT	OPEN	40-50	11,000	15,500	16'8"	8½"x26"	115	3"
42,000	140	CONMACO	SGL.-ACT	OPEN	55	14,000	30,750	13'10"	11¼"x42"	100	3"
42,000	014	VULCAN	SGL.-ACT	OPEN	59	14,000	27,500	13'8"	11¼"x32"	110	3"
40,600	3/0	RAYMOND	SGL.-ACT	OPEN	50	12,500	21,000	15'7"	10¼"x25"	120	2½"
40,000	80E5	CONMACO	SGL.-ACT	OPEN	47	8,000	17,500	17'9"	9¼"x26"	80	2½"
40,000	508	VULCAN	SGL.-ACT	OPEN	41	8,000	19,480	18'5"	9¼"x26"	65	2½"
37,500	S-14	MKT	SGL.-ACT	CLOSED	60	14,000	31,700	13'7"	—x36"	100	3"
37,375	115	CONMACO	SGL.-ACT	OPEN	52	11,500	20,830	14'2"	9¼"x32"	100	2½"
36,000	140-C	VULCAN	DIFFER.	OPEN	101	14,000	27,984	12'3"	11¼"x32"	140	3"
32,885	100-C	VULCAN	DIFFER.	OPEN	103	10,000	22,200	14'0"	9¼"x26"	140	2½"
32,500	100	CONMACO	SGL.-ACT	OPEN	55	10,000	19,280	14'2"	9¼"x32"	100	2½"
32,500	65E5	CONMACO	SGL.-ACT	OPEN	50	6,500	12,500	16'10"	8¼"x20"	95	2½"
32,500	506	VULCAN	SGL.-ACT	OPEN	46	6,500	13,025	17'5"	8¼"x20"	100	2"
32,500	2/0	RAYMOND	SGL.-ACT	OPEN	50	10,000	18,550	15'0"	10¼"x25"	110	2"
32,500	010	VULCAN	SGL.-ACT	OPEN	50	10,000	18,780	15'0"	9¼"x26"	105	2½"
32,500	S-10	MKT	SGL.-ACT	CLOSED	55	10,000	22,380	14'1"	—x30"	80	2½"
30,800	MS-350	MKT	SGL.-ACT	OPEN	40-50	7,716	10,500	15'1"	8½"x26"	105	2½"
26,000	80	CONMACO	SGL.-ACT	OPEN	56	8,000	17,280	14'2"	9¼"x32"	80	2½"
26,000	85-C	VULCAN	DIFFER.	OPEN	111	8,525	19,020	12'7"	9¼"x26"	128	2½"
26,000	08	VULCAN	SGL.-ACT	OPEN	50	8,000	16,750	14'10"	9¼"x26"	83	2½"
26,000	S-8	MKT	SGL.-ACT	CLOSED	55	8,000	18,300	14'4"	—x26"	80	2½"
25,000	50E5	CONMACO	SGL.-ACT	OPEN	48	5,000	11,000	16'10"	8¼"x20"	70	2"
25,000	505	VULCAN	SGL.-ACT	OPEN	46	5,000	11,800	17'5"	8¼"x20"	77	2"
24,450	80-C	VULCAN	DIFFER.	OPEN	109	8,000	17,885	12'7"	9¼"x26"	120	2½"
24,450	80-C (HYD)	RAYMOND	DIFFER.	OPEN	110-120	8,000	17,780	11'10"	——	5,100	——
24,450	80-C	RAYMOND	DIFFER.	OPEN	95-105	8,000	17,885	12'2"	——	120	2½"
24,375	0	RAYMOND	SGL.-ACT	OPEN	50	7,500	16,000	15'0"	10¼"x25"	110	2"
24,375	0	VULCAN	SGL.-ACT	OPEN	50	7,500	16,250	15'0"	9¼"x26"	80	2½"
24,000	C-826	MKT	COMPOUND	CLOSED	85-95	8,000	17,750	12'2"	—x26"	125	2½"
19,500	65	CONMACO	SGL.-ACT	OPEN	61	6,500	12,100	12'10"	9¼"x26"	100	2"
19,500	65-C	RAYMOND	DIFFER.	OPEN	110	6,500	14,675	11'8"	9¼"x19"	120	2"
19,500	1-S	RAYMOND	SGL.-ACT	OPEN	58	6,500	12,500	12'9"	7½"x28¼"	100	1½"
19,500	06(106)	VULCAN	SGL.-ACT	OPEN	60	6,500	11,200	13'0"	8¼"x20"	100	2"
19,500	65-C(HYD)	RAYMOND	DIFFER.	OPEN	130	6,500	14,615	12'1"	——	5,000	——
19,200	65-C	VULCAN	DIFFER.	OPEN	117	6,500	14,886	12'1"	8¼"x20"	150	2"
19,150	11B3	MKT	DBL.-ACT	CLOSED	95	5,000	14,000	11'2"	8½"x26"	100	2½"

AIR/STEAM HAMMERS - CONTINUED

RATED ENERGY (FT.-LBS.)	MODEL	MANUFACTURER	TYPE	STYLE	BLOWS PER (MIN.)	WT. OF STRIKING PARTS (lbs)	TOTAL WEIGHT (LBS.)	HAMMER LENGTH (FT.-IN.)	JAW DIMENSIONS	INLET PRESSURE (PSI)	INLET SIZE (NPT)
16,250	S-5	MKT	SGL.-ACT	CLOSED	60	5,000	12,460	13'3"	—24"	80	2"
16,000	C-5(STM)	MKT	DBL.-ACT	CLOSED	100-110	5,000	11,880	—	—x26"	100	2½"
15,100	50-C	VULCAN	DIFFER.	OPEN	117	5,000	11,782	11'0"	8¼"x20"	120	2"
15,000	50	CONMACO	SGL.-ACT	OPEN	64	5,000	10,600	12'10"	9¼"x26"	80	2"
15,000	1(106)	VULCAN	SGL.-ACT	OPEN	60	5,000	9,700	—	8¼"x20"	80	2"
15,000	1	RAYMOND	SGL.-ACT	OPEN	60	5,000	11,000	12'9"	7½"x28¼"	80	1½"
14,200	C-5(AIR)	MKT	COMPOUND	CLOSED	100-110	5,000	11,880	8'9"	—x26"	100	2½"
13,100	10B3	MKT	DBL.-ACT	CLOSED	105	3,000	10,850	9'2"	8½"x24"	100	2½"
8,750	900	BSP	DBL.-ACT	CLOSED	145	1,600	7,100	7'10"	—x20"	90	2"
8,750	9B3	MKT	DBL.-ACT	CLOSED	145	1,600	7,000	84"	8½"x20"	100	2"
7,260	30-C	VULCAN	DIFFER.	OPEN	133	3,000	7,036	8'11"	7¼"x19"	120	1½"
7,260	2	VULCAN	SGL.-ACT	OPEN	70	3,000	6,700	11'7"	7¼"x19"	80	1½"
4,700	700N	BSP	DBL.-ACT	CLOSED	225	850	6,630	7'5"	—x18"	90	2"
4,150	7	MKT	DBL.-ACT	CLOSED	225	800	5,000	6'1"	6½"x21"	100	1½"
4,000	DGH-900	VULCAN	DIFFER.	CLOSED	328	900	5,000	6'9"	VARIES	78	1½"
3,000	600N	BSP	DBL.-ACT	CLOSED	250	500	4,800	7'2"	—x15"	90	1½"
2,500	6	MKT	DBL.-ACT	CLOSED	275	400	2,900	5'3"	6½"x15"	100	1¼"
1,200	500N	BSP	DBL.-ACT	CLOSED	330	200	2,520	5'11"	—x14"	90	1¼"
1,000	5	MKT	DBL.-ACT	CLOSED	300	200	1,500	4'7"	6"x11"	100	1¼"
386	DGH-100D	VULCAN	DIFFER.	CLOSED	303	100	786	4'2"	4¼"x8¾"	60	1"
—	3	MKT	DBL.-ACT.	CLOSED	400	68	675	4'5"	—	100	1"
160	200	BSP	DBL.-ACT.	CLOSED	500	48	343	2'9"	—	90	¾"
—	2	MKT	DBL.-ACT	CLOSED	500	48	343	2'5"	3¼"x8¼"	100	¾"
350	300	BSP	DBL.-ACT	CLOSED	400	63	675	4'10"	—	90	1"
—	1	MKT	DBL.-ACT	CLOSED	500	21	145	3'3"	—	100	¾"

STEEL SHEET PILING SPECIFICATIONS CHART

General Notes

The need for a master listing of available steel sheet piling sections and their properties is well recognized. A few years ago there were two established producers offering eight shapes. Currently there are more than ten domestic and foreign manufacturers listing almost 200 sheet piling sections in their catalogs.

Pile Buck® invited these producers to supply data regarding their standard sections for inclusion in this publication. In addition to dimensional and strength properties, Pile Buck® 's charts also provide other information of interest to the user, such as the method of manufacture and type of interlocks.

Column headings in the charts are identified by numbers. Pile Buck® recommends that the charts be used in conjunction with the descriptive information provided herein for each column.

*NOTICE: The data and commentary contained within this steel sheet piling comparison chart is for general information purposes only. It is provided without warranty of any kind. Pile Buck® Inc. shall not be held responsible for any errors, omissions or misuse of any of the enclosed information and hereby disclaims any and all liability resulting from the ability or inability to use the information contained within. Anyone making use of this material does so at his/her risk. In no event will Pile Buck®, Inc. be held liable for any damages including lost profits, lost savings or other incidental or consequential damages arising from the use or inability to use the information contained within. Pile Buck® Inc. suggests contacting the manufacturer of any given steel sheet piling section to ensure its suitability for a particular application.

Ball and Socket (BS)

Double Jaw (DJ)

Single Jaw (SJ)

Double Hook (DH)

Thumb and Finger - three point contact (TF)

Thumb and Finger - one point contact (TFX)

Hook and Grip (HG)

COLUMN 1
SECTION DATA

Manufacturer's designation to identify the section. In some cases, this also describes the section. For example, PZ-27 describes a Z-shaped section with a weight of 27 pounds per square foot.

COLUMN 2
PRODUCER AND COUNTRY OF ORIGIN

"Bethlehem USA" identifies Bethlehem Steel Corp. of the United States as producer of that section, etc.

COLUMN 5
HOT ROLLED AND COLD FORMED

Hot Rolled (HR) indicates that the section shown is produced by the traditional steel hot-mill procedure. A semi-finished rough shape is reduced during a series of rolling stages to final form. Metal thickness of flanges and webs may be varied and interlocks are shaped by the flow of hot metal.

Cold Forming (CF) describes a process whereby the desired sheet piling shape is obtained by passing pre-finished sheet-steel through a series of rolls while in the cold state. While passing through these series of rolls, the interlock is also formed by bending the flange ends into a "hook and grip" type interlock. The section produced is of constant thickness.

COLUMN 6
SECTION AREA

For the Z, U and A sections, cross-sectional area is listed as square inches per foot of wall, which is according to current practice. Areas shown for flat piling are based on the single section.

COLUMN 7
NOMINAL WIDTH

Nominal width indicates the nominal dimension from the centerline of one interlock to the centerline of the second interlock of the same pile. Since a run of wall consists of a series of interlocked sheets, the theoretical number of pieces to complete a certain run can be determined. Due to mill tolerances, interlock play or installation practices, etc. actual walls may run shorter or longer than theoretical.

COLUMN 8 and COLUMN 9
WEIGHT

Column 8 lists the weight in pounds per foot of bar for each section. Column 9 indicates the weight, in pounds, of a one-square foot area of projected wall. This provides a quick calculation of the total weight in the wall. It also offers a comparison of alternative sections from an economy standpoint. Because of it's importance, this column (wt. per sq. ft.) has been printed in bold type.

COLUMN 10
WALL DEPTH

This detail indicates the maximum theoretical distance between the outboard and inboard faces of a sheet pile wall when the piling is interlocked in a conventional manner.

COLUMN 11 and COLUMN 12
WEB THICKNESS and FLANGE THICKNESS

These are nominal dimensions useful for evaluating the relative service life of the sections. Also, driving conditions may require selection of thicker sections to minimize driving damage.

COLUMN 3
SHAPE

This column describes the profile of the cross-sectional area. Sheet piling shapes in the Pile Buck® charts are grouped into four traditional shape classifications. These are:

Z-type (Z) used for intermediate to deep wall construction.

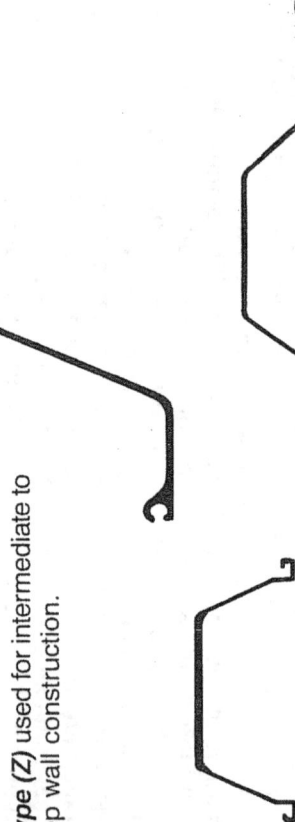

Larssen and other "U" types (U) used for applications similar to Z-piles.

Flat or straight - web types (SA), (S) with strong interlocks, and little beam strength, for filled cell construction.

Arch shaped and lightweight "gauge" sheets (A) used for shallower wall construction.

COLUMN 4
INTERLOCKS

There are no established industry standards for interlocks, however several basic designs have evolved. All manufacturers have the same objectives for their interlock designs: (1) to provide permanent connection of individual sheets in order to form a continuous, relatively water or earth - tight wall (2) the interlocks should permit reasonably free sliding to connect sheets during installation, (3) interlocks designed for tension applications should provide a guaranteed minimum pull strength and also allow some minimum swing between interlocks in order to form a circle. Following are various types of interlocks listed in the charts:

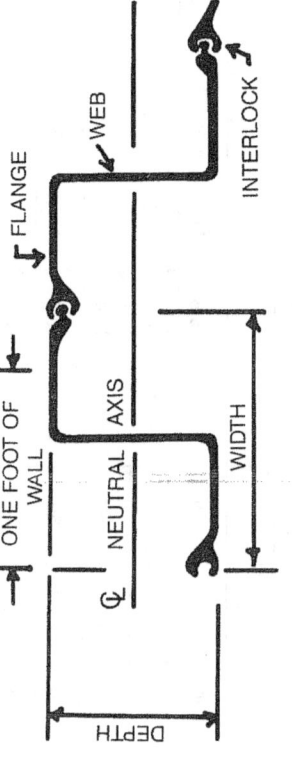

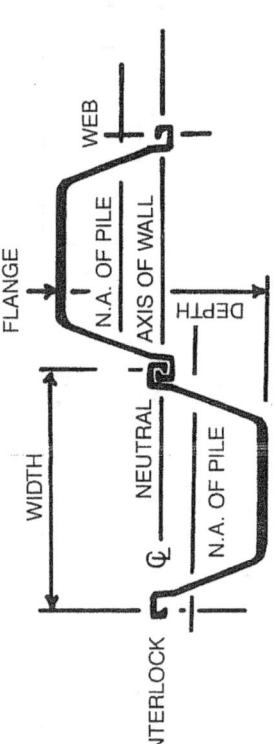

COLUMN 13

MOMENT OF INERTIA

The **Moment of Inertia** (I, inches, fourth-power) of any structural shape is obtained as the product of cross-sectional area and squared distance from a reference axis. For most structurals used as beams, including sheet piling, the important reference axis is the x-x axis which passes through the centroid of the section. Moment of Inertia is a property necessary for the proper sizing of beams so that stresses due to bending and shearing forces will not exceed safe limits.

The values for moment of inertia shown in Column 13 are based on a foot of wall, to coincide with design practice. **It should be noted that the (I) listed for the "U" shaped sections is based on the manufacturers' assumption that these sections form a continuous, solidized wall similar to Z-type, rather than a series of shapes with lesser properties. This assumption of "combined" section properties is discussed further in the commentary on Column 14, Section Modulus.**

COLUMN 14

SECTION MODULUS

The **Section Modulus** (inches, cubed) of a structural shape is the convenient expression of the ratio I/c in the equation $s = Mc/I$, the calculation for determining maximum stress due to bending moments in beams. Section Modulus is the most important property, along with weight, for sheet piling sections used to carry horizontal loads from earth, water and surcharges. Since Section Modulus is a function of Moment of Inertia (I in Column 13), it is a measure of area and depth. The heavier and deeper the section, the larger the value of Section Modulus. In the equation above, given the bending movement (M) and a maximum value for allowable stress, the required minimum Section Modulus can be calculated and a section selected which provides that value. Section Moduli for the various Z, U, and A sections have been listed in descending order for convenience, and are based on a linear foot of wall according to conventional design practice.

Larssen and Other "U" Shaped Sections

It is important to note that Column 14 - Chart 2 contains *two* values of Section Modulus for this group. The higher number in the left-hand column is based on the moment of inertia of a "combined section" as listed in Column 13. The drawings below illustrate the make-up of a wall constructed of Z-piles and one constructed of "U" type piles. The Z-pile wall contains a continuous web and the neutral axis of each pile and the wall coincide. The Section Modulus of the wall is also the Section Modulus of each pile.

The wall constructed of "U" type shapes however, contains a potential location for slippage along the neutral axis of the wall due to longitudinal shear. If full transfer of shear across this area does not occur, the maximum section modulus cannot be developed.

The manufacturers of "U" shaped sections generally assume that friction, since these areas are generally not coated. Some manufacturers do not publish this data or publish for one condition only. Check with the supplier for additional information.

COLUMN 17(a)

INTERLOCK LOCATION

For the Z-shaped piles, trough (U) and pan-shaped (A), piles, this column indicates the location of the interlocks so that consideration may be given to the correct use of section modulus. Where interlocks are shown to be on the centerline (℄), consult the manufacturer prior to and regarding the use of values in Column 14.

COLUMN 17(b)

INTERLOCK STRENGTH

For flat or straight sheet piling sections listed in Column 2 as (S) or (SA) shapes, interlock strength is of greatest importance while section modulus is of little importance. The minimum interlock strengths available for the various sections listed are shown in kips per inch of interlock. Where a range of strengths is shown, the higher values are generally obtainable through the use of higher strength steel.

pressure or artificial measures will affect this transfer and accordingly publish the combined values.

A traditional and more conservative approach in the United States has been to consider a wall constructed of this type sheet to be a series of single sheet piles with wall properties calculated accordingly. Pile Buck® has provided the Section Modulus of *single sections* for the "U" shaped piles in the right-hand column of Column 14. **It is recommended that users of this chart consult with the manufacturer regarding the proposed application and selection of a steel sheet piling section which contains a *safe* Section Modulus for design calculations.**

COLUMN 15 and COLUMN 16

SURFACE AREA

These columns provide information required to estimate protective coatings. Column 15 shows total area, two sides, in square feet per foot of pile. Column 16 shows the "nominal" coating area which excludes the interlock surfaces

Flat sheet piling shapes are used in circular structures and a certain minimum deflection or "swing" per interlock must be assured in order to close the circles. The manufacturer should be contacted regarding this important function of his interlock design.

ADDITIONAL NOTES:

* Additional sectional modulus can be obtained by welding cover plates to the flanges of conventional sections.
* Where regular grades of steel combined with section modulus provide inadequate safety factors against yield, higher strength steels such as ASTM 572 Grades may be considered. Suppliers should be consulted regarding availability.
* High section modulus walls are also obtainable from special systems such as HZ (Arbed - Luxemburg) and PSP (Peine - Salzgitter - W. Germany). Space did not permit incorporation of this data at this time.

PROPERTIES CHART 1
"Z" SHAPED SHEET PILING

1 Section Designation	2 Mfr. & Country	3 Shape	4 Type Interlock	5 Hot Rolled Or Cold Formed	6 Area sq. in. Per ft of wall	7 Nominal width, in. Per section	8 Weight In Pounds Per lin ft of pile	9 Weight In Pounds Per sq ft of wall	10 Wall Depth (height) (inches)	11 Web Thickness (inches)	12 Flange Thickness (inches)	13 Moment of Inertia, in.⁴ Per ft of wall	14 Section Modulus, in.³ Per ft of wall	15 Surface Area Total Area	16 Surface Area Nominal Coating Area	17(a) Interlock Location in wall
FSPZ45	Nippon, Japan	Z	DJ	HR	17.50	15.70	78.00	**59.4**	14.40	0.520	0.862	611.0	**84.6**	5.77	NP	Both faces
BZ42	Arbed, Lux	Z	DJ	HR	16.30	19.69	90.92	**55.4**	13.94	0.551	0.945	544.4	**78.2**	NP	5.48	Both faces
FSPZ38	Nippon, Japan	Z	DJ	HR	14.43	15.70	64.50	**49.2**	14.30	0.449	0.677	507.0	**70.7**	5.64	NP	Both faces
BZ37	Arbed, Lux	Z	DJ	HR	14.08	19.69	78.62	**47.9**	13.78	0.472	0.787	467.7	**67.9**	NP	5.45	Both faces
5RU3	Frodingham, UK	Z	SJ	HR	14.89	16.75	70.67	**50.6**	12.28	0.500	0.700	375.2	**61.1**	5.41	5.12	Both faces
PZ40	Bethlehem, USA	Z	BS	HR	11.76	19.69	65.60	**40.0**	16.10	0.500	0.600	490.8	**60.7**	5.83	5.37	Both faces
FSPZ32	Nippon, Japan	Z	DJ	HR	12.72	15.70	56.80	**43.2**	13.50	0.409	0.559	403.0	**59.5**	5.58	NP	Both faces
BZ32	Arbed, Lux	Z	DJ	HR	12.52	19.69	69.82	**42.5**	13.86	0.453	0.748	411.7	**59.4**	NP	5.25	Both faces
H215	Hoesch, W. Ger	Z	SJ	HR	12.95	20.67	75.94	**44.0**	13.39	0.472	0.740	392.2	**58.6**	NP	NP	Both faces
PZ35	Bethlehem, USA	Z	BS	HR	10.29	22.64	66.00	**35.0**	14.90	0.500	0.600	361.2	**48.5**	5.83	5.37	Both faces
AZ26	Arbed, Lux	Z	DH	HR	9.35	24.80	65.72	31.75	16.81	.0480	0.512	406.5	**48.4**	NP	5.87	Both faces
H175	Hoesch, W. Ger	Z	SJ	HR	10.54	20.67	61.76	**35.8**	13.39	0.394	0.551	323.7	**48.4**	NP	NP	Both faces
BZ26	Arbed, Lux	Z	DJ	HR	10.20	19.69	56.85	**34.7**	13.78	0.394	0.520	331.9	**48.2**	NP	5.25	Both faces

"Z" SHAPED SHEET PILING - CONTINUED

1 Section Designation	2 Mfgr. & Country	3 Shape	4 Type Interlock	5 Hot Rolled Or Cold Formed	6 Area sq. in. Per ft of wall	7 Nominal width, in. Per section	8 Per lin ft of pile	9 Per sq ft of wall	10 Wall Depth (height) (inches)	11 Web Thickness (inches)	12 Flange Thickness (inches)	13 Moment of Inertia, in.⁴ Per ft of wall	14 Section Modulus, in.³ Per ft of wall	15 Total Area	16 Nominal Coating Area*	17(a) Interlock Location in wall
							Weight in Pounds							Surface Area Sq ft per lin ft of pile		
FSPZ25	Nippon, Japan	Z	DJ	HR	11.14	15.70	49.70	37.9	12.00	0.378	0.512	280.0	46.7	5.28	NP	Both faces
4N	Frodingham, UK	Z	SJ	HR	10.30	19.00	55.39	35.0	12.99	0.410	0.550	291.7	44.9	5.28	5.05	Both faces
CZ148	Casteel, Canada	Z	HG	CF	8.88	24.02	60.68	30.3	13.39	0.433	0.433	273.9	40.9	6.92	5.90	Both faces
CZ141	Casteel, Canada	Z	HG	CF	8.48	24.02	57.92	28.9	13.39	0.413	0.413	261.4	39.1	6.92	5.90	Both faces
BZ20.7L	Arbed, Lux	Z	DJ	HR	8.86	22.64	56.85	30.2	12.80	0.394	0.520	245.9	38.4	NP	5.25	Both faces
H155	Hoesch, W. Ger	Z	SJ	HR	9.31	20.67	54.70	31.7	11.82	0.388	0.504	219.7	37.2	NP	NP	Both faces
RZ11	Unimetal, France	Z	BS	HR	8.79	19.10	47.7	30.0	11.33	0.375	0.394	221.9	37.2	NP	5.15	Both faces
CZ128	Casteel, Canada	Z	HG	CF	7.68	24.02	52.28	26.2	13.39	0.375	0.375	236.5	35.3	6.92	5.90	Both faces
SPZ26	Syro, USA	Z	HG	CF	7.42	24.00	51.62	25.8	13.38	0.375	0.375	232.9	34.8	NP	5.90	Both faces
AZ18	Arbed, Lux	Z	DH	HR	7.09	24.80	49.99	24.17	14.69	0.375	0.375	250.4	33.5	NP	5.64	Both faces
PLZ25	Bethlehem, USA	Z	BS	HR	7.30	24.00	49.60	24.8	13.50	0.375	0.375	223.3	32.8	5.98	5.52	Both faces
SZ27	Superior, USA	Z	HG	CF	7.57	21.25	46.90	26.5	12.00	0.375	0.375	206.6	32.4	6.20	5.75	Both faces
SZ27	Shoreline, USA	Z	HG	CF	7.57	21.25	46.90	26.5	12.00	0.375	0.375	206.6	32.4	6.20	5.75	Both faces
CZ114	Casteel, Canada	Z	HG	CF	6.88	24.02	46.83	23.4	13.39	0.335	0.335	211.6	31.6	6.92	5.90	Both faces
3NA	Frodingham, UK	Z	SJ	HR	7.80	19.00	42.07	26.6	12.01	0.375	0.380	188.1	31.4	5.18	4.99	Both faces
SPZ23	Syro, USA	Z	HG	CF	6.61	24.00	45.22	23.1	13.38	0.335	0.335	208.7	31.3	NP	5.90	Both faces
BZ17	Arbed, Lux	Z	DJ	HR	7.89	19.69	43.96	26.8	11.81	0.375	0.394	183.7	31.1	NP	4.86	Both faces
134N	Hoesch, W. Ger	Z	SJ	HR	7.47	20.67	45.16	26.2	11.82	0.375	0.375	181.3	30.7	NP	NP	Both faces
RZ10	Unimetal, France	Z	BS	HR	7.75	21.65	47.71	26.4	11.26	0.375	0.394	172.1	30.5	NP	5.15	Both faces
BZ16.4	Arbed, Lux	Z	DJ	HR	7.75	19.69	43.27	26.4	11.81	0.375	0.375	180.0	30.5	NP	4.86	Both faces
PLZ23	Bethlehem, USA	Z	BS	HR	6.64	24.00	45.2	22.6	13.50	0.335	0.335	203.8	30.2	5.98	5.52	Both faces
PZ27	Bethlehem, USA	Z	BS	HR	7.94	18.00	40.50	27.0	12.00	0.375	0.375	184.2	30.2	4.94	4.48	Both faces
3N(M)	Frodingham, UK	Z	SJ	HR	8.28	19.00	44.55	28.1	11.16	0.380	0.440	168.1	30.2	4.94	4.73	Both faces
SZ24	Superior, USA	Z	HG	CF	6.84	21.25	42.60	24.1	12.00	0.340	0.340	186.5	29.5	6.20	5.75	Both faces
SZ24	Shoreline, USA	Z	HG	CF	6.84	21.25	42.60	24.1	12.00	0.340	0.340	186.5	29.5	6.20	5.75	Both faces
SZ222	Superior, USA	Z	HG	CF	6.84	21.25	38.40	21.7	12.00	0.312	0.312	170.3	26.8	6.20	5.75	Both faces
SZ222	Shoreline, USA	Z	HG	CF	6.84	21.25	38.40	21.7	12.00	0.312	0.312	170.3	26.8	6.20	5.75	Both faces

AZ13	Arbed, Lux	N	DH	HR	6.47	26.38	48.38	21.92	11.93	0.375	0.375	144.3	24.2	NP	5.45	Both faces
BZ12.1L	Arbed, Lux	N	DJ	HR	6.74	22.64	43.27	22.9	10.28	0.375	0.375	115.5	22.5	NP	4.86	Both faces
H116	Hoesch, W. Ger	N	SJ	HR	6.99	20.67	40.92	23.8	9.85	0.354	0.366	109.9	22.3	NP	NP	Both faces
BZ12	Arbed, Lux	N	DJ	HR	6.76	19.69	37.70	23.0	9.57	0.335	0.335	106.4	22.3	NP	4.63	Both faces
2N	Frodingham, UK	N	SJ	HR	6.76	19.00	36.42	23.0	9.25	0.330	0.380	99.0	21.4	4.66	4.48	Both faces
2NRD3	Frodingham, UK	N	SJ	HR	7.18	19.00	34.43	21.6	9.22	0.350	0.300	92.7	20.1	4.66	4.48	Both faces
SZ22	Superior, USA	N	HG	CF	6.38	25.20	46.90	22.0	9.00	0.375	0.375	95.7	19.6	6.15	5.70	Both faces
SZ22	Shoreline, USA	N	HG	CF	6.38	25.20	46.90	22.0	9.00	0.375	0.375	95.7	19.6	6.15	5.70	Both faces
SPZ23.5	Syro, USA	N	HG	CF	6.68	24.00	46.76	23.4	8.40	.0375	0.375	81.2	19.3	NP	5.40	Both faces
CZ113	Casteel, Canada	N	HG	CF	6.80	21.65	41.70	23.1	7.88	0.375	0.375	72.7	18.4	5.53	4.78	Both faces
SPZ22	Syro, USA	N	HG	CF	6.18	24.00	43.28	21.6	8.40	0.350	0.350	77.0	18.3	NP	5.40	Both faces
PZ22	Bethlehem, USA	N	BS	HR	6.47	22.00	40.30	22.0	9.00	0.375	0.375	84.4	18.1	4.94	4.48	Both faces
SZ20	Superior, USA	N	HG	CF	5.80	25.20	42.60	20.3	9.00	0.340	0.340	86.63	17.8	6.15	5.70	Both faces
SZ20	Shoreline, USA	N	HG	CF	5.80	25.20	42.60	20.3	9.00	0.340	0.340	86.63	17.8	6.15	5.70	Both faces
CZ107	Casteel, Canada	N	HG	CF	6.44	21.65	39.58	21.9	7.88	0.354	0.354	68.8	17.5	5.53	4.78	Both faces
Z80	CMRM, Canada	N	HG	CF	5.73	22.00	35.90	19.6	8.25	0.315	0.315	69.0	16.7	5.70	5.20	Both faces
SPZ19.5	Syro, USA	N	HG	CF	5.76	24.00	39.2	19.6	8.40	0.315	0.315	69.49	16.6	NP		Both faces
SZ18	Superior, USA	N	HG	CF	5.22	25.20	38.40	18.3	9.00	0.312	0.312	86.6	16.2	6.15	5.70	Both faces
SZ18	Shoreline, USA	N	HG	CF	5.22	25.20	38.40	18.3	9.00	0.312	0.312	79.6	16.2	6.15	5.70	Both faces
BZ8.6	Arbed, Lux	N	DJ	HR	6.43	21.65	39.51	21.9	7.60	0.375	0.375	60.8	16.0	NP	4.40	Both faces
Z75	CMRM, Canada	N	HG	CF	5.40	22.00	33.7	18.4	8.20	0.295	0.295	63.9	15.6	5.70	5.70	Both faces
CZ95	Casteel, Canada	N	HG	CF	5.72	21.65	35.15	19.5	7.88	0.315	0.315	61.2	15.5	5.53	4.78	Both faces
CZ70	CMRM, Canada	N	HG	CF	5.04	22.00	31.50	17.2	8.15	0.276	0.276	58.6	14.4	5.70	5.20	Both faces
BZ7	Arbed, Lux	N	DJ	HR	5.57	21.65	34.27	19.0	7.48	0.315	0.315	52.5	14.0	NP	4.40	Both faces
H95	Hoesch, W. Ger	N	SJ	HR	5.72	20.67	33.52	19.5	7.49	0.315	0.315	52.2	14.0	5.70	NP	Both faces
CZ84	Casteel, Canada	N	HG	CF	5.05	21.65	31.05	17.2	7.88	0.276	0.276	53.6	13.6	5.53	4.78	Both faces
1N	Frodingham, UK	N	SJ	HR	5.95	19.00	32.13	20.3	6.69	0.350	0.350	44.3	13.3	4.27	4.09	Both faces
Z65	CMRM, Canada	N	HG	CF	4.68	22.00	29.20	15.9	8.10	0.256	0.256	54.1	13.4	5.70	5.20	Both faces
1BXN	Frodingham, UK	N	SJ	HR	7.87	18.75	41.73	26.7	5.63	0.500	0.500	36.0	12.8	4.10	3.91	Both faces

"Z" SHAPED SHEET PILING - CONTINUED

1	2	3	4	5	6	7		8	9	10	11	12	13	14	15	16	17(a)
						Nominal width, in.		Weight in Pounds		Wall Depth (height) (inches)	Web Thickness (inches)	Flange Thickness (inches)	Moment of Inertia, in.⁴ Per ft. of wall	Section Modulus, in.³ Per ft. of wall	Surface Area Sq ft per lin ft of pile		Interlock Location in wall
Section Designation	Mfgr. & Country	Shape	Type Interlock	Hot Rolled Or Cold Formed	Area sq. in. Per ft of wall	Per section		Per lin ft of pile	Per sq ft of wall						15 Total Area	16 Nominal Coating Area*	
SZ15	Superior, USA	Z	HG	CF	4.39	20.00		25.50	15.3	7.50	0.250	0.250	36.4	10.4	5.00	4.60	Both faces
SZ15	Shoreline, USA	Z	HG	CF	4.39	20.00		25.50	15.3	7.50	0.250	0.250	36.4	10.4	5.00	4.60	Both faces
SZ14	Superior, USA	Z	HG	CF	4.01	20.00		23.40	14.0	7.50	0.239	0.239	36.4	9.8	5.00	4.60	Both faces
SZ14	Shoreline, USA	Z	HG	CF	4.01	20.00		23.40	14.0	7.50	0.239	0.239	36.4	9.8	5.00	4.60	Both faces
SZ12	Superior, USA	Z	HG	CF	3.75	20.00		21.90	13.1	7.50	0.209	0.209	36.4	8.6	5.00	4.60	Both faces
SZ12	Shoreline, USA	Z	HG	CF	3.75	20.00		21.90	13.1	7.50	0.209	0.209	36.4	8.6	5.00	4.60	Both faces

NOTE: NP = NOT PUBLISHED

The following Z-shaped sheet piling sections were formerly offered by the two major domestic manufacturers. More recently the section designations were standardized to "PZ", a move sponsored by the American Iron and Steel Institute. Except for the ZP 27 (PZ 27) section, these sections are no longer available from the mill, having been replaced by more efficient sections by the remaining producer. Our purpose in listing these sections and their properties is to provide a continuing reference source for this information, which is no longer published.

Section Designation	Mfgr. & Country	Shape	Type Interlock	Hot Rolled Or Cold Formed	Area sq. in. Per ft of wall	Nominal width, in. Per section	Per lin ft of pile	Per sq ft of wall	Wall Depth (inches)	Web Thickness (inches)	Flange Thickness (inches)	Moment of Inertia, in.⁴ Per ft. of wall	Section Modulus, in.³ Per ft. of wall	Total Area	Nominal Coating Area	Interlock Location in wall
MZ38	US Steel, USA	Z	BS	HR	16.8	18.0	57.0	38.0	12.0	0.375	0.500	281	46.8	NP	NP	Both faces
ZP38	(1) Bethlehem, USA	Z	BS	HR	16.8	18.0	57.0	38.0	12.0	0.375	0.500	281	46.8	NP	NP	Both faces
MZ32	US Steel, USA	Z	BS	HR	16.5	21.0	53.0	32.0	11.5	0.375	0.500	220	38.3	NP	NP	Both faces
ZP32	(2) Bethlehem, USA	Z	BS	HR	16.5	21.0	56.0	32.0	11.5	0.375	0.500	220	38.3	NP	NP	Both faces
MZ27	US Steel, USA	Z	BS	HR	11.9	18.0	40.5	27.0	12.0	0.375	0.375	184	30.2	NP	NP	Both faces
ZP27	(3) Bethlehem, USA	Z	BS	HR	7.94	18.00	40.50	27.0	12.00	0.375	0.375	184.2	30.2	4.94	4.48	Both faces

(1) SAME AS AISI PZ38   (2) SAME AS AISI PZ32   (3) SAME AS AISI PZ27

PROPERTIES CHART 2
LARSSEN TYPE AND OTHER "U" SHAPED SHEET PILING
LISTED IN DESCENDING ORDER BASED ON SECTION MODULUS

1	2	3	4	5	6	7	8	9	10	11	12	13	14		15	16	17(a)
						Nominal width, in.	Weight in Pounds		Wall Depth (height) (inches)	Web Thickness (inches)	Flange Thickness (inches)	Combined Moment of Inertia, in.⁴ Per ft. of wall	Section Modulus, in.³		Surface Area Sq ft per lin ft of pile		Interlock Location in wall
Section Designation	Mfgr. & Country	Shape	Type Interlock	Hot Rolled Or Cold Formed	Area sq. in. Per ft of wall	Per section	Per lin ft of pile	Per sq ft of wall					Combined Per ft. of wall	Single Section Per ft. of wall	Total Area	Nominal Coating Area*	
6H	Frodingham, UK	U	DJ	HR	19.89	16.54	93.20	67.6	17.32	0.550	1.130	816.13	94.2	35.5	6.13	5.55	℄
6M	Frodingham, UK	U	DJ	HR	18.76	16.53	88.03	63.9	17.32	0.550	1.000	753.23	87.0	35.3	6.15	5.57	℄
6L	Frodingham, UK	U	DJ	HR	17.47	16.53	81.98	59.5	17.32	0.550	0.870	677.01	78.1	33.2	6.17	5.59	℄
L6	Unimetal, France	U	DJ	HR	17.43	16.54	81.88	59.4	17.24	0.591	0.925	673.7	78.1	34.3	NP	5.68	℄
FSP6L	Nippon, Japan	U	DJ	HR	14.46	19.70	80.60	49.2	17.72	1.090	1.090	630.0	71.1	25.3	6.00	NP	℄

LARSSEN TYPE AND OTHER "U" SHAPED SHEET PILING - CONTINUED

1	2	3	4	5	6	7 Nominal width, in. Per section	8 Weight in Pounds Per lin ft of pile	9 Per sq ft of wall	10 Wall Depth (height) (inches)	11 Web Thickness (inches)	12 Flange Thickness (inches)	13 Combined Moment of Inertia, in.^{4} Per ft. of wall	14 Section Modulus, in.^{3} Combined Per ft. of wall	14 Section Modulus, in.^{3} Single Section Per ft. of wall	15 Surface Area Sq ft per lin ft of pile Total Area	16 Surface Area Sq ft per lin ft of pile Nominal Coating Area*	17(a) Interlock Location in wall
Section Designation	Mfgr. & Country	Shape	Type Interlock	Hot Rolled Or Cold Formed	Area sq. in. Per ft of wall												
32W	Frodingham, UK	U	DJ	HR	11.91	20.67	69.62	40.4	17.87	0.410	0.670	534.6	59.8	22.2	6.56	6.04	℄
PU32	Arbed, Lux	U	DJ	HR	11.48	23.62	77.01	39.1	17.80	0.433	0.768	529.2	59.5	20.0	NP	6.07	℄
L5S	Unimetal, France	U	DJ	HR	12.76	19.69	71.23	43.4	17.72	0.453	0.811	527.3	59.5	25.5	NP	5.81	℄
FSP5L	Nippon, Japan	U	DJ	HR	12.64	16.50	70.60	43.0	15.74	0.957	0.957	461.0	58.6	19.3	5.74	NP	℄
L25	Hoesch, W. Ger	U	DJ	HR	12.37	19.70	69.22	42.2	16.55	0.453	0.787	467.9	56.5	21.2	NP	NP	℄
L24	Hoesch, W. Ger	U	DJ	HR	10.54	19.70	58.80	35.8	16.55	0.394	0.614	384.5	46.5	20.3	NP	NP	℄
L4S	Unimetal, France	U	DJ	HR	10.58	19.69	59.13	36.0	17.32	0.394	0.610	401.3	46.5	20.5	NP	5.74	℄
PU25	Arbed, Lux	U	DJ	HR	9.45	23.62	63.23	32.2	17.80	0.394	0.559	413.7	46.5	18.2	NP	6.07	℄
25W	Frodingham, UK	U	DJ	HR	10.06	20.67	59.07	34.3	17.87	0.410	0.480	415.4	46.5	21.5	6.59	6.07	℄
L64	Hoesch, W. Ger	U	DJ	HR	9.59	23.64	64.18	32.6	16.55	0.394	0.595	369.1	44.6	17.7	NP	NP	℄
4A	Frodingham, UK	U	DJ	HR	11.15	15.75	49.73	37.9	15.00	0.380	0.620	328.9	43.9	18.9	5.33	4.81	℄
FSP4	Nippon, Japan	U	DJ	HR	11.46	15.70	51.10	38.9	13.38	0.610	0.610	283.0	42.2	16.9	5.28	NP	℄
FSP4A	Nippon, Japan	U	DJ	HR	11.11	15.70	49.70	37.9	14.56	0.634	0.634	305.0	41.8	18.6	5.15	NP	℄
L63	Hoesch, W. Ger	U	DJ	HR	8.69	23.64	58.26	29.49	16.55	0.394	0.480	311.6	37.8	17.0	NP	NP	℄
20W	Frodingham, UK	U	DJ	V	8.88	20.67	51.94	30.15	15.75	0.360	0.440	294.2	37.4	16.5	6.31	5.83	℄
L23	Hoesch, W. Ger	U	DJ	HR	9.31	19.70	52.08	31.7	16.55	0.394	0.453	307.6	37.2	19.6	NP	NP	℄
PU20	Unimetal, France	U	DJ	HR	8.50	23.62	56.92	28.9	15.75	0.382	0.488	292.0	37.2		NP	5.87	℄
PU20	Arbed, Lux	U	DJ	HR	8.50	23.62	56.92	28.9	15.75	0.382	0.488	292.0	37.2		NP	5.87	℄
L3S	Unimetal, France	U	DJ	HR	9.50	19.69	53.08	32.4	14.96	0.394	0.555	278.3	37.2	16.8	NP	5.51	℄
L43	Hoesch, W. Ger	U	DJ	HR	10.02	19.70	54.77	34.0	16.54	0.473	0.473	246.8	30.9	18.0	NP	NP	℄
16W	Frodingham, UK	U	DJ	HR	7.84	20.67	45.90	26.6	13.70	0.410	0.340	204.0	29.8	12.3	5.93	5.47	℄
L2S	Unimetal, France	U	DJ	HR	8.36	19.69	46.77	28.5	13.39	0.354	0.484	201.4	29.8	12.8	NP	5.28	℄
PU16	Arbed, Lux	U	DJ	HR	7.51	23.62	50.20	25.6	14.96	0.354	0.472	223.5	29.8	12.5	NP	5.51	℄
L3N	Unimetal, France	U	DJ	V	9.35	15.75	41.66	31.7	11.42	0.354	0.512	169.9	29.8	12.0	NP	4.63	℄
FSP3A	Nippon, Japan	U	DJ	HR	8.79	15.70	39.20	29.9	11.82	0.516	0.516	167.0	28.3	11.7	4.72	4.72	℄
3	Frodingham, UK	U	DJ	HR	9.35	15.75	41.80	31.8	9.76	0.330	0.560	124.3	25.3	10.5	4.70	4.20	℄
LIII	Hoesch, W. Ger	U	DJ	HR	9.31	15.75	41.66	31.7	9.73	0.312	0.559	122.1	25.1	10.3	NP	NP	℄

LARSSEN TYPE AND OTHER "U" SHAPED SHEET PILING - CONTINUED

1	2	3	4	5	6	7	8	9	10	11	12	13	14 Section Modulus, in.³		15 Surface Area Sq ft per lin ft of pile	16	17(a)
Section Designation	Mfgr. & Country	Shape	Type Interlock	Hot Rolled Or Cold Formed	Area sq. in. Per ft of wall	Nominal width, in. Per section	Weight in Pounds Per lin ft of pile	Per sq ft of wall	Wall Depth (height) (inches)	Web Thickness (inches)	Flange Thickness (inches)	Combined Moment of Inertia, in.⁴ Per ft. of wall	Combined Per ft. of wall	Single Section Per ft. of wall	Total Area	Nominal Coating Area	Interlock Location in wall
FSP3	Nippon, Japan	U	DJ	HR	9.02	15.70	40.30	30.7	9.84	0.512	0.512	123.0	24.9	10.4	4.72	NP	℄
CU122	Casteel, Canada	U	HG	CF	7.32	22.66	41.17	25.0	14.18	0.394	0.394	164.8	23.3	14.3	5.94	5.00	℄
L22	Hoesch, W. Ger	U	DJ	HR	7.32	19.70	40.99	25.0	13.40	0.354	0.394	155.6	23.3	12.5	NP	NP	℄
PU12	Arbed, Lux	U	DJ	HR	6.61	23.62	44.28	22.5	14.17	0.354	0.386	157.8	22.3	11.3	NP	5.28	℄
PU12	Unimetal, France	U	DJ	HR	6.61	23.62	44.28	22.5	14.17	0.354	0.386	157.8	22.3	11.3	NP	5.28	℄
CU116	Casteel, Canada	U	HG	CF	6.99	22.66	44.89	23.8	14.18	0.375	0.375	158.2	22.3	13.6	5.94	5.00	℄
12W	Frodingham, UK	U	DJ	HR	6.94	20.67	40.59	23.6	12.05	0.330	0.350	134.3	22.3	10.2	5.67	5.24	℄
L62	Hoesch, W. Ger	U	DJ	HR	6.62	23.64	44.35	22.5	12.21	0.362	0.366	130.5	21.4	9.8	NP	NP	℄
CU110	Casteel, Canada	U	HG	CF	6.62	22.66	42.54	22.5	14.18	0.354	0.354	149.0	21.4	12.9	5.94	5.00	℄
L2N	Unimetal, France	U	DJ	HR	7.37	15.75	32.75	25.0	10.63	0.295	0.375	109.1	20.5	8.8	NP	4.49	℄
BU9.4	Arbed, Lux	U	DJ	HR	6.76	23.62	45.09	22.9	10.63	0.394	0.394	93.3	17.5	8.4	NP	4.99	℄
9W	Frodingham, UK	U	DJ	HR	5.86	20.67	34.27	19.9	10.24	0.250	0.350	85.9	16.8	7.2	5.31	4.87	℄
BU9	Arbed, Lux	U	DJ	HR	6.47	23.62	43.48	22.1	10.63	0.375	0.375	89.0	16.7	8.3	NP	4.99	℄
FSP2	Nippon, Japan	U	DJ	HR	7.23	15.70	32.30	24.6	7.88	0.413	0.413	64.0	16.4	7.1	4.36	NP	℄
FSP2A	Nippon, Japan	U	DJ	HR	6.50	15.70	29.00	22.1	9.44	0.362	0.362	67.6	16.2	7.5	4.40	NP	℄
L32	Hoesch, W. Ger	U	DJ	HR	7.32	17.72	36.89	25.0	9.85	0.413	0.413	77.6	15.8	9.3	NP	NP	℄
CU118	Casteel, Canada	U	HG	CF	7.10	19.69	39.66	24.2	9.45	0.375	0.375	74.5	15.8	8.5	5.25	4.10	℄
PU8	Unimetal, France	U	DJ	HR	5.48	23.62	36.62	18.6	11.02	0.315	0.315	85.0	15.4	7.2	NP	5.02	℄
L61	Hoesch, W. Ger	U	DJ	HR	5.31	23.64	35.88	18.2	12.21	0.315	0.323	94.3	15.4	9.2	NP	NP	℄
CU99	Casteel, Canada	U	HG	CF	5.96	19.69	33.28	20.3	9.45	0.315	0.315	62.8	13.3	7.1	5.25	4.10	℄
L21	Hoesch, W. Ger	U	DJ	HR	5.72	19.70	31.92	19.5	8.67	0.315	0.323	56.4	13.0	5.9	NP	NP	℄
CU104	Casteel, Canada	U	HG	CF	6.26	23.62	41.93	21.3	7.87	0.375	0.375	44.4	11.4		5.54	4.55	℄
6W	Frodingham, UK	U	DJ	HR	5.10	20.67	30.04	17.4	8.35	0.310	0.250	47.3	11.4	5.5	4.99	NP	℄
L20	Hoesch, W. Ger	U	DJ	HR	4.77	19.70	26.54	16.2	8.67	0.236	0.276	48.3	11.2	5.3	NP	4.10	℄
CU81	Casteel, Canada	U	HG	CF	4.87	19.69	27.22	16.59	9.45	0.256	0.256	52.7	11.2	5.9	5.25	4.76	℄
PU6	Unimetal, France	U	DJ	HR	4.54	23.62	30.44	15.5	8.90	0.252	0.295	49.20	11.2	4.5	NP	4.53	℄
CU94	Casteel, Canada	U	HG	CF	5.66	23.62	37.89	19.3	7.87	0.335	0.335	39.71	10.2	5.9	5.54		℄

1	2	3	4	5	6	7	8	9	10	11	12	13	14	15	16	17(a)
FSP1A	Nippon, Japan	U	DJ	HR	5.34	15.70	23.90	18.2	6.70	0.315	0.315	33.0	9.8	4.1	3.97	NP	℄
L60	Hoesch, W. Ger	U	DJ	HR	5.67	23.64	37.90	19.3	5.91	0.375	0.375	28.1	9.5	3.4	NP	NP	℄
L31	Hoesch, W. Ger	U	DJ	HR	6.00	17.72	30.24	20.5	5.91	0.375	0.375	25.3	8.6	4.2	NP	NP	℄
L31	Unimetal, France	U	DJ	HR	6.00	17.72	30.24	20.5	5.91	0.375	0.375	25.3	8.6	4.4	NP	3.87	℄

NOTE: NP = NOT PUBLISHED

The following U-shaped sheet piling sections were formerly offered by the two major domestic manufacturers. More recently, the section designations were standardized to "PDA" and "PMA," a move sponsored by the American Iron and Steel Institute. None of these sections is available from the mills now, having been displaced by the more efficient Z-type shapes. Our purpose in listing these sections and their properties is to provide a continued reference source for this information, which is no longer published.

1	2	3	4	5	6	7	8	9	10	11	12	13	14	15	16	17(a)
DP1	Bethlehem, USA	U	TF	HR	12.6	16.0	42.7	32.0	12.0	0.375	0.484	NP	NP	15.3	NP	NP	℄
MP110	(3)US Steel, USA	U	TF	HR	12.6	16.0	42.7	32.0	12.0	0.375	0.484	NP	NP	15.3	NP	NP	℄
DP2	Bethlehem, USA	U	TF	HR	10.6	16.0	36.0	27.0	10.0	0.375	0.375	NP	NP	10.7	NP	NP	℄
MP116	(2)US Steel, USA	U	TF	HR	10.6	16.0	36.0	27.0	10.0	0.375	0.375	NP	NP	10.7	NP	NP	℄
AP3	Bethlehem, USA	U	TF	HR	10.6	19.6	36.0	22.0	7.0	0.375	0.375	NP	NP	5.4	NP	NP	℄
MP115	(3)US Steel, USA	U	TF	HR	10.6	19.6	36.0	22.0	7.0	0.375	0.375	NP	NP	5.4	NP	NP	℄

(1) Same as AISI PDA32   (2) Same as AISI PDA27   (3) Same as AISI PMA22

PROPERTIES CHART 3
ARCH AND LIGHT GAUGE
(LISTED IN DESCENDING ORDER BASED ON SECTION MODULUS)

1	2	3	4	5	6	7	8	9	10	11	12	13	14	15	16	17(a)
Section Designation	Mfgr. & Country	Shape	Type Interlock	Hot Rolled Or Cold Formed	Area sq. in. Per ft of wall	Nominal width, in. Per section	Weight in Pounds Per lin ft of pile	Weight in Pounds Per sq ft of wall	Wall Depth (height) (inches)	Web Thickness (inches)	Flange Thickness (inches)	Moment of Inertia, in.⁴ Per ft of wall	Section Modulus, in.³ Per ft of wall	Surface Area Sq ft per lin ft of pile Total Area	Surface Area Sq ft per lin ft of pile Nominal Coating Area*	Interlock Location in wall
S65	CMRM, Canada	A	HG	CF	4.37	27.00	33.50	14.9	6.38	0.256	0.256	28.9	8.9	6.42	5.90	One Face
CS76	Casteel, Canada	A	HG	CF	4.57	27.58	35.62	15.6	5.91	0.276	0.276	26.3	8.9	6.37	5.86	One Face
S55	CMRM, Canada	A	HG	CF	3.70	27.00	28.30	12.6	6.38	0.217	0.217	24.5	7.5	6.42	5.90	One Face
CS60	Casteel, Canada	A	HG	CF	3.60	27.58	28.22	12.3	5.91	0.217	0.217	20.6	7.0	6.37	5.86	One Face
L65	CMRM, Canada	A	HG	CF	4.23	19.70	24.10	14.7	4.15	0.256	0.256	12.8	5.4	4.70	4.20	One Face
L60	CMRM, Canada	A	HG	CF	3.90	19.70	22.30	13.6	4.15	0.236	0.236	11.8	5.0	4.70	4.20	One Face
L50	CMRM, Canada	A	HG	CF	3.23	19.70	18.90	11.5	4.15	0.197	0.197	9.8	4.1	4.70	4.20	One Face
L45	CMRM, Canada	A	HG	CF	2.89	19.70	16.20	9.9	4.10	0.177	0.177	7.7	3.6	4.50	4.10	One Face

ARCH AND LIGHT GAUGE - CONTINUED

1 Section Designation	2 Mfgr. & Country	3 Shape	4 Type Interlock	5 Hot Rolled Or Cold Formed	6 Area sq. in. Per ft of wall	7 Nominal width, in. Per section	8 Weight in Pounds Per lin ft of pile	9 Weight in Pounds Per sq ft of wall	10 Wall Depth (height) (inches)	11 Web Thickness (inches)	12 Flange Thickness (inches)	13 Moment of Inertia, in.⁴ Per lin ft of wall	14 Section Modulus, in.³ Per ft of wall	15 Surface Area Total Area	16 Surface Area Nominal Coating Area*	17(a) Interlock Location in wall
CL57	Casteel, Canada	A	HG	CF	3.43	21.67	21.07	11.7	3.55	0.217	0.217	6.2	3.5	4.80	4.24	One Face
5GA	Contech, USA	A	HG	CF	3.41	19.69	19.10	11.6	3.20	0.209	0.209	5.7	3.4	2.24	2.00	One Face
5	Superior, USA	A	HG	CF	4.85	18.00	16.90	11.3	3.15	0.209	0.209	5.4	3.3	3.90	3.70	One Face
5	Shoreline, USA	A	HG	CF	4.85	18.00	16.90	11.3	3.15	0.209	0.209	5.4	3.3	3.90	3.70	One Face
L41	CMRM, Canada	A	HG	CF	2.68	19.70	15.00	9.2	4.10	0.164	0.164	7.0	3.3	4.50	4.10	One Face
6	Superior, USA	A	HG	CF	4.53	18.00	15.80	10.5	3.15	0.194	0.194	4.9	3.0	3.90	3.70	One Face
6	Shoreline, USA	A	HG	CF	4.53	18.00	15.30	10.5	3.15	0.194	0.194	4.9	3.0	3.90	3.70	One Face
7GA	Big R, USA	A	HG	CF	4.20	18.00	14.63	9.8	3.40	0.179	0.179	5.4	3.0	NP	4.00	One Face
CL47	Casteel, Canada	A	HG	CF	2.83	21.67	17.39	9.6	3.55	0.177	0.177	5.1	3.0	4.80	4.24	One Face
7GA	Contech, USA	A	HG	CF	2.93	19.69	16.40	10.0	3.20	0.179	0.179	4.9	2.9	2.24	2.00	One Face
7	Superior, USA	A	HG	CF	2.85	18.00	14.40	9.6	3.15	0.179	0.179	4.4	2.8	3.90	3.70	One Face
7	Shoreline, USA	A	HG	CF	2.85	18.00	14.40	9.6	3.15	0.179	0.179	4.4	2.8	3.90	3.70	One Face
L34	CMRM, Canada	A	HG	CF	2.18	19.70	12.50	7.6	4.10	0.134	0.134	5.9	2.8	4.50	4.10	One Face
8GA	Big R, USA	A	HG	CF	2.55	18.00	13.41	8.9	3.40	0.164	0.164	5.0	2.8	NP	4.00	One Face
8GA	Contech, USA	A	HG	CF	2.72	19.69	15.20	9.3	3.20	0.164	0.164	4.5	2.7	2.26	2.00	One Face
8	Superior, USA	A	HG	CF	2.53	18.00	13.20	8.8	3.15	0.164	0.164	4.2	2.6	3.90	3.70	One Face
8	Shoreline, USA	A	HG	CF	2.53	18.00	13.20	8.8	3.15	0.164	0.164	4.2	2.6	3.90	3.70	One Face
CL42	Casteel, Canada	A	HG	CF	2.53	21.67	-5.52	8.6	3.55	0.157	0.157	4.5	2.6	4.80	4.24	One Face
10GA	Big R, USA	A	HG	CF	2.09	18.00	10.97	7.3	3.40	0.135	0.135	4.0	2.2	4.50	4.10	One Face
L27	CMRM, Canada	A	HG	CF	1.72	19.70	9.91	6.0	4.10	0.106	0.106	4.7	2.2	4.50	4.10	One Face
10GA	Contech, USA	A	HG	CF	2.23	19.69	12.50	7.6	3.20	0.135	0.135	3.7	2.2	2.24	2.00	One Face
10	Superior, USA	A	HG	CF	2.07	18.00	10.80	7.2	3.15	0.135	0.135	3.5	2.2	3.90	3.70	One Face
10	Shoreline, USA	A	HG	CF	2.07	18.00	10.80	7.2	3.15	0.135	0.135	3.5	2.2	3.90	3.70	One Face
12GA	Big R, USA	A	HG	CF	1.64	18.00	8.53	5.7	3.40	0.105	0.105	3.1	1.75	NP	4.00	One Face
12GA	Contech, USA	A	HG	CF	1.77	19.69	9.90	6.0	3.20	0.105	0.105	2.9	1.71	2.24	2.00	One Face
12	Superior, USA	A	HG	CF	1.61	18.00	8.40	5.6	3.15	0.105	0.105	2.7	1.65	3.90	3.70	One Face
12	Shoreline, USA	A	HG	CF	1.61	18.00	8.40	5.6	3.15	0.105	0.105	2.7	1.65	3.90	3.70	One Face

PROPERTIES CHART 4

FLAT SHAPED SHEET PILING
(LISTED ALPHABETICALLY BY MANUFACTURER)

1 Section Designation	2 Mfgr. & Country	3 Shape	4 Type Interlock	5 Hot Rolled Or Cold Formed	6 Section Area sq. in.	7 Nominal width, in.	8 Weight Per lin ft of pile	9 Weight Per sq ft of wall	10 Wall Depth (height) (inches)	11 Web Thickness (inches)	12 Flange Thickness (inches)	13 Moment of Inertia, in.⁴ Per section	14 Section Modulus, in³ Single Section	15 Surface Area Sq ft per lin ft of pile Total Area	16 Nominal Coating Area*	17(b) Interlock Strength (Flat Web Only) Kips (3) Per Inch of Interlock
PSA23	Bethlehem, USA	SA	TF	HR	8.99	16.0	30.7	23.0	2.75	0.375	NA	5.50	3.20	3.76	3.08	12.0
PS27.5	Bethlehem, USA	S	TF	HR	13.27	19.69	45.1	27.5	3.50	0.400	NA	5.30	3.30	4.48	3.65	16.0, 21.0
PS31	Bethlehem, USA	S	TF	HR	14.96	19.69	50.9	31.0	3.50	0.500	NA	5.30	3.30	4.48	3.65	16.0, 21.0, 24.0
SW1	Frodingham, UK	S	TFX	HR	10.90	16.25	37.17	27.5	NP	0.375	NA	NP	NP	3.58	3.36	16.7 to 21.5
SW1A	Frodingham, UK	S	TFX	HR	12.60	16.25	42.85	31.6	NP	0.500	NA	NP	NP	3.58	3.36	16.7 to 21.5
FL408	Hoesch, W. Ger	S	TF	HR	10.90	15.76	37.09	28.3	3.39	0.375	NA	4.86	2.93	NP	NP	Up to 28.0
FL409	Hoesch, W. Ger	S	TF	HR	10.7	15.76	36.29	27.7	3.39	0.355	NA	4.86	2.93	NP	NP	Up to 28.0
FL409.5	Hoesch, W. Ger	S	TF	HR	10.2	15.76	34.68	26.4	3.39	0.315	NA	4.86	2.93	NP	NP	Up to 28.0
FL412.7	Hoesch, W. Ger	S	TF	HR	12.35	15.76	42.07	32.0	3.39	0.500	NA	4.86	2.93	NP	NP	Up to 28.0
FL511	Hoesch, W. Ger	S	TF	HR	12.80	19.70	45.36	27.7	3.47	0.433	NA	4.76	2.93	NP	NP	Up to 28.0
FL512	Hoesch, W. Ger	S	TF	HR	13.93	19.70	47.37	28.9	3.47	0.473	NA	4.76	2.93	NP	NP	Up to 28.0
FL512.7	Hoesch, W. Ger	S	TF	HR	14.32	19.70	48.72	29.7	3.47	0.500	NA	4.76	2.93	NP	NP	Up to 28.0
YSP F	Nippon, Japan	S	TF	HR	10.71	15.70	36.4	27.9	NP	0.374	NA	4.56	2.92	NP	NP	22.4
YSP FA	Nippon, Japan	S	TF	HR	12.01	15.70	40.9	31.1	NP	0.500	NA	4.71	2.95	NP	NP	22.4
YSP FX	Nippon, Japan	S	TF	HR	13.28	15.70	45.2	34.4	NP	0.500	NA	5.38	3.42	NP	NP	36.6
500J9.5	Unimetal, France	S	TF	HR	12.42	19.69	42.67	26.0	NP	0.375	NA	3.60	2.40	NP	3.94	16.0 to 20.0
500J12	Unimetal, France	S	TF	HR	14.42	19.69	49.05	29.9	NP	0.472	NA	3.70	2.50	NP	3.94	18.5 to 22.5
500J12.5	Unimetal, France	S	TF	HR	14.71	19.69	50.07	30.5	NP	0.492	NA	3.80	2.60	NP	3.94	28.0
500J12.7	Unimetal, France	S	TF	HR	14.90	19.69	50.73	30.9	NP	0.500	NA	3.90	2.60	NP	3.94	28.0

NOTE: NP = NOT PUBLISHED
NA = NOT APPLICABLE

The following flat-shaped sheet piling sections were formerly offered by the two major domestic manufacturers. More recently, the section designations were standardized to "PSA" and "PS," a move sponsored by the American Iron and Steel Institute. Except for PSA 23 however, these sections are no longer available from the mills. They have been replaced by more efficient sections by the remaining manufacturer. Our purpose in listing these sections and properties is to provide a continued reference source for this information, which is no longer published.

1	2	3	4	5	6	7	8	9	10	11	12	13	14	15	16	17(b)	
							Weight in Pounds		Wall Depth (height) (inches)	Web Thickness (inches)	Flange Thickness (inches)	Moment of Inertia, in.^{4} Per section	Section Modulus, in^3 Single Section	Surface Area Sq ft per lin ft of pile		Interlock Strength (Flat Web Only)	Kips (3) Per Inch of Interlock
Section Designation	Mfgr. & Country	Shape	Type Interlock	Hot Rolled Or Cold Formed	Section Area sq. in.	Nominal width, in.	Per lin ft of pile	Per sq ft of wall						Total Area	Nominal Coating Area*		
SP4	Bethlehem, USA	SA	TF	HR	9.0	16.0	30.7	23.0	2.7	0.375	NA	NA	2.4	NP	NP	12.0	
MP112	[1] US Steel, USA	SA	TF	HR	9.0	16.0	30.7	23.0	2.7	0.375	NA	NA	2.4	NP	NP	12.0	
SP5	Bethlehem, USA	SA	TF	HR	11.0	16.0	37.3	28.0	2.7	0.500	NA	NA	2.5	NP	NP	12.0	
MP113	[2] US Steel, USA	SA	TF	HR	11.0	16.0	37.3	28.0	2.7	0.500	NA	NA	2.5	NP	NP	12.0	
SP6a	Bethlehem, USA	S	TF	HR	10.3	15.0	35.0	28.0	3.5	0.375	NA	NA	2.4	NP	NP	16.0	
MP101	[3] US Steel, USA	S	TF	HR	10.3	15.0	35.0	28.0	3.4	0.375	NA	NA	1.9	NP	NP	16.0	
SP7a	Bethlehem, USA	S	TF	HR	11.8	15.0	40.0	32.0	3.5	0.500	NA	NA	2.4	NP	NP	16.0	
MP102	[4] US Steel, USA	S	TF	HR	11.8	15.0	40.0	32.0	3.4	0.500	NA	NA	1.9	NP	NP	16.0	
MP103	[5] US Steel, USA	S	TF	HR	12.9	16.5	44.0	32.0	3.7	0.453	NA	NA	2.4	NP	NP	28.0	

(1) Same as AISI PSA 23 (2) Same as AISI PSA 28 (3) Same as AISI PS28 (4) Same as AISI PS32 (5) Same as AISI PSX32

www.ingramcontent.com/pod-product-compliance
Lightning Source LLC
Chambersburg PA
CBHW080906170526
45158CB00008B/2014